Non-Destructive Evaluation of Corrosion and Corrosion-assisted Cracking

Non-Destructive Evaluation of Corrosion and Corrosion-assisted Cracking

Edited by

Raman Singh
Baldev Raj
U. Kamachi Mudali
Prabhakar Singh

The American Ceramic Society

WILEY

Published by John Wiley & Sons, Inc., Hoboken, New Jersey. Published simultaneously in Canada.

Registered Office
John Wiley & Sons, Inc., 111 River Street, Hoboken, NJ 07030, USA

Editorial Office
111 River Street, Hoboken, NJ 07030, USA

For details of our global editorial offices, customer services, and more information about Wiley products visit us at www.wiley.com.

Wiley also publishes its books in a variety of electronic formats and by print-on-demand. Some content that appears in standard print versions of this book may not be available in other formats.

Library of Congress Cataloging-in-Publication Data

Names: Singh, Raman, 1959– editor.
Title: Non-Destructive evaluation of corrosion and corrosion-assisted cracking / edited by Raman Singh, Baldev Raj, U. Kamachi Mudali, Prabhakar Singh.
Description: Hoboken : Wiley-American Ceramic Society, 2019. | Includes bibliographical references and index. | Description based on print version record and CIP data provided by publisher; resource not viewed.
Identifiers: LCCN 2018036498 (print) | LCCN 2018038143 (ebook) | ISBN 9781119428619 (Adobe PDF) | ISBN 9781119428534 (ePub) | ISBN 9781118350058 (hardcover)
Subjects: LCSH: Corrosion and anti-corrosives–Testing. | Fracture mechanics. | Nondestructive testing.
Classification: LCC TA462 (ebook) | LCC TA462 .N586 2019 (print) | DDC 620.1/1223–dc23
LC record available at https://lccn.loc.gov/2018036498

Cover Design: Wiley
Cover Image: © Rohit Mattoo/Getty Images

Set in 10/12pt Warnock by SPi Global, Pondicherry, India

Printed in the United States of America

V10008597_031119

Dedication

This unique book on nondestructive evaluation (NDE) of corrosion was conceived as a result of late Dr. Baldev Raj's world renowned expertise in NDE and Professor Raman Singh's complementary expertise in corrosion, when Dr. Raj was visiting Australia way back in 2009. Upon Dr. Raj's initiative, Dr. Kamachi Mudali and Professor Prabhakar Singh joined the editorial team. It is truly sad that Dr. Baldev Raj is not physically with us to see the first edition of the book, but his inspiration will always be there with us.

Dr. Baldev Raj was born in Jammu Tawi, India, on 9 April 1947. At the time of his death on 6 January 2018, he was director of National Institute of Advanced Studies, Bangalore. At that time, he was also chancellor of Academy of Scientific and Innovative Research (AcSIR) and chairman of Research Councils of Gas Turbine Research Establishment, DRDO, and Central Glass Ceramic Research Institute, CSIR. His pioneering research contributions in material characterization, testing, and evaluation using nondestructive evaluation methodologies have been recognized globally. He played a key leadership role for India achieving global leadership stature in fast breeder nuclear reactors with closed fuel cycles. He has made sustained contributions through research and development of nuclear materials, components and systems for water, fast spectrum, fusion reactors, and back-end recycling plants. He has chaired Strategic Committee for Materials and Manufacturing (Ministry of Defence) and Rare Earths Leadership (NITI Aayog). His contributions to science policy and science diplomacy are widely acclaimed. Diversity of contributions is evident from his deep and effective engagement in areas ranging from cultural heritage to medical technology to education. He has rare distinction of being a *fellow of all the four science and engineering academies* in the country. He is a fellow of The World Academy of Sciences, International Nuclear Energy Academy, German Academy of

Sciences, etc. He has over 1300 publications, over 80 books, over 100 articles in encyclopedia and handbooks, 22 patents, and 5 standards. He has received highest national and international honors and distinctions. He is recipient of numerous awards: Homi Bhabha Gold Medal; H. K. Firodia; Om Prakash Bhasin; Vasvik, Gujar Mal Modi Science and Technology Award, National Metallurgist; Lifetime Achievement of Indian Nuclear Society, Distinguished Material Scientist; 20th SIES Sri Chandrasekarendra Saraswati National Eminence Award for S&T; and Distinguished Alumni of Indian Institute of Science, to name a few. He also received the national honor of Padmashri. Late Dr. Baldev Raj was also known for his unique human touch and passion to interact with students and young professionals for mutual inspirations and service to society.

Contents

List of Contributors

A. Bowler
Maritime Systems Project Office (MPSPO)
RAAF Base Edinburgh
Edinburgh
South Australia
Australia

M. Dorman
Aircraft Structural Integrity – Directorate General Technical Airworthiness (ASI-DGTA)
RAAF Williams
Laverton
Victoria
Australia

Alec Groysman
Faculty of Chemical Engineering
Technion
Israel Institute of Technology
Haifa
Israel

J. Jayaraj
Corrosion Science and Technology Group
Indira Gandhi Centre for Atomic Research
Kalpakkam
India

R. Jones
Centre of Expertise in Structural Mechanics
Department of Mechanical and Aeronautical Engineering
Monash University
Clayton
Victoria
Australia

N. Jothilakshmi
Bhabha Atomic Research Centre
Mumbai
India

U. Kamachi Mudali
Corrosion Science and Technology Group
Indira Gandhi Centre for Atomic Research
Kalpakkam
India

M. Lo
Centre of Expertise in Structural Mechanics
Department of Mechanical and Aeronautical Engineering
Monash University
Clayton
Victoria
Australia

M.B. Venkataraman
Research School of Engineering
The Australian National University
ACT
Australia

Paritosh Nanekar
Bhabha Atomic Research Centre
Mumbai
India

Baldev Raj
National Institute of Advanced Studies
Bengaluru
India

V.S. Raja
Department of Metallurgical Engineering and Materials Science
Indian Institute of Technology Bombay
Mumbai
India

B.P.C. Rao
Fast Reactor Fuel Cycle Facility
Indira Gandhi Centre for Atomic Research
Kalpakkam
India

D. Roles
Maritime Systems Project Office (MPSPO)
RAAF Base Edinburgh
Edinburgh
South Australia
Australia

Prabhakar Singh
Center for Clean Energy Engineering
University of Connecticut
Storrs, CT
USA

R.K. Singh Raman
Department of Mechanical & Aerospace Engineering, Department of Chemical Engineering
Monash University
Clayton Campus (Melbourne)
Victoria
Australia

Girija Suresh
Corrosion Science and Technology Group
Indira Gandhi Centre for Atomic Research
Kalpakkam
India

B. Venkatraman
Indira Gandhi Centre for Atomic Research
Kalpakkam
India

S.A. Wade
Swinburne University of Technology
Hawthorn
Victoria
Australia

Foreword

Corrosion represents an enormous challenge across all industries and commences and even impacts each one of us in our daily lives. For example, the corrosive degradation of artificial hips and knees, as we are now experiencing, is causing the replacement of many implants because of the toxicity of the corrosion products toward human tissue. This alone will cost the medical system in the United States many billions of dollars by the time that the problem is corrected. I could cite many other examples, including repair and replacement of infrastructure (roads, bridges, and building, because these structures have been compromised by chloride-induced rebar corrosion, particularly in the northeast and in mountain regions of the country). The cause of the problem has been traced to the use of road salts and/or the use of calcium chloride as a "setting agent" when pouring concrete in freezing weather. This is a good example of an "unintended consequence." While the causes of such corrosion maladies are well known and have been well articulated, the first line of defense is to detect and characterize the developing corrosion damage without compromising the performance of the system itself. This is done through "nondestructive testing (NDT)," which is the subject of this outstanding book. Finally, the American Society for Testing and Materials (ASTM E1316, "Standard Terminology for Nondestructive Examinations") defines nondestructive examination (NDE) as the development and application of technical methods to examine materials or components in ways that do not impair future usefulness and serviceability in order to detect, locate, measure, and evaluate flaws. The reader will note that the definition is comprehensive in that it is just not detection that is at issue but it is also characterization of the damage that is important.

The book begins with an overview of NDT by the editorial team led by Professor Raman Singh, which is both comprehensive and informative. Importantly, this overview makes the case and paves the way for subsequent chapters. As noted by the editor, NDT or nondestructive evaluation (NDE) techniques are therefore essential for the early and accurate detection of corrosion and corrosion-induced damage, including general, pitting, and crevice corrosion and of hydrogen embrittlement, stress corrosion cracking, and corrosion fatigue in systems under active mechanical loading. Thus, the first chapter of this book articulates the principles of key NDE techniques and identifies which techniques are particularly suitable for detecting corrosion and corrosion-assisted cracking. This is followed by a chapter on different forms of corrosion as well as brief descriptions of their underlying mechanisms. After the two introductory chapters on NDE and corrosion, the following two present case studies on applications of NDE techniques for online corrosion detection and monitoring in critical applications, e.g. nuclear power plants. Subsequent chapters describe specific NDE techniques, such as electrochemical impendence spectroscopy (EIS), electrochemical noise analysis (ENA), acoustic emission (AE), and thermography (T). Importantly, these chapters describe in detail the application of the various techniques in detecting corrosion in specific industries (e.g. aerospace and transportation) or specific corrosion phenomenon (such as corrosion scale cracking and spallation), again with the support of case studies. The final chapter is dedicated to NDE for online corrosion monitoring.

Inspection is also an integral part of NDT or NDE and includes visual testing (VT), liquid penetrant testing (LPT), magnetic particle testing (MPT), eddy current testing (ECT), and *in situ* metallography (ISM). Inspection techniques can be divided into surface examination, volumetric examination methods, and performance testing. Thus, VT, LPT, and ISM are examples of surface inspection methods, while MPT, ECT, radiography testing (RT), and ultrasonic testing (UT) are examples of volumetric examination techniques. Performance tests include leak testing (LT) and acoustic emission testing (AET). These techniques are vital for the *in situ* characterization of the extent of damage.

The material contained in this book on EIS and ENA are close to my heart, as I have used both to detect and characterize numerous corrosion systems. Thus, EIS, which has been applied to a variety of corrosion systems, is a perturbation/response technique that is based on linear systems theory (LST) for which a firm and broad mathematical basis exists. Its principal advantages are that it is a steady-state technique, it can be applied over a very wide frequency range, and it transfers an enormous amount of information to the observer. Its principal disadvantage is that in order to take full advantage of EIS, it is necessary to possess mathematical skills. Thus, while most EIS studies end in characterizing a "equivalent electrical circuit (EEC)," its real power lies in reaction mechanism analysis (RMA). In RMA one seeks to define reaction mechanisms in terms of charge transfer and coupled chemical reactions. Fortunately, a number of excellent textbooks exist on EIS that describe in detail how RMA is affected. The application of EIS in the analysis of organic coatings, passive films, anodic oxide films, localized corrosion, atmospheric corrosion, fouling, and concrete structures is reviewed in the book.

ENA is related to EIS (in fact the impedance of a system can be determined from the natural fluctuations in potential and current) but is non-perturbative in nature. Potential and current fluctuations emanating from corroding surfaces carry information about the type of corrosion process and the corrosion rate, but its extraction requires a mathematical skill level that, like EIS, is usually above that possessed by corrosion scientists and specialists in NDE. Because it is non-perturbative, ENA is the corrosionist's stethoscope. Just like a physician useful information requires a skill level that can prize the information from the recorded noise, as noted above. This information can only be extracted by using sophisticated data analysis techniques, such as wavelet analysis (WA), maximum entropy analysis (MEA), and power spectral density analysis (PSDA). Although the fluctuations in current and potential are often stated to be random (stochastic), the noise is actually deterministic in nature because they arise from processes occurring on the metal surface and are interrelated by the impedance.

The material contained in this book is directly related to the risk equation, $R = P \times C$, where P is the probability that an event (e.g. SCC) will occur and C are the consequences. NDE primarily articulates P. The consequences, C, however, are often imposed legislatively, such as in the case of spillage from a pipeline. However, it is their coupling that defines the risk, R. Thus, the consequences for the environment on rupture of a water pipeline are generally small (and maybe even beneficial in some respects), whereas spillage from an oil pipeline can and usually does have severe consequences. Thus management of risk is usually accomplished by controlling P. It is therefore obvious that NDE is a vital part of controlling risk.

Although the benefits of NDE are clearly apparent, there does not currently exist a comprehensive book on the principles underlying NDE or their application in specific industrial systems. This book goes a long way to fulfilling that need, in spite of the fact that it focuses on environmentally induced damages (corrosion). As such, the book represents a considerable resource in detecting and characterizing corrosion across industries ranging from rail infrastructure, offshore petroleum production, and power generation to aging aircraft. I therefore rate this book to be a "must" for anyone who is involved in NDE.

Digby Macdonald

FRSC, FRSNZ, FNACE, FECS, FISE, FIC, FASM, FWIC
Khwarizmi Laureate in Fundamental science, Doctuer Honoris Causa-INSA Lyon, Faraday Memorial Trust Gold Medalist, Gibbs Award Recipient (IAPWS), De Tao Master, Frumkin Medalist, OLIN Palladium Medalist

Professor in Residence, Departments of Nuclear Engineering and Materials Science and Engineering, University of California at Berkeley.
Distinguished Professor Emeritus of Materials Science and Engineering, Penn State University

Preface

Corrosion of engineering alloys and the mitigation measures continue to cost dearly (1–4% of GDP of any developed economy, which translates to an annual loss of ~\$250 billion to the United States). The problem of corrosion is endemic to a large cross section of industries and affects our daily lives. For example, in the United States, there are more trips per day over structurally deficient bridges than there are McDonald's hamburgers eaten in the entire United States [1]. The significance of corrosion is further illustrated by the June 2007 Report to Congress by the Under Secretary of the Department of Defense (DoD) (Acquisition, Technology, and Logistics) [2]. This report estimated the cost of corrosion associated with US DoD systems alone to be between US\$10 billion and US\$20 billion annually.

Traditional approaches such as use of corrosion resistance alloys and coatings have brought significant mitigation of the age-old problem of corrosion. However, a durable corrosion resistance is still a nontrivial challenge in some critical applications, where aggressive corrosive solutions are handled (e.g. concentrated chloride solutions in desalination plants) or where corrosion resistance is required for very long durations (e.g. nuclear waste containers) or where corrosion can cause serious health problems (such as degradation of human implants). It will be socially fulfilling as well as commercially attractive to completely circumvent corrosion, which is often prohibitively challenging. This challenge emphasizes the need of an early and accurate detection and assessment of corrosion that are crucial for timely mitigation/repair.

Nondestructive evaluation (NDE) techniques are, therefore, invaluable for an early and accurate detection of corrosion and corrosion-assisted cracking. To this end, the first chapter of this book describes principles of key NDE techniques, with elaborations of those techniques that are particularly suitable for detection of corrosion and corrosion-assisted cracking. This is followed by a chapter on different forms of corrosion as well as brief descriptions of their underlying mechanisms.

On the solid footings of the two introductory chapters on NDE and corrosion, the following two chapters present case studies on applications of NDE techniques for detection and online monitoring of corrosion in some critical applications (such as nuclear power plants). Subsequent chapters present specific NDE techniques (viz. thermography, electrochemical impendence spectroscopy, electrochemical noise, acoustic emission) and their applications in detection of corrosion in specific industry (such as aerospace) or specific corrosion phenomenon (such as corrosion scale cracking and spallation), again with the support of case studies. The final chapter is dedicated to NDE for online corrosion monitoring.

The magnitude of losses and risks caused by corrosion is immense. Though it is invaluable to accomplish early detection of corrosion by NDE, there seems to be no resource/book on the topic of NDE of corrosion and corrosion-assisted cracking. As such this is a maiden attempt to accomplish a formidable resource on NDE of corrosion across industries that range from rail infrastructure, offshore, and power generation to aging aircraft. Editorial team is grateful to the true international experts from each of the component areas on the topic of the application of NDE to corrosion, who have contributed to this book.

References

1 Davis, S.L., DeGood, K., Donohue, N., and Goldberg, D. (2013). *The Fix We're in For: The State of Our Nation's Busiest Bridges*. Transportation for America. http://t4america.org/docs/bridgereport2013/2013BridgeReport.pdf (accessed 3August 2016).

2 *Efforts to Reduce Corrosion on the Military Equipment and Infrastructure of the Department of Defense*. (2007). Department of Defense Report (June). Office of the Secretary of Defense, USA.

Professor Raman Singh
Dr. Baldev Raj
Dr. U. Kamachi Mudali
Professor Prabhakar Singh

1

Nondestructive Testing

An Overview of Techniques and Application for Quality Evaluation

B. Venkatraman[1] *and Baldev Raj*[2]

[1] *Indira Gandhi Centre for Atomic Research, Kalpakkam, India*
[2] *National Institute of Advanced Studies, Bengaluru, India*

Nondestructive testing (NDT) or nondestructive evaluation (NDE) or nondestructive testing and evaluation (NDT&E) as the name implies is the science and technology of assessing the soundness, acceptability, and fitness for purpose of processes, products, plants, and systems without affecting the functional properties. It is now an inseparable part of modern society. Be it the field of engineering, technology, healthcare, security, research, or heritage, to name a few, NDE is being used right from cradle to end of service to optimize processes, manage quality, predict the life, effect conservation/preservation measures, and limit liability. It is natural to understand from appreciation of this perspective that NDE is a multidisciplinary profession with crosscutting domains where physics, chemistry, materials science, mechanics, electronics, instrumentation, etc. work with synergy to realize end objectives.

Figure 1.1a and b summarizes the history and growth of NDE science and technology. Looking back in the history, it can be observed that while NDT was practiced in ancient times too in a heuristic manner (e.g. potters used tap testing and acoustic emission [AE] to check quality of pots), it was during the seventeenth to nineteenth centuries that the physical basis and science for the NDE methods was laid through the formulations of Maxwell, Faraday, Huygen, Planck, and Kaiser and the discovery of infrared (IR) radiation and X-rays by Herschel and Roentgen, to name a few. The industrialization of economies and the realization that technologies were needed to verify the quality and fitness for purpose of components and processes led to the application of NDT techniques in the late nineteenth and early twentieth centuries. Radiography, penetrant, ultrasonic, and visual testing were the major NDT techniques that were practiced, and NDT was primarily a go/no-go technique for the detection

Non-Destructive Evaluation of Corrosion and Corrosion-assisted Cracking, First Edition.
Edited by Raman Singh, Baldev Raj, U. Kamachi Mudali and Prabhakar Singh.

(a)

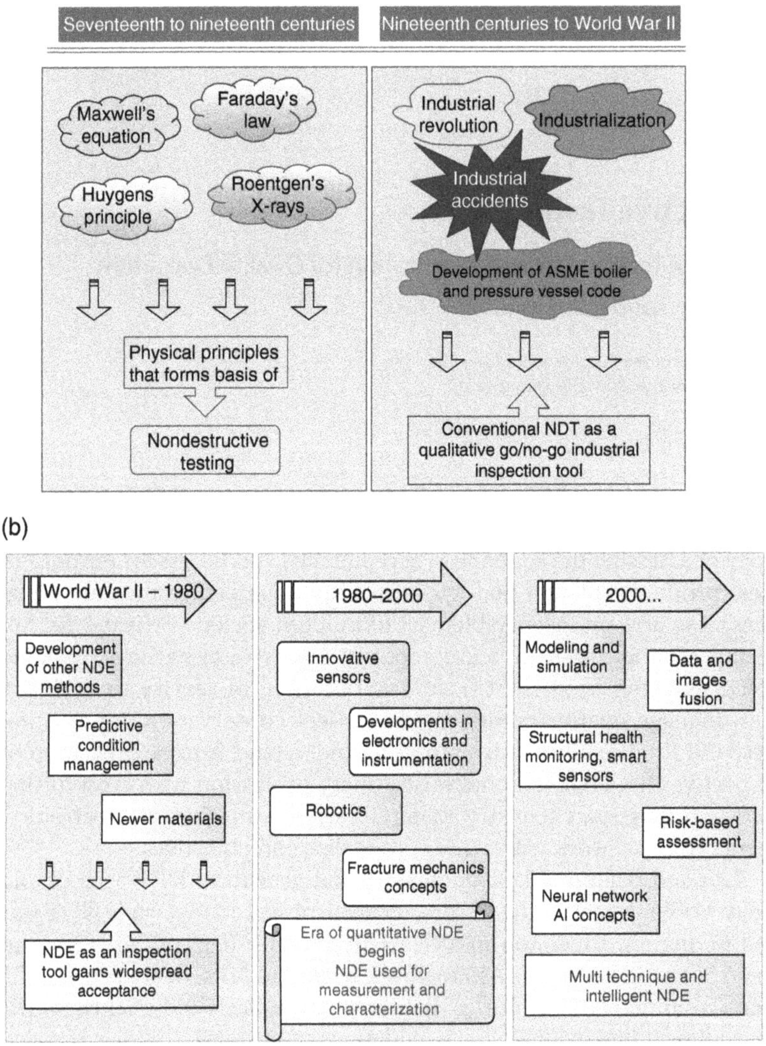

Figure 1.1 (a) and (b) NDT&E: a historical perspective.

of flaws. This is schematically indicated in Figure 1.1a. The post-World War II with rapid industrialization oversaw extensive applications of NDT in many industries. Many new methods such as infrared thermography (IRT), AE, holography, etc. were developed and applied during this period. In all these cases, NDT was primarily a qualitative method. The advent of fracture mechanics concepts in the 1980s coupled with the liberalization of economies, importance of reduced margins of safety in order to be competitive, and stringency of specifications spurred the development and growth of quantitative NDE. Parallely, innovations in sensor technologies and advances in electronics, instrumentation, computers, robotics, and automation coupled with modeling, simulation, etc. led to miniaturization of NDE and also development of smart and intelligent NDE. This is schematically depicted in Figure 1.1b.

Measurements form the heart of inspection and quantitative NDE. By making the right measurements at right points and at right times with quantified uncertainties in the product life cycle, one can ensure fitness for purpose as well as excellence in quality, extended life, and global product competitiveness and customer delight. While NDE has traditionally been applied for defect detection and evaluation during material inspection, manufacturing, fabrication, and in-service inspection, a niche area wherein it has found extensive application is corrosion detection, monitoring, and evaluation.

1.1 Corrosion Damage and NDE

While extensive understanding has been developed in the last few decades about the science of corrosion process and technology to mitigate it, corrosion damage still continues to be a challenging problem in practically all industrial sectors. Corrosion damage progresses with time and can result in leakages and structural failures and, in some cases, can be catastrophic, resulting in loss of human lives. Despite the best efforts at various stages such as selection of materials, proper design, and maintaining appropriate operating environment, corrosion degradation continues to occur and is inevitable. It thus becomes essential to monitor the performance of the installed components for assessing the progress of corrosion degradation and ensuring that it is well within the acceptable limits. It is in this context that the role of NDT&E and quality assurance (QA) becomes crucial.

The role of NDT&E in corrosion damage evaluation is twofold: (i) detection and characterization of the damage and (ii) ensure product quality level in accordance with criterion as set forth by the codes and standards or customer's specifications, thus ensuring the overall safety and reliability and also paving way for remnant life assessment and risk-based analysis. Some of the major advantages of NDE for material inspection include:

1) Ensures product quality and safety and thus fitness for purpose.
2) Provides crucial inputs with respect to flaw dimensions for fracture mechanics-based risk assessment and remnant life prediction.
3) Aids in optimizing the future product design.
4) Ensures reliability and customer satisfaction.
5) Helps in predictive condition management by revealing incipient corrosion damage areas. Preventive measures can then be taken in timely manner, thus preventing costly shutdowns.

As a diagnostic tool for corrosion damage evaluation, a wide range of industries/professions use NDT&E methods: nuclear, aerospace, automotive, chemical, defense, electronics, electrical, fabrication, fertilizers, food processing, marine, medical, metals and nonmetals, petrochemical, power, security, and surface transport, to name a few. It should be emphasized here that, the successful NDE can be achieved only through:

1) Right choice of NDE techniques (single or complementary NDE techniques).
2) Qualified and certified personnel.
3) Calibrated sensors and equipment.
4) Documented procedures with clearly defined evaluation and acceptance criteria based on standards and codes.

Conventional NDE methods – namely, visual testing (VT), liquid penetrant testing (LPT), magnetic particle testing (MPT), radiographic testing (RT), ultrasonic testing (UT), and eddy current testing (ECT) – are primarily used for corrosion detection and corrosion damage evaluation. In the last two decades, advanced techniques such as phased array, digital radiography (DR), Compton backscatter radiography, terahertz (THz) imaging, etc. are being applied for corrosion detection and evaluation. This being an important area, in this book, three chapters are devoted to NDE and corrosion evaluation including corrosion-assisted cracking. In this chapter, we provide a brief overview of the principles, advantages, limitations, and applications of the NDE methods with specific examples. A brief overview of the R&D trends in NDE for corrosion evaluation is also outlined.

1.2 Corrosion: A Brief Overview

As per ISO 8044:2015, corrosion is defined as "Physicochemical interaction (often of an electrochemical nature) between a metal and its environment that results in changes in the properties of the metal, and which may lead to significant impairment of the function of the metal, the environment, or the technical system, of which these form a part." It is to be noted here that while the term "corrosion" applies to the process, the end result is the deterioration or "corrosion damage," which needs to be detected preferably by NDE.

Corrosion of metals and materials is a highly complex phenomenon and can take many different forms (as summarized in Table 1.1). The process of corrosion is affected by several factors. Among these, material type, composition, heat treatment if any, and environmental conditions (including medium and stresses if any) in which the material resides are considered as the major ones. However, the most prominent result of all corrosion types is the loss of thickness and strength of the material, ultimately leading to failure of the component or structure itself. A fundamental understanding of the various types of

Table 1.1 Corrosion types and characteristics.

Corrosion type	Cause	Appearance	By-product
Uniform	Exposure to corrosive environment	Irregular roughening more or less uniformly distributed over the entire exposed surface of a metal	Scale, metallic salts
Pitting	Weak spots (due to the presence of impurity or chemical inhomogeneity) in the protective film/coating/passive layer or damaged surface. Pitting initiation also influenced by surface condition and by temperature >CPT (critical pitting temperature (CPT))	Localized pits or holes with jagged edges	Rapid localized dissolution of the base metal
Intergranular	Preferential/localized attack along or adjacent to the grain boundaries	Grain or phase boundaries appear uniformly damaged	Indications: scale type at smaller magnitude than stress corrosion
Crevice	Triggered by local difference in environment composition (oxygen concentration), this affects mechanical joints, such as coupled pipes or threaded connections or welds (fillet, for example) where there is a crevice or gap wherein the corroding medium can stay stagnant	Localized damage (considered a severe form of pitting)	Same as scale and pitting
Filiform	Micro/small defects or gaps in coatings or paints that can result in moisture/water/corroding medium seepage or ingress	Fine, meandering, threadlike indications that spread from the source	Similar to scale
Galvanic corrosion	Generally results due to electrochemical action between two dissimilar metals in contact with each other and exposed to a corrosive medium (electrolyte)	Corrosion buildup in the dissimilar metal region	Emission of mostly molecular hydrogen gas in a diffused form

(*Continued*)

Table 1.1 (Continued)

Corrosion type	Cause	Appearance	By-product
Stress corrosion cracking	Tensile stresses combined with corrosive medium/environment	Micro/macrocracks that are branched	Initially scale type and finally leads to cracking
Microbiologically induced corrosion (MIC)	A form of corrosion caused or accelerated by living organisms such as bacteria, algae, or fungi found in marine applications. Often associated with the presence of tubercles or slimy organic substances. MIC is usually found in aqueous environments or services where water is always or sometimes present, especially where stagnant or low-flow conditions allow and/or promote the growth of microorganisms	Usually observed as localized pitting under deposits or tubercles that shields the organisms	Damage is often characterized by subsurface cavities

Note: This table is of significance to find right and optimum NDE measurement method and technique based on corrosion mechanism and outcomes.

corrosion is thus essential to evaluate the significance and accordingly arrive at the most appropriate NDE technology and optimum technique for quantitative detection and characterization of corrosion damage. Please refer to Chapter 2 of this book for elaborate description of various corrosion types.

1.3 NDE Methods [1]

1.3.1 Visual Examination

Visual examination (VE), the first and most important testing method in the examination and evaluation of a component, is one of the oldest, simplest, and widely used methods for the inspection of defects/discontinuities open to the surface. The flaws present on the surface are more detrimental compared with those located underneath because:

1) Surface is the part of the component where maximum stress is present.
2) Corrosion often initiates at surface. Hostile environment introduce possibility of corrosion–fatigue, corrosion, erosion, etc.
3) Many studies have revealed that small imperfection on the surface can serve as a nucleus for fatigue crack initiation and such cracks may lead to catastrophic failure.

The method finds wide applications for inspection of materials during various stages of manufacture. For many noncritical parts, integrity is verified principally by visual inspection. Even when other NDT methods are used, visual inspection still constitutes an important part of practical quality control (QC). VE enhanced by measurement, sensors, and instrumentation is useful for assessing the following:

1) Dimensional accuracy of the components.
2) Conformity of components to size and contour requirements.
3) Acceptability of part appearance with regard to surface roughness and finish.
4) Detection of surface-breaking cracks and other defects in castings, forgings, and weldments.
5) Determining the weld profiles externally and internally.

A variety of aids are available to enhance the capabilities of visual inspection. These range from simple aids such as magnifying glass and optical microscopes to fiberscope, video image scopes, and robotic-based remote inspection systems for closely inspecting the internal surface of tubes, pipes, and areas with limited access. Some of the typical aids are shown in Figure 1.2a–d.

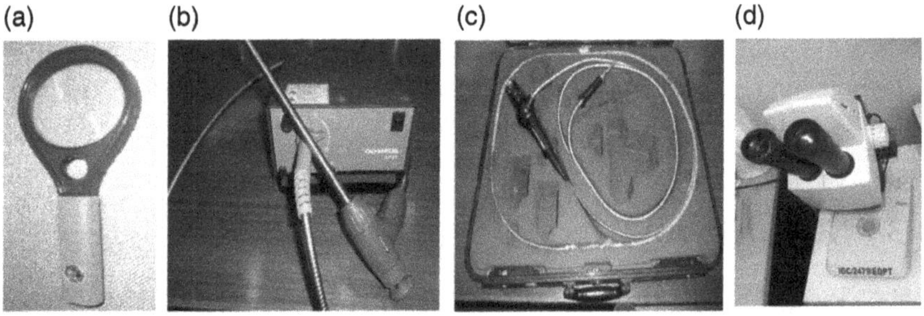

Figure 1.2 Aids to visual inspection. (a) Self-illuminated magnifying glass – 6×. (b) Rigid borescope. (c) Flexible video image scope. (d) Stereomicroscope.

Surface lighting and angle of inspection are two of the important factors to be considered during visual inspection. The surface should be inspected at an angle that is preferably at 90° to the surface of inspection. Adequate lighting should be available on the inspection surface (>200 lx). For critical inspections, more than 500 lx should be available on the inspection surface. Visual acuity of the personnel plays an important role in the efficiency of defect detection. Personnel involved in VE should undergo vision acuity tests at least once in a year.

General corrosion, pitting corrosion, crevice corrosion, corrosion in weld and heat-affected zone, and erosion corrosion can be detected by VE either through direct inspection or with optical aids such as borescope, fiberscope, etc. In the case of stress corrosion cracks, appropriate illumination at an angle and with magnification enables location and detection of the tight cracks. Typical corrosion types detected through visual inspection are shown in Figures 1.3–1.6.

1.3.2 Liquid Penetrant Testing

Liquid penetrant testing (LPT) is one of the simplest, effective, and sensitive NDT methods that reveal discontinuities and defects that are open to the surface. The method is applicable during fabrication, manufacturing, and pre- and in-service of almost any component, large or small, of simple or complex configuration. It can be employed on welds, castings, forgings, wrought products of both ferrous and nonferrous metals and alloys, ceramics, and certain glassware.

The various steps involved in penetrant inspection are schematically shown in Figure 1.7. Pre-cleaning is the first and most important step. Organic solvents, alkaline cleaners, water, soap solution, etc. are some of the typical cleaning agents often used to remove loose scales, dirt, rust, scales, oil, etc. that may adhere to the surface and block the opening of surface

(a) (b)

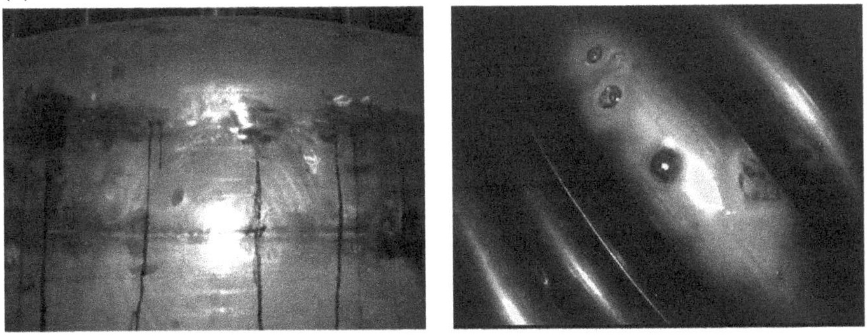

Figure 1.3 Pitting corrosion (a) in stainless steel (SS) tanks detected through normal visual examination and (b) in heat exchanger tubes detected through remote visual examination using video image scope.

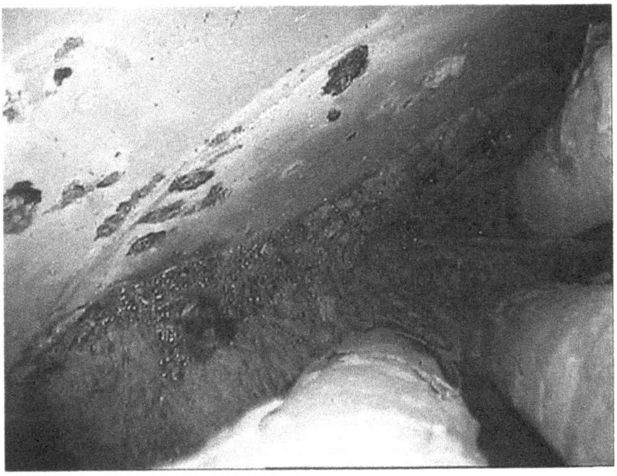

Figure 1.4 Typical crevice corrosion detected in a heat exchanger of a power station using video image scope-based remote visual inspection.

(a) (b)

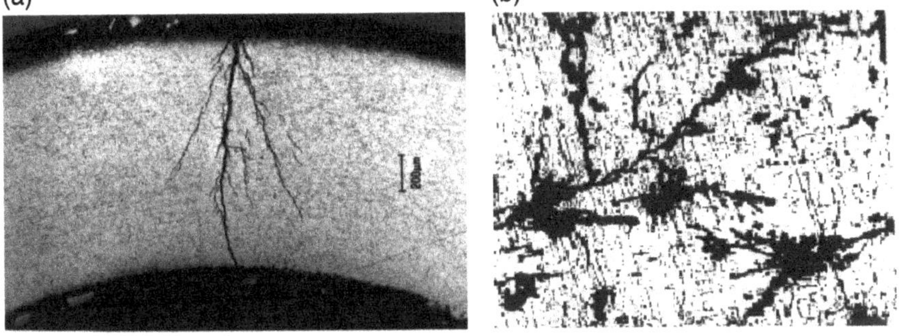

Figure 1.5 Transgranular stress corrosion cracking (TGSCC) visualized through (a) optical microscope and (b) stereomicroscope. (Note the branched nature of TGSCC. The crack initiated from corrosion pits located on the outer surface of stainless steel tube.)

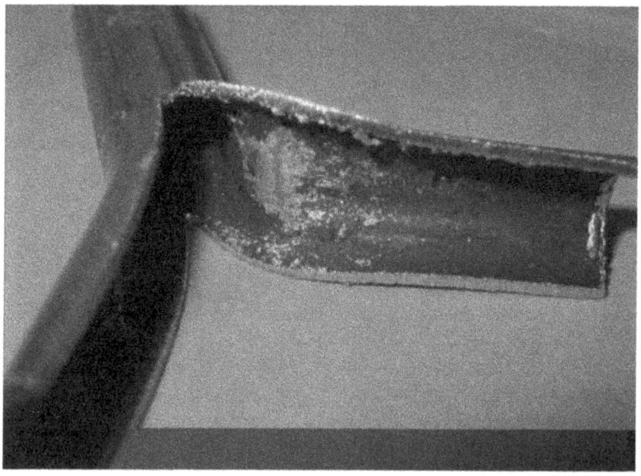

Figure 1.6 Typical erosion and pitting corrosion in cupro-nickel alloy (90 : 10) that failed prematurely in service.

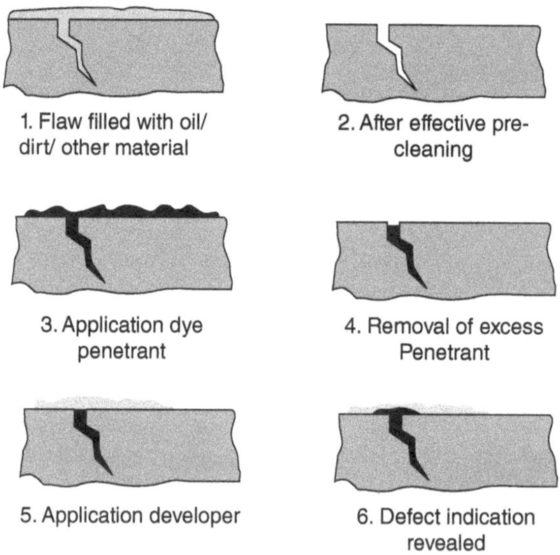

Figure 1.7 Schematic sketch of LPT methodology.

flaws. The second step is the application of visible or fluorescent solution called penetrant on the surface either by dipping, brushing, swabbing, or spraying.

The penetrant is allowed to remain on the surface for a period (referred to as dwell time) of 5–30 min depending on nature of defect to be detected. During this dwell time, the penetrant seeps into the discontinuities by capillary action. After the dwell time has elapsed, the excess penetrant is removed from the surface by gently wiping the surface with a moistened lint-free cloth or light rinse depending on nature of penetrant process.

A developer is then applied uniformly over the surface. The developer draws out the penetrant trapped inside the discontinuity and spreads it on the surface. Referred to as bleed-out, this has a magnifying effect that helps in revealing even minute defects. In the case of visible dyes, the contrasting color between the penetrant (red) and developer (white) makes the discontinuity to be seen under natural or artificial lighting. With fluorescent dyes, the surface is illuminated by ultraviolet (UV) light (also referred to as black light), which causes the bleed-out to fluoresce brightly, thus enabling the imperfections to be seen.

Penetrants are classified as visible or fluorescent depending on the nature of the dyes used. While visible penetrants use a color contrast (usually red) dye, fluorescent penetrants use a dye that fluoresces essentially in the greenish-yellow region of electromagnetic spectrum under black (UV) light. Fluorescent penetrants have higher sensitivity compared with visible dye penetrants. A number of penetrant types have been developed to cater for a wide variety of inspection conditions and sensitivity requirements. The main types of penetrants are water-washable, post-emulsifiable, and solvent-based systems. The primary difference between these is the method by which the excess penetrant is removed. While a rinse of water is used in the first case, there is a separate emulsification step using an emulsifier in the case of post-emulsified penetrant system, which renders the penetrant soluble in water. In solvent-removable system, a solvent is used to remove the excess penetrant. In terms of sensitivity, the water-washable penetrant system is the least sensitive, while post-emulsified systems have the highest sensitivity.

Applications Penetrant testing is one of the best techniques for surface-breaking defects, and hence, corrosion-related surface cracks and pits can be easily detected by this technique. Very tight stress corrosion cracks with widths of the order of a few microns can be detected by fluorescent penetrant. Figure 1.8 shows a typical intergranular stress corrosion crack detected through penetrant testing.

Apart from corrosion-assisted cracks and other types of cracks, LPT is also well suited for detection of all types of laps, porosity, shrinkage areas, laminations, and other similar defects open to the surface. As mentioned earlier, the strength of the technique lies also in the versatility of its use, as it can be applied on any material regardless of size, shape, thickness, and composition. It is a highly sensitive, simple, and cost-effective NDE method finding wide applications in practically all types of industries.

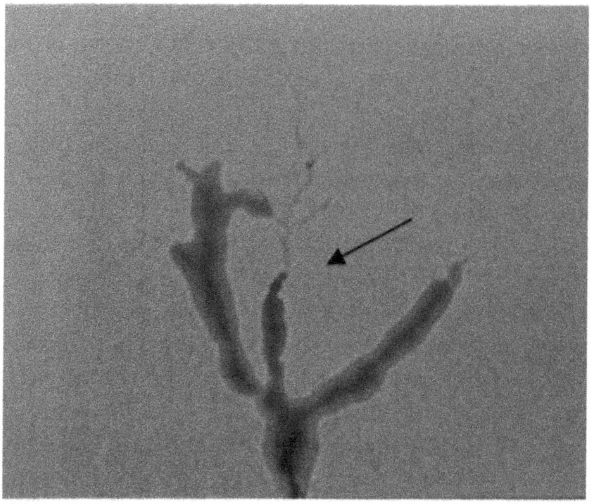

Figure 1.8 Dye penetrant testing reveals intergranular stress corrosion crack in austenitic stainless steel dished end. Fine branches are revealed by LPT.

The technique cannot detect subsurface defects, and it is difficult to apply the technique on rough and porous surfaces.

1.3.3 Magnetic Particle Testing

Magnetic particle testing (MPT) is a sensitive and widely used test method for detecting surface and subsurface defects in ferromagnetic materials and components. It is based on the principle that when a ferromagnetic material is placed in a magnetic field, the material gets magnetized. The magnetic lines of force that pass through the material are continuous from one pole to another. The presence of any surface or subsurface defect causes a deflection in the magnetic field and the associated lines of force and results in forming a leakage field. This leakage field is present at the surface of the magnetized component. When finely divided magnetic particles such as iron particles are spread onto the surface, it results in the collection of the magnetic particles at the discontinuity as shown in Figure 1.9, revealing the defect.

The "indication pattern" so formed can be clearly visualized and it reveals the location, approximate size, and shape of the discontinuity. Magnetization of a component is accomplished using one of the following: alternating current (AC), direct current (DC), half-wave DC (HWDC), and permanent magnet or electromagnets.

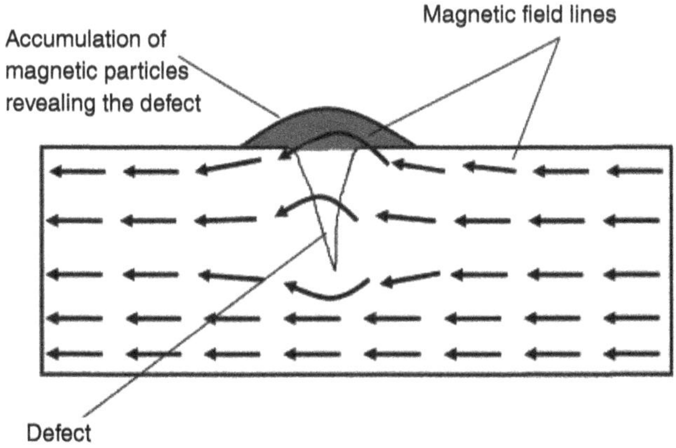

Figure 1.9 Magnetic particle testing principle.

Applications MPT is effective for the detection of near-surface corrosion and also surface cracking of ferromagnetic materials (such as ferritic steels). It is also used for detection of grinding cracks, heat treatment cracks, stringer-type nonmetallic inclusions, porosities, laps, and folds, which during service can lead to accelerated corrosion. It can be applied to finished articles, billets, hot rolled bars, castings, and forgings.

This technique can be used only for ferromagnetic materials. It is important that the magnetic fields be perpendicular to the principal plane of discontinuity so that maximum lines are intercepted. Thus, two or more sequential inspections are required to cover all orientations of defects. Demagnetization and post-cleaning following inspection is often necessary. Exceedingly high currents are required for large parts. Thus, adequate precaution needs to be taken to avoid local overheating in the component at points of electrical contact.

1.3.4 Radiographic Testing

Radiographic testing (RT) is one of the oldest and most widely used NDT methods [2]. In this method, a source of penetrating radiation is used. When these radiations pass through an object, a part of it is absorbed, a part gets scattered, and the balance gets transmitted. The amount of radiation absorbed depends on the nature of the object and its composition (density) and thickness. Variations in the transmitted beam intensity are detected by a detector, which can be the traditional sheet of photographic film or imaging plates or flat panel detectors. On processing the film, a permanent image is formed. In the case of imaging plate or flat panel detector, a digital image is formed, which can be processed and archived electronically. Film still continues to be the major detection medium though digital methods are becoming popular.

The developed film is called radiograph. Possible imperfections in the materials are indicated on the film as density changes. Thus, a void would appear darker since more radiation will pass through, while a tungsten inclusion because of higher density will appear to be white since it would attenuate more radiation. Figure 1.10 shows schematic setup for radiographic inspection. Radiography primarily involves a source of radiation, the object to be radiographed, and the detection medium. The commonly used sources are X-rays or gamma rays. X-rays are produced when fast-moving electrons are suddenly brought to rest in X-ray machines. A wide variety of X-ray machines are available in the market with energies ranging from 10 to 450 kV. A typical 450 kV X-ray unit can be used to examine about 100 mm of steel equivalent. For higher thicknesses, linear accelerators and betatrons are used. X-ray radiography gives better sensitivity as the energy of radiation can be varied to match thickness. Gamma rays are produced during decay of radioactive isotopes such as iridium-192 and cobalt-60. Radiography with gamma rays using isotopic sources is very popular especially for field applications, since they are compact and portable and do not require power supply. Cobalt-60, selenium-75, and iridium-192 are the

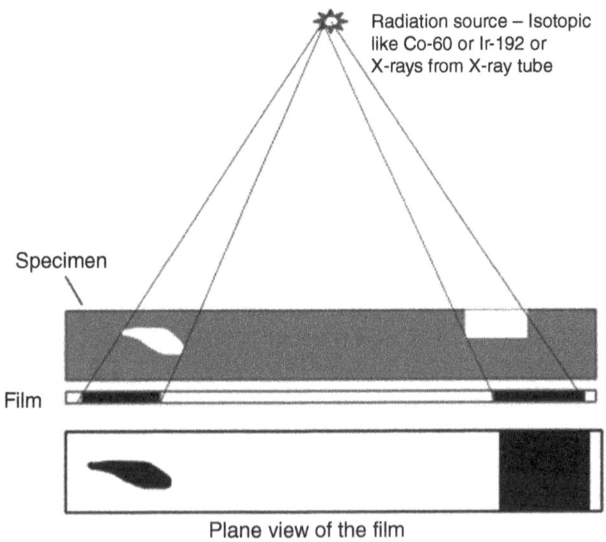

Figure 1.10 Schematic setup for radiographic inspection.

popular isotopic sources. The main disadvantage of gamma ray sources is that they decay with time and need replacement.

Applications Radiography is widely used in the process sector for detection of pipeline corrosion. It is also used for detection of pitting corrosion or localized corrosion, resulting in wall thinning. In civil structures, RT is used for detection of rebar corrosion. Apart from the above, other applications include inspection of welds, castings, assemblies and adhesively bonded structures, plastics, wood, composite materials, and electronic components. One of the much relevant advantages of radiography is that objects can vary in size from microminiature parts such as integrated circuit chips to huge components such as rocket motor casings. The technique is best suited for the detection of volumetric defects such as voids and porosities, and since a visual image is formed, interpretation is direct and has also the advantage of a permanent record.

Radiography requires access from both sides of the component. Linear flaws are difficult to detect, e.g. cracks cannot be detected unless these are oriented in line to the radiation beam. Also compared with other (NDT) methods of inspection, radiography is expensive. Since radiations are hazardous, proper precautions are to be taken while doing radiography.

The advances in radiography w.r.t. sources, digital technologies, Compton scattering, etc. remove many a limitations of RT and enhance its application.

1.3.5 Ultrasonic Testing

Ultrasonic testing (UT) is a widely used inspection method based on the application of sound waves with frequencies typically ranging from 0.5 to 10 MHz to examine materials. It is well known from basic physics that sound waves will travel through a given medium at a specific velocity and when these waves encounter an interface or boundary with a different medium, they tend to get reflected or transmitted depending on the acoustical properties of the medium. It is this principle that is used in UT. The ultrasonic waves used for testing are generated and received by small probes called ultrasonic transducers. These transducers contain piezoelectric ceramic or piezocomposite elements that convert electrical pulses into mechanical sound waves (ultrasonic transmitter) and sound waves into electrical energy (ultrasonic receiver). The transmitter probe generates ultrasonic pulses and transmits these into the materials, while the receiver probe receives the reflected pulses. The time taken by a sound pulse to travel through a material is a direct measure of the length of path traveled by it. By analyzing the received signals in time and frequency domain, one can locate the defects/flaws in the object under inspection. This is schematically depicted in

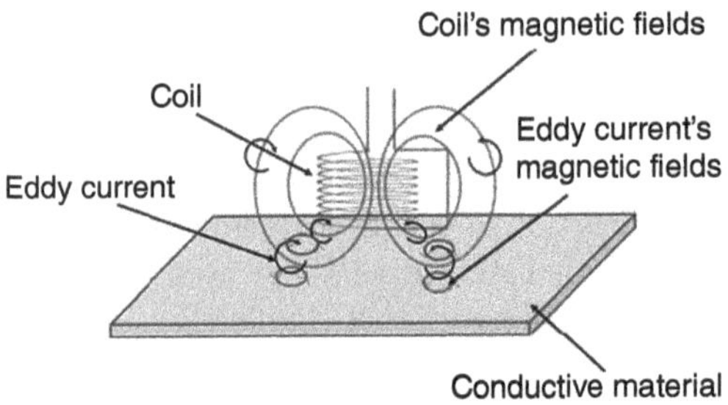

Figure 1.11 Principle of ultrasonic testing.

Figure 1.11. In ultrasonic inspection, both through transmission and pulse-echo techniques are used [3]. The last three decades has seen extensive advancements in the fields of materials science and instrumentation, which has led to the development of newer transducers (dual, multi-element, etc.) and techniques such as phased array and diffraction-based UT. The advent of robotics has resulted in enhanced applications of UT in industries, especially in areas such as corrosion mapping of pipelines, tanks, and vessels.

Applications Ultrasonics is one of the most widely used techniques for pipeline corrosion in boiler tubes, refinery, and process sector piping especially as an in-service tool. With better probability of detection for linear defects compared with radiography and capable of single-sided inspection, the technique is widely used for detection of cracks such as transgranular and intergranular stress corrosion cracking and hydrogen-induced cracking in nuclear, petrochemical, and other industries. The technique is also used for detection of under skin corrosion, localized corrosion, pits, and corrosion cracks in aircraft and marine structures.

Ultrasonics is also widely used for detecting defects such as laminations, shrinkage, cavities, flakes, pores and other discontinuities, inclusions, etc. in plates, pipes, welds, castings, and forgings in a reliable manner. The technique is widely used for QC and material inspection in all major industries. Online techniques have also been developed to monitor and classify materials as "acceptable" or "reject" and for process control applications.

Manual operation requires careful attention by experienced technicians. Unfavorable geometries, coarse anisotropic structures, and parts that are rough or irregular in shape or very small or thin are difficult to examine. Couplants are needed to provide efficient transfer of ultrasonic wave energy between transducers and parts being inspected. Reference standards are needed for calibrating the equipment and for characterizing the flaws.

1.3.6 Eddy Current Testing

Eddy current testing (ECT) is employed for examination of electrically conducting material [4]. In ECT, a coil carrying AC is brought close to the surface of the object under test. The magnetic field in the coil induces eddy currents in the object. The eddy currents generate secondary magnetic field that opposes the primary magnetic field of the coil. The change in the magnetic field is detected as a change in impedance of the test coil. The impedance change of the coil is affected by electrical conductivity, permeability and geometry of the material, test frequency, and spacing between the coil and the object. The principle of operation of ECT is shown in Figure 1.12. The test coil also referred to as the probe is the main link between the test instrument and test object and serves two main functions: the first is to establish a varying electromagnetic field that induces eddy currents within the test object, and the second is to feed the response due to the electromagnetic field to a signal analysis system. Various types of probes suitable for the inspection of flat surfaces and tubular products are chosen for testing. It can detect surface as well as subsurface flaws at certain depth. The depth of penetration for a particular material can be increased by lowering the frequency.

Applications ECT is commonly used for material sorting and measurement of coating thickness and to identify or differentiate between a wide variety of physical, structural, and metallurgical conditions in conductive materials. It is the most suitable technique for carrying out in-service inspection of heat exchanger tubes to detect pitting and cracking.

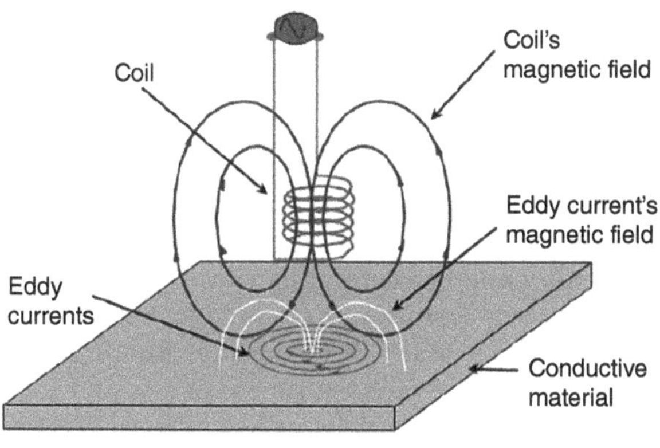

Figure 1.12 Principle of eddy current testing. *Source:* https://www.nde-ed.org/GeneralResources/MethodSummary/ET1.jpg.

In the aerospace industry it is widely used for detection of exfoliation corrosion, corrosion underneath structures that are often assembled by means of riveted lap jointing of various layers or skins or painted structures in aircrafts, and corrosion cracks around rivets and fasteners, and also in process industries it is used for detection of wall thinning due to corrosion. ECT is also used:

1) To measure or identify such properties as electrical conductivity, magnetic permeability, grain size, heat treatment conditions, hardness, and physical dimensions.
2) To detect seam, laps, cracks, voids and inclusions, and other surface discontinuities in non-ferromagnetic material.
3) To sort out dissimilar metals and detect differences in composition, microstructure, and other properties.
4) To measure the thickness of a nonconductive coating on a conductive metal.

Limitations The main limitation is its low penetration depth into parts being examined, and hence, its use is limited to thin walls or near-surface flaws. The technique is difficult to use on ferromagnetic materials. False indications are possible because of mixed variables, edge effects, and lift-off effects. Extensive technical knowledge is required for the development of inspection techniques and design of specific probes and to interpret the inspection data. The limitations of ECT are overcome by the advances such as multifrequency remote field and signal analysis technologies.

1.3.7 *In Situ* Metallography

The NDE methods elucidated earlier such as ultrasonics and ECT provide only an indirect characterization of material state such as sensitization, wall thinning, degradation, etc. However, for assessing the corrosion resistance, integrity, and health of a component, it is essential to know the actual metallurgical status of the material. It is a well-realized fact that microstructure governs properties like mechanical, metallurgical, and corrosion resistance of a metal/alloy. A microstructure also records subtle evidences of manufacturing processes like casting and also deterioration in service. Thus, visualization and analysis of microstructural signature can reveal the incipient damages and early onset of corrosion that would be beneficial in initiating corrective actions and thus assist in enhancing the life of the component.

Metallography is one of the best and most widely used techniques to assess the microstructural signature and hence the state of the component. While conventional metallography is not applicable for fabricated components and in-service inspection, field metallography practices also referred to as "*in situ* metallography" and sometimes referred to as "nondestructive metallography" are widely practiced in the industry. *In situ* metallography is an essential technique that a practicing corrosion, QA, and NDE professional should be aware of. Basic tools required to perform *in situ* metallographic technique are as follows:

1) A portable motorized handheld grinding and polishing device.
2) Disks of adhesive-backed grinding papers (120, 240, 320, 400, and 600 grit) in separate containers, disks of polishing cloths, diamond paste, slurry, suspension, or aerosol spray (6 and 1 μm).
3) Suitable etching agent (either manual or electrolytic).
4) A battery-operated, portable upright microscope with reflected light source with 5×, 10×, and 20× objective camera.
5) Cellulose acetate tape to acquire the positive replica of the etched area for assessment.

Etching plays an important role in deciding the outcome of the *in situ* metallographic examination. It can be done either manually or electrolytically. The type of etchant depends upon the type of material being examined such as stainless steel (SS), carbon steel, nonferrous alloys, etc. A convenient way of applying etchant on surface is by swabbing using cotton. Etching time depends upon the type of etchant and surface temperature of object being examined.

In situ metallography is routinely used in performing life assessment investigations of nuclear and fossil-fuel power plant equipment components such as turbine rotors, blades, steam piping, pressure vessels, etc. to ascertain the useful remaining/residual life of the component. It is also widely used to ascertain the effectiveness of solution annealing by observing the etch structure in the solution-annealed cold-formed austenitic stainless components such as dished ends and pipe bends in nuclear facilities. Figure 1.13a–c depicts a typical microstructure of AISI type 304 SS. While Figure 1.13a shows a step structure, Figure 1.13b shows a dual structure and Figure 1.13c shows a ditch structure. While step structure is a normal and preferable structure indicating effective solution annealing, dual structure is indicative of susceptibility to sensitization, while ditch structure indicates sensitization and also confirms that the material is prone to stress corrosion cracking (SCC).

In situ metallographic techniques are also widely employed in corrosion mapping of representative areas of vessel/equipment during in-service inspection/annual maintenance shutdown and wear surface mapping.

1.3.8 Neutron Radiography

Neutron radiography (NR) is a complementary NDE method to photon (X-ray or gamma ray) radiography. There are similarities as well as clear differences

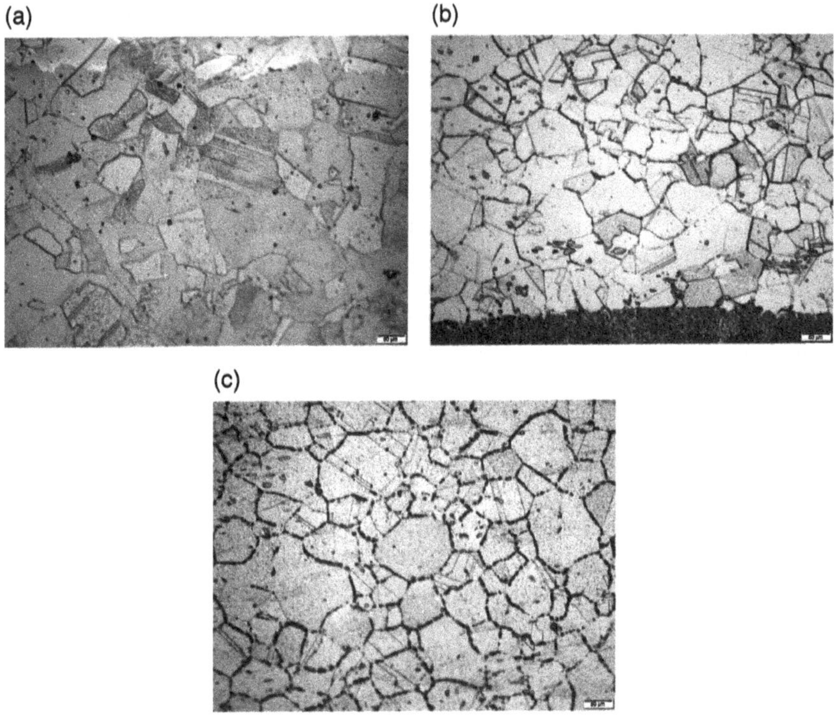

Figure 1.13 Typical microstructures of AISI type 304 stainless steel: (a) normal step structure, (b) dual structure, and (c) ditch structure. The dual structure indicates susceptibility to sensitization, while the ditch structure confirms prone for stress corrosion cracking.

between neutron and photon radiographic techniques. Similarities include the ability to produce a photographic image that records the changes in composition, density, and thickness of the object under examination. It is the degree of difference in attenuation of elements that is the advantage of NR over photon radiography. Figure 1.14 is a typical attenuation plot for X-rays (125 keV) and thermal neutrons. The attenuation differences arise because of the varied mechanisms of interactions with matter. While neutrons interact with the nuclei of atoms, photons interact with orbital electrons. Thus, NR is advantageous in imaging low atomic number material in high atomic number matrices and also in detection of changes in isotopic composition that are difficult to accomplish in photon radiography.

A typical setup for NR includes neutron source, moderator, collimator, and detection assemblies all enclosed in a suitably shielded enclosure. Three widely used neutron sources for radiographic applications are nuclear

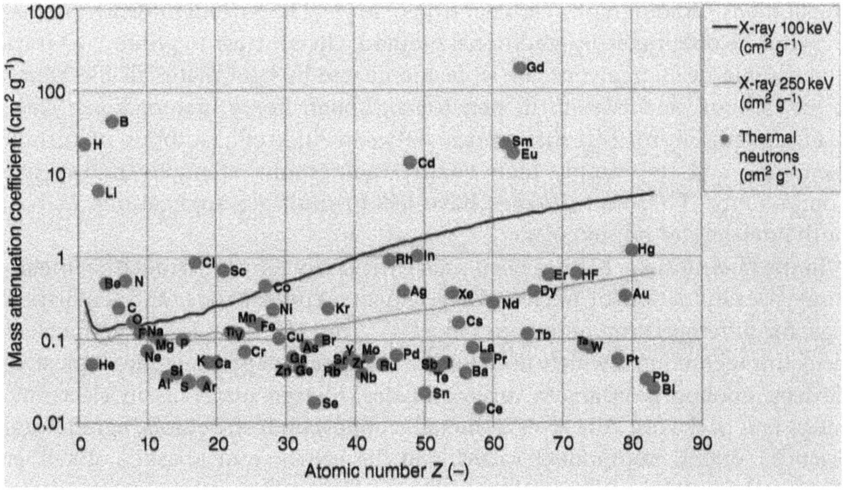

Figure 1.14 Typical attenuation plot for X-rays and neutrons.

reactors, particle accelerators, and radioisotopes. Reactor sources are best as they can offer very high neutron flux. But the cost and engineering complexity are high compared with accelerators that rank next followed by radioisotopic sources such as californium. NR is normally accomplished using thermal neutrons. The neutrons produced by the abovementioned sources are primarily fast neutrons (energy >10 keV). Hence, they must be moderated to a level of thermal energy range (<1 eV, typically 0.025 eV). Water, heavy water, graphite, etc. are some of the widely used moderator materials. The extraction of beam from core is achieved through the insertion of probe (beam) tube or collimator into the moderator, which permits only those neutrons having a direction to that of the tube axis to pass through. Two main types of collimators are the parallel wall collimator and the divergent collimator. The divergent collimator is widely used since a large inspection area can be covered. Collimators are characterized by their length to diameter ratio. The higher this ratio, the better the neutron image quality.

Neutrons are nonionizing unlike X-rays or charged particles. Hence a two-step process is used in NR to produce the radiographic image. In the first step, the neutrons are absorbed by a convertor screen. The screen gets activated and emits ionizing radiations (charged particles/photons). In the second step, a film is used in conjunction with the convertor screen to record the image produced. The techniques used for imaging can be classified as direct and indirect (also known as transfer imaging). Gadolinium foils are used for direct imaging and indium/dysprosium foils for indirect or transfer imaging.

Applications Radiography with neutrons can yield important information that may not be obtainable by traditional methods. In contrast to photons (X-rays and gamma rays), neutrons can (i) be attenuated by light materials like water, hydrocarbons, and boron; (ii) penetrate through heavy materials like steel, lead, and uranium; (iii) distinguish between different isotopes of certain elements; and (iv) supply high quality radiographs of highly radioactive components. These advantages have led to multiple applications of NR, both nuclear and nonnuclear.

In the nuclear field, NR has been used extensively for post-irradiation metallurgical examination of nuclear fuel elements, control rods, and irradiation rigs, for differentiation of isotopes like U^{235} from U^{238}, etc. In the nonnuclear field, this technique is widely used for the inspection of pyrotechnic (explosive) devices, cooling passages in turbine blades, foreign materials in electronic relays and packages, adhesives in metallic honeycomb structures, agricultural science, ceramic component valves, and the presence or absence of rubber seals, gaskets, etc. in complex assemblies.

Neutron sources are neither cost effective nor portable. Applications are limited to detection of low Z materials only. Field applications are limited and isotopic sources have low neutron yields. Hazards of radiation exist and hence adequate shielding and radiation protection measures need to be undertaken, which makes applications cumbersome.

1.3.8.1 Corrosion Detection Through NR

Corrosion products contain low Z elements such as oxides/hydroxides and NR is a good tool for the detection of corrosion. This is due to the fact that while other nondestructive methods for corrosion detection, such as X-ray imaging, ultrasonics, or eddy currents, depend on metal/material loss as the principal means for detection, NR using thermal neutrons relies on the relatively high thermal neutron scattering cross section for hydrogen. The presence of hydrogen in practically all corrosion products thus leads to the possibility of detection of hidden corrosion in assemblies at nascent stages itself.

The latest advances in this field is digital real-time NR using flat panel detectors and computed radiography (CR) using imaging plates. Real-time imaging methods offer advantages in immediate visualization, digital processing of data, and archiving for baseline comparison studies. The images also lend themselves to processing and enhancement with techniques such as frame averaging, beam flattening, and gray-level stretching, all of which are effective in bringing out details in neutron radiographic images of corrosion. Detection capability in aluminum is shown to be at least 0.18 mg H cm^{-2}, representing an aluminum metal loss of about 25 μm. Figures 1.15 and 1.16 show typical neutron radiographs of aircraft components revealing hidden corrosion.

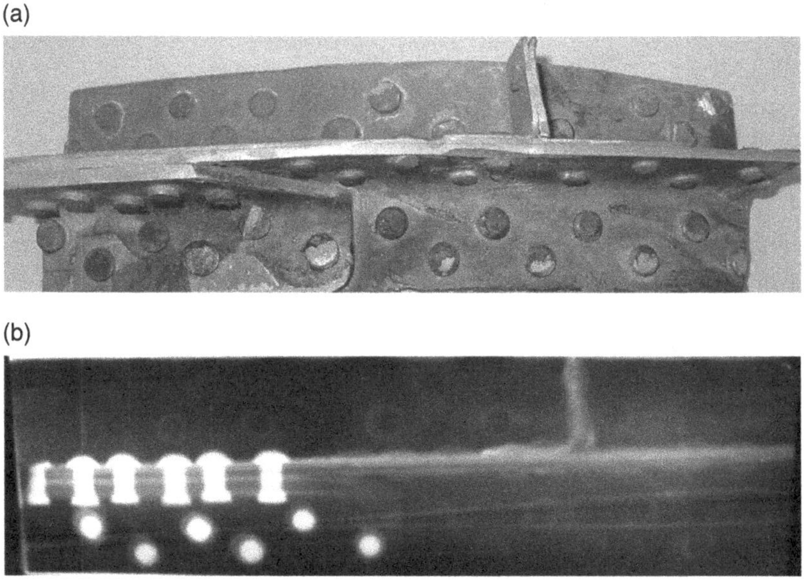

Figure 1.15 (a) Photograph of aluminum riveted plate. (b) Neutron radiograph of the above aluminum riveted plate. Hidden corrosion revealed as continuous white line in the center of the radiograph. Presence of corrosion is also seen underneath the rivet heads in the base plate.

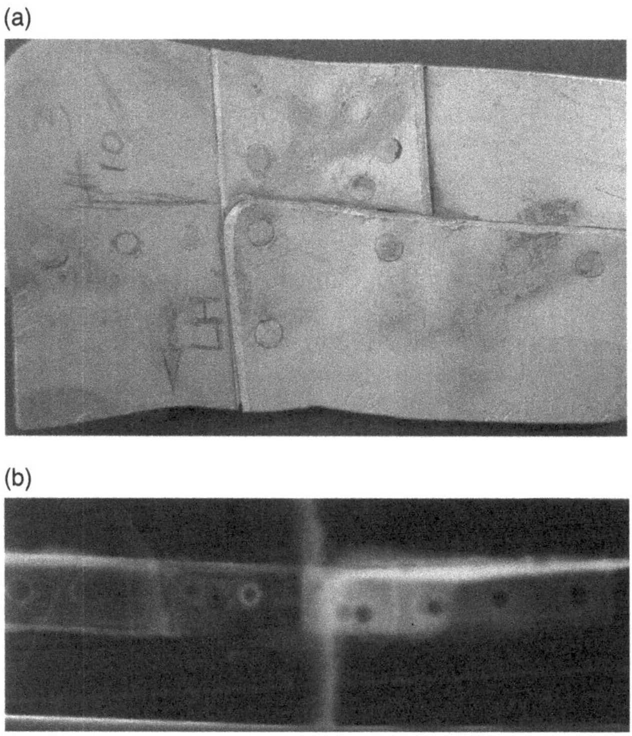

Figure 1.16 (a) Photograph of aluminum riveted plate. (b) Neutron radiograph of the above plate. Presence of corrosion product below the aluminum plate is clearly revealed due to higher attenuation of thermal neutrons.

1.4 Advances in Nondestructive Evaluation for Corrosion Evaluation and Damage Assessment

The need to have minimum plant downtimes coupled with shorter maintenance schedules and high productivity has led industries to look for alternative and advanced NDE technologies that can enhance reliability and sensitivity of testing and evaluation with reduced inspection times and cost. In process and piping industries, there has also been a need to inspect corrosion damage without removing the insulation. Some of the recent NDE techniques that have lent themselves toward all these objectives include DR, AE, and IRT.

1.4.1 Digital Radiography

While conventional radiography utilized the film as the detector medium, DR uses digital detector arrays and imaging plates. Compared with film radiography, DR offers many advantages such as lesser exposure time, wider dynamic range, environmental friendliness (with no chemicals required), cost-effectiveness, and easier archival of digital images. Digital images can also be subjected to a variety of image analysis and enhancement functions that lead to reliable quantitative inspection.

Flat panels using detector arrays and CR using imaging plates are two forms of DR that are extensively applied for corrosion evaluation and damage detection. Both these techniques provide quantitative data with respect to material wall thickness changes due to internal corrosion or erosion damages.

1.4.1.1 Computed Radiography

CR is similar to conventional radiography except for the fact that instead of film, an imaging plate is used as the detection medium. The imaging plate contains photostimulable storage phosphors, which retains the latent image. When the imaging plate is scanned with a laser beam in the scanner, the latent image information is released as visible light. Emitted light is detected by a photomultiplier tube, which is then converted to an electronic signal. The electronic signal is then converted to discrete (digital) values that are displayed on the monitor as a gray-level image. This is schematically depicted in Figure 1.17. The main advantage of CR is that the imaging plate is reusable. When handled carefully, each plate can be used more than 500 times. Since there is no chemical processing, the technique is considered eco-friendly. Since these plates have higher dynamic range, the thickness range that can be covered in a single exposure is quite large compared with film radiography. This means lesser number of exposures. Apart from these, the other major advantage is that the image can be digitally processed, enhanced, and archived.

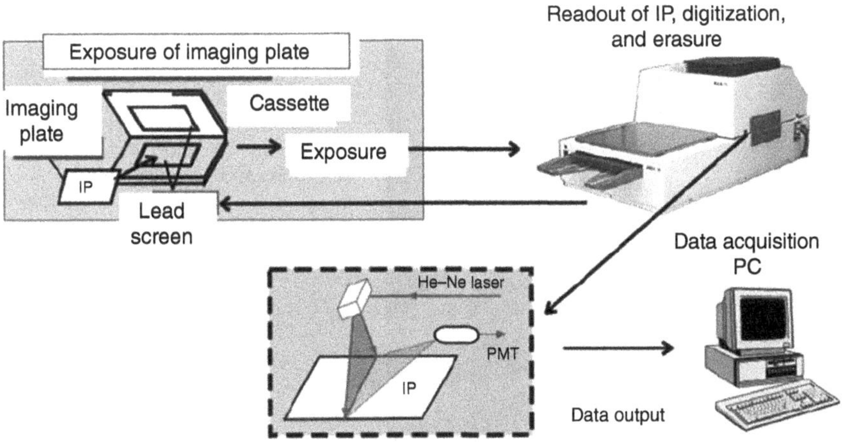

Figure 1.17 The computed radiography process. *Source:* Courtesy of Dr. Uwe Ewert and Dr. Uwe Zscherpel, IAEA presentation.

Applications

1) CR is used for evaluation of internal or external corrosion/erosion losses in process piping, pressure vessels, and valves. Figure 1.18 is a typical CR image of a pipe with insulation. Through application of contrast stretching and image processing, one can visualize the thicker and thinner portions in the same image. In the case of any internal corrosion, estimation of remaining thickness and wall losses can be accurately undertaken out using appropriate image processing and analysis techniques.
2) Evaluation of valves and their internals for internal or external corrosion and erosion.
3) Evaluation of piping and valves for buildup of corrosion products and for flow blockage.

In addition to CR, DR using flat panel detectors is increasingly being used in field applications. The better image quality and the ability to provide instant or real-time images provide attractive alternatives to CR and film radiography. Flat panel detectors are ideal solution for aboveground cross-country pipeline inspection for corrosion and erosion, where often miles of pipeline require inspection. Flat panel detectors with a robotic crawler help to optimize productivity and safety.

1.4.2 Acoustic Emission

AE is defined as a class of phenomenon whereby transient elastic waves are generated by the rapid release of energy from localized sources in a material [5]. AE occurs as a release of a series of short impulsive energy packets. The energy

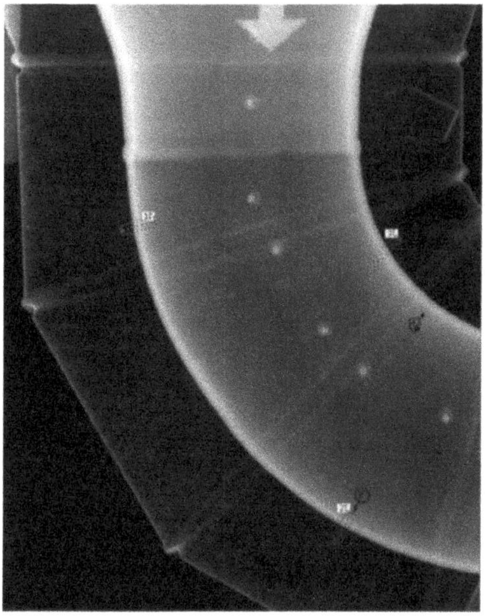

Figure 1.18 Computer-aided inspection of an insulated pipe by means of the projection radiography. The pipe wall thickness along with insulation details can be clearly seen, which is an indication of the sensitivity and versatility of CR imaging. *Source:* Courtesy of Dr. Uwe Ewert and Dr. Uwe Zscherpel, IAEA presentation.

thus released from the material travels as spherical wavefront and can be picked up from the surface of the material using highly sensitive transducers. The wave thus picked up by the transducer is converted into electrical signal that on suitable processing and analysis can reveal valuable information about the source causing the energy release. It is to be emphasized that this method is for dynamic evaluation, i.e. when the phenomenon is taking place and cannot detect degradation or phenomena after the event is completed.

Sources of AE include many different mechanisms of degradation, deformation, and fracture. The naturally occurring sources are earthquakes and rock bursts in mines. In metals, different AE sources include generation and propagation of cracks, movement of dislocations and grain boundaries, formation and growth of twins, fracture and decohesion of brittle inclusions, phase transformations, etc. In composite materials, the sources of AE include matrix cracking, debonding, and fracture of fibers. There are also secondary or pseudo sources of AE, which include leaks and cavitation, friction, realignment and growth of magnetic domains (Barkhausen effect), liquefaction and solidification, and solid–solid phase transformation. Relating to corrosion, AE is generated during SCC, hydrogen embrittlement, cracking/spallation of scales during oxidation, etc.

AE signals are broadly classified into two major types, namely, continuous type and burst type. The waveform of continuous type AE signal is similar to Gaussian random noise, but the amplitude varies with AE activity. In metals and alloys, this form of emission is considered to be associated with the motion of dislocations. Burst-type emissions are short-duration pulses and are associated with discrete release of high amplitude strain energy. In metals, the burst-type emissions are generated by twinning, micro-yielding, and development of micro- and macrocracks [6]. The difference between these two types of emissions is in the average repetition rate. Above certain values of the average repetition rate, the length of the bursts exceeds the time between them, leading to the inability of the instrumentation to discern between two successive burst emissions. The result is the superposition of bursts giving the appearance of continuous emission. Actually both are generated by discrete processes, but the amplitude of continuous emission is usually lower than that of the burst type. A typical burst-type AE signal is shown in Figure 1.19. For burst-type AE signals, certain AE parameters are commonly used for the analysis. These include threshold V_t, ring-down count (RDC), event duration (ED), peak amplitude (PA), rise time (RT), and energy (En). Apart from these, root mean square (RMS) voltage of the AE signal is used for characterizing the continuous emission.

Among various types of sensors, the most commonly used is the piezoelectric type. It has the sensitivity of the order of 10^{-14} m. Both resonant and broadband sensors are used for different applications. The frequency range of interest for AE is from 30 kHz to 2 MHz.

Advantages

- AE is a dynamic inspection method in that it provides a response to discontinuity growth under an imposed structural stress.

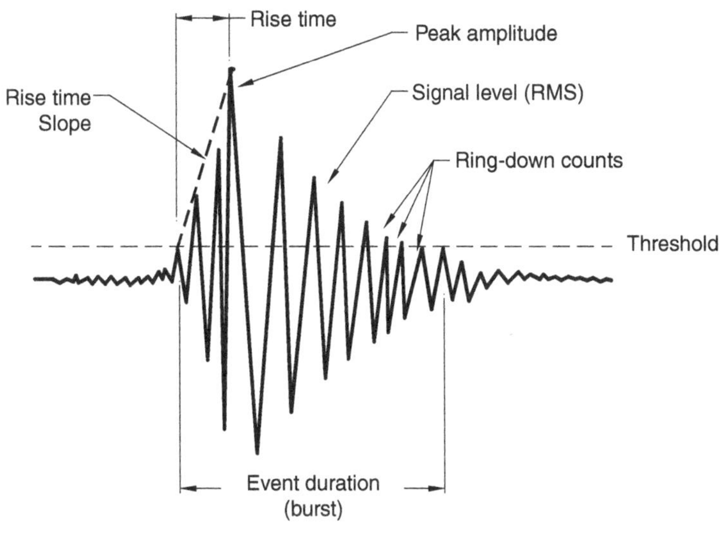

Figure 1.19 Characteristics of AE signals.

- AE can detect and evaluate the significance of discontinuities throughout an entire structure during a single test.
- Since only limited access is required, discontinuities that are inaccessible to the more traditional NDT methods may be detected.

Limitations

- Irreversible process.
- Background noise can have a major impact on AE testing (AET).
- Sensors must be placed within a reasonable distance.
- Static discontinuities will not be detected unless they propagate.

Applications of AET AET is applied in the field of materials research and in industrial applications. Various applications include mechanical property testing, study of crack growth and fracture behavior, preservice proof testing, in-service (requalification) testing, leak detection and location, study of phase transformation, structural integrity evaluation, stress corrosion and oxidation studies, online monitoring of welding, and forming processes.

1.4.2.1 AET for Corrosion Evaluation

Online monitoring ability of AE technique can be used for the detection and evaluation of corrosion. AET is used for monitoring initiation and propagation of cracks resulting from different types of corrosion including pitting corrosion [7, 8], crevice corrosion [9], erosion corrosion [10], and SCC [11, 12]. Metallic corrosion involves some chemical reactions that result in a redistribution of energy. Any corrosion process thus is a potential source of AE.

Pitting is a form of extremely localized attack, the rate of attack being greater at some areas than at others. It is the most common form of corrosion that results in failure of equipment due to formation of holes leading to leak. To establish an AE system, which could detect ongoing pitting corrosion and quantitatively monitor the pitting corrosion rates of various vessels or chemical equipment in service, a series of investigations was conducted to study the development of pitting corrosion on AISI 316L austenitic SS [7]. Tests were conducted at room temperature in 3% NaCl solution acidified to pH 2, at the free corrosion potential or with applied anodic polarization. AE signals were detected during pitting corrosion, and a good correlation was observed between AE activity and pitting rate. The exact nature of AE sources within pits could not be identified clearly, but the results demonstrated that AET can be used to detect and monitor the occurrence of such phenomena. Figure 1.20 shows the number of detected AE events versus the calculated cumulative pit size in terms of total corroded volume of pit cells or total area of the inner active surface of the cells.

In another work, AE generated from a low-carbon type 304 SS during pitting and transgranular stress corrosion cracking (TGSCC) has been reported [8].

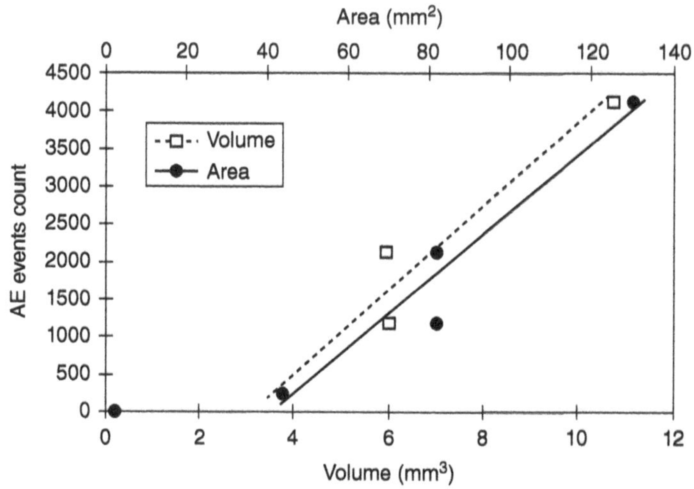

Figure 1.20 Number of detected AE events versus calculated pit size.

Tests were conducted in 0.01 and 1 M NaCl with the pH adjusted to 1 with HCl at potentials of −380 and 0 mV (SCE) for no applied stress and with a stress equal to 75% of the yield strength of the material. In this study, a correlation of AE with corrosion was suggested by higher AE rate observed for the 1 M versus the 0.01 M NaCl solutions and the increase in AE rate with increasing pit frequency and current. Both stress and potential affected the transition from low to high AE rate regimes. From the above case studies, it can be comprehended that there is a good correlation between AE activity and damage due to pitting corrosion.

Crevice corrosion occurs in interstitial zones where a passive metallic material is locally in contact with a confined corrosive environment, whereas the majority of the metallic surface remains in contact with bulk liquid. An investigation was carried out to study the ability of AE technique for detection and monitoring of crevice corrosion in 304L austenitic SS [9]. Crevice initiation, propagation, and repassivation were controlled by additions of hydrogen peroxide and hydrochloric acid and by the extent of the applied torque of the assembly, in the presence of chloride ions. The simultaneous measurements of corrosion potential of the specimen and AE activity generated during the test showed a good correlation between potential and AE activity and amplitude of crevice damage in terms of weight loss and condition of metallic surface [9]. Results demonstrated that different parameters of AE signals such as rise time, counts, duration, and cumulative energy are affected by crevice development (i.e. initiation and propagation of crevice during the test). It was proposed that formation of bubbles from cathodic reactions within the crevice is the source of AE [9].

The phenomenon of corrosion–abrasion is responsible for the premature destruction of numerous industrial installations in which corrosive fluids circulate, together with another phase, essentially a solid one. In addition to the deterioration resulting from electrochemical corrosion, a mechanical damage results from the velocity of the fluid and particularly from the interactions between the particles in suspension and the surface of the impacted material. In a noncorrosive environment, the way in which material is torn off by the impacts of particles depends on the mechanical properties of the material and the conditions of the impacts. In a corrosive environment, several factors are responsible that control the relative extent of the corrosion compared with the abrasion alone. Applications of only electrochemical techniques do not allow the separation of both the corrosion and abrasion contributions. In order to study thoroughly the interactions between the two types of damages, an investigation combining AE and electrochemical measurements has been reported [10]. This work is related with the correlation of acoustic energy and electrochemical-free corrosion potential during abrasion–corrosion tests of AISI 304L SS in saline environment. The abrasion mechanism was examined for three angles of impingement in noncorrosive conditions. Then, the synergistic effect between mechanical damage and corrosion was investigated. The results indicate that the lower the impingement angle, the lower the mechanical contribution. In corrosive conditions, it was shown that at low abrasion rate, the synergistic effect depends on the impingement angle [11].

Corrosion processes present an attractive proposition for detection and characterization by AE technique. Some of the processes during corrosion such as dissolution of metal and film formation do not emit AE since the lattice is not strained. However, some other processes such as film cracking, gas evolution, hydrogen migration, plastic zone formation, discontinuous cracking during SCC, and hydrogen cracking produce AE since all these processes strain the lattice. Several studies have been reported on application of AE for detection and monitoring of SCC in different materials. Shaikh et al. [12] have analyzed the AE signals to determine the microprocess during SCC of AISI type 316LN SS and thus the mechanism of the SCC process. Investigators used annealed AISI type 316LN SS compact tension (CT) specimens polished to 600 grit finish, pre-cracked, and subjected to SCC testing in 45% magnesium chloride solution at 413 K in the range of stress intensity factors of 13–26 MPa. The analysis of the AE signals in conjunction with fractography suggested a surge in the AE counts and energy to indicate initiation of SCC. AE was found to be continuous prior to the initiation. The time gap between AE events increased during initiation. AE events occurred in bursts during crack growth. Plastic deformation ahead of the crack tip was determined to be the major source of AE during propagation of SCC in type 316LN SS. The cracking was found to initiate and propagate in the transgranular mode.

1.4.3 Infrared Thermography

IRT has gained wide acceptance as an NDE technique due to its ability to inspect large areas in a noncontact/noninvasive manner and its fast inspection rates. It started as a condition monitoring tool in electrical and process industry. The better resolution and thermal sensitivity due to improved sensors, electronics, and imaging devices have resulted in wider range of applications such as material characterization, tensile deformation studies, online weld monitoring, thermal diffusivity measurements, and defects including corrosion detection and evaluation.

The basic principle of IRT inspection involves measurement or mapping of surface temperatures when heat flows from, to, or through a test object. All bodies above absolute zero emit electromagnetic radiation. At ambient temperatures and above, these radiations are emitted predominantly in the IR band of the electromagnetic spectrum. Using an IR detector, it is possible to convert these IR radiations into electrical signals, which can then be displayed on a monitor as a gray or color image. With appropriate calibration, it is also possible to get the absolute temperature values of any point on the surface of the object. Thus a complete surface temperature map of the object can be obtained in a noncontact way. In IRT, temperature differentials on a surface, or changes in surface temperature with time, are related to heat flow patterns and can be used to detect flaws or to determine the heat transfer characteristics of a test body or for corrosion detection and evaluation. The principles, advantages, limitations, and varied applications are extensively discussed in references [13, 14].

A typical IR imaging system essentially consists of the IR camera, control unit, image acquisition, and analysis unit. A typical schematic of IR thermal imaging system is shown in Figure 1.21. The heart of the system is the IR detector, which converts the electromagnetic thermal energy radiated from an object into electric signals. These signals are amplified and transmitted to a display monitor. To obtain the absolute temperature values, appropriate correction factors (emissivity, relative humidity, surrounding temperature, etc.) are applied, and calibration is performed using internal black body references.

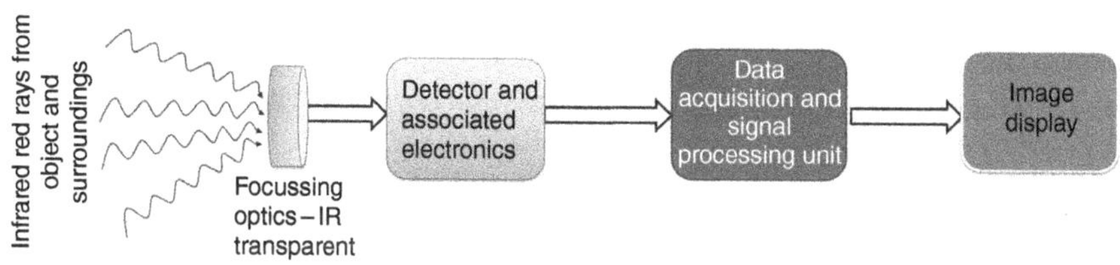

Figure 1.21 Schematic diagram of the IR imaging system.

1.4.3.1 Techniques in Infrared Thermography

There are two primary techniques in thermography, namely, passive and active. Passive technique involves applications where the material already contains its own internal source of heat. Majority of the condition monitoring applications where the component themselves get heated up due to a variety of reasons fall under this category. Active techniques involve the application of an external thermal perturbation (heating or cooling) directly or indirectly to the object as a whole or of a small area of interest within the object. While both heating and cooling can be applied, it is heating that is generally preferred. A variety of stimulation sources have been used such as hot air guns, incandescent and flash lamps, lasers, plasma arcs, inductive heating, heating strips, acoustic energy, etc. Active thermography is mainly used for defect detection and depth estimation, coating thickness evaluation, etc. Pulsed thermography (PT), lock-in thermography (LT), pulsed phase thermography (PPT), vibro-thermography, and step heating are some of the variants in active thermography. The external stimuli used to excite the material can be optical, mechanical, ultrasound, eddy current, hot and cold simulation, etc.

1.4.3.2 Pulsed Thermography

In PT, a short and high energy light pulse of a few milliseconds duration is flashed on the testing material. Qualitatively, the phenomenon is as follows. The front surface of the material absorbs the light energy, causing an increase in surface temperature. The thermal front propagates inside the material primarily through diffusion. The presence of a flaw or anomaly alters the diffusion rate. The surface temperature is recorded using an IR camera primarily from same side (reflection mode). On the thermal image, defects/anomalies appear as areas of different temperatures with respect to unflawed surrounding areas. An advantage of PT is that one can evaluate the defect depth through thermal time of flight. Defects located in the vicinity of the surface appear immediately, while deeper defects appear later and with a reduced contrast and definition. Xenon flash lamps are widely used for pulse heating.

1.4.3.3 Lock-in Thermography

In LT, a continuous sinusoidal light source of single frequency is used to heat the surface of testing object. The surface temperature of the object also gets modulated with the same frequency as the heating source. This change in temperature is monitored using an IR camera. Then the input and output waves are compared to get the amplitude and the time lag information. The defective area will have different amplitudes and phase values when compared with the area without defect. To get the amplitude and phase images, algorithms like fast Fourier transform (FFT) or four-point correlation (FPC) are used. Phase image is not sensitive to emissivity variation and surface irregularities, and it gives more precise depth information as compared with mere amplitude image.

1.4.3.4 Pulsed Phase Thermography

PPT combines the advantages of both PT and LT. The testing specimen is exposed to short pulse, like in PT, and the temperature decay is recorded using an IR camera. Mathematically, any pulse can be approximated by a sum of sinusoids. Hence, when a specimen is pulse heated, thermal waves of various amplitudes and frequencies are generated in the specimen at any point of time. Using mathematical tools such as FFT, one can analyze the various frequencies, and phase and amplitude information can be obtained for the desired frequency. PPT combines the advantage of PT (fast inspection) and LT (more depth information). The technique is also independent of emissivity variation.

1.4.3.5 Corrosion Detection Using IR Thermography

Extensive investigations have been undertaken in the area of corrosion detection and evaluation by advanced IRT. The technique has been employed for an early detection of intergranular corrosion (IGC) in AISI type 304L SS and biofouling and corrosion of rebars in concrete. A few of these are outlined below.

1.4.3.5.1 Detection of Sensitization and IGC

Austenitic SS are widely used as structural materials and for component fabrication in nuclear and other process industries. IGC is one of the major localized corrosion problems faced by austenitic SS components during service. Conventional nondestructive methods such as ultrasonic, eddy current, radiography, etc. have limitations in detecting localized corrosion, particularly at an early stage. Authors have explored the feasibility of using IRT for detection of sensitization. Sample plates of AISI types 304 and 304L SS were subjected to sensitization heat treatment at 675 °C for durations varying from 30 min to 250 h. IGC resistance was evaluated as per ASTM A262 Practice A (oxalic etch test) and double loop electrochemical potentiokinetic reactivation (EPR) method. Metallography was also carried out, which revealed that both samples had got sensitized (ditch structure observed). LT was then carried out on the samples with the frequency varying from 0.05 Hz to a few Hz. Detailed analysis of the phase images revealed that the phase angle value of the sensitized specimen increased with thermal aging time. Good correlation was obtained for 304 SS between the phase angle and the degree of sensitization (DOS) and the microstructure (Figure 1.22). This work reveals the potential for LT as a tool for early detection of sensitization and IGC. It is a significant result as chemical methods cannot be applied to manufactured component.

1.4.3.5.2 Detection of Wall Thinning in Metallic Pipes

Corrosion in metals especially in process and piping industries results in material loss and consequent wall thinnings. Thus, the quantification of corrosion in terms of metal loss is required for pipings over large lengths in nuclear fuel processing or chemical and process industries. A series of systematic studies

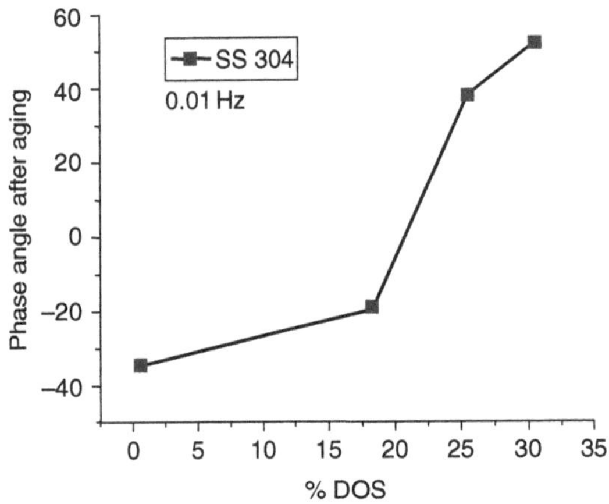

Figure 1.22 Plot of phase angle with degree of sensitization (DOS). Note the increase in the phase angle values with the quantum of sensitization.

Figure 1.23 Wall thinnings (~20%) detected in zircaloy tube.

have been carried out on the detection of wall thinning in SS, carbon steel, and zircaloy pipes by IR thermal imaging. The outer diameter of the pipes ranged from 80 to 200 mm and the wall thickness varied in range of 4–7 mm. Grooves of varying depths (corresponding to wall thinnings in the range of 12.8–90%) were made on both inner and outer surfaces. Active thermal imaging through internal and external heating was adopted for creating the required temperature differential. Analysis of the thermal images revealed that in SS pipes, it was possible to reliably detect wall thinnings of the order of 32%. In the case of zircaloy, as indicated in Figure 1.23, wall thinning of the order of 20% could be detected.

In both the above studies, the minimum size of the defect that could be detected was well above the requirements of the ISI code of practice followed in industries for wall thinning. The detailed studies also clearly revealed that apart from emissivity, the other main parameters that influence the detectability of wall thinnings include (i) size and extent of wall thinning, (ii) material thickness, (iii) thermal properties of the material such as thermal conductivity, and (iv) method of creating a thermal differential.

1.4.3.5.3 Biofouling and Microbial Corrosion Detection

A study on understanding the behavior of SS under immersed condition in the sea for evaluation of settlement and fouling characteristics of fouling organisms and their seasonal variation and the effect of the fouling organisms on the test materials in terms of (microbial) corrosion was carried out at the author's center. SS samples were exposed in the coastal seawaters of the Bay of Bengal. The samples were kept at water surface as well as at depths of 0.3 and 3 m from surface. The periods of immersion varied from 5 to 8 months.

Application of PT for the study was ruled out due to uncertainties pertaining to emissivity, surface variation, and geometry of surfaces. LT based on phase images appeared more appropriate as it reduces the influence of emissivity variation and surface problems on the measurement and provides more quantitative information on the various stages of corrosion.

Figure 1.24 shows the phase thermal images after 5, 7 and 8 months. The evolution of the corrosion pits is clearly revealed with time. Small pits observed after 5 months (Figure 1.24a) grew larger in 7 months (Figure 1.24b). An increase in the corrosion pit depth was also observed. Figure 1.24c shows the phase image of the sample exposed to the surface for 8 months. The effect of fouling organisms on the SS can be clearly observed from Figure 1.24c. The variation in depth of the pits could be correlated to the phase angle variation on the mapped surface, confirming the ability of LT to map the depth, area, and morphology of the corrosion damage and biofouling formation in marine corrosion.

1.4.3.5.4 Corrosion Detection in Concrete Reinforcement Using Infrared Thermography

Corrosion of reinforcing steel and other embedded metals is the leading cause of deterioration of reinforced concrete structures. The steel corrosion products occupy a greater volume as compared with steel. This expansion creates tensile stresses in the concrete, which can eventually cause cracking, delamination, and spalling of concrete cover, leading to further degradation. Corrosion of steel in concrete is an electrochemical process that takes place as a result of the formation of a corrosion electrochemical cell. Exposure of reinforced concrete to chloride ions in an aggressive environment is the primary cause of premature corrosion of steel reinforcement. By monitoring the corrosion of the structures and taking suitable measures at early stages, enormous saving is possible. Mapping of corroded and uncorroded reinforcement bars in the concrete samples and cracking of concrete due to accelerated corrosion and degradation have been monitored using lock-in IRT.

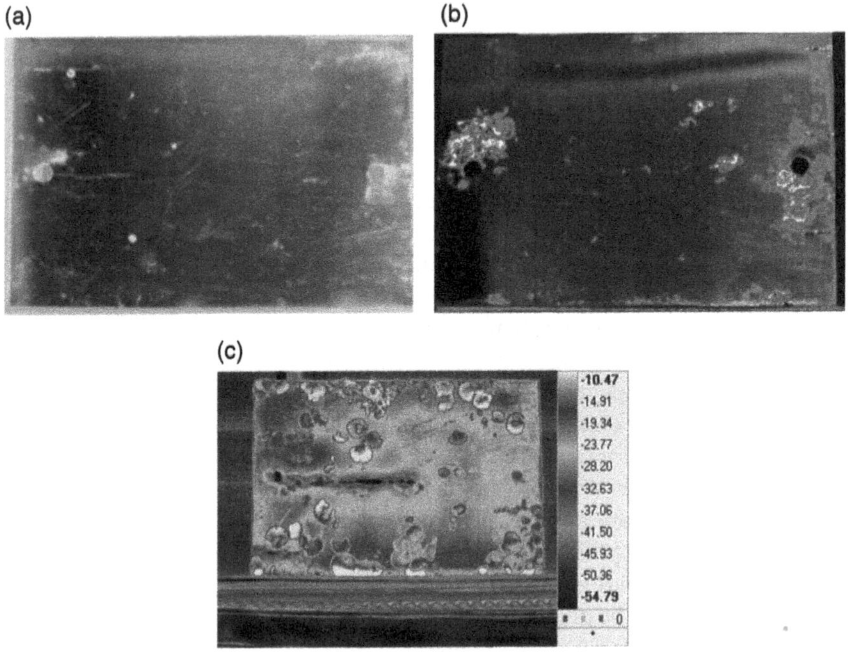

Figure 1.24 Phase image of steel coupons at a depth of 3 m exposed for (a) 5 months, (b) 7 months, and (c) 8 months. Pits are indicated by arrows. Note the progressive increase in the number of corrosion pits with time. The depth and diameter of the corrosion pits were also observed to increase with time.

Three concrete samples of size 600 mm × 300 mm × 25 mm with reinforcement bars of 8 mm diameter in each of them were fabricated and subjected to LT. An optimum frequency range was established through experiments. The distance between the bars and orientation of bars was recorded in the amplitude and phase images and analyzed with appropriate image processing. The corroded area has metal oxide that is an insulator and hence has different phase signal than uncorroded metal bars. This can be clearly seen in Figure 1.25.

The advantages offered by IRT as a fast and noninvasive tool coupled with the fact that it can be applied in service have spurred large interest in its application for corrosion detection. Liu et al. [15] have investigated the application of PT along with image and signal processing methodologies such as amplitude, phase, first and second derivative, and second principal component for characterization and quantification of pitting corrosion in small diameter ductile iron pipes. The features used by the investigators [15] clearly detected the pitting corrosion, but quantification of depth was challenging. A direct linear relationship between second principal component and corrosion depth

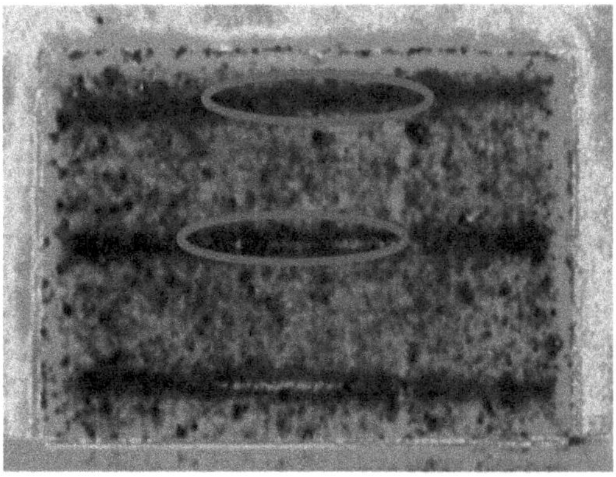

Figure 1.25 Phase image of the concrete sample clearly revealing the rebars and corroded regions (indicated by ellipse).

for the pipe diameters up to 8 mm was observed. Jonsson et al. [16] have investigated the application of PT for detection of blisters and filiform corrosion in magnesium. Han et al. [17] have investigated the application of active PT for the detection of ruptured blister, filiform corrosion, and blisters on steel substrates under organic coating. Kobayashi and Banthia [18] have combined induction heating and thermal imaging for detection of corrosion of reinforcing steel in concrete. The authors have also studied the influence of rebar diameter, cover depth, extent of corrosion, and heating time. Based on the experiments, they concluded that (i) under identical heating conditions, the temperature rise in a corroded rebar is higher than that in a noncorroded rebar and this difference increases with the extent of corrosion; (ii) the greater the extent of corrosion, the smaller is the temperature rise on the surface; and (iii) the technique is particularly effective in detecting corrosion of larger bar diameters and smaller cover depths.

Modeling of corrosion sites requires 3D numerical models. Numerous practical corrosion scenarios have been simulated, and the inversion formulas for determining material loss have been modeled for both flash and square-pulse heating by Marinetti and Vavilov [19]. They have also validated the proposed inversion formulas experimentally by analyzing two samples made of AISI type 1045 steel and observed that the accuracy of corrosion characterization varies from 0.2 to 20%. Grinzato et al. [20] have extended the classical inversion formula for detecting corrosion in plate onto cylinder and spheres and the case of pipes that are of practical interest. Apart from the above, extensive modeling studies based on 1D, 2D, and 3D heat conduction models have been undertaken and inversion functions proposed for a variety of corrosion defects by Grinzato, Vavilov, and others [21–23]. The studies have also attempted to take into account the variations in test material, pulse duration observation times, etc. on the detectability and quantification of corrosion.

1.4.4 Compton Backscatter Radiography

Compton backscatter radiography, also referred to as Compton backscattering, is one of the relatively new NDE techniques. While the principles of Compton backscatter has been well established for over five decades, its applications for materials testing and characterization have found appreciation during the last two decades.

In Compton scattering, a collimated beam of ionizing radiations (X-rays or gamma rays) is incident on the material under test. When the photons in the beam interact with free electrons in the material, part of its energy is transferred to electron, and the photon gets scattered from its original direction with reduced energy. The energy of the scattered photon is a function of incident photon energy and scattering angle. Scattered photons are detected using radiation detectors. The number of detected photons depends on the number of scattered photons, as well as on the attenuation all along the path within the material. The path of photons from the source to the detector can be divided into three stages, which determine their contribution to the measured signal. The schematic of the photon paths is shown in Figure 1.26.

In the first stage, the photons travel from the source to the scattering point (shaded region in Figure 1.26) along path α. The second stage of development for signal is the scattering toward the detector, which occurs at the scattering point. The third stage is the transport of the scattered photons back through the materials toward the detector. Then the signal is further attenuated and scattered intensity I_2 changes to I_3. The signal intensity corresponding to point P $S(P)$ for a fixed source and detector position and the collimator sizes is given by

$$S(P) = K\rho \text{AF} \tag{1.1}$$

where K is a constant for given experimental setup, AF is the attenuation factor, and ρ is the physical density.

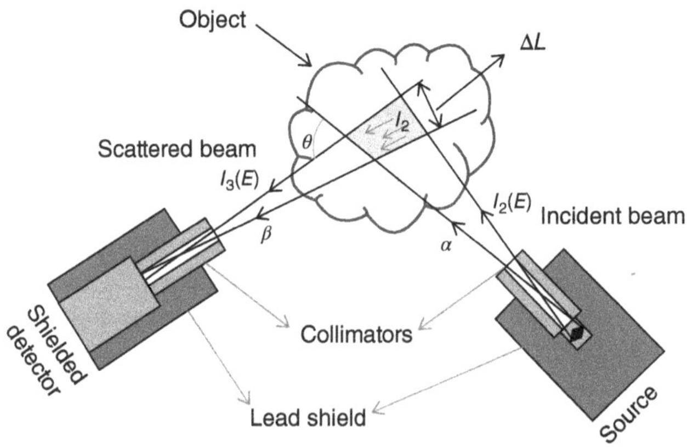

Figure 1.26 Schematic sketch of the photon paths in Compton backscatter testing.

In NDE application, the Compton scattering is observed for a reference sample without flaw and then investigated for the actual sample with flaws. During the scanning for detection of flaws, the K value is constant for a given experimental setup. By applying attenuation correction for incident beam and scattered beam, one can determine the density of the material at scanning location from the scattered intensity and correlate with the flaw.

The main advantage of Compton backscatter is that single-sided access is adequate and no surface preparation is necessary. The technique has been used for detection of fluid levels and interfaces. Its main disadvantage is that it is slow and not very field portable.

1.4.4.1 Compton Backscatter for Corrosion Detection

In corrosion detection, the primary interest is in estimation of metal loss or the thickness reduction. Hence Eq. (1.1) above needs to be modified for thickness loss. As outlined above, the intensity of scattered photon that reaches the detector from a sample is a combination of three processes (attenuation of the incident gamma photons, scattering, and attenuation of the scattered gamma photons).

At the author's center, a Compton backscatter system consisting of computer numerically controlled six-axis source detector system and a four-axis job positioning system has been indigenously designed and developed. ^{137}Cs of activity 155.4 GBq (4.2 Ci) has been used as the source. The source is shielded using lead, and beam collimation with varied collimator diameters has also been provided. The detector used is a 50% efficiency coaxial high purity germanium (HPGe) providing high-resolution energy-dispersive analysis of the scattered spectrum. Both of these are mounted separately on the source and detector subassemblies of six-axis system. Apart from other applications, this system has been successfully used to demonstrate the feasibility of detecting pitting corrosion in mild steel.

For corrosion quantification, an algorithm to reconstruct the densities of the objects under investigation and an unambiguous interpretation of the signal as a function of material thickness at any point of the thick object being inspected was specifically developed. In this algorithm only the knowledge of the target material's mass attenuation coefficients (composition) for the incident and scattered energies is enough to reconstruct the thickness of each voxel of the specimen being studied. A reference sample (uncorroded) with known thickness is selected for quantification of thickness loss. The magnitude of corrosion was quantified, and the results were correlated with those obtained from the conventional gammatography and radiography imaging measurements. The thickness losses due to corrosion of mild steel plate obtained by gamma scattering method compared well with those values obtained by gammatography and radiography techniques [24]. Figures 1.27 and 1.28 are the radiographic images and the graphical plot of radiographic results compared with Compton scatter and transmission methods.

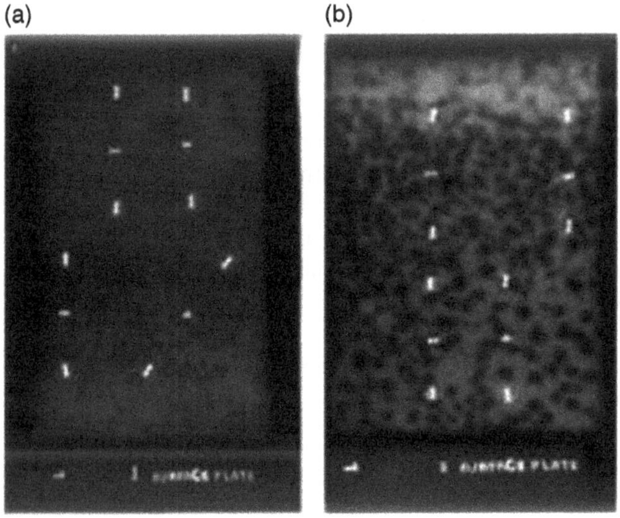

Figure 1.27 Radiographic image of (a) normal and (b) corroded mild steel plates.

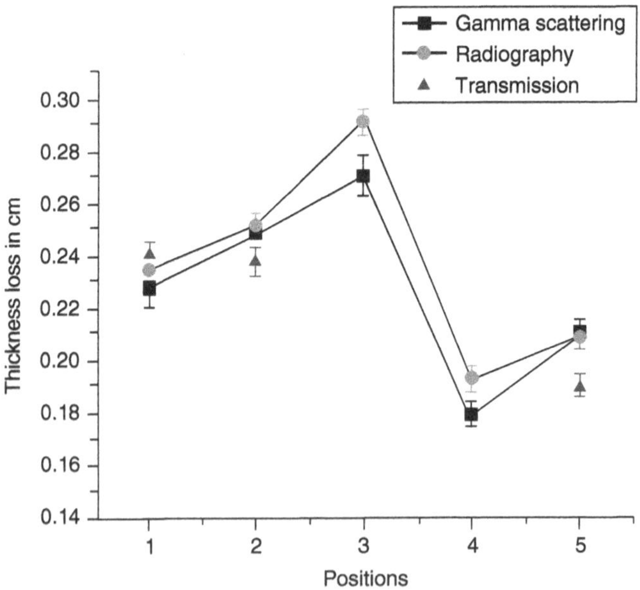

Figure 1.28 Graphical plot comparing the wall thickness loss estimated by gamma scattering, radiography, and transmission method.

1.5 R&D Trends in NDE for Corrosion Detection Technologies

Many of the technologies of NDE lend themselves to the detection, characterization, and quantification of corrosion damage. Presently, international experiences reveal that about 10–15% material loss can be detected reliably through the combination of conventional and NDE techniques coupled with image, signal processing, and analysis. However, designers, utility service providers, and the scientific community would like to be able to reliably determine material loss in the range of 0.5–1%. While this may seem to be a tall order for the present, it is considered essential as this will mean very early detection of incipient corrosion and also enable much greater precision in understanding of the mechanisms leading to rates of corrosion (and the underlying processes of corrosion, especially in newer materials for which there are not much data or field experience available). Another related aspect is the need for quantitative NDE. Corrosion metrics in quantitative terms are required by designers and plant engineers to feed into analytical models capable of assessing the effect of corrosion damage on residual strength and residual life of systems. Without quantitative evaluation, probability of detection cannot be assessed. The spin-off of this would be more accurate predictions of useful service life and judicious recommendations for periodicity of inspection and maintenance intervals.

Hence internationally, much of R&D is being devoted to improving the limits of detectability through advanced and intelligent processing techniques as well as research into smart sensor technologies with better quantum detection efficiencies. Table 1.2 outlines the trends.

A detailed literature review reveals that during the last decade, there have been rapid developments in the areas of phased array and guided wave-based ultrasonics, DR, especially, those that employ computer radiography (CR) coupled with image processing for corrosion detection, particularly in the chemical and process industries. A portable X-ray computed tomography known as TomoCAR, which stands for "Tomographic Computer Aided Radiometry," has been developed recently by the Federal Institute for Materials Research and Testing (BAM) in Berlin. The main advantage of CT has been that not only variations of the order of 1–2% can be determined but also the material densities can be determined accurately. In all these cases, modeling and simulation have contributed extensively. In fact before the actual work is undertaken, an initial attempt is made to evaluate the limits of detectability and possible pitfalls through simulation combined with computer-aided design (CAD) models of the object under investigation, thus ensuring better reliability of inspection in a cost- and time-effective manner.

Table 1.2 Corrosion detection technology R&D trend.

Technology	Observed trend
Enhanced visual inspection	• Multiplexed robotic scanning systems • Automated image acquisition and interpretation • Quantification of corrosion
Eddy current	• More sophisticated signal and data processing (pulsed eddy current, C-scan imaging) including wavelet ANN based and fuzzy logic • More sophisticated sensors (multifrequency)
Ultrasonic	• Advanced techniques such as guided wave-based UT • Efficient transducer systems including air-coupled transducers • Efficient scanning methods through robotic manipulators • Advanced signal processing tools such as wavelet, combinatorial optimization, clustering and visualization, etc.
Radiography	• Single-sided methods (backscatter) • Three-dimensional image processing (computed tomography)
Thermography	• Time domain analysis (thermal wave imaging) • Multispectral (dual-band infrared) • Three-dimensional image processing (computed tomography)
Data fusion	• Image processing (color coding, three-dimensional, etc.) • Image correlation (C-scan, etc.) • Multitechnique fusion technologies

The development of newer generation of high speed IR imaging systems with noise equivalent differential temperature (NETD) of milli-Kelvin, response times of milliseconds coupled with image processing and elements of artificial intelligence, and availability of modular and powerful flash tubes has really seen a spurt in the application of quantitative thermal imaging for corrosion detection in a variety of industries, ranging from aerospace to chemical, steel, and process. Dedicated 1D, 2D, and 3D numerical models based on heat transfer with appropriate boundary and experimental conditions have also been attempted for specific corrosion geometries. Advanced signal reconstruction techniques based on principal component analysis (PCA) are now being attempted to improve the limits of detectability and quantification accuracies. PCA is a multivariate statistical technique that is used for the reduction of a large set of variables by finding the directions of maximum variance and transforming it into a smaller subset while retaining most of the information. Mathematically, PCA depends on the eigendecomposition of positive semidefinite matrices and on the singular value decomposition (SVD) of rectangular matrices. The steps involved in PCA include standardizing the data, obtaining the eigenvectors and eigenvalues from the covariance matrix or correlation matrix or performing singular vector decomposition, sorting the eigenvalues and constructing projection matrix from the selected eigenvectors, and transforming the original data set through the projection matrix to the newer smaller data set.

THz imaging as a tool for NDE is relatively new for inspection technology. THz radiation falls between the IR and microwave bands in the electromagnetic spectrum. The frequency of the radiation is between 100 GHz and 10 THz and has a free space wavelength of 300 μm at 1 THz. It is a noninvasive and high-resolution inspection tool for detection of corrosion hidden under a dielectric layer such as nonconductive paint. This radiation, like microwaves, penetrates and inspects nonconductive material for defects. The advantage of THz is that it can provide time domain signals that are easy to interpret and has higher spatial and temporal resolution.

1.6 Quality Assurance/Evaluation Requirements for Corrosion Service

NDE is always considered as a QA tool to ascertain the extent of corrosion in a component after it is put in service. However, it is a well-realized and accepted fact that preventing/mitigating/reducing the kinetics of particular corrosion in a given service depends to a large extent on effective QA and QC checks right from material selection onward. An effective QA/QC right from raw material stage for early detection of corrosion signs is essential to do away with costly firefighting after detection of severe corrosion by NDE. Hence it was thought prudent to have a brief overview of the QA practices in material selection that would help in revealing the incipient corrosion. Austenitic SS has been chosen as a case study since this is a structural material with extensive applications worldwide in practically all industrial sectors – nuclear, process, chemical, etc.

1.6.1 Selection and Qualification of Austenitic Stainless Materials: An Example

Austenitic SS are the candidate material for many of the industrial components due to their corrosion resistance, ease of fabrication, weldability, etc. Even though austenitic SS have good mechanical and physical properties, their limitation is their susceptibility to sensitization, distortion, IGC, and SCC due to their interactions with the environment, which can become pronounced in coastal environment. The following are some of the QA and QC measures to avoid localized corrosion failures in austenitic SS:

- Material control
  - Select appropriate grade SS (preferably those with higher corrosion resistance).
  - Ensure absence of sensitization through *in situ* metallographic techniques.
  - Fine-grain size.
  - Low nonmetallic inclusions.
- Processing control
  - Low cold work (LTS, residual stress).
  - Welding parameters (low heat input and interpass temperature to avoid sensitization, stringer bead).
  - Heat treatment (removal of oil, grease).
  - No grinding of pipe internal surface after welding.
- Environmental control
  - O_2, pH, chloride controls in thermal insulation.
  - Surface contamination (Fe, C, Halides).

Table 1.3 gives a consolidated list of tests to be conducted on austenitic SS materials to detect susceptibility to IGC under given service condition. All product forms such as plates, pipes, tubes, forgings, fittings, and weldments need to undergo testing before being placed in service.

ASTM A262-14 Practice A and Practice E are the two tests generally used for nuclear reactor components. Practice A is basically a rapid screening testing using oxalic acid etching to detect susceptibility to IGC. Classification of the structure after the test will determine if the material is acceptable or if additional tests are required. Practice A is not the test for rejection.

In ASTM A262-14, Practice C, "Huey test" is used for austenitic SS samples that are susceptible to rapid intergranular attack associated with chromium carbide precipitation. In this test, samples are boiled for five 48 h periods in a freshly prepared 65% nitric acid to determine weight loss. Average weight loss over five periods is calculated and expressed in mpy (mills per year). As per US NRC guidelines, for nuclear reprocessing applications, the average value over five periods shall not exceed 15 mpy, and any of the individual period shall not exceed 18 mpy. The average mass loss over the total specimen is converted to corrosion rate that is compared with the specified value to determine its suitability.

In ASTM A262-14, Practice E, Strauss test, the specimen is boiled in copper sulfate–sulfuric acid mixture for 15 h and then bent it for 180° over an equal diameter bend to detect susceptibility to intergranular attack in austenitic SS. The bent sample is examined at 100× to detect microcracks or fissures on the tension portion of the bend to confirm intergranular attack.

Since nuclear fuel reprocessing plants handle fuming nitric acid as process fluid, a greater emphasis is given in controlling nonmetallic inclusions such as sulfide (type A), alumina (type B), silicate (type C), and globular oxide (type D)

Table 1.3 As per ASTM A262-14.

ASTM standard	Environment	Exposure	Evaluation	Species attacked
A262-14 Practice A, "Oxalic Acid Etch Test"	10% oxalic acid electrolytic etch	1.5 min. anodic at $1\,A\,cm^{-2}$	Classification of etch structure	Carbides
A262-14 Practice B, "Ferric Sulfate–Sulfuric Acid Test"	Boiling 120 h ferric sulfate–50% sulfuric acid	One 120 h period	Weight loss per unit area	Chromium-depleted area
A262-14 Practice C, "Nitric Acid Test"	65% nitric acid	Five-period 48 h duration	Average weight loss per unit area	Chromium-depleted area, sigma phase, and carbide
A262-14 Practice E, "Copper–Copper Sulfate–Sulfuric Acid Test"	Copper–copper sulfate–16% sulfuric acid	One 15 h period	Macroscopic appearance after bending	Detecting susceptibility to intergranular attack in austenitic stainless steels
A262-14 Practice F, "Copper–Copper Sulfate–50% Sulfuric Acid Test"	Copper–copper sulfate–50% sulfuric acid	120 h	Weight loss per unit area	Detecting susceptibility to intergranular attack in molybdenum-bearing austenitic stainless steels

in the raw material to avoid tunneling corrosion by nitric acid. All raw material product forms are evaluated for inclusion rating as per ASTM E45.

1.6.2 Quality Control Checks During Fabrication

During the course of fabrication, the following QC checks are suggested with prime focus on avoiding corrosion:

1) Extra-low-carbon consumables such as ER308L, ER316L, and ER316LN are selected for welding austenitic SS materials such as 304L, 316L, 316LN, etc. to avoid sensitization.
2) Contact between carbon steel and SS right from storage to during the course of fabrication to be avoided. This would eliminate the chances of localized corrosion of SS due to iron particles.
3) During welding, interpass temperature (<150 °C) and heat input control ($<1000\,J\,mm^{-1}$) shall be exercised to avoid sensitization. The same has to be demonstrated during welding procedure qualification as per ASME Section IX as well as during production.

4) Due to high coefficient of thermal expansion and low thermal conductivity, distortion and development of high residual stresses are more pronounced in austenitic SS. Proper welding sequence such as step and block and controlled heat input shall be used during welding to prevent SCC.
5) All formed components shall be subjected to solution annealing heat treatment at 1050 °C with water quenching if percentage of fiber elongation exceeds 20% to avoid SCC due to the presence of lock-in or residual stresses.
6) Since austenitic SS materials are prone to SCC under chloride environment, total halogen content (THC) shall be <25 ppm in all the testing materials such as liquid penetrant examination consumables, DM water for surface treatment, etc.
7) All formed components of austenitic SS shall be surface-treated (pickled and passivated) to get the uniform layer of chromium oxide to replace disturbed passivated layer during transport, storage, fabrication, etc.

1.7 Case Study: Zircaloy Hydriding

The previous sections provided an overview of the QA steps and the various methods for susceptibility, detection, and quantification of corrosion. It can be seen from above that no single technique provides a panacea for the complete detection and characterization of corrosion. Each method has its specific capabilities. Thus, it is a combination of NDE methods (multitechnique NDE) that offers reliable solution for corrosion characterization. Again, the choice of the NDE methods depends on material, its application, and industry type. While early detection of corrosion and its mitigation have been a focal point in the industry, parallel efforts have always been underway by academia and R&D organizations to explore the science of corrosion and formulation of combinations to inhibit corrosion. To highlight this point, we take a specific case of zirconium hydriding, an important and life-limiting phenomenon in the case of nuclear pressure/calandria tubes of pressurized heavy water reactors (PHWRs).

A PHWR is a nuclear power reactor that commonly uses natural uranium as its fuel and heavy water (deuterium oxide, D_2O) as its coolant and moderator. More than 40 reactors are in operation worldwide. In India, PHWRs form the mainstay of the first stage of the nuclear power program. Presently, 18 such reactors are under operation and 4 are under construction. Figure 1.29 is a schematic of a typical Indian PHWR. A brief description of their salient features is given below.

The reactor core consists of a low-pressure horizontal reactor vessel called calandria. The calandria is the heart of the reactor housing the fuel, moderator, coolant, and coolant channel assemblies. Each coolant channel assembly

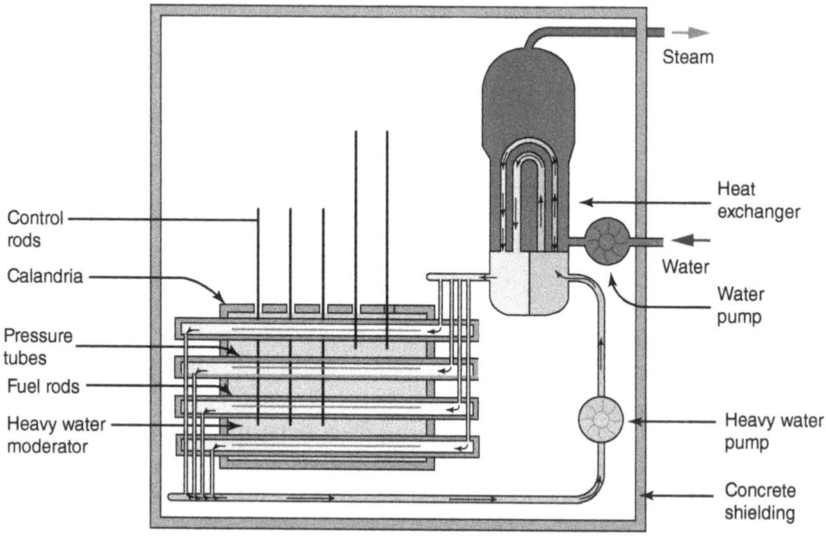

Figure 1.29 Schematic of a typical Indian pressurized heavy water reactor (PHWR).

comprises of a pressure tube and a calandria tube separated by garter springs. Inside the pressure tubes, heavy water coolant flows under a pressure of ~10 MPa and at a temperature of ~573 K. There are 306 pressure tubes in case of 220 MWe and 392 in 540 MWe PHWR. The pressure tube contains fuel bundles (19 in case of 220 MWe and 37 in case of 540 MWe) that are cooled by the pressurized heavy water. The calandria is filled with heavy water moderator at an average temp. of 334 K. The moderator is insulated from the hot reactor coolant by the calandria tube and the pressure tube. The calandria also houses all reactivity and reactor shutoff devices in the low-pressure and low temperature environment.

Zirconium alloys (zircaloy-2, zircaloy-4, Zr–Nb) find wide applications as core structural materials in nuclear reactors, especially the water/heavy water cooled ones such as light water, pressurized water reactor (PWR), boiling water reactor (BWR), and PHWR. These materials are used as cladding for fuel, wrappers for subassemblies/channels, and pressure and calandria tubes. One of the main reasons for choosing Zr alloys is that its thermal neutron absorption cross section is about 30 times less compared with that of austenitic SS, thus giving a better thermal neutron efficiency. Apart from this, it also has other desirable properties such as high mechanical strength, ductility, fracture toughness, radiation damage tolerance, good corrosion resistance at reactor operating temperatures and in aqueous environments, and good compatibility with fuel materials and acceptable creep and fatigue properties.

As mentioned above, Zr alloys are widely used as cladding and also in coolant channel assemblies. While in general the performance of the alloy has been good, the main life-limiting factor has been the hydrogen ingress into the material during corrosion, resulting in the formation of zirconium hydride and finally the hydride leading to embrittlement and failure. Coolant channels in Indian PHWRs have been replaced for this reason for the purpose of extending the life of the plant. For the sake of completeness, outlined here is the mechanism of hydriding so that readers can get a feel of this important phenomenon. Extensive investigations have been done in India and worldwide to understand the phenomenon of hydriding. IAEA-TECDOC-996, *Waterside Corrosion of Zirconium Alloys in Nuclear Power Plants*, January 1998 [25], is a comprehensive treatise.

Corrosion of zirconium alloys in an aqueous environment is principally related to the oxidation of zirconium by oxygen in the coolant, dissolved or produced by irradiation of water. All zirconium components normally have a thin oxide film (2–5 nm) on their surface in their as-fabricated state. This oxide film is continuous and adherent and acts as a protective barrier limiting the access of oxidizing species to the bare metal and thus inhibiting corrosion. When the component is in service in a reactor, with the passage of time and under the influence of high temperature, pressure, and irradiation, the oxide layer grows in thickness. It is also reported to recrystallize, resulting in micropore formation. During normal reactor operating condition, flow of hot heavy water coolant through the pressure tube causes slow oxidation of the inside surface of the tube as represented by the following chemical reaction:

$$\mathrm{Zr} + 2\mathrm{D_2O}\left(\text{or } 2\mathrm{H_2O}\right) \rightarrow \mathrm{ZrO_2} + 4\mathrm{D}\left(\text{or } 2\mathrm{H_2}\right) \tag{1.2}$$

A part of hydrogen (H) or its isotope deuterium (D) produced in the above reaction is picked up by pressure/calandria tube. It is reported that once the oxide layer thickness exceeds ~15–20 μm, oxidation rate is enhanced significantly as a consequence of the formation of microcracks and micropores in the oxide. Much evidence exists to indicate that Zr oxidation occurs by inward migration of oxygen ions through the oxide layer, either through grain boundaries or through the bulk [26–28]. This accelerated oxidation results in accelerated hydrogen/deuterium pickup, thus leading to hydriding. Zirconium has a very low solubility for hydrogen (about 80 wt ppm at 300 °C and 200 wt ppm at 400 °C), and once the solubility limit is exceeded, the hydrogen precipitates as a zirconium hydride phase:

$$\mathrm{Zr}\left(\mathrm{H}, \text{ solution}\right) + \mathrm{H_2} \rightarrow \mathrm{ZrH_{1.6}} \text{ or } \mathrm{ZrH_2} \tag{1.3}$$

A typical micrograph revealing hydrides in Zr-2.5 Nb pressure tube material is shown in Figure 1.30.

As a result of hydriding, the following effects have been reported (although not all confirmed) to occur: hydrogen embrittlement due to excess hydrogen or

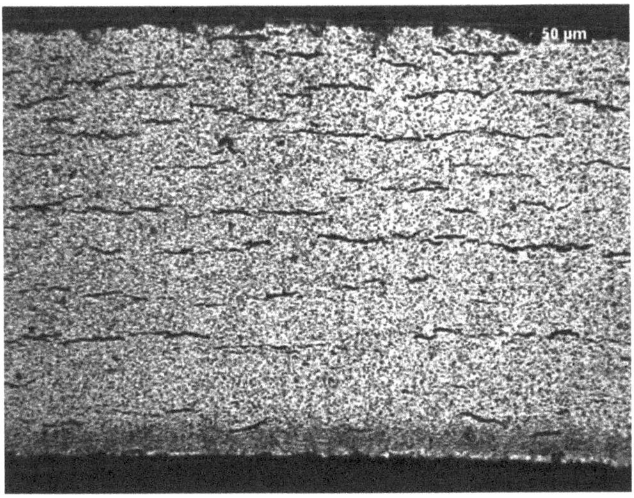

Figure 1.30 Optical micrograph of zircaloy-2 cladding material containing 270 ppm of hydrogen. Dark lines are traces of circumferential hydrides.

its localization into a blister or rim [29, 30] causing loss of fracture toughness, delayed hydride cracking (DHC), and acceleration of corrosion and of irradiation growth.

Several types of corrosion morphologies have been observed in nuclear reactors and in autoclave experiments, of which the most important are:

1) Uniform: The formation of a thin uniform layer of zirconium dioxide on the surface of a zirconium alloy component.
2) Nodular: The formation of local, small, circular zirconium oxide blisters.
3) Shadow: The formation of local corrosion regions that mirror the shape (suggestive of a shadow) of other nearby noble reactor core components.

1.7.1 Detection and Characterization of Zircaloy Hydrides

Hydrogen-induced degradations like blisters, embrittlement, cracking, etc. limit the safe operating life of Zr alloy-based components in the reactors. Hence these are required to be studied for (i) understanding their mechanisms and (ii) quantification of the extent of damage. Several experimental and analytical studies have been undertaken internationally on small specimens primarily with blisters and a few of them on actual components. These studies not only help in improved understanding of the mechanisms and refining boundary conditions of the numerical models but also help in evolving quantitative measures to develop better alloys with improved corrosion resistance. Ultrasonic and eddy current technologies based on modeling and experimental

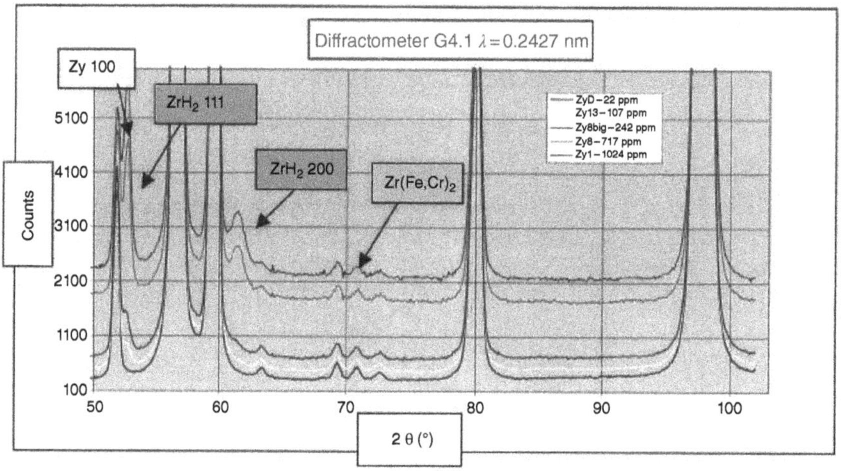

Figure 1.31 Powder diffraction diagrams measured for several zircaloy samples. One observes the increase of background with the hydrogen content, and the weak diffraction lines of δ-Zr H_2 hydride and Laves phase.

optimization have been used for finding hydrogen damage in zircaloy components in the reactor (Figure 1.31). A brief outline of some of the analytical and NDE techniques is given below.

1.7.1.1 High Energy Synchrotron X-ray Diffraction

Synchrotron radiation is a powerful NDE tool for material characterization [31]. High energy synchrotron X-ray diffraction has been used to investigate the isothermal precipitation of δ-hydride platelets in zircaloy-4 at a range of temperatures relevant to reactor conditions, during both normal operation and thermal transients. The propensity of a synchrotron to produce a highly collimated, high-flux beam yields a significant degree of angular resolution, allows for rapid diffraction pattern acquisition, and generates well-defined peaks with minimal artificial broadening. This is especially valuable when studying the hydride precipitation process that takes place when hydrogen saturates zirconium, as measurements of very weak reflections, that may otherwise be overlapped by dominant zirconium peaks.

1.7.1.2 Electron Backscattered Diffraction

Understanding the processes behind zirconium hydride precipitation is critical for the safety of nuclear reactors. Although the hydrides can usually be revealed by polishing and etching techniques, these techniques do not give information about their crystallographic texture and its relationship to the matrix zircaloy. Electron backscattered diffraction (EBSD) is an ideal technique allowing not only the identification of the hydrides but also an analysis of their crystallographic orientation. Characterization of hydrided zirconium alloys in terms of the orientation relation of hydrides with the matrix, local misorientation, and internal structure can be carried out using EBSD.

1.7.1.3 Neutron Scattering Techniques

Neutron scattering is the only experimental technique that allows measuring nondestructively the total hydrogen content in zircaloy tubes [32]. The large coherent and incoherent cross sections of hydrogen make this tool powerful to determine the hydrogen content and characterize the hydrides. The incoherent cross section results in a continuous background in diffraction spectra, which is proportional to the hydrogen content. The coherent cross section contributes significantly to the structure factor of the hydride phase, i.e. to the intensity of the diffraction peaks. Zirconium is a specially favorable element for this study because its neutron absorption coefficient and incoherent scattering cross sections are very weak: the diffraction spectra of H-free samples show very low background. But more than just a simple hydrogen content determination at room temperature, the technique, when a combined use is made of the incoherent background and of the diffraction peaks, is also sensitive, reliable and quite appropriate to follow in situ high temperature precipitation/dissolution phenomena of zirconium hydrides, even when the quantity of precipitates is very small.

1.7.1.4 Neutron Radiography

NR is known to be capable of detecting inclusions containing low atomic number elements in a matrix of relatively higher atomic number materials. The presence of zirconium hydride blisters/platelets in zircaloy matrix can be detected by NR. Zirconium hydride blister of nearly 0.35% of wall thickness could be detected using NR. While NR has been used for the identification of the areas of blistering, other NDE and analytical tools have been used to characterize the size, shape, integrity, depth, etc. of the blisters in experimental and actual specimens. Figure 1.32 shows a typical blister detected and characterized using NR, video microscopy, 3D surface profiling, scanning electron microscopy, and optical microscopy.

1.7.1.5 Neutron Diffraction and Quasi-elastic Neutron Scattering

Neutron diffraction is one of the techniques to identify the zirconium hydride phase and to find the extent to which the sample has hydrided [33]. Neutron diffraction measurements are made with the neutron spectrometer.

Quasi-elastic neutron scattering (QENS) measurements help to understand the diffusion mechanism of hydrogen in zircaloy. In QENS, the process of diffusion of hydrogen is monitored by measuring the line broadening of a monoenergetic neutron beam. QENS spectra show no quasi-elastic peak, indicating the absence of free hydrogen in the interstitial sites of Zr lattice.

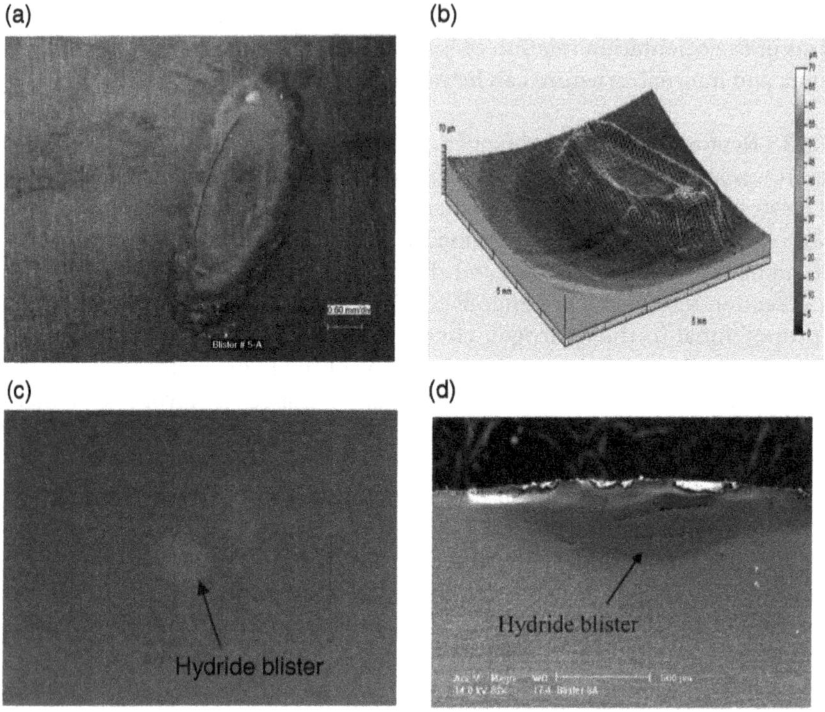

Figure 1.32 Investigations of hydride blisters grown on Zr-2.5Nb pressure tube spool piece under stimulated condition of in-reactor pressure and temperature: (a) video image, (b) 3D surface profile, (c) neutron radiograph, and (d) cross section of blister. Source: From http://barc.ernet.in/publications/eb/golden/reactor/toc/chapter13/13_5.pdf.

The appearance of elastic peak confirms the presence of hydrogen in the hydride phase. QENS spectra are recorded under the same conditions; the intensity scale in each spectrum indicates proportionately the amount of hydrogen present in each sample.

1.7.1.6 Ultrasound Backscattering Technique

Ultrasound backscattering technique is one of the means used for the investigation of the structural properties of metals [34]. Kazys et al. [34] have used this method for the investigation of hydrides. Using focused high frequency ultrasonic transducer (50 MHz central frequency), they investigated zirconium alloy tube samples with different concentrations of zirconium hydride. One test sample, in which the hydrogen was not added, was used as reference sample. Their investigations revealed that strongest backscattered signals occurred in the focal zone of the transducer. For quantitative estimation, signal processing using Hilbert transform was carried out. Analysis indicated that as the hydride concentration increased, the scatter factor also increased. This method enabled estimation of the concentration of the zirconium hydrides in the range of very low levels (<100 μm).

1.8 Conclusion

The ability to make, shape, and treat metals and materials has shaped the pathways of human civilization. Nations with superior and competitive materials and manufacturing technologies and with innovative efforts and robust techniques combined with smart technologies for predictive and risk-based inspection and maintenance (judiciously) have reaped rich rewards through better performing assets. NDE combined with robust QA has been the critical and catalytic enabler for expensive and strategic assets. Corrosion of materials and its early and efficient evaluation have always been a challenge to the engineering and scientific community. A wide spectrum of NDE methods based on ionizing and nonionizing radiations and acoustic, electrical, and magnetic methods combined with signal/image processing and artificial intelligence have lent themselves very ably for detection, characterization, and quantification of corrosion damage and also enhancing the understanding of the mechanisms so that better mitigating technologies can be adopted. This chapter basically outlined the fundamental principle, application, and limitation of the fundamental and widely used NDE methods by the industries. The developments of advanced NDE methods especially with respect to corrosion damage detection and the R&D trends have also been highlighted. For the sake of completeness, QA requirements from corrosion point of view for putting a material into service especially in nuclear facilities are aptly outlined.

Acknowledgments

The authors are thankful to colleagues Smt M. Menaka, Dr. C.K. Mukhophadyay, Shri M.V. Kuppuswamy, Shri N. Raghu, and Shri R. Ramar of Indira Gandhi Centre for Atomic Research for their valuable inputs.

References

1 Raj, B. and Venkatraman, B. (2013). Chapter 4. Overview of diagnostics and monitoring methods and techniques. In: *Handbook of Technical Diagnostics: Fundamentals and Application to Structures and Systems* (ed. H. Czichos). Berlin: Springer-Verlag.
2 Raj, B. and Venkatraman, B. (2004). *Practical Radiography Testing*. Pangbourne: Alpha Science International Limited.

3 Subramanian, C.V. (2006). *Practical Ultrasonics* (ed. B. Raj and B. Venkatraman). Pangbourne: Alpha Science International Limited.
4 Rao, B.P.C. (2004). *Practical Eddy Current Testing* (ed. B. Raj and B. Venkatraman). Pangbourne: Alpha Science International Limited.
5 Miller, R.K. and McIntire, P. (1987). *Nondestructive Testing Handbook, Volume 5, Acoustic Emission Testing*. Columbus, OH: American Society for Nondestructive Testing.
6 Liptai, R.C. and Tatro, C.A. (1976). *Metals Handbook*, vol. 11, 234. Materials Park, OH: ASM.
7 Mazille, H., Rothea, R., and Tronel, C. (1995). An acoustic emission technique for monitoring pitting corrosion of austenitic stainless steels. *Corrosion Science* 37: 1365.
8 Jones, R.H. and Friesel, M.A. (1992). Acoustic emission during pitting and transgranular crack initiation in type 304 stainless steel. *Corrosion* 48: 751.
9 Kima, Y.P., Fregonesea, M., Mazillea, H. et al. (2003). Ability of acoustic emission technique for detection and monitoring of crevice corrosion on 304L austenitic stainless steel. *NDT&E International* 36: 553–562.
10 Ferrera, F., Idrissia, H., Mazillea, H. et al. (2000). A study of abrasion–corrosion of AISI 304L austenitic stainless steel in saline solution using acoustic emission technique. *NDT&E International* 33: 363–371.
11 Kusanagi, H., Nakasa, H., Ishihara, T., and Ohashi, S. (1978). *Proceedings of the 4th Acoustic Emission Symposium*, Tokyo (18–20 September 1978).
12 Shaikh, H., Amirthalingam, R., Anita, T. et al. (2007). Evaluation of stress corrosion cracking phenomenon in an AISI type 316LN stainless steel using acoustic emission technique. *Corrosion Science* 49: 740–765.
13 Maldague, X.P. (2001). *Theory and Practice of Infrared Technology for Non Destructive Testing*. New York: Wiley.
14 Maldague, X.P.V. and Moore, P.O. (eds.) *Non Destructive Testing Handbook*, 3e, vol. 3. New York: ASNT.
15 Liu, Z., Genest, M., and Krys, D. (2012). Processing thermography images for pitting corrosion quantification on small diameter ductile iron pipe. *NDT&E International* 47: 105–115.
16 Jonsson, M., Rendahl, B., and Annergren, I. (2010). The use of infrared thermography in the corrosion science area. *Materials and Corrosion* 61 (11): 961–965.
17 Han, J.-S. and Park, J.-H. (2004). Detection of corrosion steel under an organic coating by infrared photography. *Corrosion Science* 46 (4): 787–793.
18 Kobayashi, K. and Banthia, N. (2011). Corrosion detection in reinforced concrete using induction heating and infrared thermography. *Journal of Civil Structural Health Monitoring* 1 (1–2): 25–35.
19 Marinetti, S. and Vavilov, V. (2010). IR thermographic detection and characterization of hidden corrosion in metals; general analysis. *Corrosion Science* 52: 865–872.

20 Grinzato, E., Vavilov, V., Bison, P.G., and Marinetti, S. (2007). Hidden corrosion detection in thick metallic components by transient IR thermography. *Infrared Physics & Technology* 49 (3): 234–238.

21 Grinzato, E. and Vavilov, V. (1998). Corrosion evaluation by thermal image processing and 3D modelling. *Revue Générale de Thermique* 37 (8): 669–679.

22 Vavilov, V., Grinzato, E., Bison, P.G. et al. (1996). Surface transient temperature inversion for hidden corrosion characterisation: theory and applications. *International Journal of Heat and Mass Transfer* 39 (2): 355–371.

23 Vavilo, V.P. (2014). Modeling thermal NDT problems. *International Journal of Heat and Mass Transfer* 72: 75–86.

24 Priyada, P., Margret, M., Ramar, R. et al. (2011). Intercomparison of gamma scattering, gammatography and radiography techniques for mild steel nonuniform corrosion detection. *Review of Scientific Instruments.* 82: 035115–035118.

25 IAEA-TECDOC-996 (1998). *Waterside Corrosion of Zirconium Alloys in Nuclear Power Plants.* Vienna, Austria: International Atomic Energy Agency.

26 Cox, B. (1968). Effects of irradiation on the oxidation of zirconium alloys in high temperature aqueous environments: a review. *Journal of Nuclear Materials* 28: 1.

27 Cox, B. (1969). Processes occurring during the breakdown of oxide films on zirconium alloys. *Journal of Nuclear Materials* 29: 50.

28 Cox, B. and Pemsler, J.P. (1968). Diffusion of oxygen in growing zirconia films. *Journal of Nuclear Materials* 28: 73–78.

29 Garde, A.M., Smith, G.P., and Pirek, R.C. (1996). Effects of hydride precipitate localization and neutron fluence on the ductility of irradiated Zircaloy-4. In: *11th International Symposium on Zr in the Nuclear Industry, STP 1295, Garmisch-Partenkirchen*, 407–430. West Conshohocken, PA: ASTM.

30 Pierron, O.N., Koss, D.A., Motta, A.T., and Chan, K.S. (2003). The influence of hydride blisters on the fracture of Zircaloy-4. *Journal of Nuclear Materials* 322: 21–35.

31 Daum, R.S., Chu, Y.S., and Motta, A.T. (2009). Identification and quantification of hydride phases in Zircaloy-4 cladding using synchrotron X-ray diffraction. *Journal of Nuclear Materials* 392 (3): 453–463.

32 Couvreur, F. and André, G. (1997). In situ neutron scattering study of hydrogen-containing Zircaloy-4 alloys. *LLB Scientific Report* 1998: 72–73.

33 Shaikh, A.M., Vaidya, P.R., Shah, B.K. et al. (2006). Application of neutron radiography and neutron diffraction techniques in study of zirconium hydride blisters in zirconium based pressure tube materials. *BARC New Letter* 273: 104–114.

34 Kazys, R., Sliteris, R., Mazeika, L. et al. (2006). Ultrasonic estimation of hydride degradation of zirconium pressure tubes of RBMK fuel channel. *European Conference Non-destructive Testing*, Berlin (September 2006), pp. 25–29.

2

Corrosion: An Overview of Types, Mechanism, and Requisites of Evaluation

U. Kamachi Mudali[1], J. Jayaraj[1], R.K. Singh Raman[2] and Baldev Raj[3]

[1] *Corrosion Science and Technology Group, Indira Gandhi Centre for Atomic Research, Kalpakkam, India*
[2] *Department of Mechanical & Aerospace Engineering, Department of Chemical Engineering, Monash University, Clayton Campus (Melbourne), Victoria, Australia*
[3] *National Institute of Advanced Studies, Bengaluru, India*

2.1 Metallic Corrosion

Degradation/loss of metals and alloys through chemical/electrochemical reaction with the environment is called corrosion. The rate of corrosion depends on the severity of the environment. For example, the famous Delhi iron pillar in India was constructed almost 1600 years ago but has not undergone very slow insignificant corrosion, presumably due to the protective surface scale and the relatively less aggressive natural atmospheric condition in Delhi. On the other hand, if an iron piece were to be exposed to marine environment, it would undergo severe corrosion. Thus by the nature of the environment, corrosion can be slow or fast. It should be understood that both structure of the metals/alloys and the parameters of environment control the kinetics of a corroding metal. Indeed, it is the structure and property of the interface that the metal/alloy establishes with the environment that decides the rate at which corrosion can proceed. Hence, the nature and severity of the environment and the microstructure and surface characteristics of metals/alloys play crucial roles in governing the rate and type of corrosion.

2.2 Cause of Corrosion

The obvious fundamental question is why do metals and alloys corrode, i.e. what is the driving force for the metallic corrosion? All metals and alloys are produced through extractive metallurgy of their oxides or sulfides of metal ores (i.e. the stable state in which metals have stayed for centuries). In the extractive

Non-Destructive Evaluation of Corrosion and Corrosion-assisted Cracking, First Edition. Edited by Raman Singh, Baldev Raj, U. Kamachi Mudali and Prabhakar Singh.

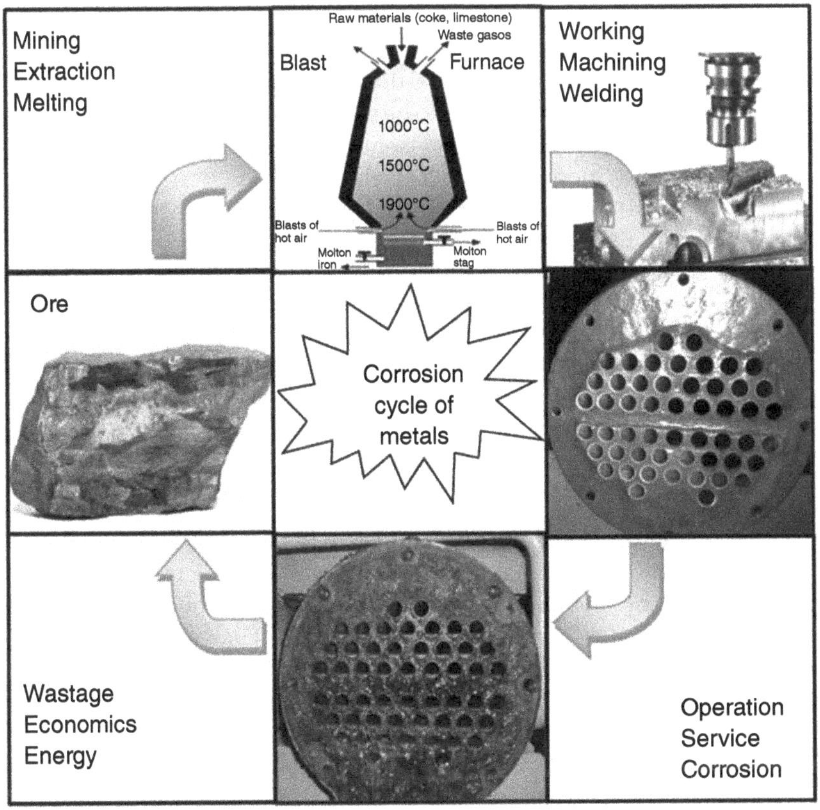

Figure 2.1 Life cycle of a metal from its formation to corrosion stage in service.

metallurgical process (e.g. in the blast furnaces, smelters, etc.), these ores (oxides/sulfides) are transformed into pure metals, i.e. a thermodynamically more active state that has inherent characteristics to react. When exposed to the environment, due to their high energy state, the metallic components have a strong driving force to return to its native, stable, and low energy state, i.e. corrosion process. For instance, the most common iron ore is hematite (Fe_2O_3), which is also the common corrosion product. Thus, the life cycle of a metal is completed when the metal/alloy corrode and return to its stable compound state (Figure 2.1).

2.3 Electrochemical Nature of Corrosion

Corrosion of metals and alloys in an aqueous environment is generally an electrochemical process that requires four essential components: (i) an aqueous electrolyte, (ii) an anode, (iii) a cathode, and (iv) a current carrying pathway (i.e. circuit).

As mentioned earlier, the nature of corrosion largely depends on the electrolyte. The question that often arises is how acidic or alkaline the environment is. In an electrolyte, if the concentration of the H^+ is high, it is acidic, and if the concentration of OH^- is high, it is said to be alkaline. If the concentration of H^+ and OH^- is same, then the electrolyte is neutral with pH of 7. If the pH is less than 7, it is acidic, and more than 7, it is alkaline.

The anode is where the corrosion action takes place by the chemical/electrochemical reaction of the metal surface, which is accompanied by removal of electrons from metal that is referred to as oxidation. For instance, formation of Fe^{2+} (ferrous ion) and Fe^{3+} (ferric ion) during corrosion denotes that a Fe atom has been stripped off 2 and 3 electrons, respectively, so the anodic reaction in any aqueous corrosion is the oxidation of a metal to its own ions. The number of electrons produced equals the valence of the metal; accordingly, the oxidation reaction at anode is given below:

$$\mathrm{M} \rightarrow \mathrm{M}^{+n} + ne^- \tag{2.1}$$

In electrochemical corrosion, the anodic oxidation reaction should be compensated by the reduction reaction at cathode. The electrons generated by the formation of metallic ions at the anode pass through the current carrying pathway to the cathodic areas. The electrical balance of the system is restored by both the anodic and cathodic reactions. The electrons produced and consumed are balanced due to the simultaneous anodic and cathodic reactions.

The most common cathodic reaction in metallic corrosion is reduction of H^+ or O_2 as given below:

$$2\mathrm{H}^+ + 2\mathrm{e}^- \rightarrow \mathrm{H}_2\,(\text{acidic}) \tag{2.2}$$

$$\mathrm{O}_2 + 4\mathrm{H}^+ + 4\mathrm{e}^- \rightarrow 2\mathrm{H}_2\mathrm{O}\,(\text{acidic}) \tag{2.3}$$

$$\mathrm{O}_2 + 2\mathrm{H}_2\mathrm{O} + 4\mathrm{e}^- \rightarrow 4\mathrm{OH}^-\,(\text{alkaline or neutral}) \tag{2.4}$$

It is worth mentioning that for corrosion to occur, it is not essential to have two different metals as anodes and cathodes. It is common that anodic (corrosion) and cathodic (reduction) reactions could occur at the same surface on which several microlevel anodic and cathodic sites can be generated (Figure 2.2).

The arrangements and locations of these anodic and cathodic sites are different for the different forms of corrosion. Environmental effects such as temperature, the presence of oxygen and other oxidizers, flow rates (velocity), reactant concentrations, and pH would influence rates of anodic and cathodic reactions. The most common measure of corrosivity of environment for a given metal/alloy is pH that has a very low magnitude (zero) for

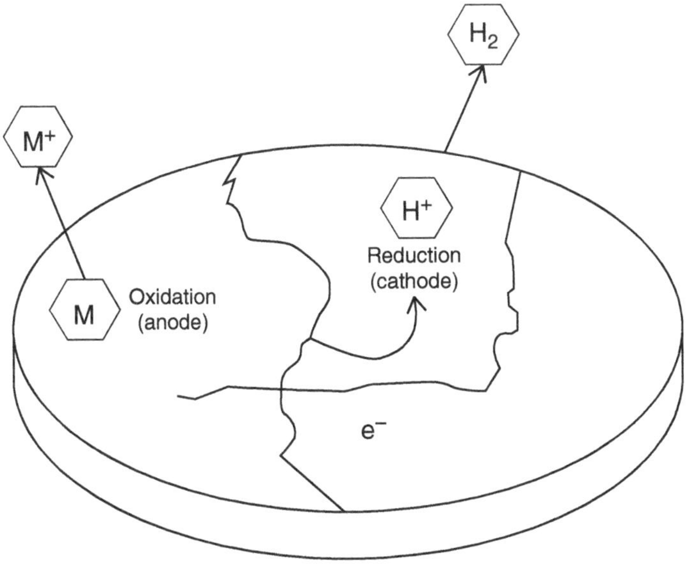

Figure 2.2 Anodic (corrosion) and cathodic (reduction) reaction occurring at the same surface in which microlevel anodic and cathodic sites can develop.

an acidic solution and high (max 14) for an alkaline solution, whereas for a neutral solution, pH is 7.

2.4 Types of Corrosion

The basis for the types of corrosion is their appearance and propagation mode. The eight well-known forms of corrosion listed below are unique, but they can be interrelated. They are:

1) Uniform or general corrosion.
2) Galvanic corrosion.
3) Pitting corrosion.
4) Crevice corrosion (that includes filiform corrosion).
5) Intergranular corrosion (that includes exfoliation).
6) Dealloying or selective leaching.
7) Erosion corrosion (that includes cavitation corrosion and fretting corrosion).
8) Environmentally assisted cracking (EAC) (including stress corrosion cracking [SCC], corrosion fatigue [CF], and hydrogen damage).

Apart from these eight forms, the microbiologically influenced corrosion (MIC) is also considered as one of the important corrosion forms.

2.4.1 Uniform Corrosion

Uniform corrosion is characterized by both anodic and cathodic reactions taking place uniformly and simultaneously over the entire surface, i.e. with no preferential location for cathodic or anodic reaction. The cathodes and anodes are located randomly, and as the corrosion process proceeds, the position of these cathodes and anodes may change.

The surfaces of some metals and alloys like weathering steel, stainless steel (SS), nickel, and titanium alloys are protected from uniform corrosion by an extremely thin oxide film that naturally and quickly develops. Weathering steel undergoes uniform corrosion to form an oxide layer that protects the outdoor sculptures as well as provides rustic antique appearance. These steels were developed to eliminate the need for painting, as they form the protective stable rust-like appearance upon their extended exposure to the atmospheric conditions. SS develops as robust protective layer of chromium oxide (Cr_2O_3). The minimum amount of chromium required is 10.5%; however, conventionally, SS contains ~18% of Cr to form the self-healing, tenacious, thin, and protective passive film of Cr_2O_3, with thickness of a few tens of angstroms.

The rate of uniform corrosion is determined by the oxidizing power of the electrolyte. Highly aggressive environment will cause severe metal loss at a rapid rate leading to failure. During such process, the component/sample becomes thinner as it corrodes until its thickness is reduced such that failure would occur. The uniform corrosion rate expressed in thickness loss is mils per year (mpy) or millimeter per year ($mm\,yr^{-1}$). The uniform corrosion should be taken care of at the design stage itself by proper material selection and by providing a suitable corrosion allowance for the designed service life of the component. Such properly planning, inspection, and maintenance mitigate the risk of uniform corrosion. However, in some instances even a low rate of uniform corrosion cannot be tolerated, such as in food processing and packaging industry where corrosion product may be unacceptable contamination. Paintings, coatings, and superior alloys selection are a few common approaches to avoid/control uniform corrosion.

2.4.2 Galvanic Corrosion

The galvanic or bimetallic corrosion commonly refers to electrochemical corrosion as a result of two different metals coming in contact with a corrosive electrolyte. The anodic metal corrodes, whereas the "noble" component of the dissimilar couple will support the cathodic reaction. Different metals and alloys have different electrochemical potentials (or the tendency to act as anode) in the same electrolyte. The difference between the potential between two dissimilar metals is the driving force for corrosion on the active metal. Galvanic series in specified environments are available, which could be used to predict the galvanic behavior of the materials. The relative electrochemical potential of different engineering metals and alloys is presented in Table 2.1.

Severe galvanic corrosion could be expected if the highly active material is coupled with the noble material, i.e. those with wide the separation in the galvanic series. Further, if the area of the active metal is smaller to the area of the noble metal, a severe galvanic corrosion takes place at the anodic site (i.e. "area effect"). Further, the galvanic corrosion is higher at the junction of bimetal, and the intensity of corrosion attack decreases with increasing distance from the point of contact. Apart from bimetal, it should be noted that galvanic corrosion can occur due to differences in electrochemical potentials of alloy matrix and secondary precipitates in multiphase alloys containing precipitates.

Table 2.1 Galvanic series for the galvanic behavior of different engineering metals and alloys in seawater.

Noble end
Titanium
Monel
Passive stainless steel
Inconel alloy
Nickel
Copper nickel (70/30)
Copper nickel (90/10)
Aluminum bronze
Copper
Alpha brass (70Cu–30Zn)
Aluminum brass
Muntz metal (60Cu–40Zn)
Tin
Lead
Active stainless steel
Cast iron
Mild steel
Aluminum
Zinc
Magnesium
Active end

Notably, the principle of galvanic effect is very commonly used for corrosion protection by cathodic protection, such as by using sacrificial anode. For example, galvanizing is the process of coating steel structures with zinc. In the event of corrosion of such structures, zinc acts as a sacrificial anode and corrodes in preference to steel, thus protecting the steel.

2.4.3 Pitting Corrosion

Pitting is a localized form of corrosion that refers to the formation of small cavities/holes at the surface of passivating metals and alloys. Pitting occurs when a thin passive film is chemically or mechanically disrupted and does not instantaneously repassivate. Local breakdown of passive film of SS is usually by the interaction of aggressive halide species, such as chloride in marine environment. Breakdown of passive film occurs at the sites of inhomogeneities in the materials such as inclusions, grain boundaries, and second-phase precipitates. The unattacked areas of the surface that are covered with passive film act as a larger cathode when compared with the small anodic corroding/pitting region, thereby accelerating the localized dissolution. The aggressive halide anions react with the metal at the location of film disruption to form metal chlorides that hydrolyze to form metal hydroxide at the cost of increasing the acidity. This increase in acidity at the pitting site enhances the metal dissolution. The electrons released during the dissolution are consumed by the cathodic reaction occurring at the large remaining surface with passive film away from the pit site. Due to the autocatalytic nature of the pitting, rapid dissolution of the metal takes place within the pit (anodic), and reduction of oxygen takes place at the adjacent surface (cathodic) as shown in Figure 2.3. The resulting pits can rapidly propagate and can often perforate the wall thickness of a metal. Therefore, such localized loss of metal is more dangerous when compared with the uniform corrosion.

Depending on the metallurgical features such as secondary precipitates of the alloy and chemistry of the environment, pit morphologies may be narrow or deep, elliptical, wider or shallow, subsurface, undercutting, and vertical and horizontal. There are several ways by which pits can be quantified including pits/unit area, diameter of pits size, depth of pits, and pitting factor. "Pitting factor" is the ratio of the maximum of pit depth to the average penetration of the total pits at the surface. The larger the pitting factor, the greater the depth of penetration. Common approaches to avoid pitting include the use of superior alloys or modification of alloy surface.

2.4.4 Crevice Corrosion

Crevice corrosion refers to the localized attack on a metal surface at or immediately adjacent to the gap between two joining surfaces where stagnation of

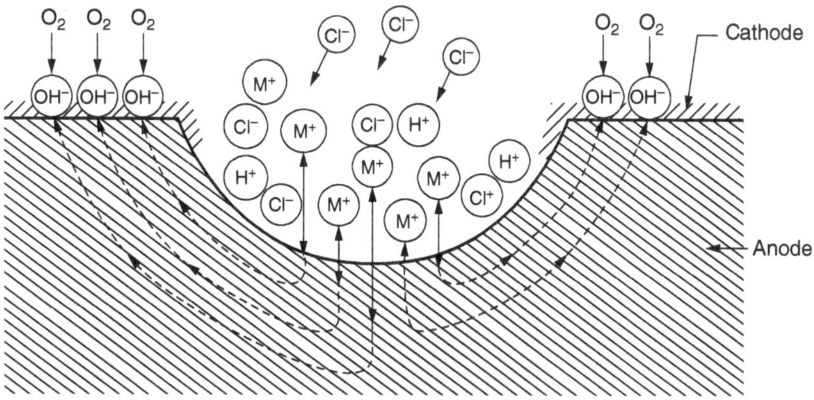

Anodic	$M \rightarrow M^{n+} + ne$
	$Fe \rightarrow Fe^{2+} + 2e$
	$Cr \rightarrow Cr^{3+} + 3e$
	$Ni \rightarrow Ni^{2+} + 2e$
Cathodic	$O_2 + 2H_2O + 4e \rightarrow 4OH^-$
Acidification of pits	$M^+Cl^- + H_2O \rightarrow MOH + H^+Cl^-$

Figure 2.3 Autocatalytic nature of pitting, with rapid dissolution of metal taking place in the pit (anodic) and reduction of oxygen taking place at adjacent surface (cathodic).

electrolyte solution is possible. Such a gap or crevice could be formed on gasket fittings, riveted connections, spot-welded lap joints, stacked metal sheets, etc. These situations are often unavoidable in practice. The geometry and configuration crevice leads to the poor access of electrolyte to it. Initially the oxidation and reduction reactions occur uniformly in the crevice and the surrounding area. However, after short duration, the oxygen within the crevice is depleted because of the restricted convection, thereby stifling the oxygen reduction reaction, but still the metal dissolution continues within the crevice. This produces an excessive metal ion concentration in the crevice, which is necessarily balanced by the migration of chloride ions into the crevice. Thus the crevice solution becomes increasingly aggressive causing breakdown of the passive film and active dissolution of metal within the crevice (Figure 2.4). Thus crevice corrosion propagates in a self-sustaining manner.

The factor affecting crevice corrosion is crevice geometry that includes the gap, depth, bulk to crevice area ratio, number of crevices, and type of crevices (metal or nonmetal). In addition, bulk solution composition (chloride ion, oxygen, and pH), bulk solution environment (temperature, agitation, and volume), alloy composition, and passive film characteristics have significant role on crevice corrosion.

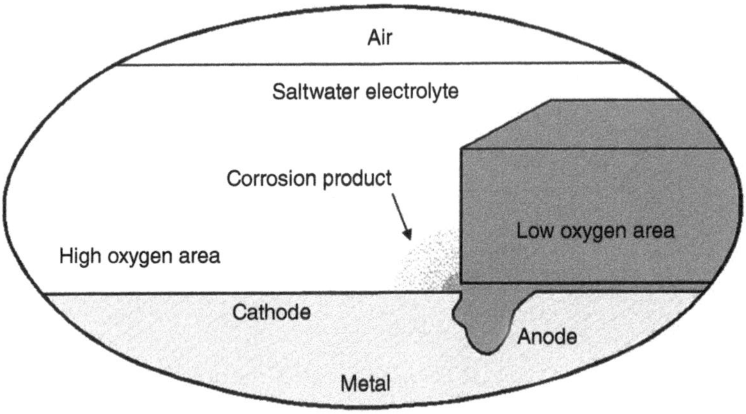

Figure 2.4 Metal within the crevice dissolves as anode, while reduction takes place away at cathode during crevice corrosion.

The most effective approach to eliminating crevice corrosion would be by suitable design and fabrication of components and structures to avoid crevice formation. However, when crevices are unavoidable, cathodic protection and inhibitors may provide adequate corrosion control. In the worst case, a suitable replacement with high corrosion resistance material may be required to avoid crevice attack. High nitrogen and high molybdenum containing stainless steels have been recommended for severe conditions where crevice corrosion is unavoidable.

2.4.5 Intergranular Corrosion

Intergranular corrosion is the preferential corrosion along the grain boundaries of metals/alloys, while the grain interiors are virtually unattacked. The intergranular corrosion starts at the exposed grain boundaries and can penetrate to the extent of grain falling out. Intergranular corrosion of sensitized austenitic stainless steels is well known. When austenitic stainless steels experience sufficient duration in the temperature range of 550–850 °C, they suffer the intergranular precipitation of chromium-rich carbide ($Cr_{23}C_6$) and the concomitant depletion of chromium in the regions adjacent to carbides along grain boundaries, i.e. sensitization of the SS (Figure 2.5).

The chromium-depleted areas adjacent to the grain boundaries act as anodic, whereas the rest of the areas within austenite grains become the cathodic. A galvanic cell with a small anode to larger cathode area ratio is set up between the chromium-depleted regions immediately next to the boundaries and the high chromium areas in the bulk of the grains. The areas of small anode, i.e. chromium-depleted zone near the grain boundary, suffer rapid dissolution as

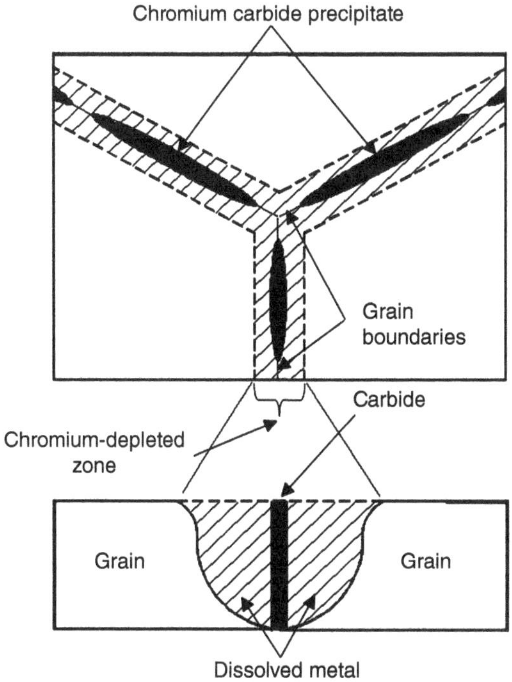

Figure 2.5 Chromium-rich carbide formations along grain boundary leading to Cr-depleted zone formation.

they do not contain sufficient corrosion resistance. The chromium carbide in the grain boundary is not attacked. Weld decay is a form of intergranular corrosion, usually of SS welds that occur as a result of sensitization in the heat-affected zone during the welding operation. The corrosive attack is restricted to the heat-affected zone that experiences the required temperature of 550–850 °C zone (Figure 2.6).

To avoid or minimize intergranular corrosion of austenitic SS, there are several ways such as lowering the carbon content in the alloy (<0.03%), adding stabilizers like Ti and Nb to form strong carbide formers, adding nitrogen to delay carbide formation, and performing solutionizing treatment. Knife-line attack (KLA) is a form of intergranular corrosion of stabilized SS. For stabilized stainless steels and alloys, carbon is bonded with stabilizers (Ti or Nb), and no weld decay occurs in the heat-affected zone during welding. But during subsequent heat treatment or welding, precipitation of chromium carbide is possible, and this leaves the narrow band adjacent to the fusion line susceptible to intergranular corrosion. The corrosion attack is restricted to extremely narrow line that appears as knife line, adjoining the fusion line.

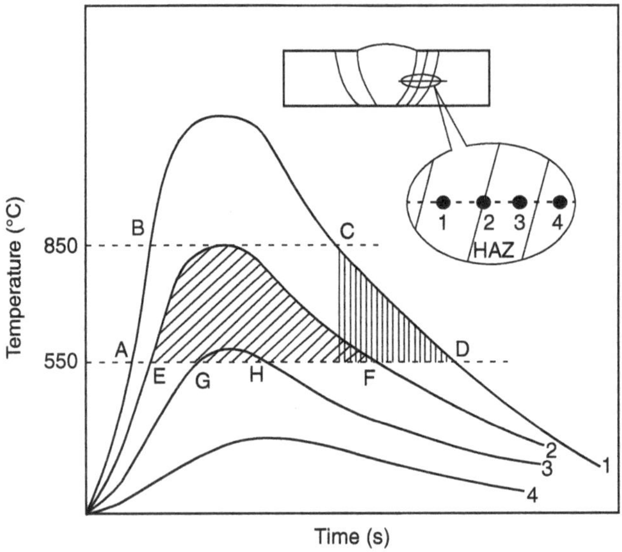

Figure 2.6 Time–temperature diagram showing formation of heat-affected zone (HAZ) where sensitization occurs.

2.4.6 Dealloying

Dealloying is a corrosion process during which an alloy is "parted" by the selective dissolution of the electrochemically more active element of the alloy. For the dealloying process to occur, the important parameters to control are the alloy composition, the electrolyte, and the electrochemical potential of the constituents of the alloy. The potential difference between the electrochemically more active elements and the noble element in the alloy is the driving force for the dealloying to occur. In a typical Au–Cu solid solution alloy, the Cu is selectively leached to leave behind a porous Au layer. In the case of dezincification of brass, zinc is preferentially leached out of the copper–zinc alloy, leaving behind a copper-rich surface layer that is porous and brittle. Selective removal of iron from cast-iron leaves porous graphite structure that is a specific type of dealloying process called graphitization. The mechanical properties of a porous layer of dealloyed layers are different from those of the bulk alloy, rendering this layer susceptible to brittle crack propagation.

2.4.7 Erosion Corrosion

Erosion corrosion refers to the combined action of erosion and corrosion of a metal/alloy component in contact of a moving fluid. The flow or velocity (mechanical effect) of the fluid combined with the corrosive action (chemical effect) of the fluid causes accelerated loss of metal. The severity of erosion corrosion depends on the fluid velocity, turbulence, and the tenacity of the protective layer formed on the metal/alloy component. Such attack is severe in high velocity slurries because of abrasive action of the suspended solids. Erosion corrosion often occurs at elbows, turbines, pumps, tube constrictions, and other structural features that alter flow directions or velocity and increase turbulence. Attack may be very serious in two-phase flow in which steam and water vapor condensate droplets are present together. Hard, strong metals, although resistant to wear, are not resistant to erosion corrosion if their corrosion resistance is low or if they form a very weak passive film. Erosion corrosion takes the form of grooves, waves, gullies, teardrop-shaped pits, and horseshoe-shaped depressions in the surface. In general, alloys with high toughness (stainless steels, nickel, and titanium alloys) are resistant to erosion corrosion when compared with the carbon steel, aluminum, and copper alloys. Erosion corrosion can be prevented by design to reduce surface velocity and turbulence and careful material selection. Inhibitor or passivator addition can also control erosion corrosion.

2.4.7.1 Cavitation Corrosion

Cavitation corrosion occurs due to the combined action of cavitation and corrosion. Components where large pressure changes are present (such as turbine blades, pump impellers, ship propellers, and pipes) can undergo cavitation corrosion during service. Cavitation corrosion is caused by the formation and collapse of vapor bubbles. Collapsing of the vapor bubbles produces shock waves with pressures ~414 MPa. These shock waves with high force can create plastic deformation in metals and even can remove metal particles away from the surface. In the early stage of cavitation corrosion, the damaged surface appears as closely spaced pits on the metal surface, and as time progress it looks like a heavily ground surface. Those roughened surfaces further increase nucleation for collapsing of vapor bubbles and create more cavitation damage. These processes are repetitive and result in the formation of deep holes. Cavitation damage can be minimized by proper control of the hydrodynamic pressure differences in the flow stream. Also smooth surfaces minimize the bubble nucleation. Selection of corrosion resistance alloy and proper operation of pumps and other equipment minimize the cavitation corrosion.

2.4.7.2 Fretting Corrosion

Fretting corrosion occurs at dry contact surface of metal components that is subjected to load and repeated relative motion against each other. The load and relative motion of the interface must be sufficient to produce slip or deformation on the surfaces. For fretting the displacement of relative motion is of very small magnitude of about 10^{-8} cm. Fretting corrosion is expected in many tight fitting parts that are designed not to slip against each other, but in the presence of vibration and pressure fluctuation, these parts do slip against each other. Fretting often appears in press-fitted bearings on shaft, bolted tie plates on railroad rails, and at other contact points during shipping of components. The mechanism of fretting is a combination of oxidation and wear as in the following sequence: metal surface exposed to air to form a thin oxide layer, vibration of the surfaces against one another ruptures oxide at the contact points, bare metal rubbing against bare metal from cold welds, continued vibration breaks these microscopic welds, and the fractured surface immediately oxidizes in the air and the process repeats. Thus, fretting corrosion leads to production of oxide debris and destruction of metal surfaces. Fretting may be mitigated by lubrication of contacting surfaces to prevent the wearing action of fretting, insertion of insulating or cushioning material between surfaces as sealants and gaskets, and substitution of more wear-resistant materials.

2.4.8 Environmentally Assisted Cracking

EAC is the expedited cracking of a susceptible material under the simultaneous influence of corrosion and mechanical loading. The EAC or corrosion cracking process leads to unexpected and premature failures of the components. Corrosion cracking in a broad term includes SCC, CF, and hydrogen embrittlement. EAC crack propagation is generally attributed to one of two main mechanisms: anodic dissolution at the crack tip or hydrogen-assisted embrittlement ahead of the crack tip. A simple description of anodic dissolution mechanism implies that the stress gets intensified at an existing crack tip and locally disrupts the protective film, allowing rapid dissolution at film-free crack tip. This localized dissolution amounts to crack extension, before the crack tip repassivates and is again ruptured. Repetition of the film rupture–dissolution–repassivation processes is responsible of SCC crack propagation. In the most common of the hydrogen-assisted embrittlement mechanisms, hydrogen atoms accumulate just ahead of the crack tip under the influence of stress (which has been shown to intensify ahead of crack tip). Accumulated hydrogen atoms embrittle the area ahead of crack tip, when their concentration exceeds the threshold for the given material. Hydrogen atoms may be generated external to the metal surface through cathodic reaction or may be internally present in the metal.

2.4.8.1 Stress Corrosion Cracking

Cracking of a susceptible metal under the combined action of tensile stress and corrosive environment is termed stress corrosion cracking. The tensile stress can be residual stress (such as those caused by welding) and/or an applied stress. Load-bearing capacity and time to fracture of the metal are lowered by SCC when compared with those properties when stress or corrosion was to act alone. During SCC, the metal or alloy is virtually unattacked over most of its

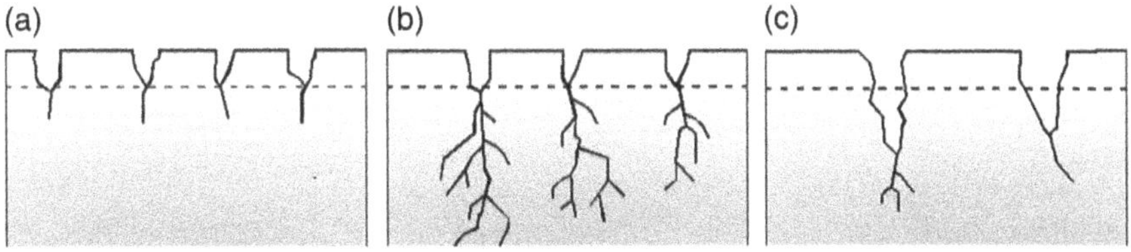

Figure 2.7 (a) Initiation of SCC and corrosion fatigue cracks at pits, (b) highly branched SCC cracks, and (c) corrosion fatigue cracks with less branching.

surface. The two main stages of SCC are initiation and propagation. The cracks can initiate and propagate to the failure of component without significant dimensional changes. Typical cracks often initiate at pits or crevices that can concentrate the stress (Figure 2.7a). These cracks run into the metal through the grains (transgranular) or along grain boundaries (intergranular) depending on the conditions during propagation. The final fracture is by mechanical overload leaving dimples on fracture surface. Highly branched cracks (Figure 2.7b) in the direction perpendicular to the tensile stress axis are developed during SCC.

A usual countermeasure against SCC is stress-relief annealing, where the residual stresses are removed. Another way is to introduce compressive stress by shot peening.

SCC is among the most dangerous forms of corrosion-assisted failures, because:

1) It can occur at stresses much below the design stress of the material/component. Even residual stresses generated during fabrication such as welding may often be sufficient to cause SCC in a suitable environment if present. For example, chloride SCC of weldments in austenitic stainless steels is well known.
2) It is intriguing that even a mildly corrosive environment can cause severe SCC. Small concentrations of harmful component can cause SCC.
3) The surface of a metallic component may appear virtually free from corrosion, while a localized crack may propagate undetected to failure.

In addition to those listed above, the most fundamentally detrimental feature of SCC is that a ductile material that would have undergone considerable elongation before undergoes serious embrittlement under the simultaneous action of stress and corrosive environment and suffers a premature brittle fracture (i.e. without much ductility). As shown in Figure 2.8, mild steel suffered caustic embrittlement and intergranular SCC over a large fraction of the fracture surface, followed by mechanical failure (i.e. ductile dimples over rest of the area) due to overloading of the cross section reduced by intergranular SCC.

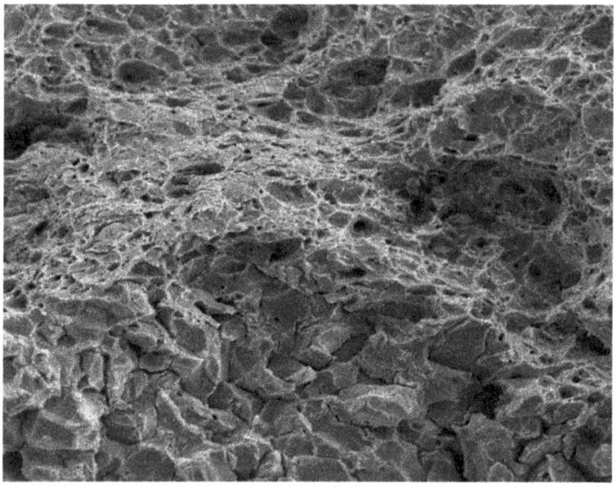

Figure 2.8 SEM fractograph of a caustic SCC of mild steel, showing a transition from corrosion-assisted intergranular cracking (lower part of the micrograph) to the pure mechanical failure (ductile dimples). *Source:* Singh Raman and Saxena [1]. Reproduced with permission of The Electrochemical Society.

Because of its dangerous features as listed above, SCC is a serious consideration while selecting materials for design of equipment operating in corrosive environment. SCC is highly environment specific, which is also a distinct characteristic of SCC. Not all alloy–environment combination produces SCC, whereas any environment providing sufficient atomic hydrogen can cause hydrogen embrittlement of an alloy. Hydroxides (caustic), nitrates, carbonates, phosphates, $CO–CO_2–H_2O$, and aqueous chlorides are among common environments that can cause SCC to steels. Other variables that govern SCC are material chemistry, microstructure, cold work, stress level, concentration of corrosive species, additives to the environment, pH, temperature and electrochemical conditions, and the vast literature on SCC of low and medium strength steels and are associated with the study of one or more of these variables.

2.4.8.2 Corrosion Fatigue

CF results from the combined effect of cyclic loading and corrosion to cause premature brittle fracture failure of metals and alloys. Due to the presence of corrosive environment, the fatigue life of a metal component is reduced. CF failure happens at stresses lower than those for failure in the absence of corrosion. For example, SS possesses good CF resistance in normal water, but in sea water, the fatigue resistance decreases to 70% of their normal fatigue resistance. The CF is prevalent in the environment that causes pitting as the pits act as stress raisers and initiate crack. A CF crack usually begins at a pit or surface irregularities produced due to corrosion (Figure 2.7a). The crack further propagates to fail the component under the simultaneous action of corrosion and cyclic stress. The predominant fracture surface is covered with corrosion products and a smaller rough area due to brittle fracture. CF cracks generally propagate transgranularly and are usually unbranched (Figure 2.7c) as compared with SCC cracks that are highly branched. It can be prevented by retarding the corrosion with inhibitors or cathodic protection, reducing chances of fatigue by selecting ductile materials with low internal stress, minimizing cyclic stress and vibration, eliminating factors that increase stress, or applying surface compressive stress by methods such as shot peening or sandblasting.

2.4.8.3 Hydrogen Damage

Hydrogen damage refers to the damages caused by interaction of hydrogen with metal. There are a few types of hydrogen damages such as hydrogen blistering, hydrogen-induced cracking (HIC), and hydride embrittlement. Hydrogen ions present in the electrolyte reduce to hydrogen atoms at the metal surface. The hydrogen atoms shall diffuse through the metal and accumulate in the voids or at inclusions. The trapped hydrogen atoms will recombine within the voids to form hydrogen molecules that become immobile due to the size of the molecule. The accumulation of hydrogen gas molecules in voids will create internal hydrogen pressure sufficient to create hydrogen damage via blistering or cracking. Adjacent hydrogen blisters on different planes in the metal shall be connected to form large cracks, i.e. stepwise internal cracks to form HIC. In other form of hydrogen damage, the diffused hydrogen atom will react with certain metals such as zirconium and titanium to form metal hydrides. These metal hydrides are brittle in nature and deteriorate the mechanical properties by initiating cracks. This phenomenon is called hydride embrittlement. In the absence of residual stress or external loading, hydrogen embrittlement that is manifested in various forms, such as hydrogen blistering, HIC, and hydride formation, will reduce the ductility of the metal. Hydrogen damage can be prevented by the modification of environment or selection of materials that are more resistant to embrittlement. Elimination of cathodic protection system and galvanic coupling is also helpful. Removal of hydrogen poisons like cyanides in the environment and elimination of inclusions and other inhomogeneities in the material will also reduce the propensity to hydrogen damage.

2.5 Microbiologically Influenced Corrosion

Microorganisms like bacteria, algae, and fungi in natural waters adhere to metal/alloy surfaces, secrete metabolic products, and reproduce and form biofilms. The biofilms (~20–80 nm thick) are composed of 90% water molecules and its metabolites. The biofilms would be able to modify the electrostatic charge and wettability of the metal surface facilitating its further colonization by microorganisms, leading to biofouling. A dynamic system is formed at the biofouled interface, and mass transport processes are governed by the biofilm matrix. Degradation of metals/alloys influenced by the presence of either microorganisms or their metabolites is called MIC. Microbes can induce corrosion by creating differential aeration/concentration cells, changing pH, depleting oxygen, utilizing cathodic hydrogen or producing hydrogen, removing metals, oxidizing or reducing corrosion inhibitors, setting up galvanic cells, and disturbing or feeding upon protective organic coatings and their metabolic products. In addition, reactions between metabolites produced by microbes and the metal take place within the biofilm. Thus, these microorganisms influence the corrosion phenomenon either directly by accelerating the electrochemical reactions or indirectly by virtue of their metabolic products. MIC is accused for failures in cooling water systems, industrial process plants, underground sub-tanks, oil storages, alcohol industries, nuclear power stations, marine tankers, offshore concrete storage cells, and several other industries.

Bacterial groups involved in MIC fall in four categories: sulfate-reducing bacteria (SRB), sulfur-oxidizing bacteria (SOB), iron-oxidizing/depositing bacteria (IOB), and manganese-oxidizing/depositing bacteria (MOB). SRB utilize sulfate from the environments and produce H_2S that could directly attack metal/alloy surfaces such as carbon steels, stainless steels, and copper alloys. SOB shall oxidize iron that will either oxidize sulfur compounds to form H_2SO_4 or induce anodic reaction directly. Such H_2SO_4 formation is very damaging for sewage pipelines made of concrete. IOB when attached to mild steel components have the capability to disrupt/remove the passive γ-Fe_2O_3 film and replace it with a biofilm under which pitting corrosion can take place. MOB are capable of oxidizing manganese in the MnS inclusion present in SS, leading to biomineralization. The presence of biomineralized manganese dioxide would act as a cathodic reactant and increase the corrosion rate.

2.6 Evaluation and Monitoring of Various Types of Corrosion by Nondestructive Technique (NDT)

Though utmost care may be taken in the selection of material, proper design, and control of environment, corrosion-related degradation process is still inevitable. Nondestructive technique (NDT) plays a vital role for damage assessment and life extension of the component as it is suitable in detection of the early stages of corrosion so that corrective measures can be taken before damage becomes severe. NDT implies that the testing does not alter the properties of the materials/components tested; it leaves the component identical to the prior state. As discussed earlier in this chapter, the physical manifestations of various types of corrosion are different. Thus, detection and evaluation of

Table 2.2 NDT techniques for evaluation of various types of corrosion.

Corrosion mechanism	NDT techniques
Uniform	UT, ECT, RT, OPTICAL/LASER, IRT
Localized	UT, ECT, RT, OPTICAL/LASER, IRT
Stress corrosion cracking	PT, AE, UT, ECT
Corrosion fatigue	PT, AE, UT, ECT
Hydrogen damage	UT
High temp. oxidation	AE, ECT

UT, ultrasonic testing; ECT, eddy current testing; RT, radiography testing; IRT, infrared thermography; PT, penetrant testing; AE, acoustic emission.

those various forms of corrosion may necessitate the use of different NDT techniques. Depending on the type of component and the type of corrosion to be monitored, there may be a choice of an appropriate NDT technique or their combination as presented in Table 2.2.

Corrosion monitoring becomes essential to assess the progress of the corrosion-related degradation during service in order to evaluate the performance of a component. According to National Corrosion Service of the National Physical Laboratory (NPL, UK), corrosion monitoring is defined as "any method used to observe or measure the progress of corrosion." However, corrosion monitoring provides deeper commitment to quantifying the rate of corrosion and determining underlying cause of corrosion damage. The function of corrosion monitoring is to provide an early warning of the damaging process conditions that may result in a corrosion-induced failure. By studying the correlation of changes in process parameters and their effect on system corrosivity, an effective evaluation of the corrosion control/prevention technique such as chemical inhibition and the determination of optimal process parameter shall be incorporated. Corrosion monitoring also provides opportunity to assess if the degradation is in the expected rate or tending toward accelerated corrosion stage. This approach is aimed to predict the life of the component and to extend the life component by avoiding failure or adopting strategies for reducing the corrosion degradation process. The effective corrosion monitoring (i) leads to excellence in plant operation, (ii) improves the confidence and credibility, (iii) plays a vital role in controlling the production losses and environmental damage, and (iv) provides safety of plant and personnel life.

A brief description of various NDT and corrosion monitoring techniques for the detection, evaluation, and monitoring of corrosion and corrosion cracking processes is given in subsequent chapters.

Reference

1 Singh Raman, R.K. and Saxena, A. (2007). Role of imposed potential in expanding the regime of strain rates for caustic cracking. *Journal of the Electrochemical Society* 154 (9): C451–C457.

Further Reading

Cramer, S.D. and Covino, B.S. Jr. (eds.) (2003). *ASM Handbook, Volume 13A, Corrosion – Fundamentals, Testing, and Protection*. Materials Park, OH: ASM International.

Fontana, M.G. (1986). *Corrosion Engineering*, 3e. New York: McGraw-Hill.

Jones, D.A. (1992). *Principles and Prevention of Corrosion*, 2e. Upper Saddle River, NJ: Prentice Hall.

Little, B.J. and Lee, J.S. (2007). *Microbiologically Influenced Corrosion*. Hoboken, NJ: Wiley.

Raj, B., Jayakumar, T., and Thavasimuthu, M. *Practical Non-destructive Testing*. New Delhi: Narosa Publishing House.

Roberge, P.R. (2008). *Corrosion Engineering Principles and Practice*. New York: McGraw-Hill.

Sedriks, A.J. (1996). *Corrosion of Stainless Steel*, 2e. New York: Wiley.

Winston Revie, R. and Uhlig, H.H. (2008). *Corrosion and Corrosion Control*, 4e. Hoboken, NJ: Wiley.

Workman, G.L. (Technical ed.) and Moore, P.O. (ed.) (2012). *Nondestructive Testing Handbook, Volume 10, Nondestructive Testing Overview*. Columbus, OH: American Society for Non-Destructive Testing.

3

Nondestructive Evaluation of Corrosion

Case Studies I

Paritosh Nanekar[1], *N. Jothilakshmi*[1] *and Baldev Raj*[2]

[1] *Bhabha Atomic Research Centre, Mumbai, India*
[2] *National Institute of Advanced Studies, Bengaluru, India*

3.1 Introduction

Corrosion is one of the most common mechanisms of failure for many industrial components. Corrosion can manifest as uniform attack or localized attack, viz. pitting, intergranular corrosion (IGC), crevice corrosion, stress corrosion cracking (SCC), hydrogen attack, etc. Corrosion failures can be avoided by monitoring the extent or rate of corrosion periodically during in-service inspection (ISI). There are many factors that influence the selection of corrosion monitoring methods. These include the type of corrosion, environmental conditions, time required for measurement, response to change in corrosion rate, ease of interpretation, and so on. One important consideration is the access to the process stream or equipment to be monitored. If access is available, methods involving probes and coupons become more feasible; otherwise nondestructive examination (NDE) methods are employed. From NDE point of view, corrosion monitoring involves detection of (i) loss of material thickness (uniform corrosion), (ii) local loss of material thickness (pitting), and (iii) cracking (SCC). This chapter deals with the commonly used NDE methods for corrosion monitoring. Some of the advanced NDE methods that are now being increasingly used in the industry for corrosion monitoring are also discussed.

Non-Destructive Evaluation of Corrosion and Corrosion-assisted Cracking, First Edition. Edited by Raman Singh, Baldev Raj, U. Kamachi Mudali and Prabhakar Singh.

3.2 Nondestructive Examination: A Perspective

The American Society for Testing and Materials (ASTM E1316, "Standard Terminology for Nondestructive Examinations") has defined nondestructive examination as follows:

> The development and application of technical methods to examine materials or components in ways that do not impair future usefulness and serviceability in order to
> - detect, locate measure and evaluate flaws
> - to assess integrity, properties and composition
> - to measure geometric characteristics

Typically, NDE involves a probing input, which interacts with the material under test. The output is analyzed to determine the presence of flaws, for assessment of material properties, and for dimensional measurement. During flaw detection, any deviation in the output from what is normally expected is referred to as an indication. Indication is defined as the evidence of a discontinuity that requires interpretation to determine its significance. The indication can be broadly classified into three categories, viz. relevant, irrelevant, or false. A relevant indication is caused by a condition or type of a discontinuity that requires evaluation for its acceptance in the component. For example, an indication from flaws like porosity, inclusion, lamination, and crack is termed as relevant indication. A nonrelevant indication is caused by a condition or type of discontinuity that is not rejectable. For example, an indication from a groove or a keyway in the component, which is present in the component as part of design, is termed as nonrelevant indication. A false indication is the one that is interpreted as a discontinuity at a location where no real discontinuity exists. Such indications are encountered during the examination if proper testing procedures are not followed or if the test equipment is not set up correctly.

NDE is applied at various stages during the life of the component, what is referred to as "cradle-to-grave" approach. When a component is being fabricated, it undergoes various processing treatments like casting, forging, rolling, extrusion, machining, heat treatment, welding, and so on. Even with the most stringent process control, there is always a possibility that the processing operations can introduce flaws in the component being manufactured. It is desirable to eliminate these flaws before the component is subjected to the next processing step. NDE plays an important role by finding out the presence and the location of these flaws. The examination of the component carried out during its manufacturing stage is called as in-manufacture inspection (IMI). Once the component is completely fabricated, it is examined by various NDE techniques to rule out the presence of any flaw more than the acceptable size, which is specified in the NDE specifications. The component after fabrication and final testing is shipped to the site where it is installed and commissioned. Prior to putting the component into service, it is examined by various NDE techniques. This inspection is referred as preservice inspection (PSI). The objective of PSI is to detect and locate the flaws in the installed component and generate a baseline data, so that their growth can be monitored periodically during service. During operation, there may be factors that can adversely affect the integrity of a component. Such factors include unanticipated stresses, operation outside designed limits, environmental effects, degradation of material properties in service, etc. These factors lead to the initiation and growth of new flaws as well as the growth of existing flaws in the component. It is difficult to predict the effect of these factors on the useful life of the component during the design stage itself. Hence the component is periodically examined during service to assure its structural integrity and avoid any catastrophic failure. Such inspection is called as in-service inspection.

NDE techniques can broadly be classified into three categories, viz. surface examination techniques, volumetric examination techniques, and performance tests. Some of the commonly used surface examination techniques are visual testing (VT), liquid penetrant testing (LPT), magnetic particle testing (MPT), eddy current testing (ECT), and *in situ* metallography (ISM). Radiography testing (RT) and ultrasonic testing (UT) fall in the category of volumetric examination techniques, while the performance tests include leak testing (LT) and acoustic emission testing (AET).

A brief overview of the commonly used NDE techniques, which are relevant for corrosion monitoring, is given in Chapter 1, while the references [1–5] provide a detailed overview.

3.3 Factors Determining Selection of NDE Method

Various NDE methods have matured over the years and successfully employed for field examination for variety of purposes, including the monitoring of corrosion damage. The choice of a particular NDE technique(s) for a given task depends on several factors. Some of these are given below.

3.3.1 Location of the Flaw to Be Detected

The location of the flaw, i.e. surface or subsurface, decides whether to use a surface examination technique like VT, LPT, eddy current testing etc. or a volumetric examination technique like UT or radiography. If the flaw is expected to originate from the surface, which is directly accessible, then surface examination techniques are easier to use. If the flaw is embedded deep in the material or if the flaw is on the opposite surface for which there is no direct access (for example, the inside surface of a closed pipe or vessel), then one would have to choose the volumetric examination techniques.

3.3.2 Type of Material

Some of the NDE techniques are suitable for a particular type of material. For example, eddy current testing and MPT can be used only on conducting and ferromagnetic materials, respectively. Similarly, LPT cannot be used for components where the examination surface is extremely rough. UT and radiography can be used on almost all kinds of materials. Some materials like austenitic stainless steel casting may pose challenges for carrying out a meaningful ultrasonic test.

3.3.3 Geometry of the Component to Be Examined

Intricate shapes may prevent the use of a particular NDE technique. The thickness of a component is also an important factor while selecting a particular NDE technique, when it is required to detect a flaw located within the volume of the component. For example, although UT and radiography are both volumetric examination techniques, the former is selected during the examination of thicker components, while the intricate shapes are more likely to be examined by radiography.

3.3.4 Accessibility

For the application of the surface examination techniques, the surface to be examined should be directly accessible for the examination. For example, it is not possible to detect pits on the inside surface of a pipe or a vessel while doing examination from the outside surface by employing any of the surface examination techniques. Radiography needs access to both the surfaces of the component, one for the placement of the source and the other for the placement of the film or any other recording media. Many times, this condition cannot be satisfied, especially during ISI. In such cases, UT is preferred over all the other NDE techniques as it needs access to only one surface of the component.

3.3.5 Sensitivity of the Flaw to Be Detected

Sensitivity refers to the smallest flaw that can be detected during examination. Although there are several NDE techniques in the category of surface and volumetric examination, some of these are more sensitive than others. For example, a very fine stress corrosion crack in the heat-affected zone may go undetected by visual examination, but can be detected by fluorescent LPT. Similarly, radiography may be able to detect fine pores in the weld joint, which may be missed by UT. On the other hand, a tight crack, like stress corrosion crack, may be missed during radiography but is likely to be detected by UT.

3.3.6 Cost

Cost of the examination is an important factor in deciding which NDE technique(s) is employed for a given task. If the intent of examination can be accomplished by an inexpensive NDE technique without comprising on the test results, then it is preferred over the expensive NDE techniques. For example, detection of pits on the outside surface of a pipe can be accomplished by VT, which is less costly as compared with the other NDE techniques like eddy current testing or UT.

3.4 NDE for Monitoring Uniform Corrosion

Uniform corrosion refers to the loss of material over a significant area that is exposed to the corrosive environment. The extent and rate of uniform corrosion can be estimated by periodically monitoring the remaining thickness of the component. Direct visual examination of the exposed surface can readily reveal whether the component is under uniform corrosion attack. However, it may not be possible to estimate the damage quantitatively in terms of loss of thickness. Also, in cases where the surface undergoing corrosion is not accessible for direct visual examination, one will have to employ other NDE techniques for detection and quantification of the uniform corrosion attack.

The monitoring of uniform corrosion by ultrasonic test is based on measurement of wall thickness. The wall thickness measurement is carried out by using a normal beam transducer and a thickness gauge [6]. The ultrasonic transducer is coupled to the surface of the component. Sound waves travel inside the component and are reflected from the back surface to the transducer, producing a signal, called as backwall echo. The thickness gauge measures the time of travel of the backwall echo and multiplies it by the velocity of sound in the material (that is given as an input to the gauge by the user). The calculated value is divided by 2 (since the sound beam travels twice the thickness while producing the backwall echo) and is displayed on the screen of the gauge as a thickness reading. For arriving at the thickness reading, one can measure the time of travel of the first backwall echo or the time-of-travel difference between the first and the second backwall echo.

The principle of thickness measurement by ultrasonic technique is shown in Figure 3.1. Figure 3.1a shows that the transducer is placed at the location where the thickness is required to be determined. The flaw detector screen shows the first and the second backwall signals. The ultrasonic instrument measures the time of flight between the two successive backwall signals to determine the thickness. Figure 3.1b shows the test setup for thickness measurement on an aluminum sample of ~50 mm thickness. The transducer is placed on a sample whose sound velocity is known and is fed to the

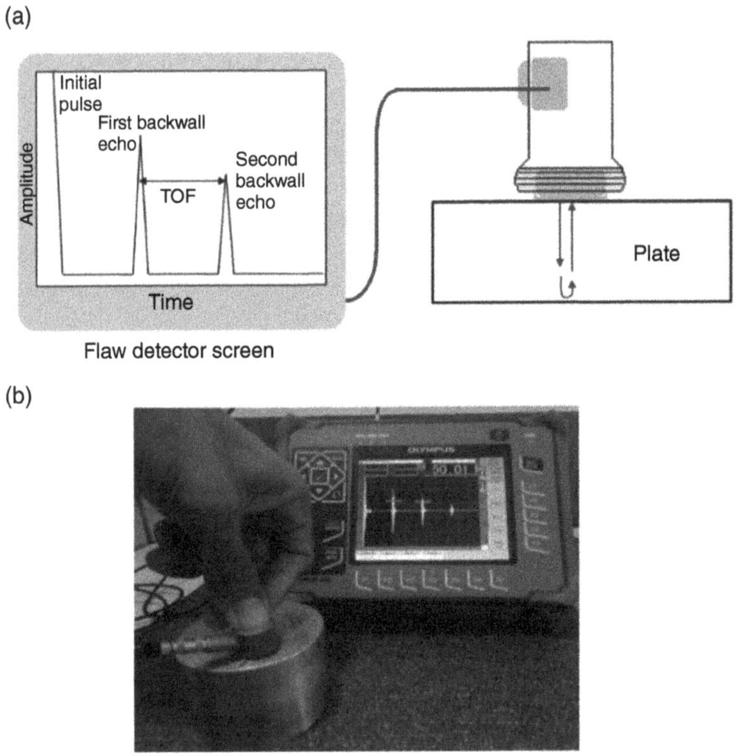

Figure 3.1 (a, b) Principle of thickness measurement by ultrasonic thickness gauge.

instrument by the user. The flaw detector shows three backwall echo signals. The gates are placed on the first backwall and the second backwall signal. The instrument measures the time of flight between these two backwall signals, and the thickness of 50.01 mm is displayed.

Ultrasonic-based thickness measurement to determine the wall loss caused by uniform corrosion is commonly used in many industries. One of the problems associated with the UT is the “dead zone,” which is the region near the test surface for which no meaningful inspection or measurement can be done. The dead zone can be order of few millimeters and can introduce difficulty during thickness measurement on thin sections. In such cases, transducers with delay line (plastic wedge) or transducers with separate transmitter and receiver (also called as dual crystal or twin crystal transducer) are used. Figure 3.2 shows the sketch of delay line and dual crystal transducers. With the delay line transducer, the dead zone gets shifted from the material under investigation to the delay line. As a result, smaller thicknesses can be measured. With the dual crystal transducers, the transmitting and receiving function for the sound

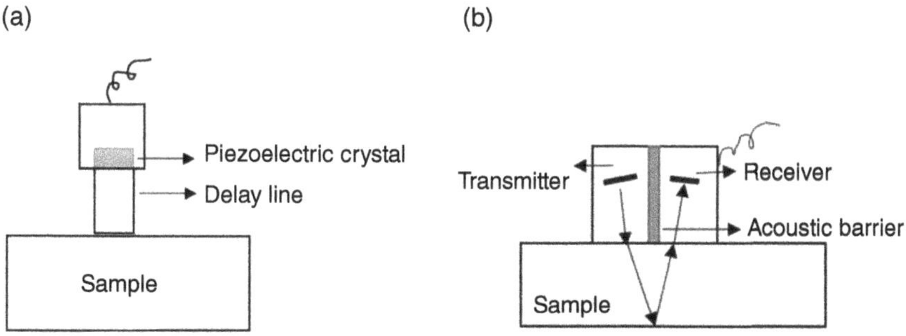

Figure 3.2 Sketch of (a) delay line and (b) dual crystal transducer.

waves are performed by two separate piezoelectric crystals. As a result, the dead zone is significantly reduced.

The condition of the test surface on which the ultrasonic transducer is placed plays an important role in determining whether the thickness of the component can be measured or not. If the test surface is very rough, there will not be proper coupling and the sound beam cannot be transmitted inside the component. Hence, rough and badly pitted scanning surfaces are cleaned and made smooth before the measurement to permit the transducer to make proper contact. Similarly, loose scales and paint are removed from the location where the transducer is required to be placed. If the paint is tightly adhering to the surface, it need not be removed. However, the paint thickness will introduce errors in the thickness measurement unless the compensation is made for this.

The accuracy of thickness measurement depends on the resolution with which the time of travel of the backwall(s) is measured. Modern ultrasonic gauges can measure the time of travel with a resolution of 10 ns or even less and hence provide an accuracy of the order of a fraction of a millimeter in the thickness measurement. Advanced approaches using digital signal processing are also used to further improve the accuracy of thickness measurement using ultrasound [7].

Many a times, it is required to get the thickness map of a large area of a vessel or a pipe. Manual measurement using thickness gauge may become cumbersome. In such cases, manipulators are used to move the ultrasonic transducer over the surface of the component. The time-of-travel data for the backwall at each location is acquired by the ultrasonic machine. This data is then processed by a software and plotted in the form of a color-coded image. The color coding is based on the time of travel of the backwall signal, which in turn is related to the thickness of the component. This kind of data presentation is referred as a C-scan image for corrosion monitoring. A typical ultrasonic corrosion map showing the C-scan (plan view), the B-scan (cross-sectional view), and the A-scan is shown in Figure 3.3.

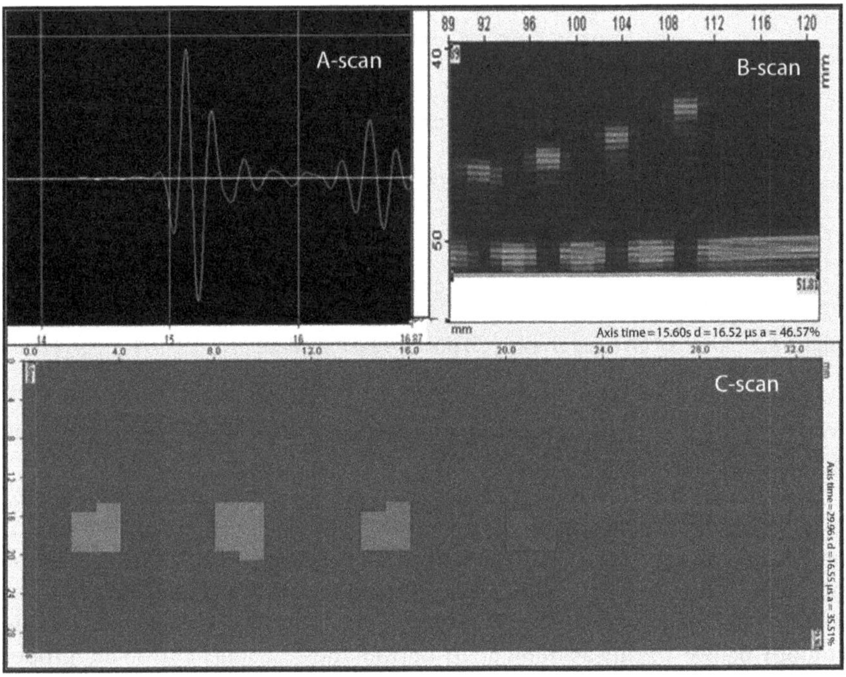

Figure 3.3 Ultrasonic corrosion map showing C-scan, B-scan, and A-scan.

The C-scan in Figure 3.3 shows four patches of different colors. These colors correspond to different thicknesses in the area that has been examined. The thickness at each location can be found from the color map. The B-scan shows the cross-sectional view, where one can see four different thicknesses, each corresponding to a patch in the C-scan image. The A-scan shows the waveform display at one of the four patches.

One of the major drawbacks of using piezoelectric transducers for UT is that one needs to have a smooth surface for proper coupling of the transducer with the job. This cannot be always guaranteed under field conditions, especially when the examination surface is corroded or heavily pitted. One way to overcome this problem is to use electromagnetic acoustic transducer (EMAT) [8]. Using EMAT, high frequency sound waves can be generated within the component using the principle of Lorentz force. With piezoelectric crystals, high frequency sound waves are generated outside the component, and hence a thin layer of couplant is required to transmit this sound within the material. In contrast to this, with EMAT, the high frequency sound waves are generated by a very thin layer of material, which is in close proximity to the EMAT. As a result, while using EMAT, no couplant (oil, grease, water) is required between the transducer and the surface of the job, and it can work effectively on rough or corroded surfaces. EMATs are extensively used for corrosion monitoring in the boiler tubes of power plants [9].

3.5 Monitoring Pitting Corrosion

Pitting corrosion is a form of localized corrosion that leads to the creation of small pits on the surface of the metal. The attack can penetrate the metal very rapidly at localized spots, while the remaining part of the metal surface remains free from corrosion.

Visual examination is one of the simplest NDE methods that can be employed for the detection of pits. When the pits are on the outside surface, direct visual examination with or without optical aids such as magnifying glass can be used to determine the presence of pits and their extent. When the pits are on the inside surface of components such as tubes, pipes, and vessels, visual examination using borescope and fiberscope is performed for detection of pitting corrosion. Current generation videoscopes can accurately characterize the pits in terms of their width and depth.

There are instances when the pitting is expected on the inside surface of a pipe, which is not accessible for visual examination, either directly or by using borescope or fiberscope. In such cases, UT using normal beam and angle beam is very effective for detection of pitting corrosion. During ultrasonic test, the transducer is kept on the outside surface, and the sound beam is sent inside the component. While using normal beam, a reflected signal from the bottom surface of the pit appears before the backwall echo signal. This signal is an indication of the presence of pit. By measuring the time of travel of this signal, one can estimate the depth of the pit. While using angle beam, the pit surface reflects the incident sound beam to produce an indication. Once calibrated, one can estimate the depth of a pit from the amplitude and time of flight of the reflected ultrasonic signal. Figure 3.4 shows the principle for detection of pits on the inside surface of the component by carrying out normal beam ultrasonic test from the outside surface.

Pitting corrosion is one of the commonly observed degradation mechanisms in tubes of heat exchangers. Eddy current test (ECT) is an effective NDE method for this purpose. It uses a differential bobbin coil, which is inserted inside the tube. The eddy currents travel along the circumference of the tube and get disturbed by the presence of pit on the inside or the outside surface of the tube as shown in Figure 3.5.

The differential coil consists of two coils wound together in the opposite direction so that the signal affecting both the coils simultaneously gets canceled. As a result, gradual variations in conductivity, dimension, and distance between the coil and the tube do not produce any net signal. However, if there is any localized flaw like a pit, strong indication is produced by both the coils. Prior to eddy current test, the instrument is calibrated using a reference defect

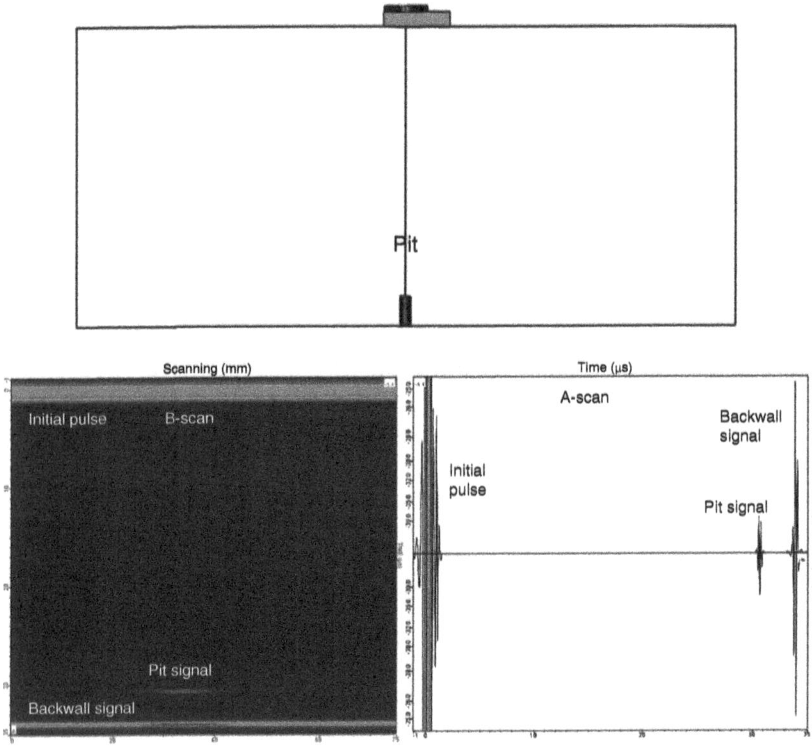

Figure 3.4 Detection of pit by normal beam ultrasonic test.

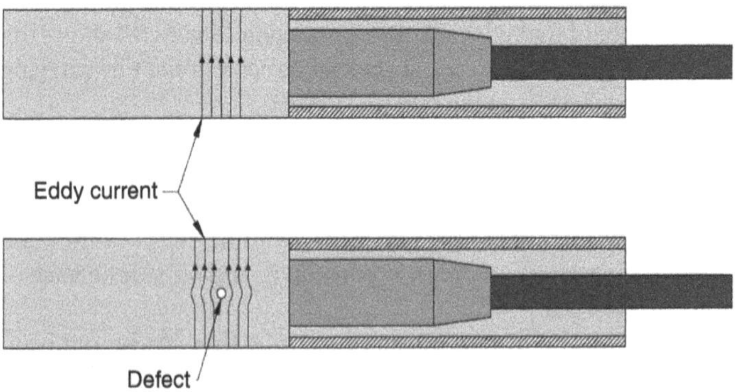

Figure 3.5 Pitting corrosion detection using eddy current test.

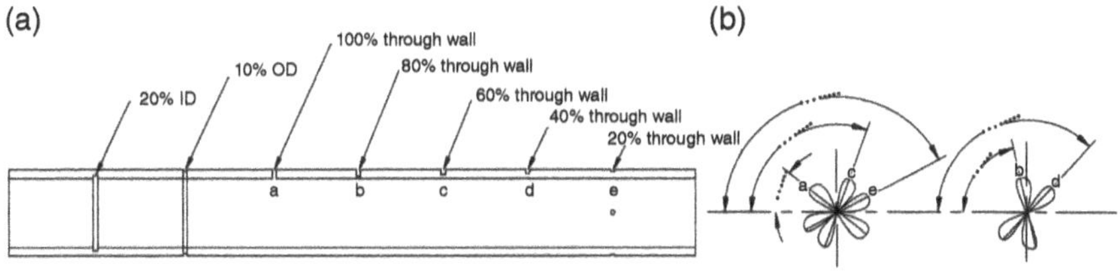

Figure 3.6 (a) Reference standard for eddy current examination of heat exchanger tubes and (b) ECT signal obtained from reference defects.

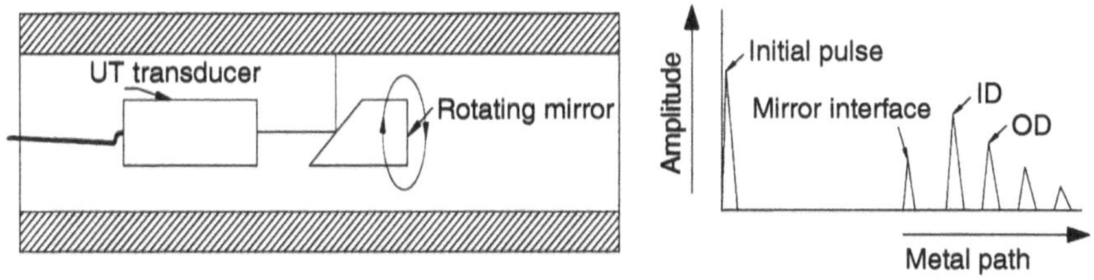

Figure 3.7 Internal rotary inspection system for tube inspection.

standard. The reference standard recommended by ASME Boiler and Pressure Vessel Code Section V Article 8 for eddy current testing of tubes is shown in Figure 3.6a. The frequency of inspection is selected in such a manner that the standard depth of penetration is more than the thickness of the tube so that pits starting from OD can also be detected. The phase of the signal from the through-wall hole in the reference standard is set to ~40°. Signals that show phase <40° represent ID defects, while signals that show phase >40° represent OD defects. Typical signatures for the reference defects of various depths in the tube are shown in Figure 3.6b.

During the inspection of tube, the phase and amplitude of the eddy current signal are recorded. From this information, the location of the pit (ID or OD) and its approximate depth are estimated. Tubes made from ferromagnetic materials need special provision for performing eddy current test to take care of the permeability variation. During eddy current testing, the tube material near the coil is magnetically saturated so that the effect of permeability variation does not affect the eddy current signal from the pit [10].

Another useful technique for detection of pits in heat exchanger tubes is the UT using internal rotary inspection system (IRIS) [11]. IRIS consists of an assembly comprising an immersion ultrasonic transducer and a reflecting mirror that is placed in front it. The assembly is inserted inside the tube to be inspected that is flooded with water. The sound beam from the transducer is reflected by the mirror toward the inside surface of the tube. The sound enters the tube material and gets reflected from the outside surface as shown in Figure 3.7.

The presence of pit on the inside and the outside surface of the tube cause the shift in the ID and OD signals, respectively. The mirror is rotated continuously so that full circumference of the tube is inspected. The data is generally presented in the form of B-scan and C-scan images.

3.6 Monitoring Crevice Corrosion

Crevice corrosion is a form of localized attack that occurs within crevices of metallic components. The attack is caused by a change in the environment within the crevice relative to the bulk solution. The region of tube-to-tubesheet weld joint in heat exchanger is prone to crevice corrosion. Eddy current test using bobbin coil is an effective NDT method for its detection. The method is similar to the one that is used for the detection of a pit. Test frequency is chosen such that the standard depth of penetration is more than the thickness of the tube. Another crucial factor that needs to be considered is the signal from the tubesheet that can mask the signal from the crevice. This is overcome by using multifrequency eddy current technique [12]. Typically, data is acquired at two frequencies, and the signal from the tubesheet is canceled by mixing of eddy current responses at these frequencies [13]. The resultant eddy current response from the crevice flaw is analyzed in terms of the amplitude and the phase to assess the depth of the attack.

UT using IRIS is also used for detection of crevice attack in to tube-to-tubesheet weld joints. Figure 3.8 shows B-scan image of crevice corrosion in heat exchanger tube. The ultrasonic transducer is placed on the inside surface of the tube. The figure shows the backwall signal obtained from the OD surface of the tube and the signal from tip of the flaw created by crevice. By finding out the difference in the time of travel of these signals, one can accurately estimate the depth of crevice attack [14].

3.7 Monitoring Intergranular Corrosion

IGC is a form of localized corrosion that proceeds along grain boundaries. Localized corrosion at grain boundaries is caused by the anodic dissolution of (i) regions depleted of alloying elements, (ii) second-phase precipitates, or (iii) regions with segregated alloying or impurity elements. Sensitization is frequently used to describe microstructure sensitivities to intergranular attack. In austenitic stainless steels, for example, sensitization can be caused by the precipitation of chromium-rich carbides at grain boundaries, leading to chromium-depleted zones. A major characteristic of IGC is the development of a relatively homogenous and uniform depth of attack. The dissolution of grain boundaries causes the dislodgement of grains, often referred to as grain dropping.

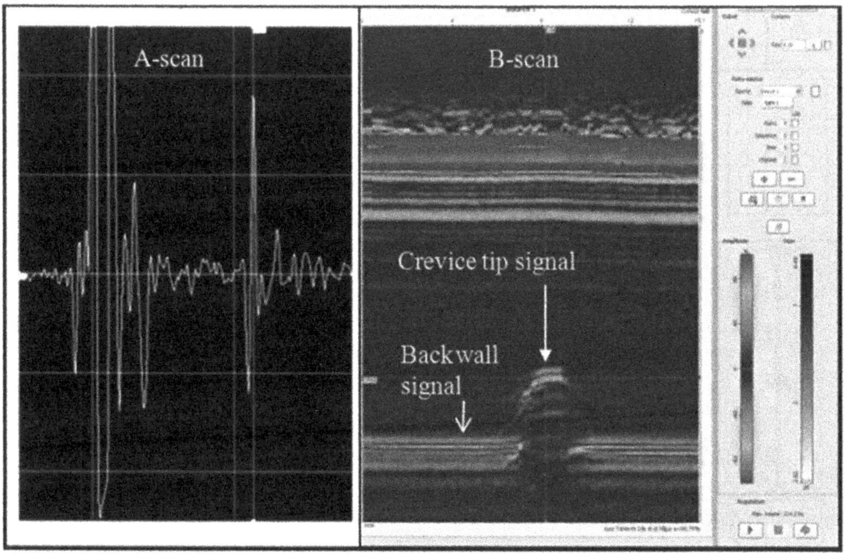

Figure 3.8 Ultrasonic imaging for crevice corrosion detection.

ISM is an effective tool to detect the IGC at an early stage [15]. During ISM, the surface is polished and etched to reveal the grain boundary attack, if any. Either the etched surface is directly seen under a portable microscope, or a replica is taken, which is subsequently seen under microscope. Although ISM can find out the IGC attack at any early stage, it cannot be used to assess the depth of IGC attack. Also, if the surface under attack is not directly accessible, ISM cannot be used.

At an early stage of damage, IGC does not result in the formation of a macroscopic flaw. As a result, conventional NDE methods that are used for flaw detection are not effective for detection of IGC. Some of the NDE techniques, like ultrasonic or eddy current test, hold a good potential for the detection of IGC attack at an early stage [16]. UT is more effective for field applications, since it can detect the IGC attack on the inside surface of a pipe or a vessel, which may not be directly accessible for eddy current test or ISM [17].

UT for detection of IGC attack is based on measuring the sound velocity and the sound attenuation. The sound velocity in a medium is related to the elastic modulus and the density [18]. The IGC attack produces numerous microcracks at the grain boundaries, which reduces the elastic modulus of the material in the attacked region. As a result, the presence of IGC attack is expected to change the velocity of sound in the material. The intergranular attacked regions also scatter the incident ultrasonic beam, especially at high frequency. As a result, the sound attenuation in the IGC attacked material is expected to be higher as compared with the unattacked material.

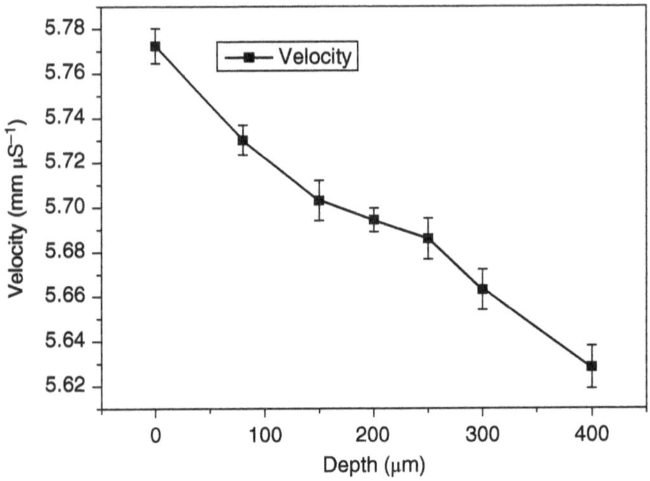

Figure 3.9 Variation of velocity with depth of IGC attack.

Figure 3.9 shows the variation in sound velocity with varying depth of IGC attack [19]. The sound velocity measurement shows that the velocity decreases with increase in the depth of IGC attack.

Using sound velocity measurement, the depth of IGC attack as shallow as 50–100 μm can be detected. The sensitivity of this technique depends on the accuracy with which the sound velocity is found out, which in turn depends on the resolution with which the time of travel of the backwall echo signals is measured. The modern ultrasonic instruments can measure the time of travel with a resolution of 1 ns, which means that sound velocity can be measured with an accuracy of ±5 m s^{-1}. As a result, a very shallow depth of IGC attack can be detected with good reliability.

Figure 3.10 shows the variation of sound attenuation with depth of IGC for different frequencies. Attenuation coefficient is calculated from the amplitude of two successive backwall echoes using Eq. (3.1):

$$\text{Attenuation coefficient}\left(\text{dB mm}^{-1}\right) = \frac{20\log\left(A_i / A_j\right)}{2d} \tag{3.1}$$

where

A_i *and* A_j = amplitudes of consecutive backwall signals i and j, respectively
d = specimen thickness (mm)

The figure shows that for a given frequency, the attenuation coefficient increases with the depth of IGC attack. The attenuation coefficient also

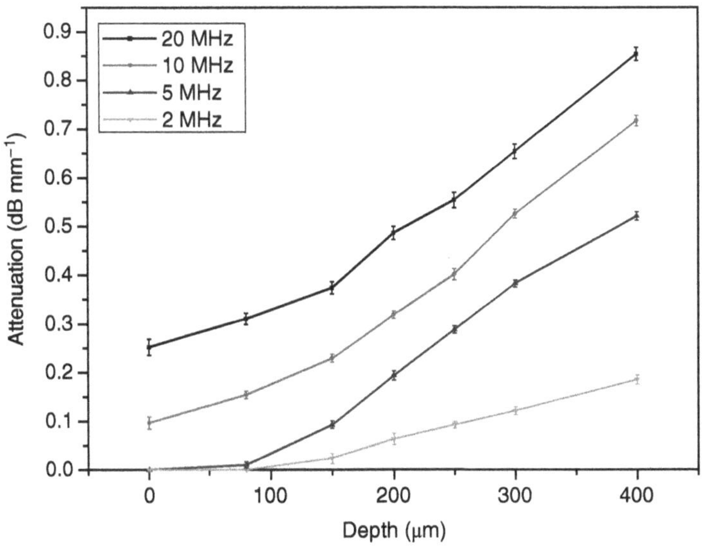

Figure 3.10 Variation of sound attenuation with depth of IGC attack.

depends on the frequency of incident sound wave, higher frequency leading to higher attenuation. By using ultrasound in the frequency range of 10–20 MHz, one can detect IGC attack as shallow as 50–100 μm deep.

One of the major limitations in the use of sound velocity and sound attenuation as the parameters for detection and characterization of IGC attack is that one needs to know the thickness of the component at the location where the IGC attack is being monitored. This information is seldom available under field conditions. In such cases, another parameter, viz. velocity ratio, is found to be very useful. Velocity ratio refers to the ratio of velocity of sound in longitudinal to shear wave mode in a given medium. When the longitudinal wave emitted by the normal beam transducer travels in the material, it gets reflected from the backwall, producing the first backwall signal. For this signal, the sound wave has traveled in the longitudinal mode for both the onward and the return path (FBE signal in Figure 3.11). At the backwall, a part of the incident longitudinal wave gets mode converted to transverse (shear) wave. This shear wave also travels back to the transducer, producing a backwall echo signal. For this backwall echo signal, the sound wave has traveled in the longitudinal mode during onward path and shear mode during the return path (SE signal in Figure 3.11). Since the velocity of sound in the transverse mode is much less than the velocity of sound in the longitudinal mode, the backwall signal SE appears later in time than the first backwall signal (FBE) but before the second backwall signal (SBE). Typical waveform in the steel sample obtained using a normal beam longitudinal wave is shown in Figure 3.10.

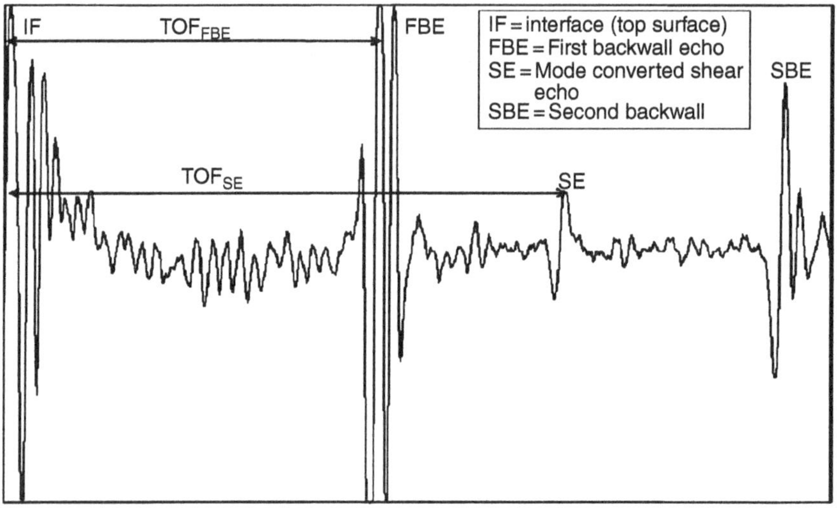

Figure 3.11 Ultrasonic velocity ratio measurement for IGC detection.

The longitudinal to shear velocity ratio is found in Eq. (3.2):

$$\frac{V_L}{V_T} = \frac{2\text{TOF}_{SE}}{\text{TOF}_{FBE}} - 1 \tag{3.2}$$

where

V_L = velocity of longitudinal wave
V_T = velocity of shear wave
TOF_{SE} = time of flight for mode converted backwall signal (longitudinal shear)
TOF_{FBE} = time of flight for first backwall signal due to longitudinal wave

Figure 3.12 shows the variation in longitudinal to shear velocity ratio (V_L/V_T) with the depth of IGC attack. The ratio decreases with the increasing depth of IGC attack. Since the velocity ratio is independent of thickness, this parameter can be used for *in situ* measurement of IGC attack.

The sensitivity of this technique depends on the accuracy with which the velocity ratio is determined, which in turn depends on the accuracy of time-of-flight measurement for FBE and SE backwall signals. Considering the accuracy of 1 ns for the measurement of backwall echo signals, the velocity ratio can be measured with an accuracy of ±0.0007 in steel. With this accuracy of measurement, one can find out the depth of IGC attack of the order of 50 μm by velocity ratio.

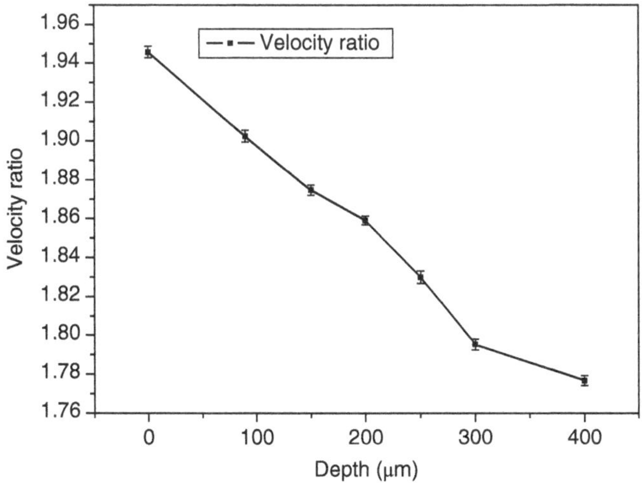

Figure 3.12 Variation of velocity ratio with depth of IGC attack.

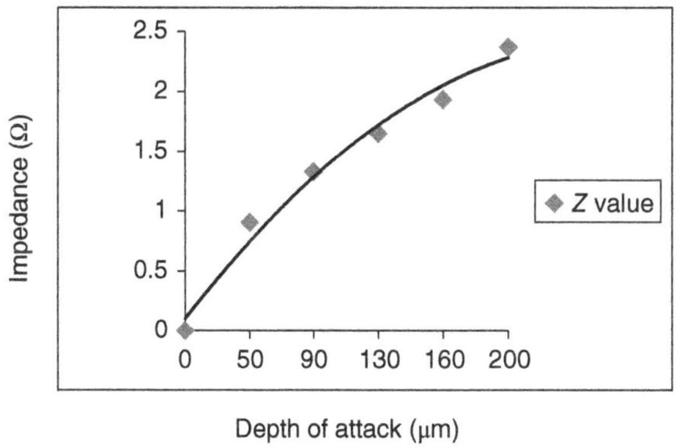

Figure 3.13 Variation of impedance with depth of intergranular attack.

Eddy current testing can also be used to detect the IGC attack, provided that the attacked surface is directly accessible for eddy current test. IGC attack changes the conductivity of the sample. As a result, there is a change in the impedance of the eddy current test coil in the IGC attacked region as compared with the virgin sample.

Figure 3.13 shows the change in the impedance of the test coil with the depth of IGC attack. The figure shows that as the depth of IGC attack increases, the impedance also increases.

3.8 Monitoring Stress Corrosion Cracking

SCC results due to the synergistic effect of tensile stress and a specific corrosion environment. The crack can either be intergranular (IGSCC) or transgranular (TGSCC) in nature. The stress corrosion cracks are generally much tighter (small opening) as compared with the cracks generated by purely mechanical effects like fatigue. Moreover, these cracks are often filled with corrosion products. Both these factors make it difficult for reliable detection and characterization of SCC by NDE techniques.

If the surface from where the SCC has initiated is accessible for direct examination, then surface examination NDE techniques can be employed for its detection. Direct visual examination may not be sensitive, and one may have to use penetrant testing using a visible or a fluorescent dye to improve the sensitivity.

Eddy current testing using high frequency surface coil is also an effective NDE technique for detection of SCC. Eddy current array testing offers several advantages over dye penetrant test. It is faster and does not need much surface preparation. Moreover, the data can be analyzed to get the information on the depth of the crack, which is seldom available with penetrant testing.

When SCC is expected to initiate from the inside surface of a pipe or vessel, which is not directly accessible for inspection, the surface examination NDE techniques like penetrant testing and eddy current testing cannot be used. In such cases, one will have to rely on volumetric examination technique like UT. With UT, the detection of SCC involves sending the sound beam at an angle so that the sound beam gets reflected from the corner that is formed by the crack and the ID surface. The principle of ultrasonic test is schematically shown in Figure 3.14.

Prior to the test, the ultrasonic test instrument is calibrated using a notch of a certain depth as a reference defect standard. The amplitude from the SCC is compared with that from the notch to estimate the depth of SCC [20].

One of the limitations of conventional ultrasonic test for depth sizing of SCC based on the amplitude of the reflected signal is its poor reflectivity to the

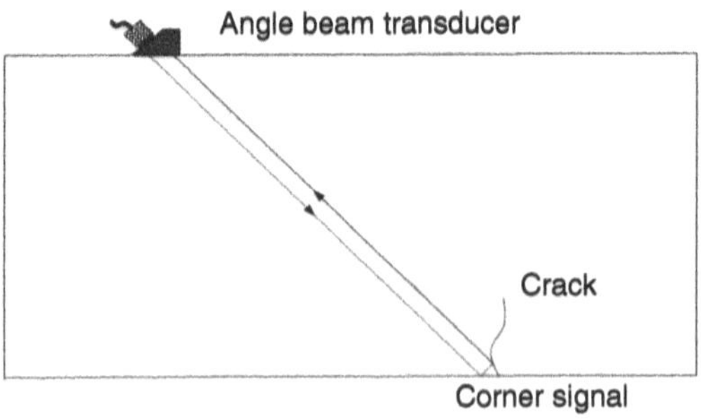

Figure 3.14 Angle beam ultrasonic technique for detection of SCC.

incident sound wave as compared with that from a notch. This is because of the fact that SCC exhibits extensive branching, is very tight, and is filled with corrosion products. As a result, the reflected signal amplitude from SCC of a certain depth is much less than that from a notch of same depth. Hence the conventional UT based on signal amplitude undersizes the depth of SCC.

This limitation can be overcome by using UT techniques that are based on time of flight instead of amplitude. One such technique is called as tip-diffraction technique [21, 22]. In this technique, the sound beam is sent toward the crack at an angle, similar to the one that is done in the conventional technique. The incident sound gets reflected from the corner that is formed by the crack and the ID surface. Additionally, the tip of the crack diffracts the incident sound, and this diffracted signal also reaches the transducer kept on the surface. The schematic of this technique is shown in Figure 3.15.

By measuring the time-of-flight difference between the diffracted signal from the crack tip and the reflected signal from the crack corner, one can accurately calculate the depth of the crack. Since the time of flight is dependent only on the depth of the crack and not on other factors like orientation, branching, etc. the tip-diffraction technique is very accurate for depth sizing of SCC as compared with the amplitude-based ultrasonic technique.

Table 3.1 shows the results of the laboratory study carried out to evaluate the effectiveness of tip-diffraction technique over the conventional amplitude-based technique for depth sizing of IGSCC [23]. For this purpose, IGSCC of various depths were generated in the laboratory on a 25 mm thick stainless steel plate.

The depth sizing of IGSCC was carried out by using two approaches: (i) amplitude comparison and (ii) tip-diffraction technique. With amplitude comparison, the depth sizing of IGSCC was carried out by comparing the signal amplitude from the crack with that from 10% thickness deep (2.5 mm deep) surface notch and 7.7 mm deep IGSCC. The table indicates that with notch as a reference defect standard, the amplitude comparison technique severely undersizes IGSCC. When the notch is replaced by a known depth IGSCC as a

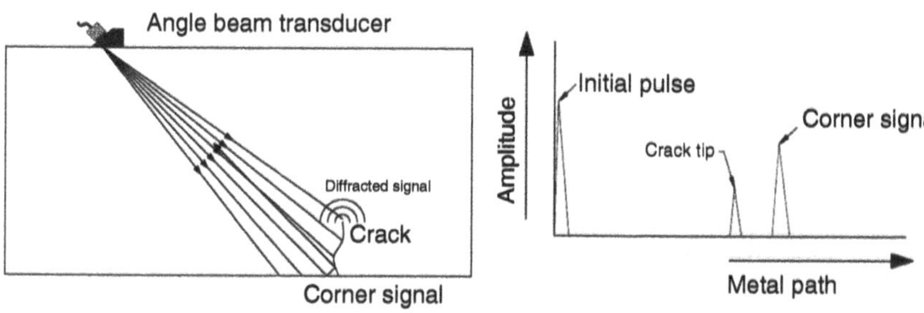

Figure 3.15 Tip-diffraction technique for depth sizing of SCC.

Table 3.1 IGSCC sizing by various methods.

	Measured depth (mm)		
	Amplitude comparison		
Actual depth by metallography (mm)	10% deep notch as reference	IGSCC as reference	Tip-diffraction technique
7.7	2.4	7.7[a]	7.7
6.7	2.1	4.8	6.6
7.0	1.5	3.0	7.0
4.3	1.9	3.2	4.9
5.8	1.2	2.4	5.6

[a] Known depth IGSCC used as reference defect.

reference defect, there is some improvement in the sizing accuracy. This is because of the fact that the difference in reflectivity from the real IGSCC and that of a reference standard is taken care of. However, the best depth sizing accuracy is obtained by the tip-diffraction technique.

3.9 Advanced NDE Methods for Corrosion Monitoring

Over the years, there have been many advances in the field of NDE. These advances are primarily aimed at improving the reliability and sensitivity of inspection and improving the accuracy of flaw characterization. Some of these advances have also been targeted to improve the speed of inspection, so that one can have shorter plant outages, and the damage can be detected at an early stage, so that the plant operators can take appropriate measures to slow down or mitigate the suspected damage. Many advanced NDE techniques have been successfully employed in the field for monitoring corrosion damage. Some of these are described below.

3.9.1 Long-range Ultrasonic Technique

Corrosion monitoring in pipelines over a large area is of major interest in nuclear and petrochemical industry. Conventional ultrasonic thickness gauging is widely used for this purpose. This technique measures the thickness at a localized spot where the transducer is placed. In order to measure the thickness over a large area, the transducer has to be moved manually or using a manipulator, which is a time-consuming process. If the component is insulated, then the insulation has to be removed for coupling the transducer to the surface, which adds to the cost of inspection. Moreover, it is not possible to employ conventional thickness gauging to measure the thickness of the component, which is not accessible, such as a buried pipeline.

Long-range ultrasonic technique (LRUT) using guided waves provides an effective solution to the above problems faced with the conventional ultrasonic thickness gauging. LRUT differs from the conventional ultrasonic thickness gauging in many ways. The wave mode employed with LRUT is the guided wave (or Lamb wave), unlike conventional ultrasonic, which uses longitudinal bulk wave. In conventional ultrasonic test, the sound beam is transmitted in the thickness direction and gets reflected multiple times between the back surface and the front surface, producing a train of backwall signals. While with LRUT, the low frequency guided waves that are generated in the pipe under inspection travel in the axial direction along the length of the pipe. These waves are guided by the inner and the outer diameter of the pipe and are confined within the geometric boundary of the structure. As a result, guided waves travel over a considerable distance along the pipe length, from few meters to tens of meters, depending on the pipe material, fluid inside the pipe, presence of weld joints, and openings in the pipe. When these waves encounter a wall loss over a significant area, they get reflected back to the transducer, producing an indication. From the time of travel of this indication, one can find out the location of the wall loss from the point where the guided wave transducer is placed. Once the location of the wall loss is identified, the actual thickness at this spot can be found out by conventional ultrasonic thickness gauging [24, 25].

In the industry environment, a ring containing an array of ultrasonic transducers is wrapped over the pipe to be examined. The transducers are excited simultaneously so that the guided waves are generated over the entire circumference. These waves travel along the length of the pipe and are reflected back from cavity formed by metal loss and other features such as weld joints, flanges, openings, etc. The reflected signals arriving at various transducers are processed, and a plot showing the locations of various signals from the ring position along with their amplitudes is generated. Since the location of geometrical features (weld joint, flange, etc.) from the ring is known, the presence of any additional signal is an indication of metal loss due to uniform corrosion.

LRUT offers several advantages over the conventional ultrasonic thickness gauging method. First and foremost is the fact that a large area of the pipeline can be screened for any possible damage in a very short time. Secondly, one needs to remove the insulation only at the location where the ring containing guided wave transducer(s) is wrapped over the pipe. The remaining pipeline can remain insulated throughout the test. This significantly saves the cost of inspection. Thirdly, with guided waves, it is possible to carry out corrosion monitoring of inaccessible areas such as buried pipelines.

Guided wave has also been used for the detection of corrosion under pipe support. During this, two transducers, one transmitted and the other receiver, are coupled to the pipe in such a way that the sound beam emitted by the transmitter travels along the circumferential direction of the pipe before it reaches the receiver. When these waves encounter corrosion under pipe support, there is a change in the amplitude of signal received at the receiver. The presence of corrosion and the approximate depth can be assessed from this data. The transducer assembly is moved along the length of the pipe using a suitable manipulator as the data at each cross section is recorded [26].

One of the major limitations of LRUT using guided waves is its poor sensitivity to detect shallow wall loss as compared with the conventional thickness gauging method. While with conventional thickness gauging one can detect the wall loss of the order of a fraction of a thickness, with guided wave, a wall loss of the order of 10% or more of the pipe cross-sectional surface area can only be detected reliably. The other limitation is that it may not be possible to accurately know the remaining thickness from the LRUT data and additional measurement by conventional thickness gauging at the suspected location is required to get this information. Nevertheless, LRUT is increasingly used in petrochemical and nuclear industry as a rapid screening tool for detection of uniform corrosion in pipelines.

3.9.2 Magnetic Flux Leakage

In magnetic flux leakage (MFL) testing, the leakage field due to the presence of metal loss, pit, or crack present in the component is detected by Hall effect sensors [27]. The leakage field is analyzed for identification of defect location and estimation of defect size in terms of length, width, and through-wall depth. A detailed description of the basic principle is given in Chapter 4.

MFL is predominantly used in pipeline inspection gauging (PIG) tool [28]. The PIG tool consists of an array of MFL and ultrasonic sensors that are mounted along the circumferential direction. The tool also contains brushes to remove any scale or sedimentation present inside the pipeline and electronics to collect and store the data. The tool is inserted inside the pipeline and propelled by the pressure of the liquid present inside the pipeline itself.

Another important application of MFL for corrosion monitoring is the ISI of tank floor. The MFL tool for this purpose consists of a magnetic bridge. A sensor to detect the flux leakage is placed between the north pole and the south pole of the magnet. At the location of wall loss due to corrosion, there is a leakage of magnetic field, which is detected by the sensor.

3.9.3 Remote Field Eddy Current Technique (RFECT)

Remote field eddy current technique (RFECT) is an electromagnetic NDE technique that is used for the detection of metal loss due to corrosion in conducting material. It is specifically used for ferromagnetic tubes and pipes,

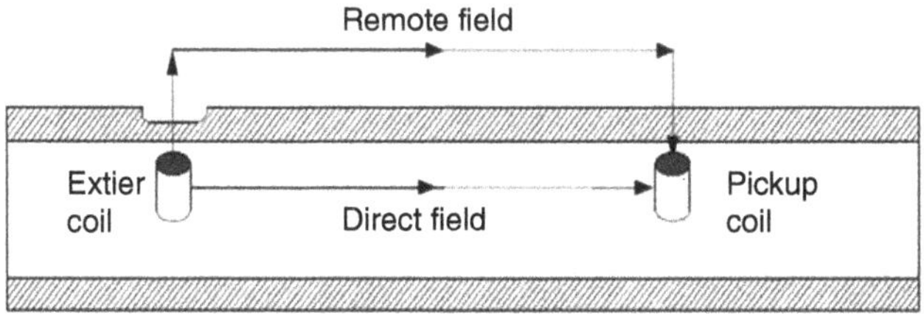

Figure 3.16 Remote field eddy current testing.

which are difficult to examine by the conventional eddy current testing. RFECT probe consists of an exciter coil (or transmitter) and a receiver coil (or detector) separated by the distance of two to five times the diameter of the pipe. The schematic of this technique for detection of flaws is shown in Figure 3.16.

The alternating current (AC) is supplied to the exciter coil, which generates eddy current in the circumferential direction and the magnetic field in the pipe thickness direction. The generated magnetic field travels along the axis of the pipe and is detected by the receiver coil, which is kept in the remote field, away from the exciter. At the area of metal loss, the magnetic field reaches the receiver earlier and with high amplitude as compared with the sound region, due to the reduced travel path. The extent of corrosion attack along the length, circumference, and through-wall direction is calculated from the phase of the remote field signal. Though this technique can detect the metal loss present at both ID and OD of the tube or pipe, it cannot differentiate between the two [29].

3.9.4 Pulsed Eddy Current Testing

The conventional eddy current testing uses an AC of certain frequency to excite the test coil. The AC induces alternating magnetic field in the conducting material, which in turn produces alternating eddy current of the same frequency as the excitation frequency. With the pulsed eddy current testing, the eddy current test coil is excited by a "pulse" of current of short duration instead of a continuous AC. As a result, eddy current of varying frequencies is generated. Since the depth of penetration of eddy current in the material depends on the frequency, information from various depths is obtained at one instance using pulsed eddy current testing. Also, the total amount of energy dissipated within a given period of time is considerably less for the pulsed waves than for the continuous waves of the same intensity. Hence higher input voltages can be applied to the exciting coil while using pulsed eddy current testing as compared with the conventional eddy current testing. One of the biggest advantages of pulsed eddy current testing is that one need not have direct access to the component under test [30]. Testing can be carried out through coating and insulated material. As a result, pulsed eddy current testing is suitable for detection of corrosion under insulation, without physically removing it. The system monitors the rate of decay of the eddy current pulse within the wall and gives an average thickness value by comparison of the observed signal with the calibrated signal.

3.9.5 Tangential Radiography

Radiography is very effective in determination of corrosion under insulation and offers some distinct advantages over other NDT techniques. First and foremost is the fact that the insulation is not required to be removed for the detection and characterization of corrosion damage. From the radiograph one can make an assessment about the nature and degree of corrosion attack. Moreover, radiography offers a permanent record, and by comparing the radiographs taken periodically, one can assess the rate of corrosion attack. The International Atomic Energy Agency (IAEA) conducted a coordinated research program titled "Validation of Protocols for Corrosion and Deposits Determination in Small Diameter Pipes by Radiography." The results of the round-robin test carried out by the participating countries on pipe samples with simulated and real corrosion damage using two approaches, viz. density measurement and tangential radiography, have been brought out in the form of IAEA-TECDOC-1445 [31]. The TECDOC recommends the use of tangential radiography for the detection of uniform corrosion and density measurement of the radiograph for the detection of pits and localized corrosion damage.

3.9.6 Infrared Thermography

Infrared thermography is used as a NDE technique for condition monitoring of electrical and mechanical equipment [32]. In the field of corrosion, it is a useful technique for detection of corrosion under insulation, especially for overhead pipelines, which are difficult to approach for performing other NDE examinations. The leakage of fluid through the pipeline changes the temperature of insulation at the point of leakage. By obtaining the thermal image of the insulation covering the pipeline, one can locate the leakage area and take appropriate corrective action.

3.10 Conclusions

NDE plays a crucial role in detection and characterization of corrosion damage in engineering components. NDE thus provides vital inputs to the plant operators and the regulators on the assessment of structural integrity of critical components, which are prone to corrosion attack. While conventional NDE techniques are routinely employed in industries for corrosion monitoring, several advanced NDE techniques are increasing being used for field inspection. These advanced techniques have helped in improving the reliability of inspection, overcoming the inaccessibility constraints, reducing the plant shutdown time, and detection of corrosion damage at an early stage.

References

1 Hellier, C.J. (2013). *Handbook of Nondestructive Evaluation*. New York: McGraw-Hill Education.
2 ASM International (1989). *ASM Handbook, Volume 17, Nondestructive Evaluation and Quality Control*. New York: ASM International.
3 Mix, P.E. (2005). *Introduction to Nondestructive Testing: A Training Guide*. Hoboken, NJ: Wiley.
4 NDT Education Resource Center (2001–2014). *The Collaboration for NDT Education*. Iowa State University. www.ndt-ed.org (accessed September 2017).
5 Raj, B., Jayakumar, T., and Thavasimuthu, M. (2014). *Practical Non-Destructive Testing*. New Delhi: Narosa Publishing House.
6 Blitz, J. and Simpson, G. (1995). *Ultrasonic Methods of Non-destructive Testing*. Berlin: Springer.
7 Hull, D.R., Kautz, H.E., and Vary, A. (1984). Measurement of ultrasonic velocity using phase-slope and cross-correlation methods. *ASNT Spring Conference*, Denver, Colorado (May 1984).
8 Innerspec. *EMAT Technology*. https://www.innerspec.com/en-EU/knowledge/emat-technology (accessed August 2018).
9 Dhanasekaran, R., Borja, L., and Arora, M. (2015). In-service application of EMAT in boiler water wall tubes at high temperature components. *Proceeding of NDE 2015*, Hyderabad (November 2015).
10 Lee, H.J., Cho, C.H., Nam, M.W. et al. (2012). Eddy current testing of type-439 stainless steel tubing using magnetic saturation technique. *Journal of Mechanical Science and Technology* 26 (7): 2081–2085.
11 Eddyfi. *Internal Rotary Inspection System (IRIS)*. http://www.eddyfi.com/technologies/internal-rotary-inspection-system (accessed July 2017).
12 Rodriguez, P. and Raj, B. (1997). Development of in-service inspection techniques for nuclear power plants in India. *International Journal of Pressure Vessels and Piping* 73 (1): 59–68.
13 Hura, D.H., Choi, M.S., Leea, D.H. et al. (2010). A case study on detection and sizing of defects in steam generator tubes using Eddy current testing. *Nuclear Engineering and Design* 240: 204–208.
14 Jothilakshmi, N., Nanekar, P.P., Kumar, A. et al. (2011). Development of non destructive techniques for characterization of crevice corrosion in tube-to-tubesheet joint. *Proceedings of the National Seminar & Exhibition on Non-Destructive Evaluation, NDE 2011*, Chennai, India (8–10 December 2011).
15 Salgadolopez, J.M. (2011). In situ metallography as non-destructive test to analyze the microstructural damage in the petrochemical industry. *5th Pan American Conference for NDT*, Cancun, Mexico (October 2011).

16 Nanekar, P.P. and Shah, B.K. (2004). Characterization of material properties by ultrasonics. *BARC Newsletter* (249): 25–38.
17 Temple, J.A.G. (1989). Estimating the depth of intergranular attack in piping by ultrasonic time delay: a theoretical analysis. *International Journal on Pressure Vessel and Piping* 36: 21–55.
18 Stella, J., Cerezo, J., and Rodrıguez, E. (2009). Characterization of the sensitization degree in the AISI 304 stainless steel using spectral analysis and conventional ultrasonic techniques. *NDT&E International* 42: 267–274.
19 Jothilakshmi, N., Nanekar, P.P., and Kain, V. (2013). Assessment of intergranular corrosion attack in austenitic stainless steel using ultrasonic measurements. *Corrosion Journal* 69 (4): 388–395.
20 ASME (2012). *Boiler and Pressure Vessel Code, Section XI Rules for In-service Inspection of Nuclear Power Plant Components*. New York: ASME.
21 Sinclair, A.N., Fortin, J., Shakibi, B. et al. (2010). Enhancement of ultrasonic images for sizing of defects by time-of-flight diffraction. *NDT&E International* 43: 258–264.
22 Ravenscroft, F.A., Newton, K., and Scruby, C.B. (1991). Diffraction of ultrasound by cracks: comparison of experiment with theory. *Ultrasonics* 29: 29–37.
23 Nanekar, P. and Shah, B.K. (2008). Non-destructive examination for ageing management of Indian nuclear pwer plants. *17th World Conference on Nondestructive Testing*, Shanghai, China (25–28 October 2008).
24 Rose, J.L. (2014). *Ultrasonic Guided Waves in Solid Media*. Cambridge: Cambridge University Press.
25 Lowe, M.J.S., Alleyne, D.N., and Cawley, P. (1998). Defect detection in pipes using guided waves. *Ultrasonics* 36: 147–154.
26 Shivaraj, K., Balasubramaniam, K., Krishnamurthy, C.V., and Wadhwan, R. (2008). Ultrasonic circumferential guided wave for pitting-type corrosion imaging at inaccessible pipe-support locations. *Journal of Pressure Vessel Technology* 130: 021502/1–021502/11.
27 Blitz, J. (1997). *Electrical and Magnetic Methods of Nondestructive Testing*, Nondestructive Evaluation Series. New York: Springer.
28 Sukhorukov, V.V. (2013). MFL technology for diagnostics and prediction of object condition. *12th International Conference of the Slovenian Society for Non-Destructive Testing*, Portoroz, Slovenia (September 2013), pp. 389–402.
29 Rao, B.P.C. (2007). *Practical Eddy Current Testing*, Indian Society for Non-destructive Testing (ed. B. Raj and B. Venkatraman). New Delhi, India: Narosa Publication.
30 Zergoug, M., Lebailia, S., and Kamel, G. Characterization of the corrosion by eddy current. http://www.iaea.org/inis/collection/NCLCollectionStore/_Public/36/105/36105061.pdf (accessed July 2017).
31 IAEA-TECDOC-1445 (2005). *Development of Protocols for Corrosion and Deposits Evaluation in Pipes by Radiography*. Vienna: IAEA.
32 Maldague, X.P.V. (1993). *Nondestructive Evaluation of Materials by Infrared Thermography*. New York: Springer Verlag.

4

NDE Methods for Monitoring Corrosion and Corrosion-assisted Cracking

Case Studies II

B.P.C. Rao[1] *and Baldev Raj*[2]

[1] *Fast Reactor Fuel Cycle Facility, Indira Gandhi Centre for Atomic Research, Kalpakkam, India*
[2] *National Institute of Advanced Studies, Bengaluru, India*

4.1 Introduction

Corrosion is one of the most common degradation mechanisms observed in many industrial components. Most of the corrosion-related failures result in loss in production, leakages, and contamination, in addition to loss of human lives. Despite the best efforts made for selection of materials, proper design, and operating environment to avoid corrosion-related degradation of components, corrosion degradation is inevitable. It becomes essential to monitor the performance of installed components for assessing the progress of corrosion degradation to be within the expected and acceptable limits. Corrosion failures can be minimized by periodically monitoring corrosion and rate of corrosion using nondestructive evaluation (NDE) methods that ensure reliable, safe, and economic operation of plant components and structures [1]. Corrosion monitoring when integrated with preventative maintenance and safety management programs can minimize plan downtime, enhance profitability, and effectively aid residual life assessment/extension strategies. Elaborate discussion on various corrosion types has been given earlier in Chapter 2.

4.2 NDE Methods for Corrosion Detection and Monitoring

Depending upon the nature of the component, type and location of corrosion, and access to the location, NDE methods such as visual, liquid penetrant, magnetic flux leakage (MFL), eddy current, infrared thermography (IRT),

Non-Destructive Evaluation of Corrosion and Corrosion-assisted Cracking, First Edition.
Edited by Raman Singh, Baldev Raj, U. Kamachi Mudali and Prabhakar Singh.

radiography, and ultrasonic methods are used for detection of corrosion. In most situations, corrosion damage starts from surface of a component. If the surface is accessible, visual, liquid penetrant, and eddy current methods can be used; otherwise volumetric methods such as radiography and ultrasonic methods are preferred. In multilayered structures, corrosion can occur in the hidden layers, and appropriate NDE techniques are necessary for detection of such hidden corrosion. While the methods like eddy current, ultrasonic, and radiography are useful to detect the static damage, acoustic emission (AE) technique offers the possibility for evaluation of dynamic corrosion damage in real time. As nearly all the NDT methods are indirect, the use of calibration specimens with known artificial reference flaws is essential for instrument calibration and for sizing purposes [2].

NDE focuses detection of loss of overall material thickness (uniform corrosion), local loss of material thickness (e.g. pitting corrosion), and cracking (e.g. stress corrosion cracking [SCC]) [3]. The interaction of a corrosive environment and tensile stress produces failure in the form of SCC in susceptible metallic components. SCC cracks are usually located on tube bend regions, and high stress concentration areas near welds and heat-affected zones (HAZ). These cracks can grow in either transgranular or intergranular mode, usually with crack branching. They can initiate from tube inner surface or outer surface or along circumferential or longitudinal directions. In stress-assisted cracking, crack initiation and propagation occur under cyclic loading in a corrosive environment, and this usually leads to brittle failure that appears as multiple and transgranular cracks that initiate from the inner surface of boiler tubes [4].

NDE instruments and probes in majority of the situations are portable, and the widely used NDE techniques include internal rotating ultrasonic inspection, guided wave ultrasonics, ultrasonic time of flight diffraction, ultrasonic backscatter, ultrasonic C-scan, remote field, pulsed and low frequency eddy current testing (ECT), real-time and digital radiography, IRT, MFL, filed signature mapping, videoendoscopy, and microwave methods. Implementation of signal and imaging methods and artificial intelligence approaches to NDE data enhances the detection and sizing capabilities. Multitechnique, multisensor, and multiparametric approaches are being used for comprehensive evaluation of corrosion-related damage. For example, for detection of pipe corrosion under insulation, saturation eddy current, long-range guided wave, and remote visual inspection techniques have to be used in a holistic manner to assess the health of the piping. Intelligent pigs using visual, MFL, and ultrasonic methods have come to stay for inspection of long pipelines. Table 4.1 gives the list of NDE methods generally applied for monitoring different types of corrosion [1]. Selection of methods is very important. This is usually done depending on the detection capability of the method and the available access.

Table 4.1 NDT methods suitable for monitoring different types of corrosion damage.

Type of corrosion	NDT methods suitable
Uniform	MFL, UT, ECT, RT, VT, IRT
Localized	UT, ECT, RT, IRT
Stress corrosion cracking	LPT, AE, UT, ECT, IRT, MFL
Corrosion fatigue	LPT, AE, UT, ECT
Pitting	AE, ECT, UT, RT
Hydrogen-induced cracking	AE, UT
Erosion corrosion	ECT, UT, VT
Stress-assisted cracking	ECT, UT, VT
Microbiologically induced corrosion	RT, UT, ECT, IRT

AE, acoustic emission; ECT, eddy current testing; IRT, infrared thermography; LPT, liquid penetrant testing; MFL, magnetic flux leakage; RT, radiographic testing; UT, ultrasonic testing; VT, visual testing (including optical).

An elaborate description on various NDT/NDE methods and techniques can be found in Chapter 1. The following sections provide the applications of eight of the most commonly employed important NDE methods for detection and evaluation of uniform corrosion, localized pitting corrosion, erosion, SCC, and microbiologically induced corrosion. This chapter places emphasis on a few novel and recent applications of NDE techniques for detection of SCC and pitting corrosion, with the help of case studies to bring out the recent advances in NDE methods for corrosion monitoring. NDE techniques for corrosion monitoring of pipes, aircraft structures, and concrete structures are covered.

4.2.1 Surface NDE Methods

Visual testing and liquid penetrant testing are the two simplest and widely used NDE methods for corrosion detection. In visual inspection, mirrors, borescopes, flexible fiber-optic instruments, or small video cameras are used together with light sources to look inside small pipes or narrow spaces as well as inaccessible regions in components for detection of stress corrosion cracks or corrosion fatigue, apart from general health [1]. Experienced inspector can readily determine the type of corrosion such as uniform corrosion, pitting corrosion, crevice corrosion, SCC, weld and HAZ corrosion, and erosion corrosion from visual inspection. For detection and characterization of surface-exposed corrosion, Moire, electronic speckle pattern interference, digital speckle correlation, laser interferometry, and holography techniques are useful, e.g. in thin aluminum aircraft structures.

Remote visual inspection is possible with rigid borescopes, industrial fiberscopes, and videoscopes attached to remote-controlled crawlers. Remote visual inspections are used for detection of corrosion and mapping of corrosion. They can rapidly confirm exact location and severity of boiler tubes for ID damage; condenser tube leaks as a result of hydrogen damage. They are valuable for assessment of turbine blade erosion and blade condition. Visual inspection process can be made effective through the incorporation of robotic technologies. The robots can be equipped with cameras that can provide images of the different areas of corrosion damage to be inspected.

Liquid penetrant inspection method is useful for detection of cracks open to surface in any type of nonabsorbent or nonporous materials based on the surface tension and capillary mechanisms and exploiting the optical recognition covering the full range of light spectrum from visible to ultraviolet light. Of the three widely adopted liquid penetrant systems water washable, post-emulsifiable, and solvent removable using visible or fluorescent dye [1], it is the fluorescent dye penetrants that have high sensitivity and are particularly useful for detecting very tight cracks (width less than a micron) such as stress corrosion cracks and other service-induced defects open to the surface [2]. The main advantage of this method is rapid inspection of large areas and large volumes of simple as well as complex geometries at low cost.

4.2.2 X-radiography

X-radiography method can be used to detect pitting corrosion and material thinning that are visible as dark areas on the film. Pit depths and the extent of thinning can be estimated by density measurements of the film and the use of proper geometrical correction factors. The minimum variation of the thickness in the direction of X-ray beam that can be detected is better than 2% of the thickness [1]. Tight or narrow cracks are difficult to detect by radiography especially if they are oriented in directions other than parallel to the X-ray beam. One important requirement of radiography method is the availability of both sides access in many situations. Real-time and digital radiography is useful for efficient and cost-effective detection of corrosion. The main advantages of digital radiography include less time for high quality imaging, no film processing, and easier storage and retrieval of images. Another important application of real-time and digital radiography is detection of corrosion under insulation and fireproofing, without the need for stripping or abatement required by visual and ultrasonic methods. Iridium-192 or selenium-75 isotopes are quite commonly used with suitable collimation for corrosion inspection in pipes.

4.2.3 Infrared Thermography

IRT method uses two imaging modalities – active imaging and passive imaging. Passive imaging is performed on machinery or components to detect hot spots, while active thermography uses external heating or cooling. Resolution up to

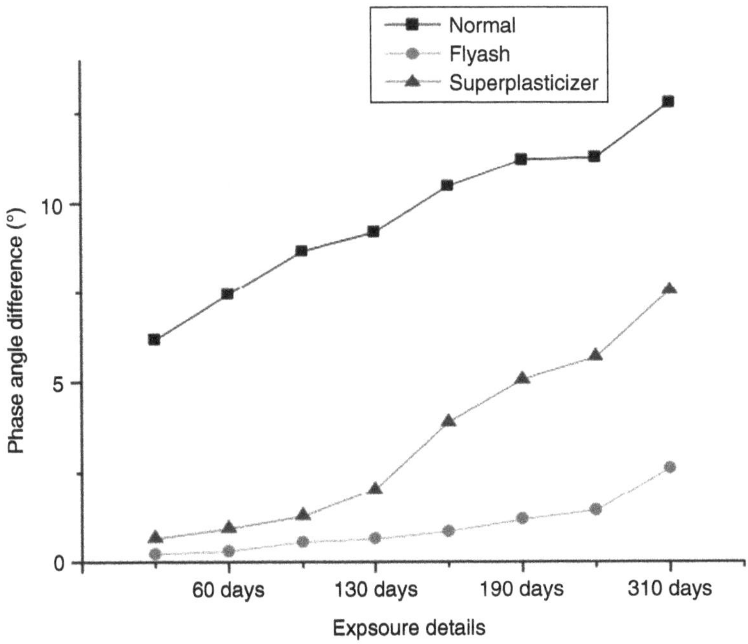

Figure 4.1 Phase angle variation with seawater exposure time for three types of concretes.

0.1 °C can be achieved, and hot spots of 5 mm and under can be detected from 5 m distance. The flow of heat in the object after the application of energy is used to study material properties and boundaries. For detection of corrosion in power plant boiler sections, IRT is used by heating the front surface using standard flash tubes and by measuring the temperature using an IR camera without and with water inside the pipes. For dynamic applications, IR cameras with a thermal sensitivity of 0.02 °C are useful. Flash, lock-in, and pulse thermography techniques can be used to detect subsurface corrosion damage and possibly identify its location [5]. Further, microbiologically induced corrosion (MIC) can be detected using IRT method. The presence of MIC can be confirmed through testing of the biofilm formed. Using the phase angle difference between unexposed and exposed to number of days to sea water, degradation of concrete due to biofouling could be compared effectively for normal concrete, fly ash, and superplasticizer as shown in Figure 4.1. Results revealed that highest phase angle variation is for the normal concrete indicating deterioration and the least variation is for fly ash indicating minimum of deterioration [6].

4.2.4 Acoustic Emission

AE method can detect localized corrosion. AE method provides a nonintrusive way to measure the corrosion activity. For real-time detection of pitting corrosion, AE method is very useful. AE systems consist of an AE sensor, an amplifier and filter, and a PC for data acquisition and analysis [3]. In the corrosion process, AE parameters such as hits, energy, etc. are extracted and correlated with corrosion damage. For example, in aluminum 2024 alloy exposed to NaCl solution, AE hit rate shows a strong correlation with corrosion rate that is a function of NaCl concentration. Three types of AE sources, viz. hydrogen bubbles, breakage of passive film, and pit growth, can be detected using AE parameters. It is also possible to determine the location of the pitting by using the difference in the arrival time of group of AE sensors. Periodic measurement of AE parameters at regular intervals will allow one to take action to mitigate the deterioration and identify the triggering condition.

Detection of corrosion by AE is well studied, and continuing developments have helped the technique become practical and in some cases even quantitative. Noise elimination is a challenge due to the small amplitude of the AE signals due to corrosion process. Pattern recognition, FFT- and wavelet transform-based signal analysis, and artificial neural networks allow early identification and classification of AE activity and also triangulation-based determination of source location.

4.2.5 Ultrasonics

Ultrasonic method is well suited for detection of isolated pitting and corrosion in the absence of surface roughness and nonparallel surfaces. Handheld ultrasonic thickness gauges that use dual element and single element transducers are employed for thickness gauge applications. For pipes and tubes, internal rotating inspection systems (IRIS) are very useful for mapping corrosion and for assessment of corrosion wall thinning, even under the support structures [3].

Long-range guided wave techniques are useful to locate internal or external corrosion and metal loss anywhere in the entire pipe wall, thus maximizing the efficiency of a corrosion management program. This technique can allow 100% volumetric screen of pipe wall from a single position of an array piezoelectric transducers. It can detect severe pitting corrosion and local gradual corrosion and is useful for monitoring corrosion growth with time. Phased array ultrasonic testing offers one possibility of identifying and analyzing corrosion. This method uses ultrasound probes made up of a number of array elements that can function separately as transmitter and/or receiver. By introducing a time delay in the excitation of the array elements, it is possible to steer and focus the ultrasound beam and display the test data as A-scans, B-scans, C-scans, and three-dimensional (3D) displays. The phased array systems are portable and can be effectively used for detection of corrosion in, e.g. storage tanks as depicted in Figure 4.2. Phased array testing can collect higher resolution data with fully encoded C-scan imaging for detection and classification of wall thickness loss due to corrosion, abrasion, erosion, and mid-wall damage such as hydrogen-induced blistering. Electromagnetic acoustic transducer (EMAT)

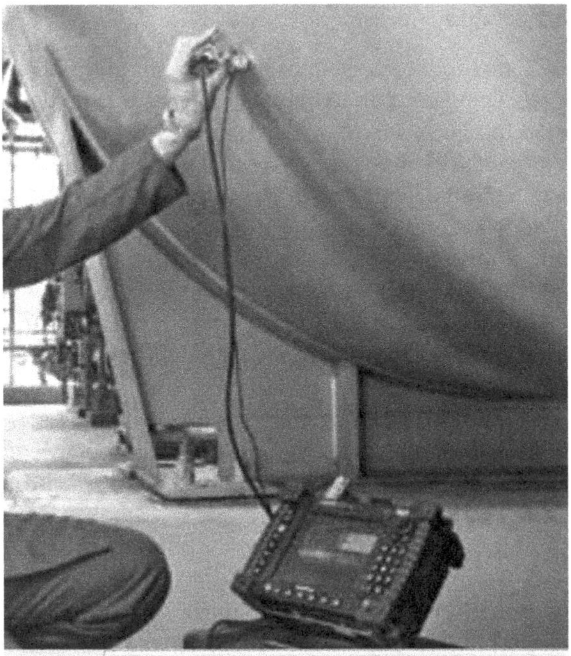

Figure 4.2 Phased array ultrasonic testing of a storage tank.

and laser-generated ultrasonic-based techniques are useful for corrosion monitoring as they are noncontact and couplant-free and work at high temperatures. Phased array ultrasonic technique has been used for detection and sizing of microbiologically induced corrosion of service water systems and shut down coolers heat exchangers in CANDU reactors [7]. Recently, low frequency ultrasonic waves that travel lengthwise over tens of meters along a pipe are gaining popularity for screening of in-service pipes with wall thickness variations. The propagated torsional waves detect variations in the total cross-sectional area of the pipe and also any changes in the material properties.

Phased array ultrasonic testing technique is preferred for detection of hydrogen-induced cracking that is a phenomenon associated with the dimerization of nascent atomic hydrogen at metallurgical defects, such as nonmetallic inclusions, large precipitates, or other discontinuities in the steel. The molecular hydrogen produces blisters, and adjacent blisters subsequently are connected on different planes in the metal by a transgranular plastic shear mechanism (stepwise cracking). Phased array technique, through beam steering and beam focusing, can readily identify a blister, as the shape of the defect can be clearly seen on the sector scans. The use of the conventional ultrasonic method as a screening tool to identify the areas with possible blisters and then detailed phased array inspection to validate and size the affected areas is one approach that is gaining popularity.

4.2.6 Eddy Current

ECT method is very useful for detection of surface as well as subsurface corrosion. The most popular applications of this method include inspection of aircraft fuselages and wings for corrosion and inspection of heat exchanger tubes for detection of pitting corrosion, wall thinning, and flaws [8]. It is a noncontact technique and does not require removal of paint coatings.

Eddy current method can be used to assess and quantify sensitization and intergranular corrosion in thermally aged AISI-type 316 stainless steel. The eddy current amplitude increases with increasing degree of sensitivity (DOS) in as-aged and Strauss-tested steel [9]. Thus, empirical relationship can be established between ECT signal amplitude and DOS parameters to assess the propensity for intergranular corrosion nondestructively without the need for chemical and electrochemical tests specified by ASTM A262 practices A and E (Strauss test) and G108. Saturated low frequency eddy current (SLOFEC) technique is a potential technique for detection of pitting corrosion and localized microbiologically induced corrosion [10].

Conventional ECT is applied for non-ferromagnetic materials. For ferromagnetic steel tubes, saturation or remote field eddy current techniques are employed whether the access available is outside or inside, respectively. In these techniques, for detection of corrosion damage under support plate and baffle plate regions, multifrequency techniques are employed following the relevant ASME standards. Apart from rotating probe coils, eddy current array probes are finding increased use. They are capable of producing images of defects, for possible application of signal and image processing methods for automated detection and sizing. Recently it has been found that significant improvements in speed, accuracy, and presentation of results can be achieved by using pulsed eddy current technique incorporating array probes. The pulsed eddy current images are easier to interpret than conventional impedance plane representations.

Eddy current pulse thermography is a potential technique for detection of corrosion blister. This technique combines the eddy current and thermal effects for detection of corrosion [11]. Basic principle of this technique comprising induction heating, heat conduction, and infrared radiation is shown in Figure 4.3. The heat is directly generated in the conductive material below the excitation coil. The temperature response is measured after a certain time. It is possible to distinguish blister region, ruptured blister, and intact regions based on the time response and parameters derived from it. Further, tone burst eddy current thermography technique allows detection of cracks, corrosion pits, and cracks in pits in stainless steels.

4.2.7 Magnetic Flux Leakage

MFL method uses sensor for detection of leakage magnetic fields in comparison with the ferromagnetic particles/powders used in magnetic particle testing. Use of sensors such as pickup coils, Hall sensors, magnetodiodes, fluxgates,

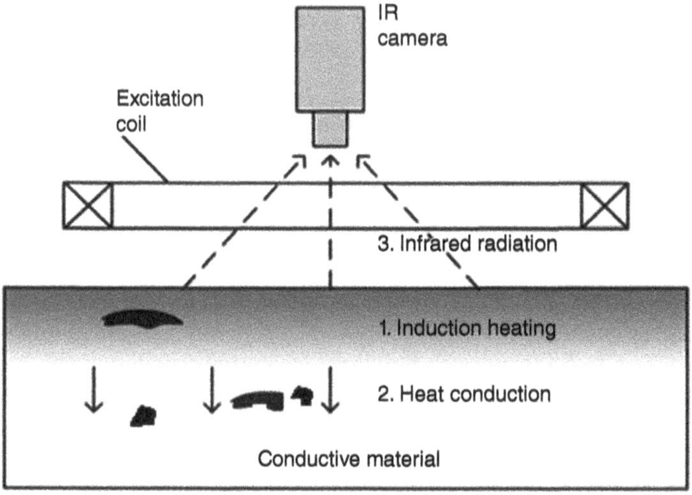

Figure 4.3 Principle of eddy current pulsed thermography.

anisotropic magnetoresistance, and giant-magneto-resistive (GMR) sensors is common in MFL method [12]. The main advantage that comes with the use of sensor is automated testing of symmetric components such as steel rods, tubes, pipes, and bars. Also use of sensor and sensor arrays enables enhanced detection and sizing of surface as well as subsurface corrosion damage by exploiting the signal and image processing methods. This method is widely used in industry for monitoring corrosion in ferromagnetic components such as underground pipelines, oil storage tank floors, rail tracks, ropes, etc. MFL method is capable of detecting pitting corrosion in materials up to 20 mm thick at speeds up to 0.5 $m s^{-1}$. MFL method is the most commonly used method to detect metal loss due to corrosion and, sometimes, gouging in nearly 80% of gas transmission pipelines.

With the availability of novel sensor, digital devices, and microelectronic systems, the capabilities and applications of the MFL techniques have increased significantly in the last two decades. Further, numerical models have enabled design of optimized magnetizing units and sensor locations, apart from better understanding of the phenomenon and the field–defect interactions. Stress variations in the test object and changes in velocity of the MFL unit produce disturbing MFL signals, and this demands the use of signal processing techniques. The amplitude and shape of MFL signal from a sensor-side defect is significantly different from an equivalent far-side defect. For such situations, employing signal processing, pattern recognition, and artificial intelligence methods is expected to ensure reliable detection and sizing of defects. Pulsed MFL technique is finding applications in crack detection and sizing of surface and subsurface cracks in thick components.

4.3 NDE Methods for Detection of Stress Corrosion Cracking

4.3.1 Eddy Current Arrays for Monitoring Stress Corrosion Cracking

Eddy current array imaging systems are useful for SCC detection. They are handheld and require much less time for inspection as compared with liquid penetrant and magnetic particle methods. When properly calibrated using known reference standard flaws, it is possible to produce 3D images of defects. The number of array elements can be 32 or 64 or 128 depending on the dimensions of the structure to be inspected. Similarly, excitation frequency can be suitably changed to cover different depths of penetration depending on the thickness of the structure [8]. They are a substitute to magneto-optical imaging; such systems can reliably detect subsurface corrosion and fatigue cracks in fasteners in aircraft structures without the need to remove paint.

4.3.2 Laser-generated Ultrasonic Technique for Detection of Stress Corrosion Cracking

Lasers, e.g. Nd:YAG, can be used for generating ultrasound waves noncontact in a small focused region of <500 μm diameter. This area is rapidly heated by the laser beam, thus expanding and generating stress. The resultant pulse of material expansion and contraction generates ultrasonic modes such as bulk longitudinal, shear, Rayleigh wave, and guided wave modes. The smaller footprint of the laser on the material allows resolution of close-by defects that are separated by a few mm or close to the plate edges that otherwise not resolved by the conventional backwall reflection-based techniques. For resolving the spatial extent of cracks in 3.9 mm thick AISI 304 stainless steel pipes, the use of laser-generated ultrasonic technique was reported [13].

4.3.3 MFL Technique Using Flux Gate Sensors for Detection of SCC

Flux gate sensors measure the absolute strength of magnetic field. Due to their advantages in comparison with other sensors, e.g. high field sensitivity, high linearity, dynamic range, and ruggedness, they are attractive for detection corrosion, especially SCC. Flux gate sensors have been used for detecting the changes of magnetic flux densities in ferromagnetic materials such as Inconel 600 alloy material due to SCC crack initiations. In an Inconel 600 alloy tube sensitized to SCC by heat treatment at 600 °C for 20 h and SCC crack induced by applying tensile stress in the axial direction in the presence of 0.5% polythionic acid solution, the variation in the vertical component of the magnetic flux density due to crack could be measured using, e.g. flux gate sensor. The flux gate sensors are used in metal magnetic memory (MMM) method to measure the magnetic field changes in the geomagnetic field without the need for external magnetization and demagnetization to identify the location of defects for automatic detection of microcracks and corrosion in coiled tubing [14]. It has been demonstrated that the method can rapidly image tubing defects and is not affected by either the surface condition of the tubing or minute changes in gap between the sensor and the tube surface.

4.3.4 Nonlinear Ultrasonic Technique for Stress Corrosion Cracking Detection

It is proven that using ultrasonic methods microstructural changes can be detected before the appearance of macroscopic damage due to the fact that microstructural changes distort the ultrasonic waves that produces higher harmonics in a monochromatic wave. As SCC starts from the surface of a component, Rayleigh waves can more appropriately detect SCC as compared with bulk ultrasonic waves. Nonlinear ultrasonic measurements can be performed before and after the SCC damage is induced. The amplitudes of the fundamental and second harmonic waves are measured with contact wedge transducers at varying propagation distances to obtain the acoustic nonlinearity of the specimens as a function of SCC damage. This normalized amplitude varies linearly with propagation distance, where the slope of the linear fit is proportional to the acoustic nonlinearity parameter, β. The percentage increase in the measured β shows a linear correlation with the defined damaged parameter due to SCC damage. Typical measurement setup for nonlinear ultrasonic testing is shown in Figure 4.4. The variation in β with propagation distance is measured

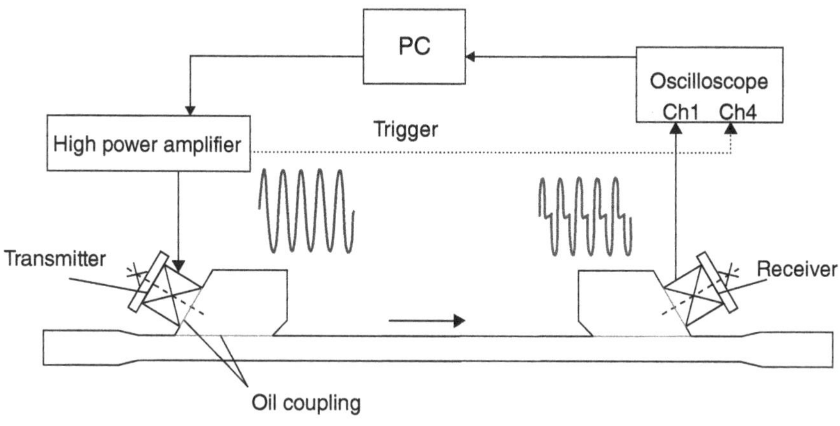

Figure 4.4 Setup for nonlinear ultrasonic measurements.

experimentally and used to determine the SCC damage. In 1018 carbon steel buried fuel pipelines at 2.25 MHz, a nearly 128% change was observed over damaged specimens as compared with undamaged specimens [15].

4.4 NDE Methods for Detection of Pitting Corrosion

4.4.1 Ultrasonic Lamb Wave-based Technique for Monitoring Pitting Corrosion

For rapid localization and parametric identification of corrosion-type damage in aluminum plates, a circular array of piezoelectric wafer active sensor that generates Lamb waves was reported [16]. Different extents of pitting corrosion were introduced in a plate by applying 0.05 ml concentrated hydrochloric acid (35% conc.) on a circular area of 20 mm diameter on an aluminum plate twice daily. This introduced pitting corrosion penetrating along thickness with time without forming soluble salt, but by giving progressive reaction. Two actuators and four receiver sensors were placed around the corroded region, and the sensor signals were acquired for a period of 8, 16, and 24 days. Wavelet time-frequency maps of the sensor signals were analyzed, and the effect of corrosion growth on the Lamb wave transmission at 20 and 60 kHz excitation was studied in time-frequency scale. For a sensor that is juxtaposed to the actuator, a decrease in signal amplitude and wavelet coefficients was observed. On the other hand, for sensor that is orthogonal to the actuator and corroded region, an increase in wavelet coefficients was noticed. Likewise, for sensors that are equidistant to the actuator, identical response was observed. The studies confirmed the possibility of circular array of sensors for monitoring the corrosion damage.

4.4.2 DC Potential Drop Technique-based Monitoring of Pitting Corrosion

For monitoring of pit from inaccessible side of a joint of type 304L stainless steel welded with 308L, the use of DC potential drop method was reported [17]. To form pits, potentiostatic polarization in 10% NaCl solution for 15 min at +400 mv was used to form four pits in base metal (BM), two in HAZ, and one in weld metal (WM) region. The difference in potential between pitted and non-pitted weld joints was analyzed. An increase in potential was observed in corroded zones in the pitted weld joints, due to increase in resistance because of the presence of pits in these zones. In the HAZ, an increase in potential by 1.7 times was observed, while in the WM zone an increase in potential by 1.5 times was observed. Studies clearly established that potential drop method can be used for corrosion monitoring in BM, HAZ, and WM regions is possible *in situ* by measuring the potential drop.

4.4.3 Eddy Current Sensors for Far-side Corrosion Detection

An integrated sensor comprising an eddy current exciter coil and a GMR sensor receiver has been developed for detection of far-side corrosion in austenitic stainless steel [18]. GMR sensors are very attractive as they offer very high sensitivity to low magnetic fields of the order of a few nT. The integrated sensor is able to reliably detect far-side corrosion at 8.0 mm below surface as compared with 4 mm that is achieved by the conventional coil-based eddy current probes. Recently, for detection of inside corrosion in 12 mm thick stainless steel waste vault storage tanks from outer surface, a high sensitivity eddy current instrument with send–receive-type cup-core-based probe has been developed. A variable current source and a digital lock-in amplifier-based phase angle measurement implemented in a LabVIEW environment have been used [19]. Experimental studies have confirmed that the higher excitation current of 550 mA and the high sensitivity lock-in amplifier-based measurements have enabled the reliable detection of subsurface notches simulating corrosion damage located 8 mm below the surface as typically shown in Figure 4.5.

4.5 NDE Techniques for Monitoring of Corrosion in Piping Systems

Monitoring pipe wall erosion/corrosion thinning rates is an important issue in petrochemical and power generation industries. Cross-correlation technique and polynomial curve fitting based estimation of time delay between front wall and backwall echoes is useful for accurate estimation of accelerated thinning rate based on the ultrasonic pipe wall thickness data collected over a short period of time [20]. Remote field eddy current technique is a potential technique for detection of corrosion in small diameter pipes. For detection of corrosion in oil and

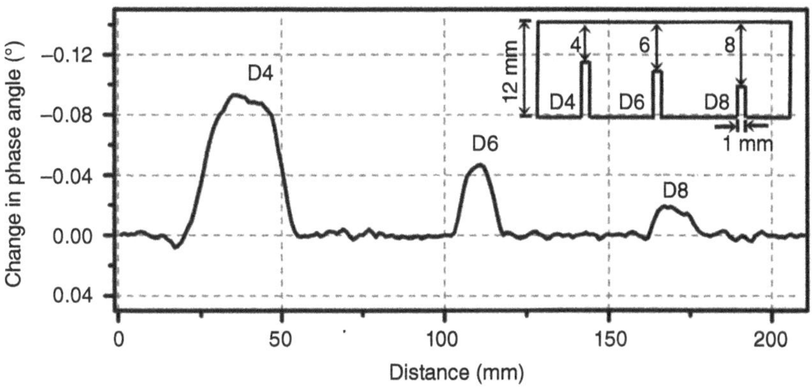

Figure 4.5 Eddy current signal of subsurface notches of width at 1 mm, length at 20 mm, and depths at 4.0, 6.0, and 8.0 mm present in a 12.0 mm thick stainless steel plate.

gas pipelines to avoid potentially catastrophic consequences, intelligent pigs (or smart pigs) were employed. These tools allow for internal inspection of pipelines using NDE methods for the early detection and sizing of corrosion damage. MFL method is used for corrosion inspection, and ultrasonic method is used for crack detection. The ultrasonic technique offers superior sizing capabilities. However, it is limited to inspection of oil pipelines in which the medium acts as couplant. Recently, this limitation has been overcome with the use of EMATs. Further, by a combined sensor, EMAT, eddy current, and MFL inspections can be performed at the same time. This multitechnique approach offers enhanced detection and sizing of defects for effective monitoring corrosion in oil and gas pipelines [21]. As large amounts of data are acquired over tens of kilometers of distance, dedicated and efficient signal processing and image processing algorithms, inversion methods, and data fusion methods are essential to demonstrate and enhance the probability of detection of defects.

4.6 NDE Techniques for Corrosion Detection in Aircraft Structures

Pitting, intergranular corrosion attack and SCC are the common problems in aircraft structures. Multifrequency eddy current and magneto-optic imaging techniques are commonly employed for their detection as the structures are made of aluminum alloys and also as it is possible to detect and size corrosion damage. Phased array ultrasonic technique is finding increased use for inspection of structural members, because of the possibility of electronic steering of the beam without the need to move the probe, imaging diagnostics, wide scanning area, and better probability of detection [22]. In a number of cases, if the components are accessible for visual inspection, it is may be sufficient if pitting corrosion and SCC are detected to initiate maintenance actions as intergranular attack commonly nucleates at pits on exposed end-grains, and in some situations, the material bulges around the fastener due to the corrosion product between multiple intergranular cracks, forming exfoliation damage. This type of damage can be detected and imaged real time using visual (miniature video cameras), ultrasonic, and thermography methods. More advanced ultrasonic techniques can interrogate the bottom of the fastener hole and locate the bottom layer. Sizing is possible by ultrasonics, while the sensitivity of IRT method is affected by the depth at which the exfoliation occurs.

4.7 NDE Techniques for Detection of Rebar Corrosion Damage

The corrosion damage of reinforced concrete structures is often a severe and invisible damage. For its detection, impact-echo and microwave techniques are very useful, apart from electrochemical methods. The displacement spectrum obtained from the impact-echo method is converted as acceleration spectrum, and the degree of corrosion damage is predicted from relative amplitude obtained from acceleration spectrum [23]. In the case of microwave technique, ground-penetrating radar, working at a fixed frequency in the range of 500 MHz to 2.6 GHz, is moved over a concrete structure, and the resulting images are observed for identifying rebars and corrosion associated with rebars. The wave speed and wavelength in concrete depend on the dielectric constant of the concrete. Typically the wavelength in concrete is usually more than the rebar diameter. When the antenna is scanned across a concrete structure containing a rebar, reflected information is obtained from the rebar along with echo arrival time. For inspection of thick concrete structures, low frequency antenna is used, while high frequency antenna is selected for high-resolution imaging in thinner concrete sections.

Formation of microcracks in concrete structures generates AE as a result of degradation and load-induced damage. Monitoring corrosion of steel rebars in concrete is very important, and its corrosion is dependent upon factors such as temperature, internal moisture content, resistivity, and the availability of oxygen. The rate and type of corrosion (e.g. chloride or carbonation induced) may influence the magnitude and rate of emissions as different corrosion products are formed [22]. Wireless AE sensor systems and wireless sensor networks are being explored for continuous surveillance of large and critical concrete structures.

Electrochemical noise technique is a potential technique for monitoring rebar corrosion. In this technique a low amplitude variation of the corrosion potential of rebar steel concrete is measured to obtain noise data as a record of changes in potential. The noise power spectra are analyzed to gain information on the mechanism and rate of corrosion process at rebar locations. Embedded sensors with suitable data logging system are finding applications as they continuously measure data of linear polarization resistance, open circuit potential, resistivity, temperature, and chloride ion concentration [24]. Apart from them, vibration strain gauges, electrical strain gauges, and fiber-optic sensors can be embedded in concrete structures for continuous monitoring of corrosion damage [25].

4.8 Case Studies

4.8.1 Eddy Current Testing for Inspection of Steam Generator

For detection of stress corrosion cracks in steam generator tubes of pressurized water reactor, eddy current technique was used because of its high inspection speed and high sensitivity for surface cracks zones of the tube [26]. ECT array probes with four pickup coils and a helical exciter were developed, as depicted in Figure 4.6 with both a speed as high as the bobbin coil probe and detectability as high as rotating pancake coil.

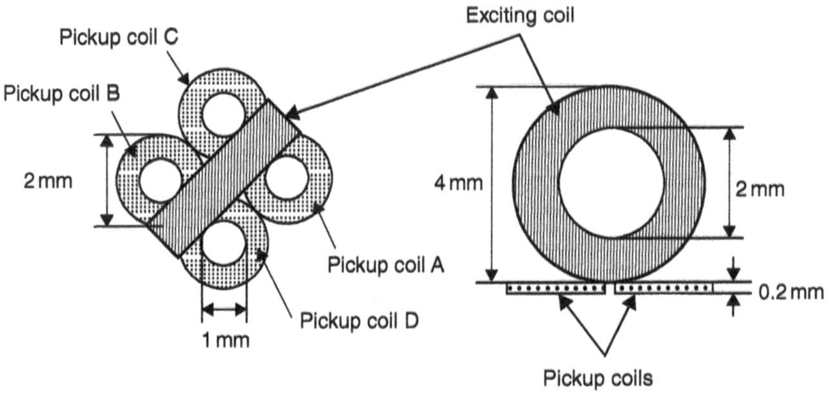

Figure 4.6 Eddy current probe for inspection of steam generator tubes.

ECT method was also developed for inspection of penetration zone of reactor pressure vessel cap of the pressurized water reactor to detect cracks and to size their length. For this, AISI 316L stainless steel plate was used, and EDM notches were fabricated and, then, subjected to cyclic loading to introduce fatigue cracks at the bottom of the notches. The plate was then loaded with four-point bending and soaked into polythionic acid to introduce stress corrosion cracks. Finally, the surface of the plate was ground to remove the notches and fatigue cracks so that the plate would only contain SCC. The lengths of the cracks were measured as 13, 15, and 27 mm. Using a lower excitation frequency of 10 kHz, weld cracks are reliably detected without much background noise. However, when a differential type plus point probe excited with a frequency of 400 kHz was used, the amplitudes of the signals were measured as 1.2, 1.8, and 3.5 V for the 13, 15, and 27 mm long cracks [27].

4.8.2 Magnetic Flux Leakage Imaging Technique for Corrosion Detection in Water Wall Tubes

In water wall tubes of power plants, erosion on OD and corrosion on ID are the major causes that leads to the puncture of tubes. Periodic inspection during overhauling period of plant using NDE methods is important to assess the condition of the tubes and to take corrective actions. For this, ultrasonic thickness measurement is usually employed. Recently, MFL technique that uses GMR sensors has been developed [28]. The technique, as shown in Figure 4.7, uses two saddle coils of 90 turns each separated by 75 mm for uniform axial magnetization (2 A magnetizing current) of the tube and 12 element GMR array sensor for measuring the axial component of leakage flux from defects. Using this technique, 1.0 mm deep OD erosion has been detected, and the MFL signals of OD notch are seen sharp and localized as compared with that of

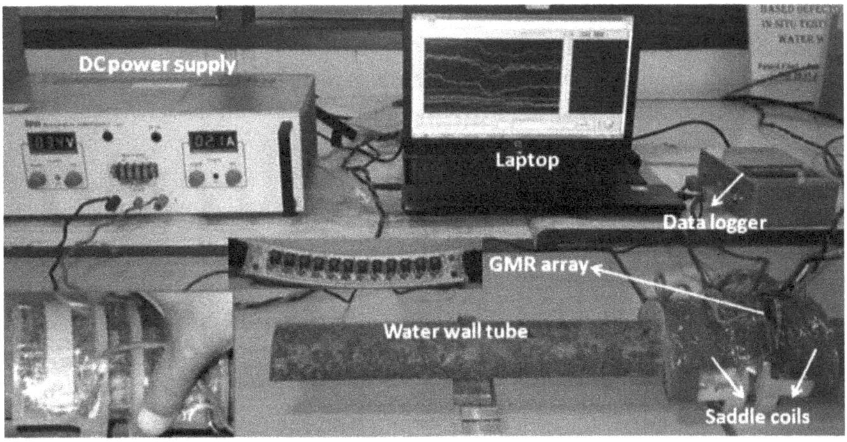

Figure 4.7 GMR array sensor-based MFL technique for testing of water wall tubes.

erosion. 2.0 mm deep ID gouging in water wall tube has been reliably detected using the GMR array probe. Images of corrosion damage have been generated by line scan of the array probe.

4.8.3 Acoustic Emission for Detection of Crevice Corrosion

Crevice initiation and propagation associated with bubbles formation from cathodic reactions within the crevice is the main source of AE. AE technique is useful for detection and monitoring of crevice corrosion on 304L austenitic stainless steel. To establish the technique, crevice initiation, propagation, and repassivation are controlled by additions of hydrogen peroxide and hydrochloric acid and by the extent of the applied torque of crevice assembly in the presence of chloride ions. The parameters of the AE signal such as rise time, counts number, duration, and cumulative energy are correlated to the potential and amplitude of crevice damage in terms of weight loss and metallic surface affected. A good correlation is reported between the cumulative % of AE signals and crevice development [29]. With a number of AE sensors covering the whole structure, AE technique is very effective for online monitoring of crevice corrosion, provided structural noise and other sources of AE are properly addressed.

4.8.4 Phased Array Ultrasonic Detection and Sizing of Microbiologically Induced Corrosion

Phased array ultrasonic technique has been developed for detection and sizing of MIC of service water systems and shut down coolers heat exchangers in OPG CANDU Stations [7]. This technique is able to detect and size different types of MIC attack, with ligaments up to 1.2 mm. This technique is also very visual, allowing easy interpretation of the data, and has been well received from the inspection technicians. The ligament measurement was improved for near-surface pin holes by using high-resolution filters and delay wedges. This technique is efficient for corrosion inspection of short-range piping, between the valves and on elbows and tees. This technique has contributed to dose reduction and reliable MIC attack detection on shut down coolers heat exchangers.

4.9 Multi-NDE and Data Fusion Techniques for Corrosion Monitoring

Recent trends in NDE for corrosion monitoring include multisensor fusion, embedded sensors, and wireless sensor networks. Often single NDE technique data is sufficient to reliably detect corrosion, and the use of multiple NDE sensors has numerous advantages over single sensor technique. Reduction in inspection time, increase reliability, improved defect visualization, and information redundancy are some of the benefits that accrue with multi-NDE modalities. For rapid inspection of non-ferromagnetic and ferromagnetic pipes and tubes of different diameters of boilers, feed water heaters, air coolers, and heat exchangers, systems that combine eddy current, MFL, remote field, and ultrasonic techniques are very attractive for detection and sizing corrosion, erosion, wear, pitting, fretting, wall loss, and cracks. Using such systems, data from same regions in a pipe can be obtained by different techniques and can be compared and also can be combined for effective decision-making.

In one typical example application, guided wave inspection, edge of light, ultrasonic pulse-echo, and eddy current techniques are combined using computer-controlled scanners for cost-effective corrosion detection in aircraft structures [30]. It was demonstrated that using guided wave defects in layered aircraft structures are detected fast and effectively. Exfoliation under the fastener heads was detected using ultrasonic guided waves launched from outside the exfoliated region, and B-scans were produced to assist in result interpretation. It was demonstrated that ultrasonic and eddy current techniques were extremely cost-effective solutions in terms of inspection implementation, efficiency, and resolution, while guided wave approach was a rapid inspection solution for large and inaccessible aircraft structures. Thus, large area defects are located by guided wave technique, and they are quantified using ultrasonic and eddy current techniques. The resulting images from the three techniques are used to extract quantitative information related to corrosion or disbonds in aircraft structures [31]. Further, data fusion, Bayesian approach, and other methodologies can be adopted for combining the information from different images and to generate a comprehensive image of the inspected regions.

4.10 Summary

The above sections provided the applications of eight of the most commonly employed important NDE methods for detection and evaluation of uniform corrosion, localized pitting corrosion, erosion, SCC, and microbiologically induced corrosion. A few novel and recent applications of NDE techniques for detection of SCC and pitting corrosion, with the help of case studies for bringing out the advances in NDE methods for corrosion monitoring, are highlighted. Selection of NDE methods is very important, and in this context, a table giving the NDE method suitable for detection of different types of corrosion damage is given. Knowing the importance, corrosion monitoring of pipes, aircraft structures, and concrete structures using various NDE methods is covered in detail. Recent trends in corrosion monitoring such as array imaging, multisensor fusion, embedded sensors, and wireless sensor networks are highlighted and a few example case studies discussed. NDE methods are continually improving by incorporating the advances in microelectronics, sensors, digital circuits, and automation. The instruments and sensors are getting miniaturized, and all the NDE equipment are nearly portable. Phased array ultrasonic technique is finding widespread use in corrosion monitoring. High sensitive electromagnetic instruments and GMR array sensors are able to detect corrosion 8.0 mm below surface and also in water wall tubes. AE and IRT techniques have proven their capability for online monitoring of corrosion damage. Signal and image processing methods and artificial intelligence approaches are enhancing the detection capability of the NDE techniques. All these developments are pushing the developments toward embedded NDE sensing for *in situ* surveillance of components and structures to be able to detect corrosion at early stages and also to be able to monitor corrosion damage toward avoiding failures.

References

1 Raj, B., Jayakumar, T., and Thavasimuthu, M. (2015). *Practical Non-Destructive Testing*, 3e. New Delhi: Narosa Publishing House.
2 Raj, B., Jayakumar, T., and Rao, B.P.C. (1995). Review of NDT techniques for structural integrity. *Sadhana, Indian Academy Proceedings in Engineering Sciences* 20: 5–38.
3 Rao, B.P.C. (2017). Non-destructive testing and damage detection. In: *Aerospace Materials and Material Technologies*, vol. 2, 209–228. Singapore: Springer.
4 Roberege, P.R. (2007). *Corrosion Inspection Monitoring*, 189–364. Hoboken, NJ: Wiley.

5 Grinzato, E., Vavilov, V., Bison, P.G., and Marinetti, S. (2007). Hidden corrosion detection in thick metallic components by transient IR thermography. *Physics & Technology* 49: 234–238.
6 Vinita, V., George, R.P., Ramachandran, D. et al. (2012). Biofilm formation and thermographic evaluation of fly ash concrete in sea water. *Corrosion Letters* 3 (2): 426–439.
7 Ciorau, P., Pullia, L., Hazelton, T., and Dak, W. (2008). Phased array ultrasonic technology (PAUT) contribution to detection and sizing of microbially influenced corrosion (MIC) of service water systems and shut down coolers heat exchangers in OPG CANDU stations. *NDT.net – The e-Journal of Nondestructive Testing* (October). http://www.ndt.net/article/v13n10/ciorau1.pdf (accessed 15 March 2017).
8 Rao, B.P.C. (2011). Eddy current testing: basics. *Journal of Non-Destructive Testing & Evaluation* 10 (3): 7–16.
9 Shaikh, H., Rao, B.P.C., Gupta, S. et al. (2002). Assessment of intergranular corrosion in AISI Type 316L stainless steel weldments. *British Corrosion Journal* 37 (2): 129–140.
10 Shaikh, H., Sivaibharasi, N., Sasi, B. et al. (2006). Use of eddy current testing method in detection and evaluation of sensitisation and intergranular corrosion in austenitic stainless steels. *Corrosion Science* 48: 1462–1482.
11 He, Y., Tian, G.Y., Pan, M. et al. (2014). An investigation into eddy current pulsed thermography for detection of corrosion blister. *Corrosion Science* 78: 1–6.
12 Rao, B.P.C. (2012). Magnetic flux leakage testing: basics. *Journal of Non-Destructive Testing & Evaluation* 11 (3): 7–17.
13 Hernandez-Valle, F., Clough, A.R., and Edwards, R.S. (2014). Stress corrosion cracking detection using non-contact ultrasonic. *Corrosion Science* 78: 335–342.
14 Zhaoming, Z., Taihe, S., Runqiao, Y. et al. (2014). Metal magnetic memory-based fluxgate sensor for imaging detection of coiled tubing. *Insight – Non-Destructive Testing and Condition Monitoring* 56 (12): 691–696.
15 Zeitvogel, D.T., Matlack, K.H., Kim, J.-Y. et al. (2014). Characterization of stress corrosion cracking in carbon steel using nonlinear Rayleigh surface waves. *NDT&E International* 62: 144–152.
16 Rathod, V.T. and Roy Mahapatra, D. (2011). Ultrasonic lamb wave based monitoring of pitting corrosion type of damage in plate using a circular array of piezoelectric transducers. *NDT&E International* 44: 628–636.
17 Atlam, A., Omar, A.A., Waheed, A.F. et al. (2010). Monitoring pitting corrosion using DCPD techniques. *Proceedings of the 4th Environmental Physics Conference*, Hurghada, Egypt (10–14 March 2010), pp. 133–139.
18 Sasi, B., Arjun, V., Mukhopadhyay, C.K., and Rao, B.P.C. (2018). Enhanced detection of deep seated flaws in 316 stainless steel plates using integrated EC-GMR sensor. *Sensors and Actuators A: Physical* 275: 44–50.

19 Soni, A.K., Thirunavukkarasu, S., Sasi, B. et al. (2015). Development of a high-sensitivity eddy current instrument for the detection of sub-surface defects in stainless steel plates. *Insight – Non-Destructive Testing and Condition Monitoring* 57 (9): 23–28.

20 Honarvar, F., Salehi, F., Safavi, V. et al. (2013). Ultrasonic monitoring of erosion/corrosion thinning rates in industrial piping systems. *Ultrasonics* 53: 1251–1258.

21 Willems, H., Jaskolla, B., Sickinger, T. et al. (2010). Advanced possibilities for corrosion inspection of gas pipelines using EMAT-technology. *ECNDT 2010, NDT of Pipelines*. http://www.ndt.net/article/ecndt2010/reports/1_11_16.pdf (accessed 15 March 2017).

22 Zaki, A., Chai, H.K., Aggelis, D.G., and Alver, N. (2015). Non-destructive evaluation for corrosion monitoring in concrete: a review and capability of acoustic emission technique. *Sensors* 15: 19069–19101.

23 Liang, M.-T. and Su, P.-J. (2001). Detection of the corrosion damage of rebar in concrete using impact-echo method. *Cement and Concrete Research* 31: 1427–1436.

24 Song, H.-W. and Saraswathy, V. (2007). Corrosion monitoring of reinforce concrete structures – a review. *International Journal of Electrochemical Science* 2: 1–28.

25 Breysse, D. (2012). *Non-destructive Assessment of Concrete Structures: Reliability and Limits of Single and Combined Techniques, State-of-the-Report of the RILEM Technical Committee 207-INR*. New York: Springer.

26 Chen, Z., Yusa, N., and Miya, K. (2008). Enhancements of eddy current testing techniques for quantitative nondestructive testing of key structural components of nuclear power plants. *Nuclear Engineering and Design* 238: 1651–1656.

27 Yusa, N., Uchimoto, T., Takagi, T., and Hashizume, H. (2012). An accurately controllable imitative stress corrosion cracking for electromagnetic nondestructive testing and evaluations. *Nuclear Engineering and Design* 245: 1–7.

28 Singh, W.S., Rao, B.P.C., Mukhopadhyay, C.K., and Jayakumar, T. (2015). Detection of localized damage in water wall tubes of thermal power plants using GMR sensor array based magnetic flux leakage technique. *Journal of Nondestructive Evaluation* 34 (3): 1–7.

29 Kima, Y.P., Fregonesea, M., Mazillea, H. et al. (2003). Ability of acoustic emission technique for detection and monitoring of crevice corrosion on 304L austenitic stainless steel. *NDT&E International* 36 (8): 553–562.

30 Brassard, M., Chahbaz, A., Pelletier, A., and Forsyth, D.S. (2000). Combined NDT inspection techniques for corrosion detection of aircraft structures. *Proceedings of 15th World Conference on NDT*, Rome. http://www.ndt.net/article/wcndt00/papers/idn534/idn534.htm (accessed 15 March 2017).

31 Forsyth, D.S. and Komorowski, J.P. (2000). The role of data fusion in NDE for aging aircraft. *Proceedings of SPIE – The International Society for Optical Engineering* (5 May 2000). https://doi.org/10.1117/12.385034.

5

Lock-in Thermography for the Wide Area Detection of Paint Degradation and Incipient Corrosion

R. Jones[1], M. Lo[1], M. Dorman[2], A. Bowler[3], D. Roles[3] and S.A. Wade[4]

[1] *Centre of Expertise in Structural Mechanics, Department of Mechanical and Aeronautical Engineering, Monash University, Clayton, Victoria, Australia*
[2] *Aircraft Structural Integrity – Directorate General Technical Airworthiness (ASI-DGTA), RAAF Williams, Laverton, Victoria, Australia*
[3] *Maritime Systems Project Office (MPSPO), RAAF Base Edinburgh, Edinburgh, South Australia, Australia*
[4] *Swinburne University of Technology, Hawthorn, Victoria, Australia*

5.1 Introduction

The last two decades has seen significant progress into the use of infrared (IR) thermography as a tool for the nondestructive inspection (NDI) of structures and components [1–35]. Thermographic techniques have several advantages over other forms of NDI. One of the main advantages is its ability to rapidly examine large areas. The thermographic techniques are relatively noninvasive in nature; however a key requirement is that the components examined should have thermal stability when subjected to nonuniform radiation heating. The major shortcoming is that the technique is primarily limited to inspecting thin structures and for detecting surface corrosion and paint anomalies. This is due to the diffusive nature of thermal signals in materials such that there is a non-linear drop in the detectability of anomalies below the surface. Marinetti and Vavilov [5] suggested that flash heating was preferable to lock-in thermography. However, one problem with flash heating is that less energy is imparted to the structure. Hence, as stated in [5], there is a "compromise between producing necessary temperature contrasts and ensuring detectable temperature signals."

In this context it should be noted that Schonberger et al. [7] have shown that lock-in IR thermography was able to detect both corrosion and paint degradation in steel and galvanized steel panels that were coated with an organic paint. Similarly Jönsson et al. [25] revealed that this method holds great promise for

Non-Destructive Evaluation of Corrosion and Corrosion-assisted Cracking, First Edition.
Edited by Raman Singh, Baldev Raj, U. Kamachi Mudali and Prabhakar Singh.

detecting corrosion in simple panels, and Cramer et al. [22] reported a similar finding for the NDI of material disbonds and material loss due to corrosion in aircraft fuselage skins. As a result IR thermography is now in use by Northwest Airlines for detecting corrosion in civil aircraft [35] and has been shown to be useful in detecting corrosion in both military and civil aircraft [32–34].

Lock-in thermography, which involves heating the structure using a sinusoidal input of heat energy at a frequency ω, produces both amplitude and phase difference images both at the (first harmonic) ω and other harmonics, the most relevant of which (to this study) is the second harmonic 2ω. The amplitude image represents the amplitude (peak-to-peak) thermal response of the structure. The phase image is generated by phase difference measured between the input and the measured signal. The phase image is often more useful than the amplitude image for detecting subsurface defects [5]. In this paper we focus on the use of lock-in IR thermography, and in particular the combined use of phase images that are associated with the first and second harmonics, to detect incipient corrosion and paint degradation associated with 7075-T6 wing skins on RAAF AP-3C (Orion) aircraft. The breakdown of paint, particularly around fastener holes, can lead to internal corrosion in the fastener hole. As such, the sensitivity of the lock-in system to detect paint degradation is significant for the detection of incipient corrosion. In each case there is no special treatment given to the surface of the wing skins, i.e. the paint is not sprayed black to enhance emissivity and special paints are not used, since this will not be possible in the field. In this context it is shown that by using a combination of the first and second harmonic, lock-in IR thermography can be used to identify the breakdown of paint around fastener holes, paint delamination, and the extent that surface corrosion has spread beneath the paint.

5.2 Lock-in IR Thermography Background

IR thermography is a process by which the surface temperature of a structure is measured and used to form a thermal map of the structure. The information in the image is then used to assess the condition of the body. Lock-in thermography is a particular form of IR thermography that works by applying a sinusoidal heat input at a frequency ω to the structure and measuring the components of the temperature field at both the input frequency ω (the first harmonic) and, in some instances, the second harmonic. Lock-in thermography yields both amplitude and phase images. The amplitude images generate a picture of the peak-to-peak thermal response at that frequency. This image reveals changes in thickness and material variations. It has the drawback that it can be influenced by nonuniform heating of the surface and by lateral diffusion effects. The phase refers to the measured phase difference, $\Delta\phi$, between the lock-in signal from the lamps used to heat the structure and the thermal signal seen by

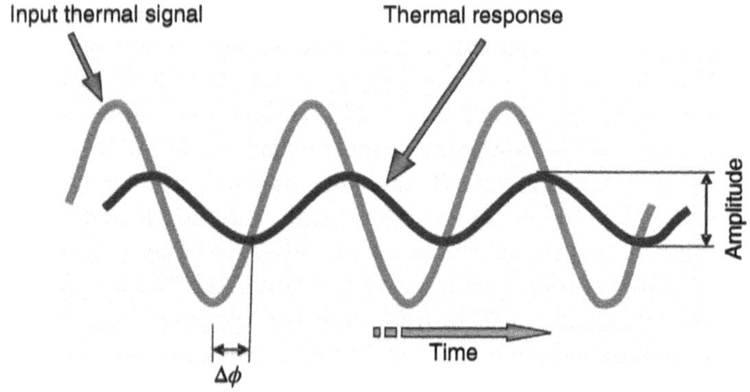

Figure 5.1 Basic lock-in thermography theory. *Source:* Wallbrink et al. [28]. Reproduced with permission of American Institute of Physics.

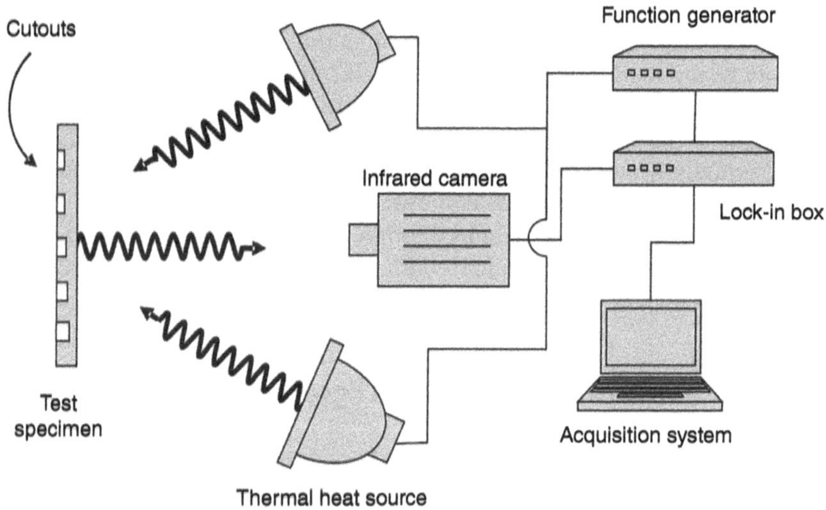

Figure 5.2 Schematic diagram of the lock-in system. *Source:* Wallbrink et al. [28]. Reproduced with permission of American Institute of Physics.

the IR camera (see Figure 5.1 and [28] for details). The phase image is particularly useful as it is generally invariant to nonuniform heating.

Figure 5.2 presents a schematic diagram of a typical lock-in thermography system. A thermal source such as a heater lamp is modulated with a sinusoidal signal to heat the structure. This is the sinusoidal harmonic heat source given by

$$Q = \frac{Q_0}{2}\left[1 + \cos(\omega t)\right]$$

where Q_0 is the amplitude of the input thermal signal. Defects in the structure alter the flow of heat through and hence create a phase shift in the emitted thermal signal at the surface of the target sample compared with the input signal. The emitted thermal signal is normally written in the form[1]

$$T(x,t) = T_0 \exp\left(\frac{-x}{\mu}\right)\cos\left(\omega t - \frac{x}{\mu}\right)$$

where x is the depth, t is the time, $T_0 = Q_0 / 2e\sqrt{\omega}$, $\mu = \sqrt{2\alpha / \omega}$ is the thermal diffusion length, and α is the thermal diffusivity. The frequency of the input heat source affects the resulting phase measurements. In general when investigating deeper defects, a lower input (heat) frequency is preferable since the thermal diffusion length is larger as shown by the expression above. Here $\Delta\phi = x/\mu$ constitutes the phase difference between the lock-in heat signal from the lamps used to heat the structure and the thermal signal seen by the IR camera.

In lock-in thermography a number of frames are taken during the acquisition of the thermal signal. The signal is processed via temporal averaging techniques to improve the thermal resolution of the signal and reduce the signal noise. To obtain the phase and amplitude data, numerous algorithms exist, and the most commonly used algorithm is four-point correlation (FPC) method. The averaging procedure reduces the number of points (frames) to only four points that are equidistant (90° out of phase) in one modulation cycle, viz. S_1, S_2, S_3, and S_4. Hence if less number of frames is acquired, information will be lost.

The amplitude[2] or change in temperature over one modulation cycle can be obtained using the FPC method and is given by

$$A = \sqrt{[S_1 - S_3]^2 + [S_2 - S_4]^2}$$

and the phase ϕ is given by

$$\phi = \tan^{-1}\frac{S_1 - S_3}{S_2 - S_4}$$

As the equation above shows that the phase information is a ratio of the heat signal, it is not impacted by factors such as nonuniform heating, optical absorption, and variation in the emissivity of the material.

---

1 In this chapter the response at the frequency ω is termed the first harmonic. However, there is also a response at the frequency 2ω that we will term the second harmonic. This response is shown in the various RAAF AP-3C-related wing panel tests presented in this chapter. The importance of the second harmonic in thermoelasticity was first outlined in [36].

2 Thermoelastic stress analysis can also be performed in order to obtain the stress amplitude map. The analysis involves thermoelasticity principles that relate thermal stress and strain in an isotropic, elastic element. The relationship between the amplitude of the principal stress and the change in temperature is essentially proportional if the element exhibits reversible, adiabatic behavior.

5.3 Detection of Incipient Corrosion and Paint Degradation

The lock-in thermography tests reported in this section were performed using a Cedip Jade 3 mid-wavelength infrared (MWIR) camera on a section of an RAAF AP-3C Orion wing that was located at DSTO. These tests involved cutting specimens out of a relatively small section of the wing, and as such there was a significant amount of the wing that could be used for other testing. A photograph of the relatively complete wing taken before it was cut up into sections is shown in Figure 5.3. It is believed that the wing has been stored outdoors for a significant period that was evidenced by extensive surface corrosion, paint degradation, and fouling by birds.

The section of wing that was used in this work is shown in Figure 5.4. Prior to the tests the wing section was cleaned to remove excess dirt and bird excrement.

The lock-in IR thermography test apparatus consisted of a bank of heat lamps and an IR camera connected via lock-in hardware manufactured by Cedip. The IR system used in this work was a Cedip Jade 3 MWIR camera, which operates in the MWIR spectral range (3.7–4.8 μm). This system uses an MCT ($Hg_xCd_{1-x}Te$) detector array producing images with a resolution of 320 × 240 pixels and has a quoted thermal sensitivity of less than 25 mK for temperatures above 25 °C. A total of six Source-Four Parcan lamps were set up (see Figure 5.5). However, not all six lamps were used in all of the tests.

IR images were taken at a number of locations on the specimen that are marked on Figure 5.4b. The test locations were chosen to cover a range of areas with different types of either visible corrosion or paint breakdown. The effect

Figure 5.3 The P-3C wing before cutup at DSTO.

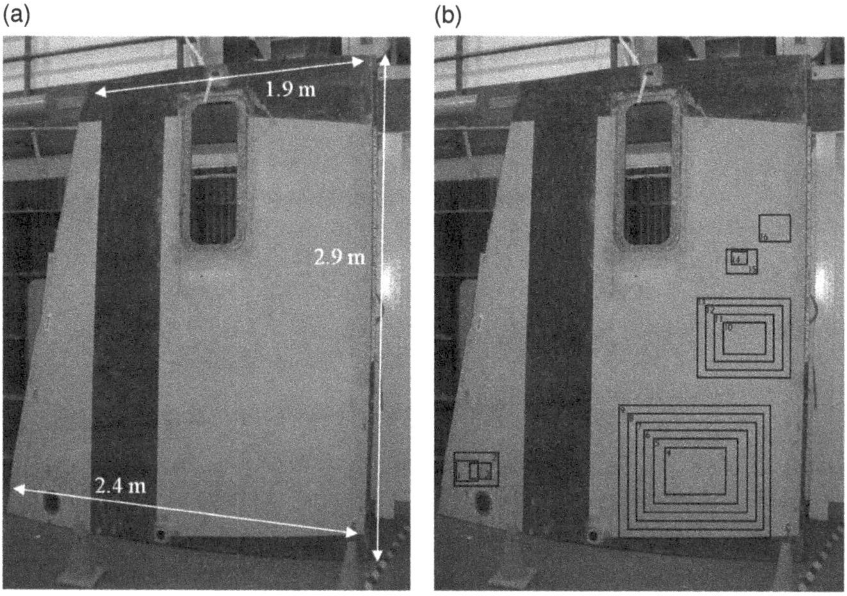

Figure 5.4 Section of P-3C wing used in the tests showing (a) dimensions and (b) locations where IR images were obtained.

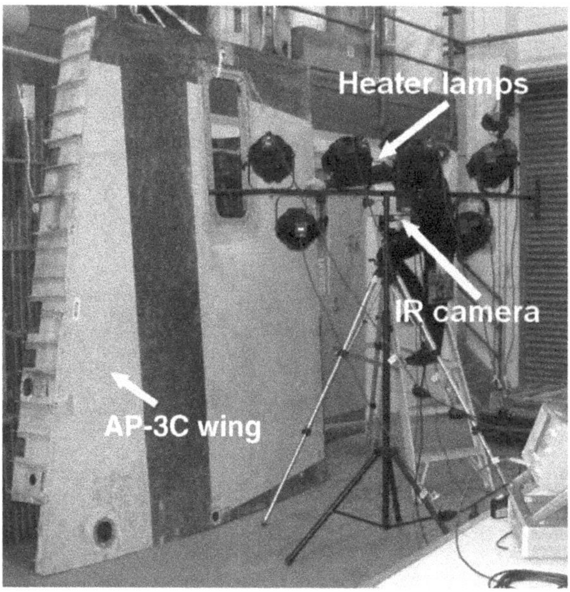

Figure 5.5 Photo of the P-3C wing specimen taken during IR thermographic tests.

of camera distance and lock-in frequency was evaluated at a number of the locations chosen, details of which are provided in Section 5.3.1.

5.3.1 Results

As mentioned previously, lock-in IR images were taken at a number of locations across the surface of the wing using different test parameters. The results below are broken up into the locations in which the tests were carried out. A digital camera (5 mega pixels) was used to take photos at each test location to allow comparison. For each test location masking tape was placed on the wing section to roughly designate the corners of the area captured by the IR camera.

Location 1: Fasteners (Site #1)

Test details:

Lock-in frequencies (Hz): 0.2, 0.5, and 0.9
IR camera to specimen distance: ~700 mm
IR camera lens: 50 mm

This location was chosen due to what would appear to be cracking of the paint around the circumference of several fasteners. Examples of a digital photo (where the black box corresponds to the IR image area) and a lock-in thermography phase image of test location 1 are shown in Figure 5.6. The fastener holes and the joint between the two wing planks are easily seen in the IR lock-in image.

Cropped and rescaled versions of sections of the photo and IR images of Figure 5.6, highlighting the fasteners and the plank joint, are shown in Figure 5.7.

Locations 2 and 3: Exfoliation Corrosion

Test details:

Lock-in frequencies (Hz): 0.2, 0.5, 0.9, and 5
IR camera to specimen distance (mm): 600 (location 2) and 300 (location 3)
IR camera lens: 25 mm

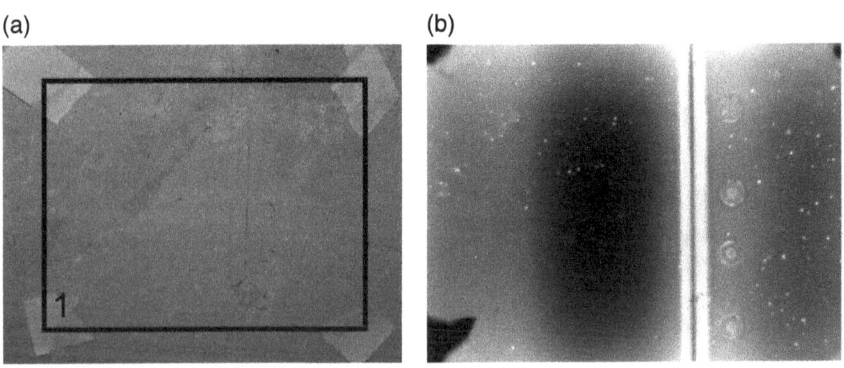

Figure 5.6 Location 1. (a) Digital camera image. (b) IR image at 0.2 Hz.

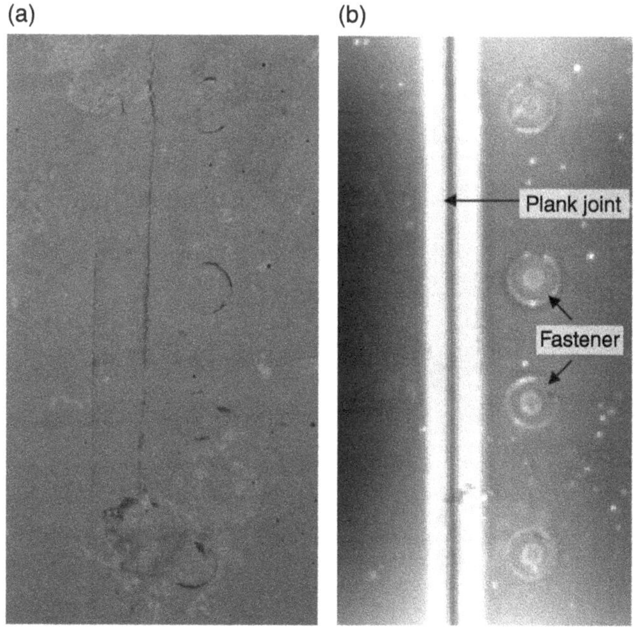

Figure 5.7 Close-up of location 1. (a) Digital camera image. (b) IR image at 0.2 Hz.

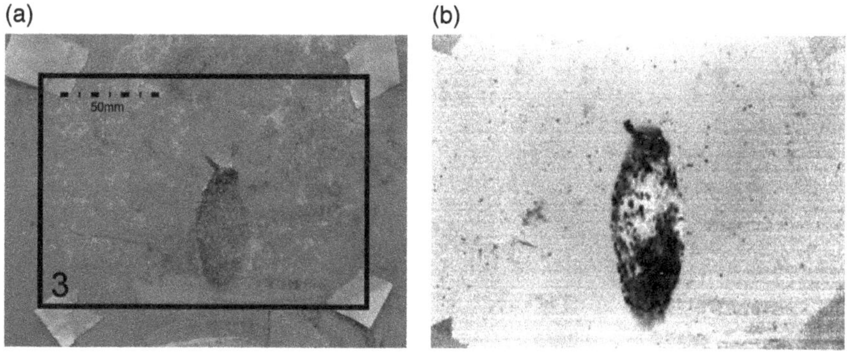

Figure 5.8 Location 3. (a) Digital camera image. (b) IR image at 5 Hz.

These locations were chosen because there was visible exfoliation corrosion. This area is of interest to determine if the lock-in IR thermographic method can pick up the extent of any corrosion that has spread beneath the painted section. Examples of a digital photo (where the black box corresponds to the IR image area) and a lock-in thermography phase image of test location 3 are shown in Figure 5.8.

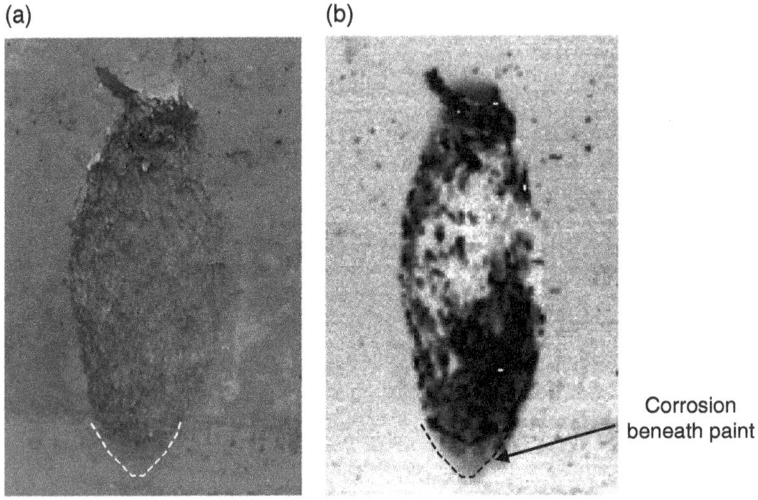

Figure 5.9 Close-up of location 3. (a) Digital camera image. (b) IR image at 5 Hz. The dashed lines indicate the possible extent of corrosion beneath the paint.

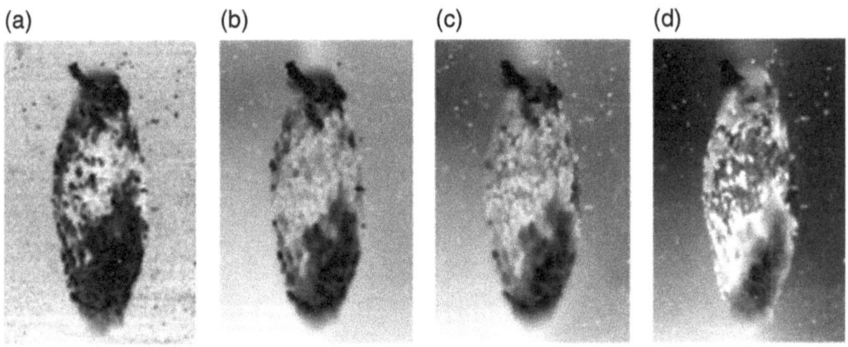

Figure 5.10 IR images of location 3 at various lock-in frequencies. (a) 5 Hz, (b) 0.9 Hz, (c) 0.5 Hz, and (d) 0.2 Hz.

Zoomed-in versions of the sections of the photo and the IR images shown in Figure 5.8, highlighting the corroded section, are shown in Figure 5.9. The darkening below the lower edge of where the paint has detached on the IR image is indicative of corrosion extending beneath the paint.

The effect of the test frequency used to obtain the IR images is shown in Figure 5.10, in which cropped images taken at location 3 are compared. Here we see that each frequency gave images that were consistent with the damage. However, in this instance the 5 Hz test frequency gave a slightly better view of extent of the corrosion beneath the paint. The lower test frequencies provided views of the extent and the depth of the corrosion underneath the corroded surface, and the lowest frequency of 0.2 Hz appeared to show the extent of the deepest detectable corrosion underneath the surface.

The previous examples presented information obtained from the component of the signal at the frequency ω corresponding to the frequency of the heater lamps. However, [31] revealed that the second harmonic is also useful for detecting delamination damage in composites. As such this study also examined the response at the second harmonic, i.e. at a frequency of 2ω, and the associated phase image, at an input frequency of 0.9 Hz, is shown in Figure 5.11 where we see that the various paint and corrosion-related defects are clearly evident as is the corrosion that lies beneath the paint (at the bottom of the picture). (The importance of the second harmonic in thermoelasticity was first outlined in [36].) This is in contrast to the first harmonic, where it is difficult to see the corrosion that lies beneath the surface of the paint and as such illustrates the advantage of using both the first and second harmonics.

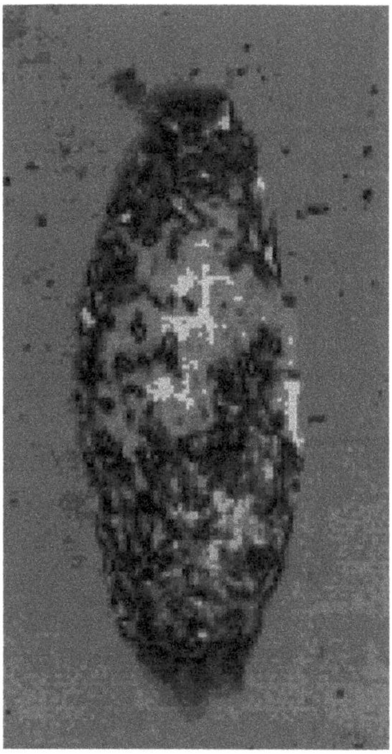

Figure 5.11 Close-up of the second harmonic.

Locations 4–9: Fasteners (Site #2). Effect of Camera Distance

Test details:

Lock-in frequencies (Hz): 0.2, 0.5, 0.9, and 5
IR camera to specimen distance (mm): 400, 600, 800, 1000, 1250, and 2000
IR camera lens: 25 mm

This location was chosen as it contained a variety of sites of interest including areas of paint breakdown and also a joint between wing planks and hence fasteners. For this location IR images were taken with the IR camera initially at a distance of 400 mm from the wing and then at increasing distances from the wing up to 2 m. The aim of this was to show how moving the camera back from the test object increases the area of inspection but at the same time reduces the ability to pick up small defects.

Figure 5.12 shows a digital photo (where the black box corresponds to the IR image area) and a lock-in thermography phase image taken with the IR camera at a distance of 400 mm from the wing (location 4). There are four areas

(a)

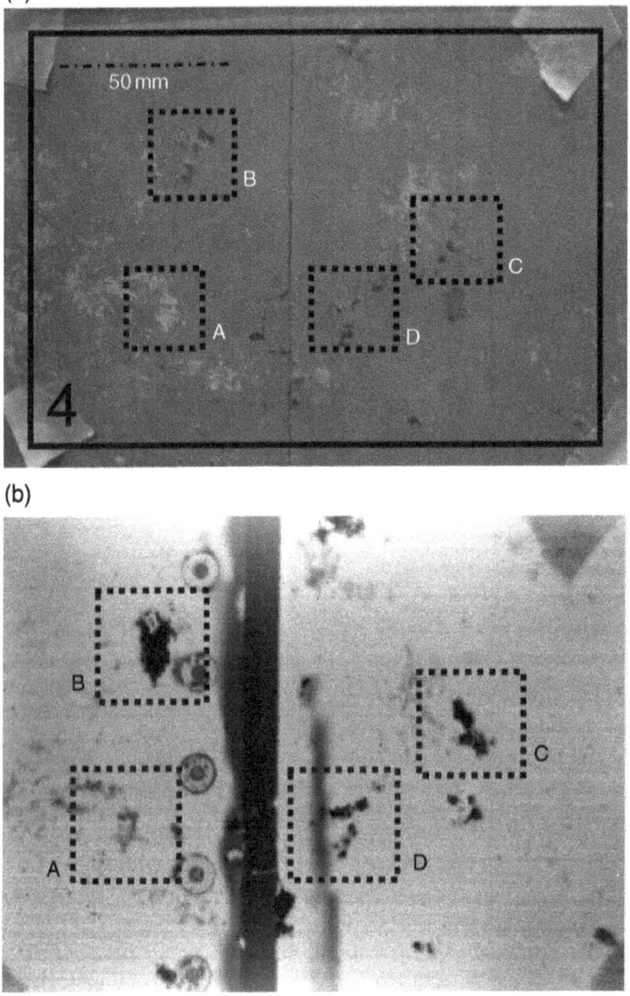

Figure 5.12 Location 4 with individual areas as denoted in the text. (a) Digital camera image. (b) IR phase image at 5 Hz.

highlighted in the images shown in Figure 5.12. The area designated A appeared to have a buildup of additional muck (possibly bird excrement) on the surface, while at areas B, C, and D, the painted surface was slightly raised indicating that paint breakdown and possibly corrosion has occurred underneath. Each of these areas can be seen as dark areas in the IR phase image shown in Figure 5.12b. It is interesting to note that the areas corresponding to paint breakdown (i.e. B, C, and D) are more easily distinguished than area A.

The phase image associated with the first harmonic and amplitude and phase images associated with the second harmonic, i.e. at a frequency of 2ω, measured at location 5 at an input frequency of 0.9 Hz, is shown in Figure 5.13. Here we see that the fasteners and the various paint and corrosion-related defects, which in these figures show as white, are clearly evident. Location 5 is a subset of the region designated location 7 and shown in Figure 5.15. Examining Figure 5.7 reveals that prior to these tests the extent of the paint degradation/corrosion evident between the fasteners in Figure 5.13 was not readily apparent. However, close visual examination of the sites subsequently confirmed the IR findings. This finding again illustrates the advantage of using both the first and second harmonics.

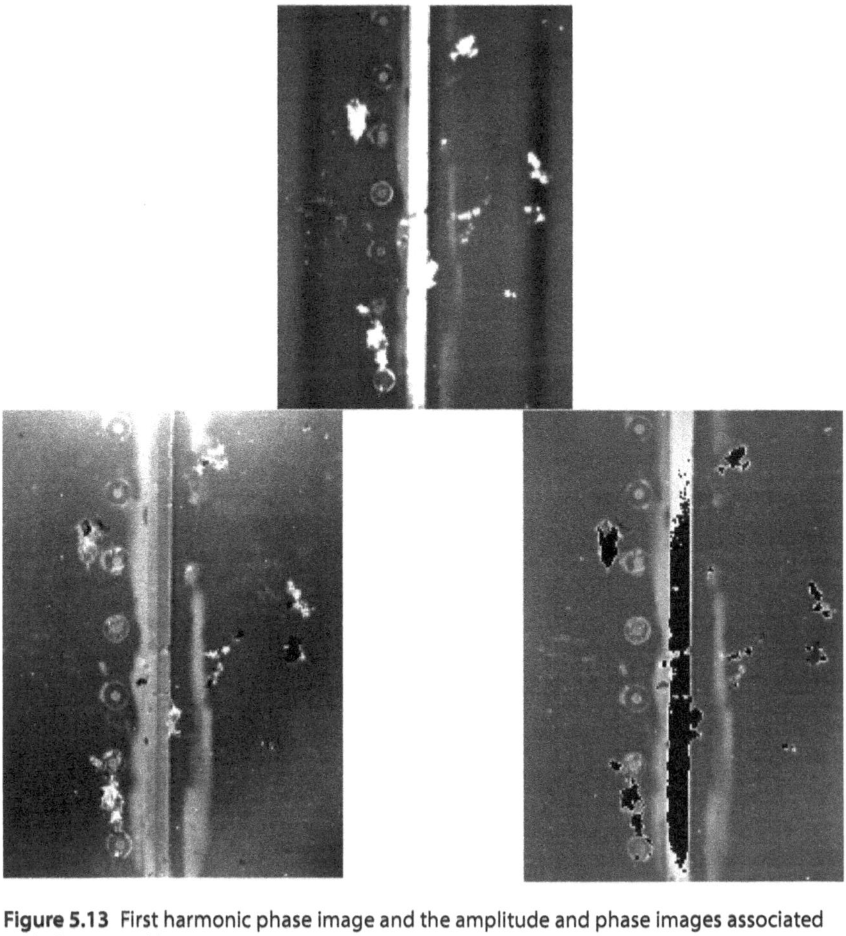

Figure 5.13 First harmonic phase image and the amplitude and phase images associated with the second harmonic at 0.9 Hz.

The effect of moving the camera away from the test specimen is shown in Figure 5.14 for images that were taken with the camera 400 mm, 1 m, and 2 m from the wing. Again digital camera images are shown for each area where an IR image was taken. The images shown were taken with a lock-in frequency of 5 Hz. The area of wing captured by the IR image is listed in Table 5.1 as a function of the distance of the IR camera from the wing.

It is easy to see from Figure 5.14 that the area captured in the IR image increases with camera distance from the wing; on the flip side however the increased area captured results in a reduction in the ability to resolve small

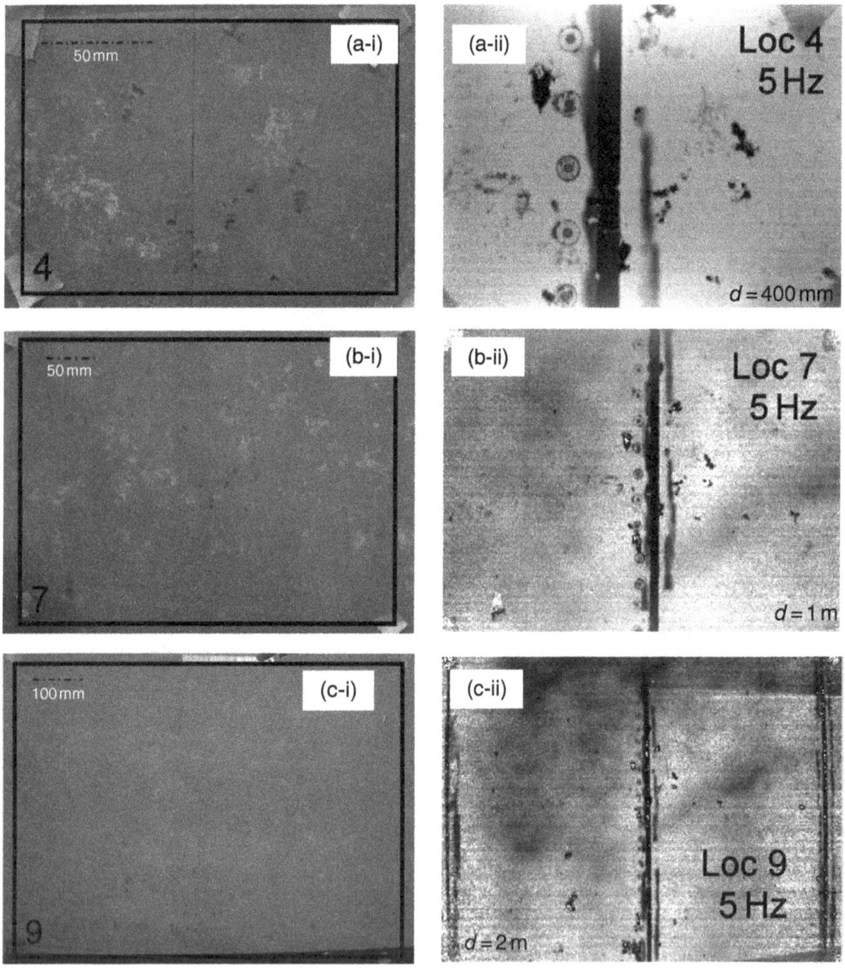

Figure 5.14 Effect of moving camera away from the test specimen, where the IR camera was (a) 400 mm, (b) 1 m, and (c) 2 m from the wing.

Table 5.1 IR image area as a function of IR camera distance from the wing.

IR camera distance from wing (m)	IR image area (mm)
0.4	158 × 118
1.0	395 × 296
2.0	780 × 585

defects. This is highlighted in Figure 5.14 where lock-in IR images taken at different camera distances to the wing have been cropped to show the same area of wing. Nevertheless we see that as the distance from IR camera to wing increases, the defects still remain visible in the lock-in IR image.

The minimum detectable feature can be correlated to the resolution of the IR camera. For the image area as a function of camera distance shown on Table 5.1 and assuming that the scale of the image in the vertical and horizontal directions is approximately similar, the height of the area and the scale can be correlated to the camera distance X (m) by the following expressions:

$$\text{Area height}(\text{mm}) = 291.58(X) + 2.5408$$

$$\text{Scale}\left(\text{mm pixel}^{-1}\right) = \frac{\left(291.58(X) + 2.5408\right)}{R_2}$$

where the resolution of an IR camera is $R_1 \times R_2$ in pixels. Thus, a minimum detectable feature of size D_{min} can be correlated to a minimum number of detectable pixels P_{min} by

$$D_{\text{min}} = P_{\text{min}} \frac{\left(291.58X + 2.5408\right)}{R_2}$$

Since the Cedip IR camera used here has a resolution of 320 × 240 pixels, the scale of the image is given by

$$\text{Scale}\left(\text{mm pixel}^{-1}\right) = \frac{\left(291.58(X) + 2.5408\right)}{240}$$

and the minimum detectable feature as a function of minimum detectable pixels is

$$D_{\text{min}} = P_{\text{min}} \frac{\left(291.58X + 2.5408\right)}{240}$$

Here, as an example, if the camera distance is 0.4 m away from a structure under examination and the required number of pixels to confidently detect any feature is 10 pixels, then the size of the minimum detectable feature is ~5 mm. The number of pixels required to confidently detect any feature can vary and is dependent on the usage conditions and the nature of the anomalies. In general, the minimum detectable feature of many IR cameras can be determined using a similar methodology as outlined here.

Location 10: Fasteners (Site #3). Effect of Lock-in Frequency and IR Camera Lens

Test details:

Lock-in frequencies (Hz): 0.05, 0.1, 0.5, 0.7, 0.9, 2, and 5
IR camera to specimen distance: 400 mm
IR camera lens: 25 and 50 mm

This location was chosen as it was another area with what would appear to be cracking of the paint around the circumferences of several fasteners. The effect of changing the lock-in frequency on the IR phase image was investigated for this location and is shown in Figure 5.15. The choice of the best lock-in frequency for testing depends upon a number of factors including the type of defect of interest. For example, in the image taken with a 5 Hz lock-in frequency, the areas around many of the fasteners appear to be obscured by what is thought to be sealant. While it is somewhat subjective, the 2 Hz lock-in frequency would appear to provide the best contrast if the areas around the fasteners were of interest.

A comparison of the images taken at location 10 with the 25 and 50 mm IR camera lenses is given in Figure 5.16 that presents examples of digital camera photos showing the area captured by each of the IR camera lenses and the corresponding IR images. The IR images were taken with the IR camera 400 mm from the wing.

Location 14: Close-up of Fasteners (Site #4)

Test details:

Lock-in frequencies (Hz): 0.05, 0.5, 0.9, 2, and 5
IR camera to specimen distance: 270 mm
IR camera lens: 50 mm

This location was chosen as it was another area in which the paint covering fasteners appears to have delaminated. Relatively closeup images were taken to evaluate the level of detail that could be obtained using lock-in thermography. Examples of a digital photo and a lock-in IR image covering the same area are shown in Figure 5.17.

Figure 5.18a and b presents cropped and rescaled versions of the images shown in Figure 5.17. Here we again see that there are a number of dark areas in the IR image above the top of the bottom fastener and at the bottom of the

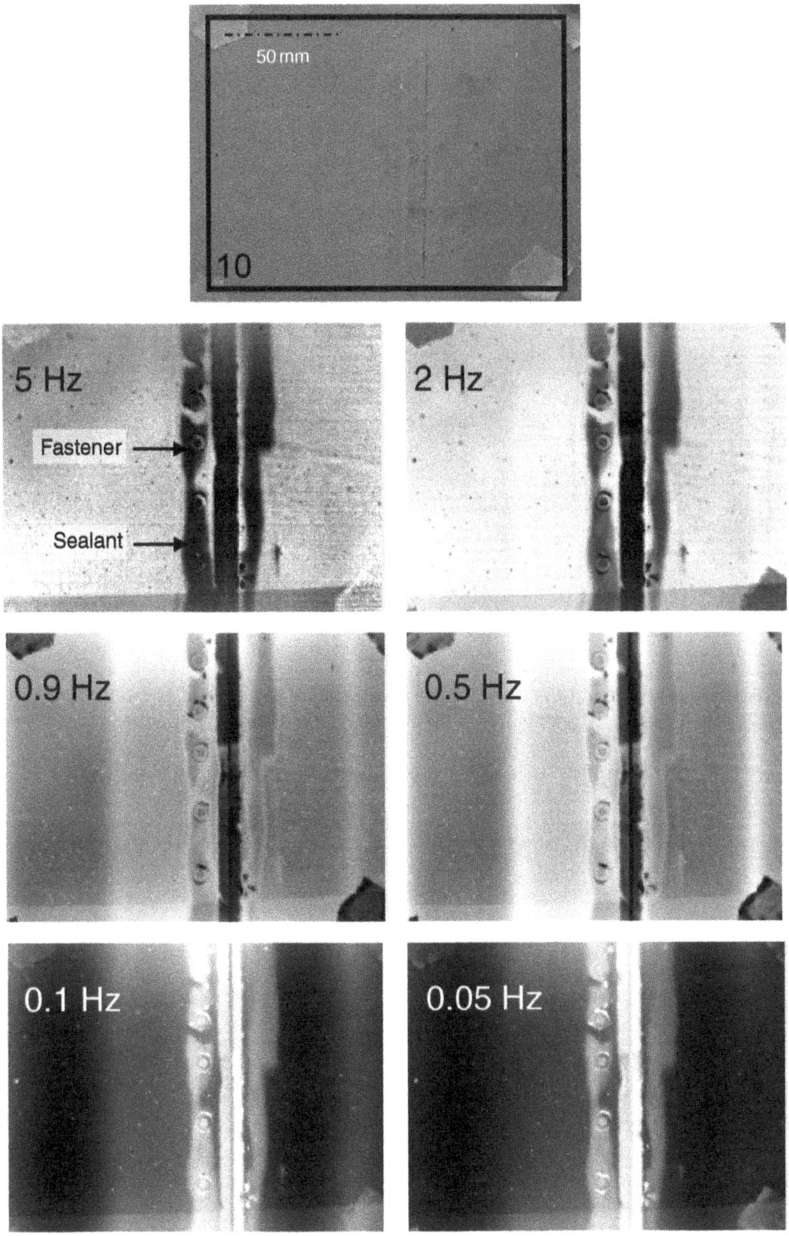

Figure 5.15 Effect of lock-in frequency on IR phase image for location 10 together with digital camera photo taken of the same location.

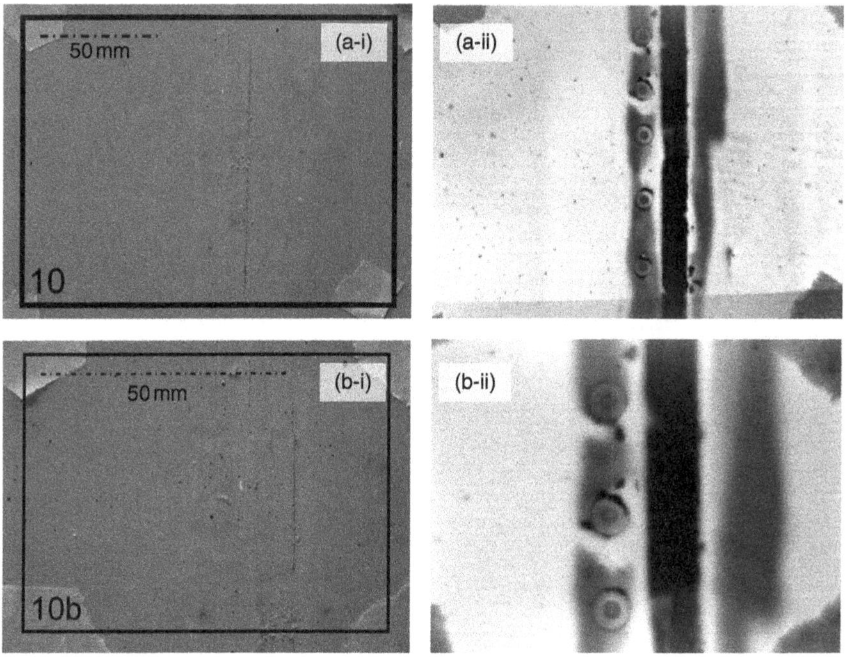

Figure 5.16 Digital camera photos showing the area and IR images taken at location 10 using (a) a 25 mm lens and (b) a 50 mm lens taken at a lock-in frequency of 2 Hz.

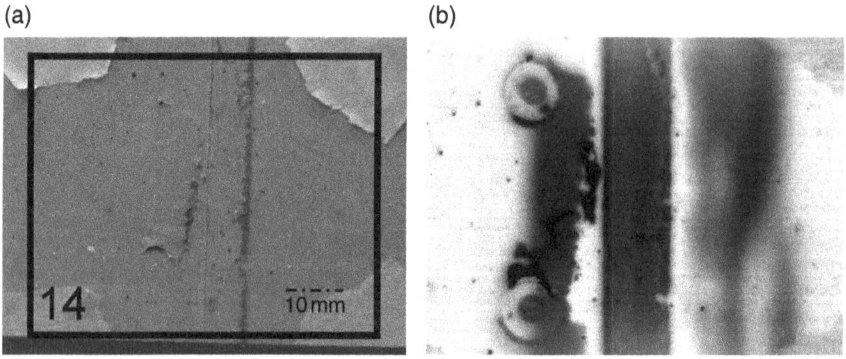

Figure 5.17 (a) Digital camera photos showing the area of interest and (b) IR image taken at location 14 using a lock-in frequency of 5 Hz.

top fastener. These relatively dark areas contrast with the surrounding lighter areas indicate that the material in the dark regions has different properties to the surrounding areas. This was due to paint delamination and/or corrosion underneath the paint.

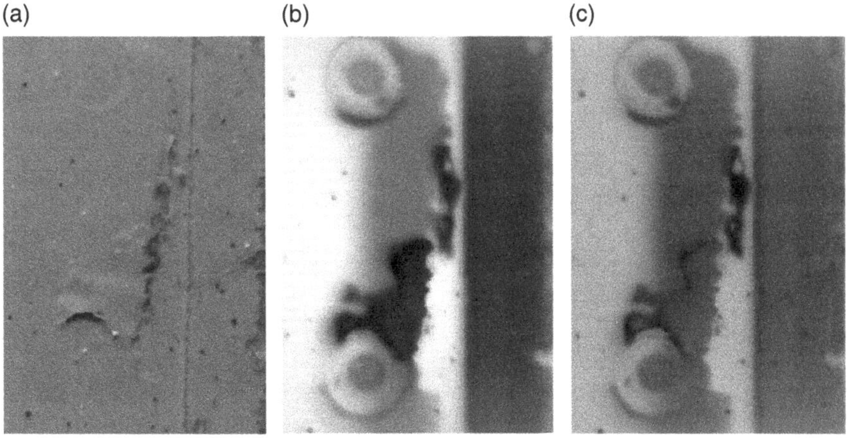

Figure 5.18 Closeup images of location 14. (a) Digital camera image. (b) IR camera image with auto gain control. (c) IR camera image without auto gain control.

It is important to note that the images taken by the IR camera can be presented in ways that can affect the interpretation of the image. An example of this is shown in Figure 5.18b and c where the same image is presented but the magnitude of the gray scales has been altered using the auto gain control (AGC) function of the Altair software that was used to analyze the IR images. The AGC function can be useful as it often allows a clearer image of the region of interest. However, care needs to be taken so as to ensure that its use does not result in parts of the image being washed out.

Locations 16: Paint Delamination

Test details:

Lock-in frequencies (Hz): 0.05, 0.5, 0.9, 2, and 5
IR camera to specimen distance: 500 mm
IR camera lens: 50 mm

This location was chosen as it contained paint blistering (see Figure 5.19). To highlight the blistered paint section, Figure 5.20 presents cropped and rescaled versions of the images shown in Figure 5.19. The digital photos and the IR image appear to be relatively similar in both shape and size.

5.3.2 Discussion

This section has presented the results of a lock-in IR thermographic inspection of a section of an RAAF AP-3C Orion wing. A range of areas were inspected including sites with visible surface corrosion, areas with fasteners that had localized paint degradation, and areas where the paint had blistered. The

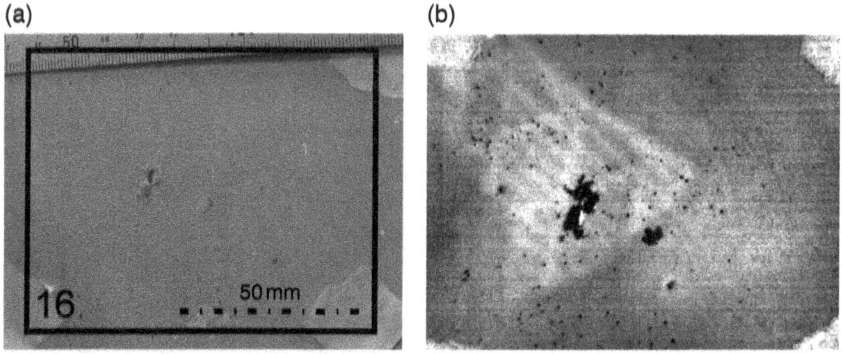

Figure 5.19 Location 16 containing paint blistering. (a) Digital camera image. (b) IR image taken with a 5 Hz lock-in frequency.

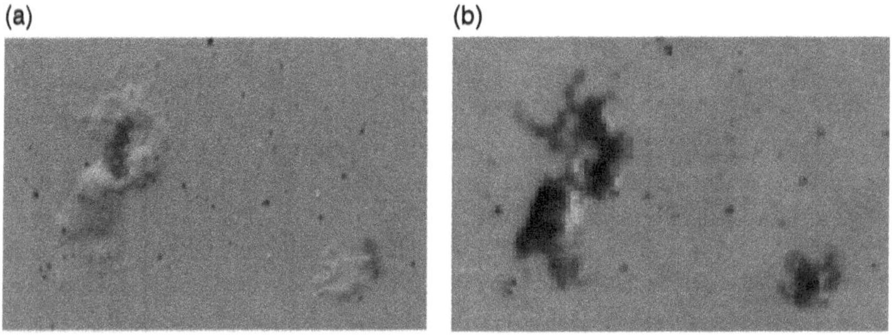

Figure 5.20 Closeup images of the blistered paint section. (a) Digital camera image. (b) IR image at 5 Hz with no ACG.

results of this study revealed that by evaluating both the first and the second harmonics in the resultant IR signal, it was possible to pick up the locations of fasteners beneath the paint, incipient corrosion, paint blistering, and also the extent that corrosion has traveled beneath a painted surface.

5.4 Infrared Thermography and Paint Degradation

To continue this study the RAAF Directorate General Technical Airworthiness (DGTA) and the RAAF Maritime Platform Systems Project Office (MPSPO) provided four AP-3C wing panels, viz. LHW Upper Panel #08 (P/N 938808-101), RHW Upper Panel #07 (P/N 938807-102), RHW Upper Panel #08 (P/N 938808-102), and RHW Upper Panel #09 (P/N 938809-102). These panels contained extensive multilayer intergranular cracking (MLIC) (see Figure 5.21)

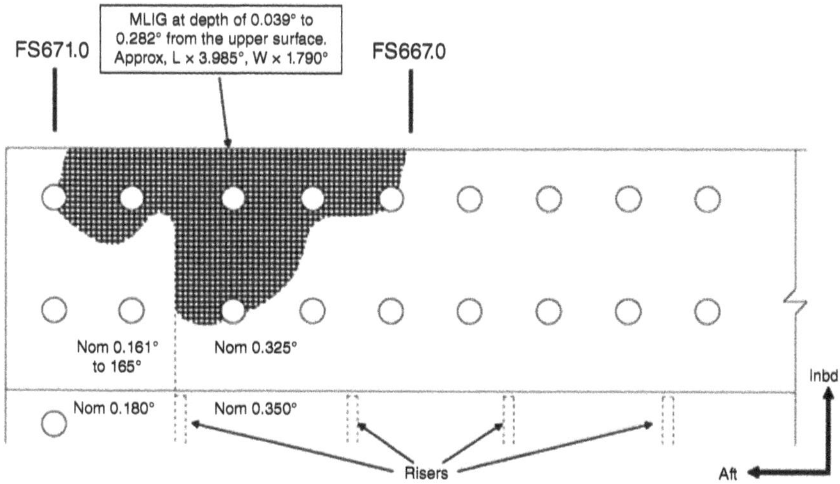

Figure 5.21 Detail of intergranular corrosion on the RH Wing Upper Panel #07 viewed from above.

that was associated with the lap splice region(s) of the panels. Unfortunately, lock-in IR thermography was unable to detect the MLIC. Following the MLIGC analysis an investigation was undertaken into the level of paint degradation on all four of the AP-3C wing panels.

The region along the lap joint of the RHW Upper Panel #07 was originally not painted; thus a fairly fine layer of black spray paint was applied to enhance material emissivity in the IR range and to prevent undesired reflections of background radiation. The other regions of the (wing skin) panels, which all had the original aircraft paint, were not spray-painted black since it was anticipated that this would not be done in the fleet. In these tests both the first and second harmonics associated with the sinusoidal heat (input) signal were measured.

The exterior and interior surfaces of RHW Upper Wing Panel #07 are shown on Figure 5.22. As mentioned previously, the tests were conducted on sections of the wing panels. Locations 1–4 of RHW#07 are shown in Figure 5.23.

5.4.1 Paint Degradation Test of RHW Upper Wing Panel #07 Locations 2, 3, and 4

Figures 5.24–5.26 present the thermal images obtained during paint degradation tests on RH Wing Upper Panel #07 at locations 2–4. These images were obtained at a pulse frequency of 0.5 Hz over 90 periods. In this study, and subsequent studies, the images presented are for an input frequency of 0.5 Hz. However, tests performed at higher frequencies and yielded almost identical

Figure 5.22 Exterior (a) and interior (b) views of RHW#07 Upper Wing Panel (P/N 938807-102).

results. Some pen markings on the skin surface of RHW#07 are visible on both the amplitude and the phase images as are the regions covered with masking tape. Note how the internal stiffeners are visible in the phase images. The anomalies in the paint seen in Figure 5.24a are also clearly evident in Figure 5.24c. Similarly anomalies (including pencil marks) in the paint seen in Figure 5.26a are also visible in Figure 5.26b and d. There was little damage seen at location 3 (see Figure 5.25). The images on Figures 5.24–5.26 show that a combination of first and second harmonic thermography images plus the amplitude image provides greater levels of details of the anomalies on the painted wing surface than a stand-alone lock-in thermography image from the first harmonic signal.

The exterior and interior surfaces of RHW Upper Wing Panel #08 are shown in Figure 5.27. The positions of thermal imaging sections are labeled as locations 1–5.

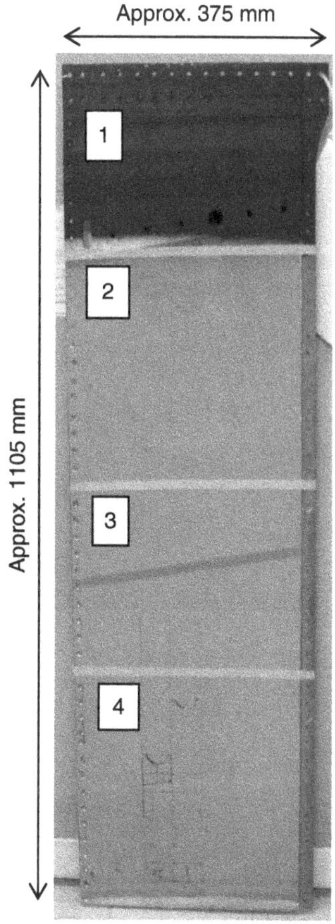

Figure 5.23 A photograph of RHW#07 with black spray paint on the lap joint region. Masking tapes mark out locations 1–4 for thermal imaging.

5.4.2 Paint Degradation Test of RHW Upper Wing Panel #08 Locations 1, 2, 3, 4, and 5

Figures 5.28–5.32 present the thermal images for the paint degradation tests on RH Wing Upper Panel #08 at locations 1–5. These images were taken at a pulse frequency of 0.5 Hz over 90 periods. Note how the internal stiffeners are visible in the phase images. Similarly marks and scratches seen in Figure 5.28a are also visible in Figure 5.28b and d. The marks and scratches seen in Figure 5.29a are also clearly visible in Figure 5.29b–d. There was little apparent damage in the paint associated with locations 3, 4, and 5 (see Figures 5.30–5.32). Again, here the images show that a combination of first and second harmonic images plus the amplitude image exemplifies the details of the anomalies such as marks and scratches on the painted wing surface when compared with a single first harmonic image.

The exterior and interior surfaces of LHW Upper Panel #08 are shown on Figure 5.33 where the various sections examined are labeled as locations 1–5.

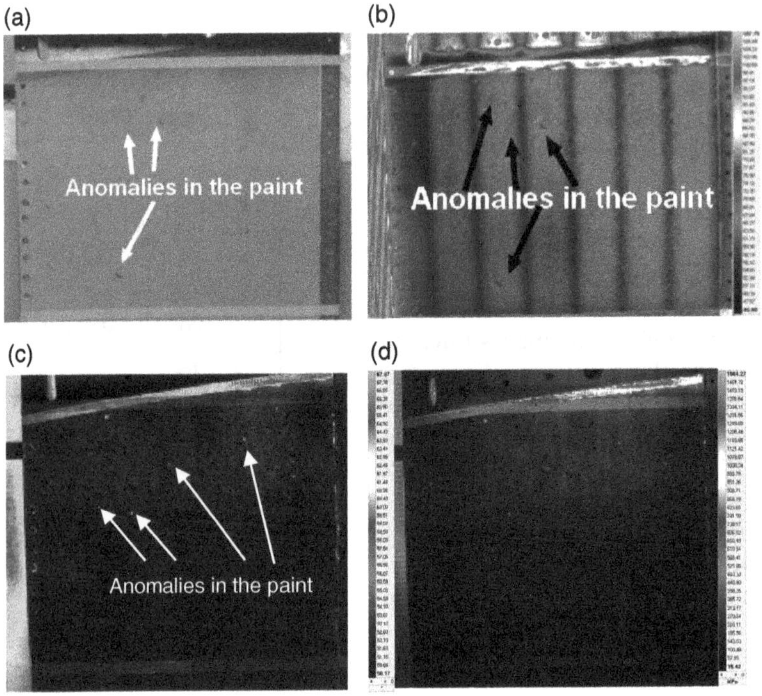

Figure 5.24 Location 2 of RHW Upper Panel #07 at 0.5 Hz with six lamps. (a) Digital camera image. (b) and (c) First and second harmonic phase images, respectively. (d) An amplitude image.

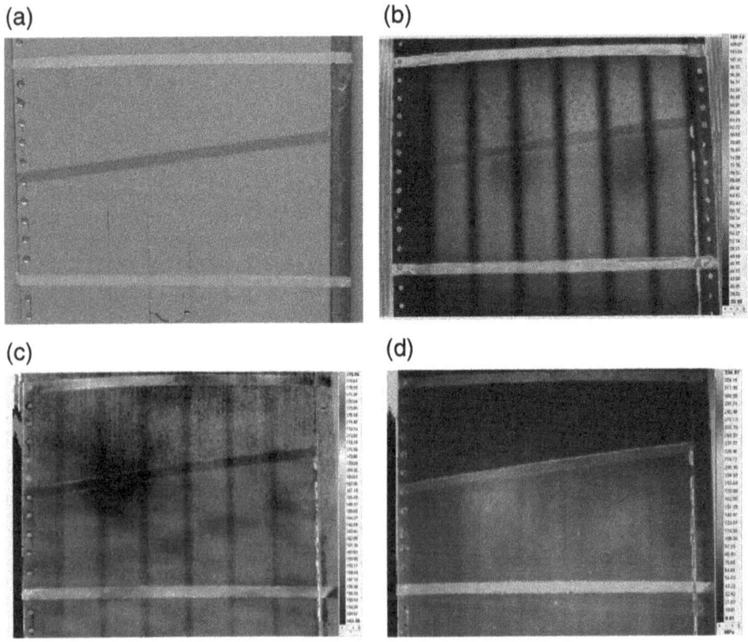

Figure 5.25 Location 3 on RHW Upper Panel #07 with six lamps. (a) Digital camera image. (b) First harmonic phase image at 0.5 Hz. (c) Second harmonic phase image at 0.5 Hz. (d) Amplitude image at 0.5 Hz.

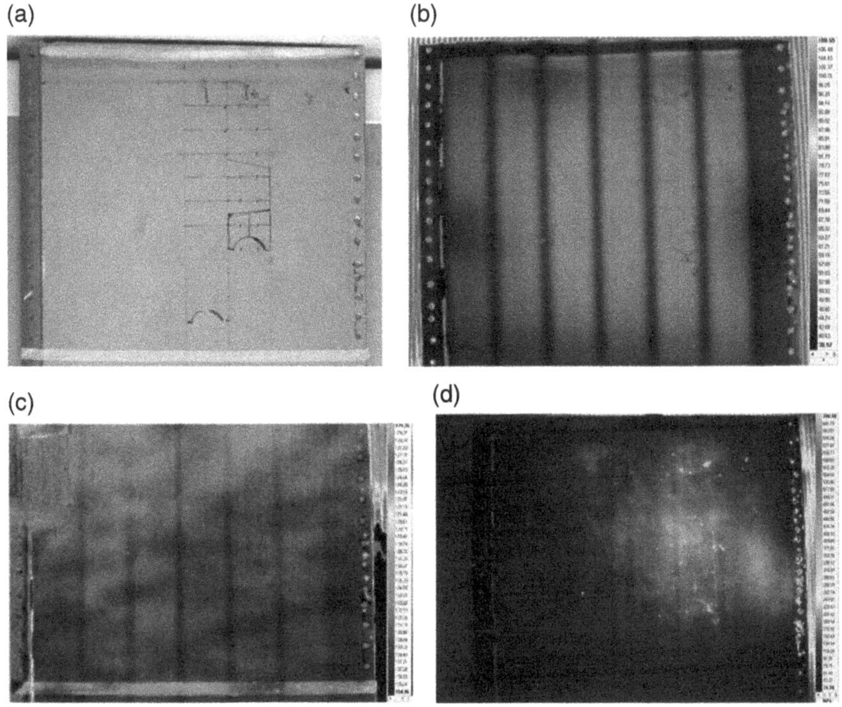

Figure 5.26 Location 4 on RHW Upper Panel #07 with six lamps. (a) Digital camera image. (b) First harmonic phase image at 0.5 Hz. (c) Second harmonic phase image at 0.5 Hz. (d) Amplitude image at 0.5 Hz.

5.4.3 Paint Degradation Test of LHW Upper Wing Panel #08 Locations 1, 2, 3, 4, and 5

Figures 5.34–5.38 show the thermal images for paint degradation tests on LH Wing Upper Panel #08 at locations 1–5 at a pulse frequency of 0.5 Hz. Note that the marks in the paint in Figure 5.34a are visible in Figure 5.34b and c.

A range of scratches and other surface anomalies shown in Figure 5.35a are also visible in Figure 5.35b–d.

The marks and paint anomalies seen in Figure 5.36a are also apparent in Figure 5.36b–d in location 3. Similar minor paint anomalies are seen in Figure 5.37a–d. There were minimal paint anomalies associated with location 5 on Figure 5.38. As previously shown, the images on Figures 5.34–5.38 show the first and second harmonics, and the amplitude images in concert provide finer details of the anomalies on the painted wing surface.

The exterior and interior surfaces of RHW Upper Wing Panel #09 are shown on Figure 5.39. The thermal imaging sections are labeled as locations 1–2.

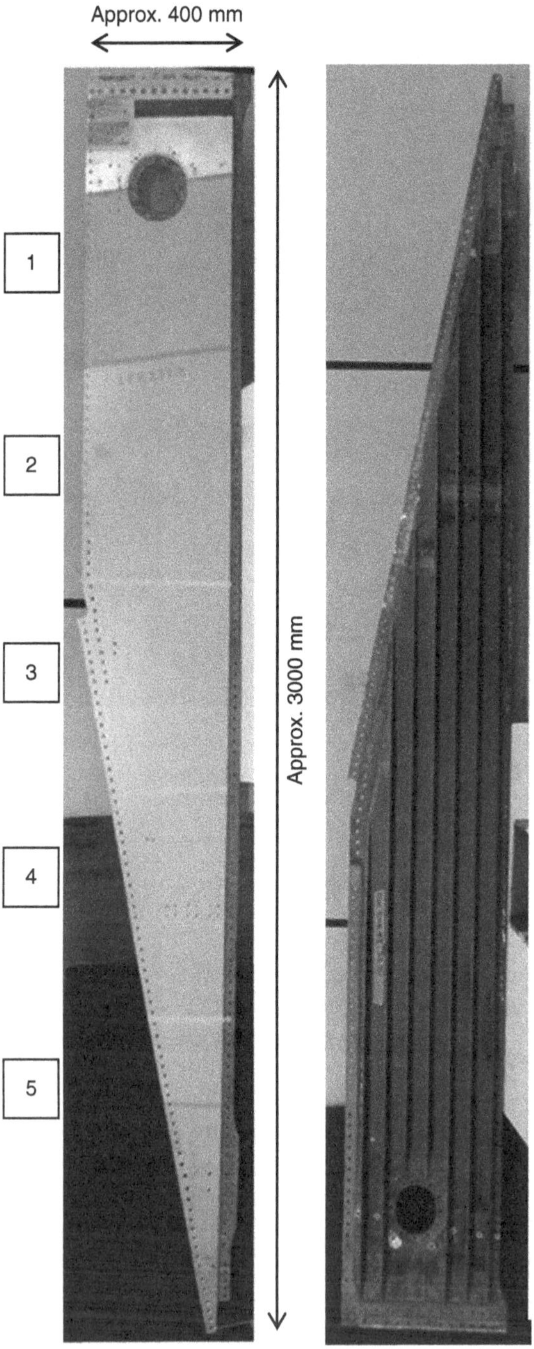

Figure 5.27 The exterior (left photo) and interior (right photo) surfaces of RHW Upper Wing Panel #08 (P/N 938808-102). Masking tapes mark out thermal imaging locations 1–5.

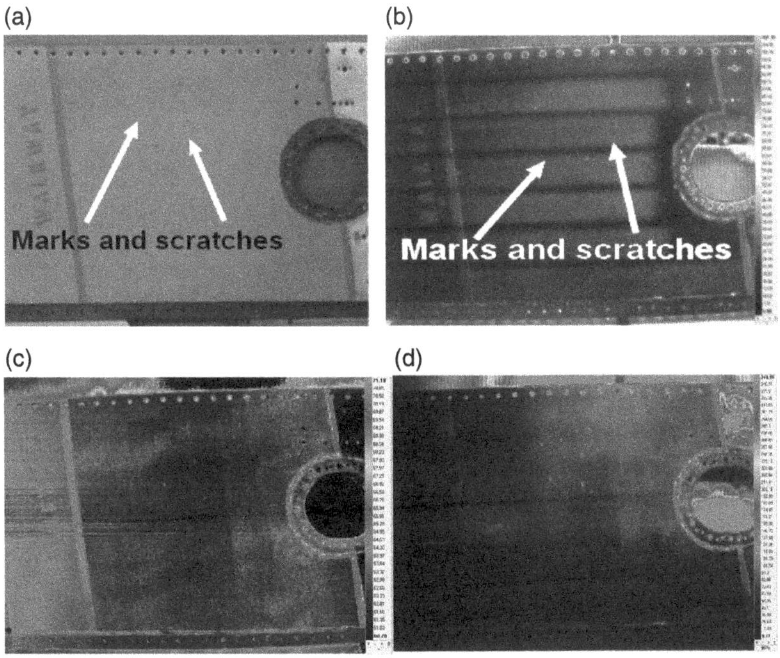

Figure 5.28 Location 1 on RHW Upper Panel #08 at 0.5 Hz with six lamps. (a) Digital camera image. (b) and (c) First and second harmonic phase images, respectively. (d) An amplitude image.

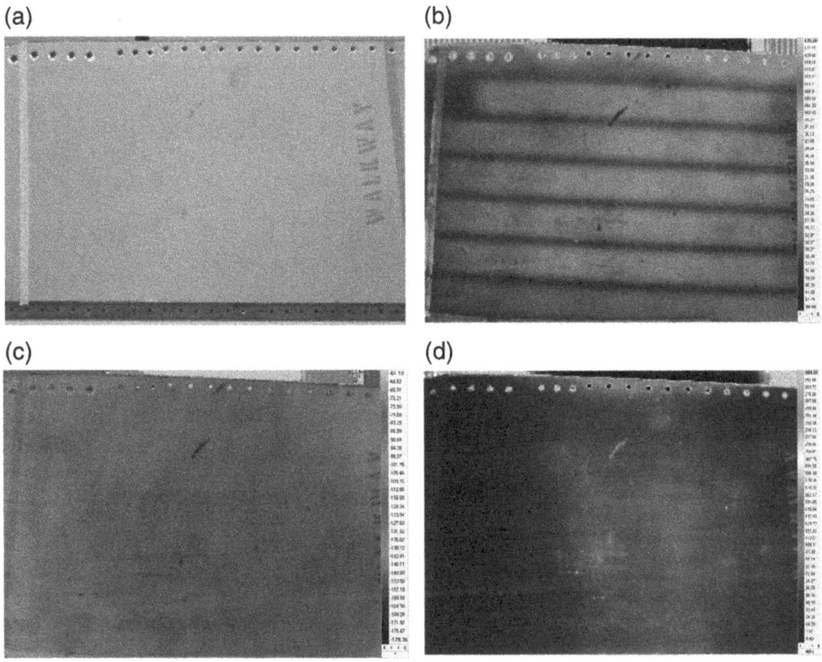

Figure 5.29 Location 2 on RHW Upper Panel #08 at 0.5 Hz with six lamps. (a) Digital camera image. (b) and (c) First and second harmonic phase images, respectively. (d) An amplitude image.

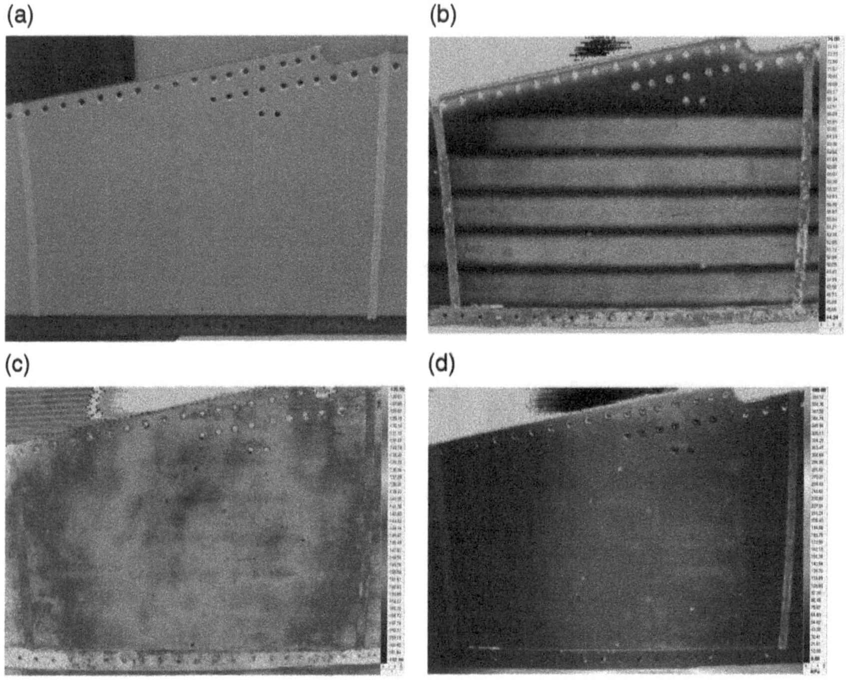

Figure 5.30 Location 3 on RHW Upper Panel #08 at 0.5 Hz with six lamps. (a) Digital camera image. (b) and (c) First and second harmonic phase images, respectively. (d) An amplitude image.

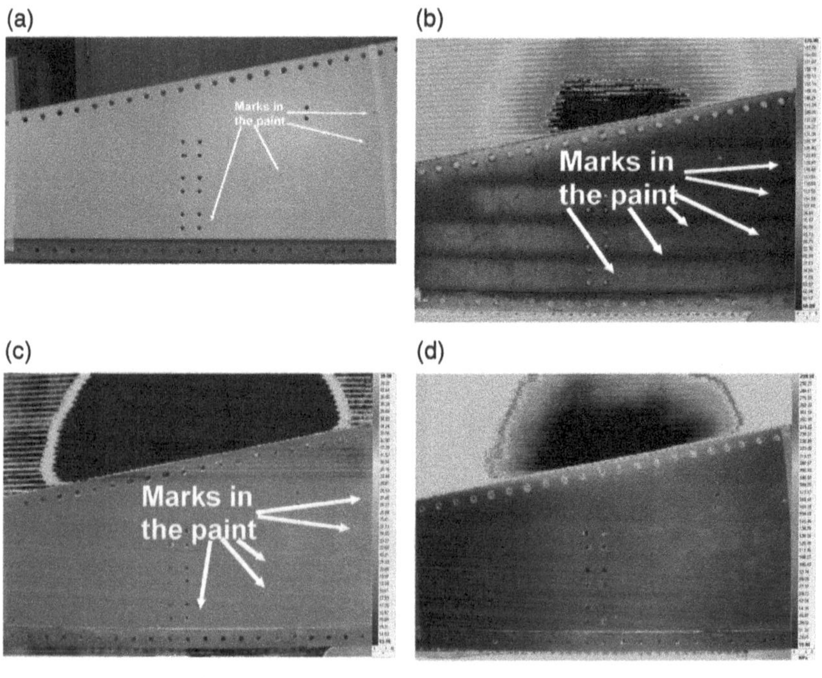

Figure 5.31 Location 4 on RHW Upper Panel #08 at 0.5 Hz with six lamps. (a) Digital camera image. (b) and (c) First and second harmonic phase images, respectively. (d) An amplitude image.

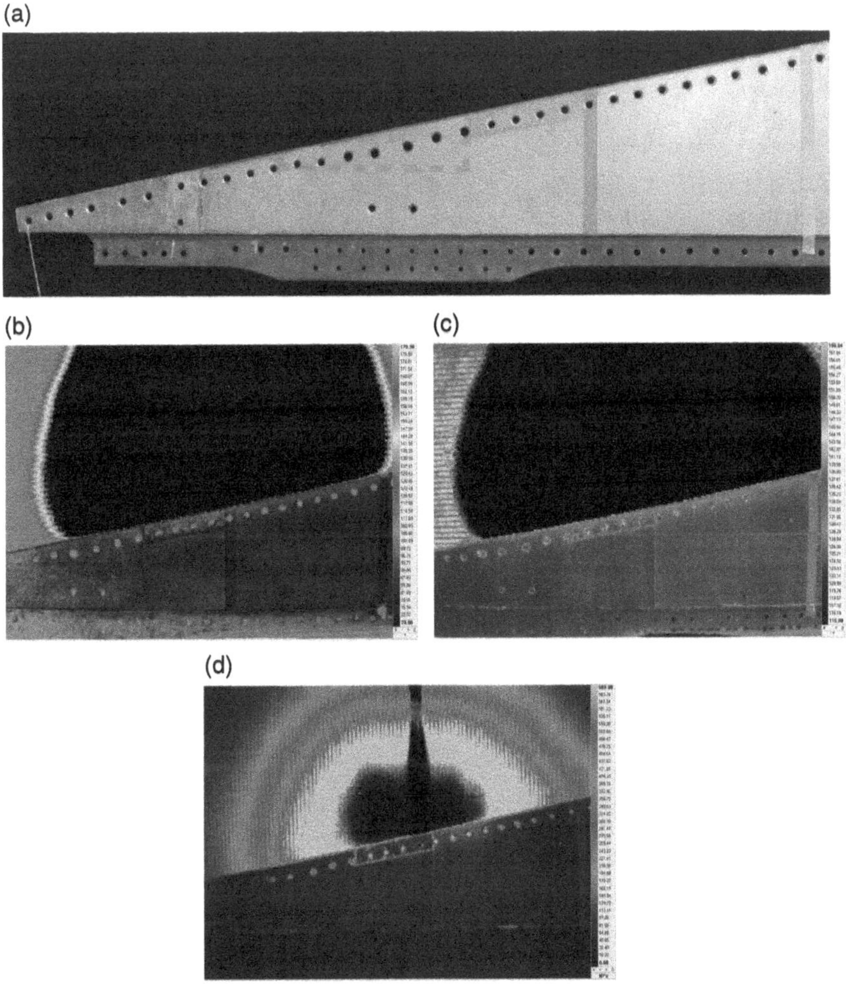

Figure 5.32 Location 5 on RHW Upper Panel #08 at 0.5 Hz with six lamps. (a) Digital camera image. (b) and (c) First and second harmonic phase images, respectively. (d) An amplitude image.

5.4.4 Paint Degradation Test of RHW Upper Wing Panel #09 Locations 1 and 2

Figures 5.40 and 5.41 show the thermal images for the paint degradation tests on RH Wing Upper Panel #09 at locations 1 and 2. Scratches and marks in the paint shown in Figure 5.39a are also clearly visible in Figure 5.40b and c. There were minimal anomalies at location 2 (see Figure 5.41). Again we see that the images on Figures 5.40 and 5.41 show that the combination of first and second

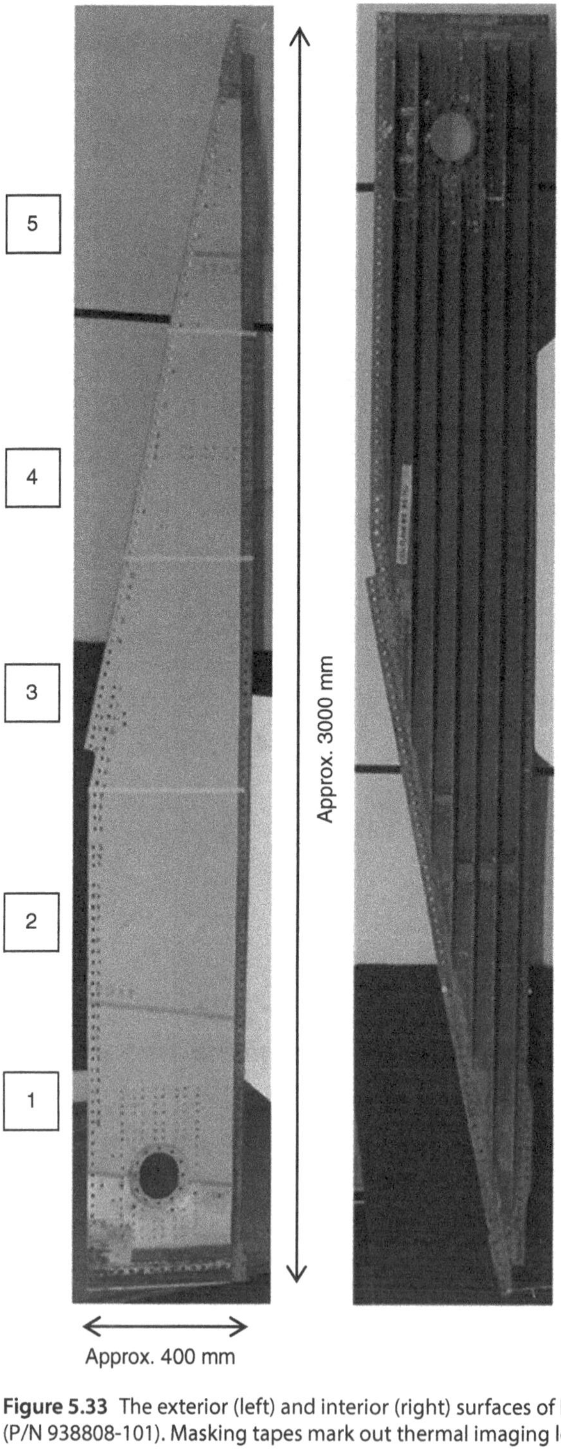

Figure 5.33 The exterior (left) and interior (right) surfaces of LHW Upper Wing Panel #08 (P/N 938808-101). Masking tapes mark out thermal imaging locations 1–5.

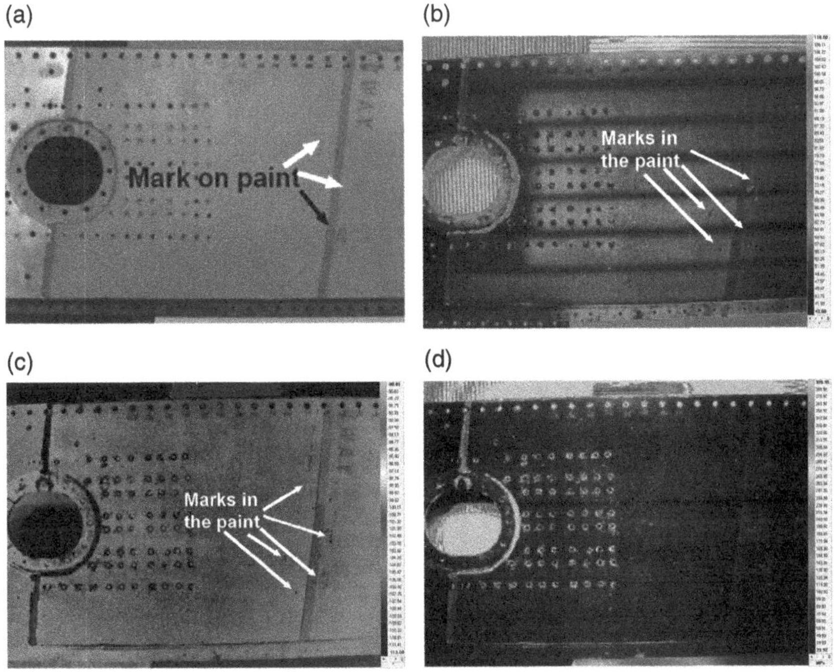

Figure 5.34 Location 1 on LHW Upper Panel #08 at 0.5 Hz with six lamps. (a) Digital camera image. (b) and (c) First and second harmonic phase images, respectively. (d) An amplitude image.

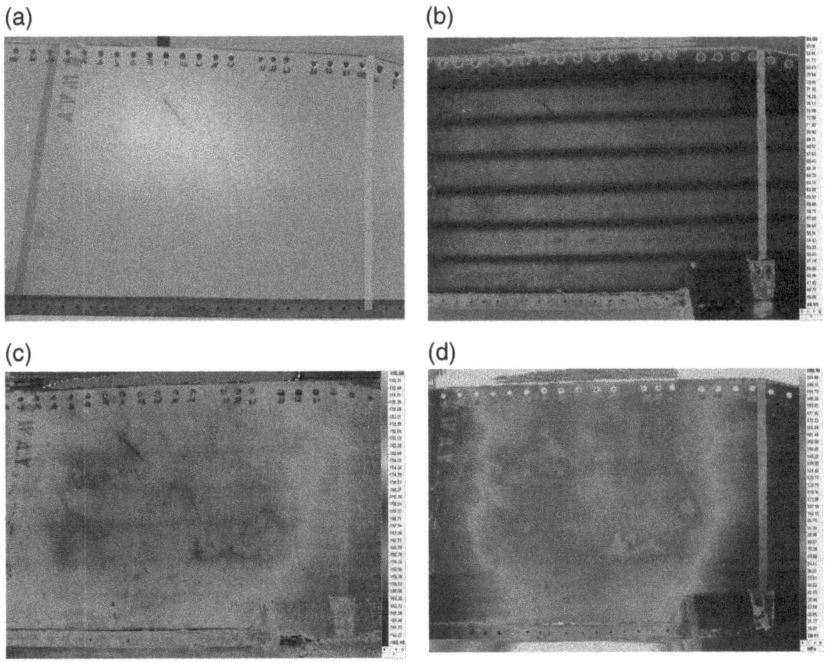

Figure 5.35 Location 2 on LHW Upper Panel #08 at 0.5 Hz with six lamps. (a) Digital camera image. (b) and (c) are first and second harmonic phase images, respectively. (d) An amplitude image.

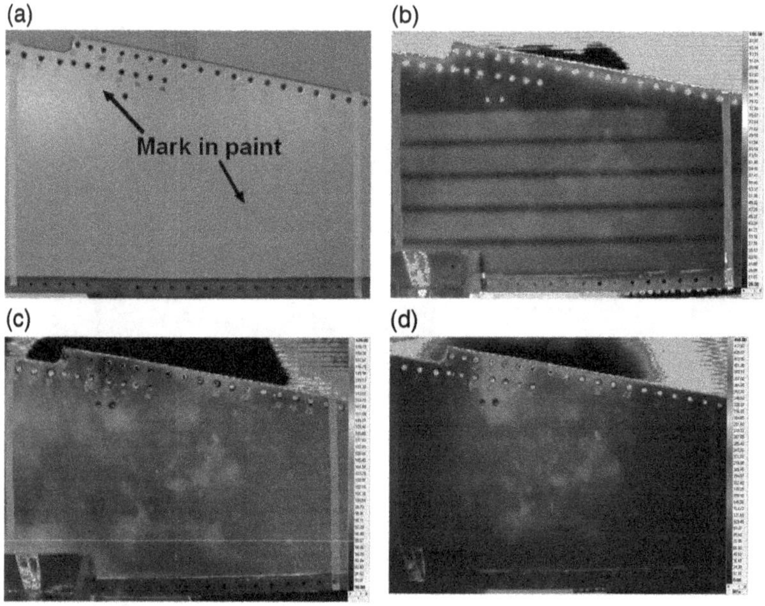

Figure 5.36 Location 3 on LHW Upper Panel #08 at 0.5 Hz with six lamps. (a) Digital camera image. (b) and (c) First and second harmonic phase images, respectively. (d) An amplitude image.

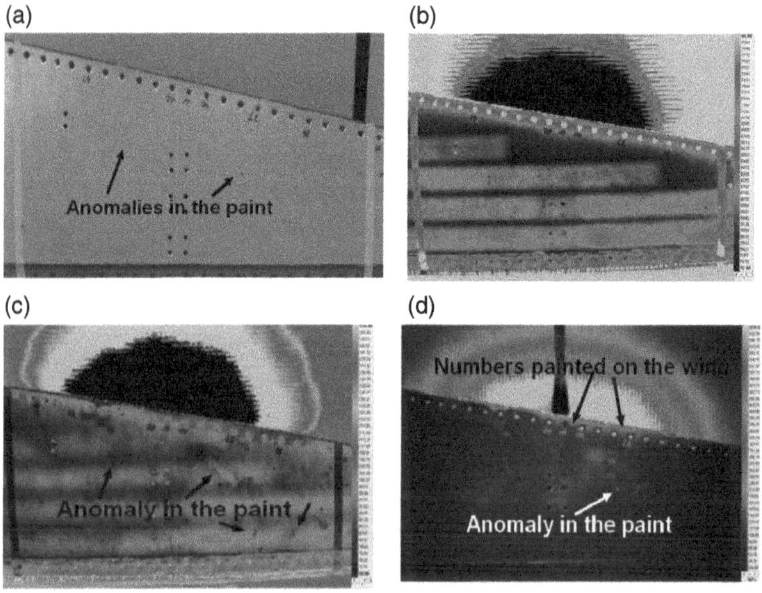

Figure 5.37 Location 4 on LHW Upper Panel #08 at 0.5 Hz with six lamps. (a) Digital camera image. (b) and (c) First and second harmonic phase images, respectively. (d) An amplitude image.

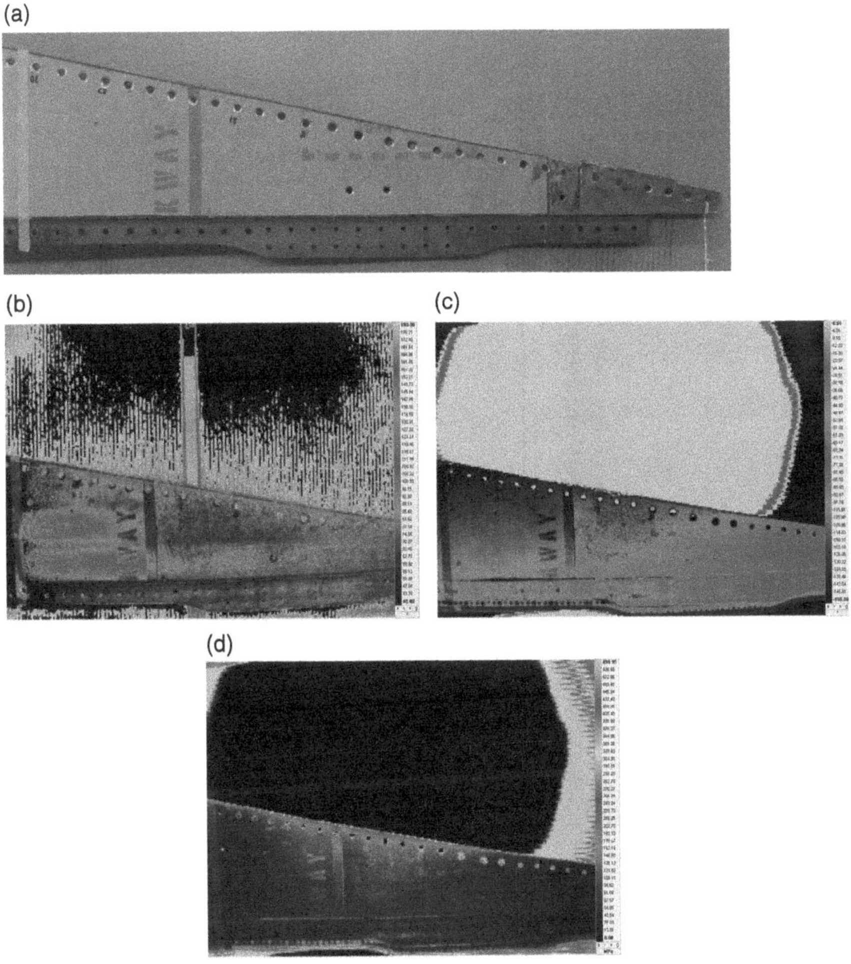

Figure 5.38 Location 5 on LHW Upper Panel #08 at 0.5 Hz with six lamps. (a) Digital camera image. (b) and (c) First and second harmonic phase images, respectively. (d) An amplitude image.

harmonics and the amplitude images gives improved information on the anomalies on the painted wing surface compared with an image sourced from first harmonic signal only.

5.4.5 Discussion

These results presented in Section 5.4 and the NDI of the AP-3C wing described in Section 5.3 show that the first and second harmonic phase images generally yield similar information. The first harmonic phase image is often more useful.

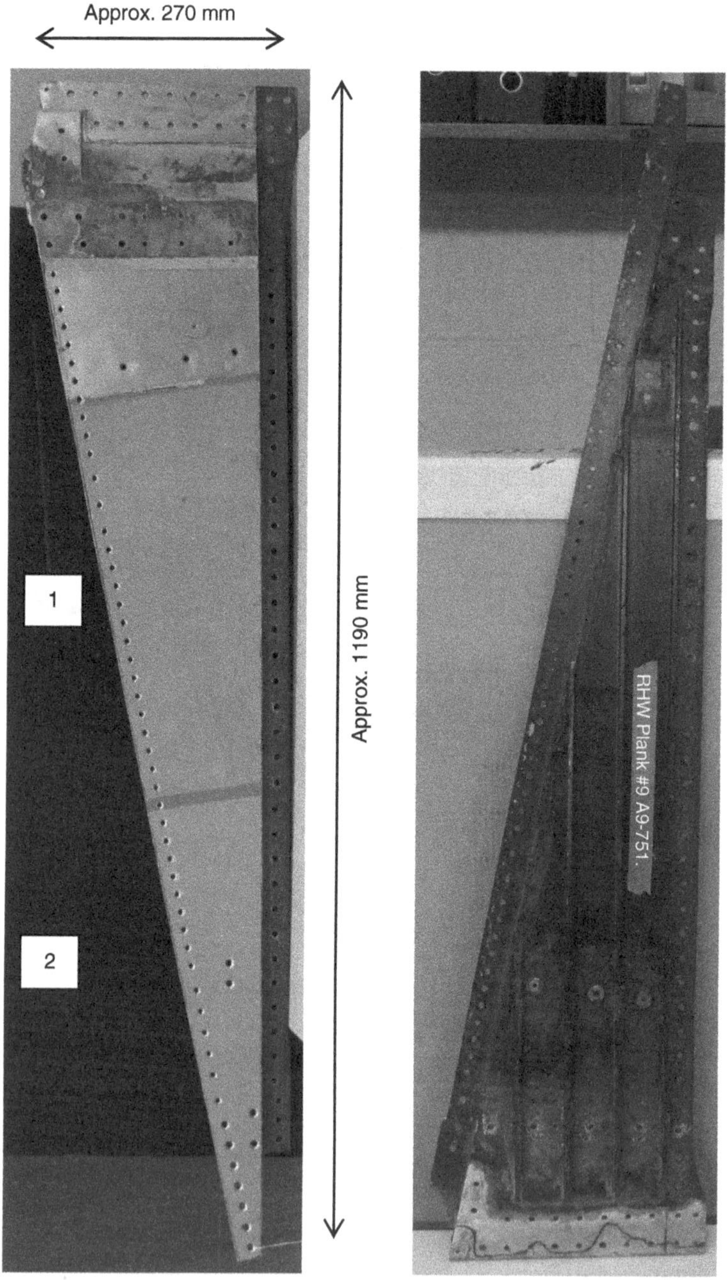

Figure 5.39 The exterior (left) and interior (right) surfaces of RHW Upper Wing Panel #09 (P/N 938809-102). Thermal imaging locations 1 and 2 are as displayed.

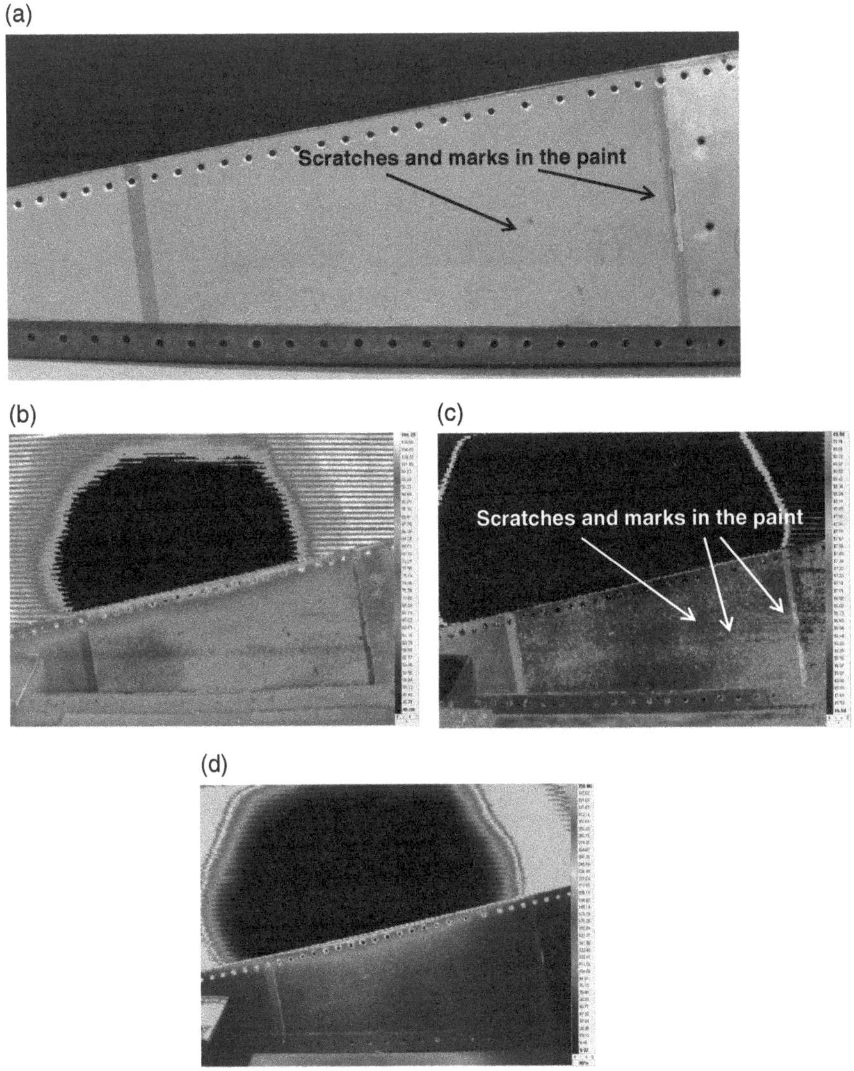

Figure 5.40 Location 1 on RHW Upper Panel #09 at 0.5 Hz with six lamps. (a) Digital camera image. (b) and (c) First and second harmonic phase images, respectively. (d) An amplitude image.

However, there are instances when this is not true, and hence it is recommended that both frequency responses, i.e. both the first and second harmonics, be measured.

The amplitude images were sometimes diffused due to nonuniform heating on the surface of the wing panels. This effect impacted the temperature contrasts seen in the images and made distinction of surface and paint degradation

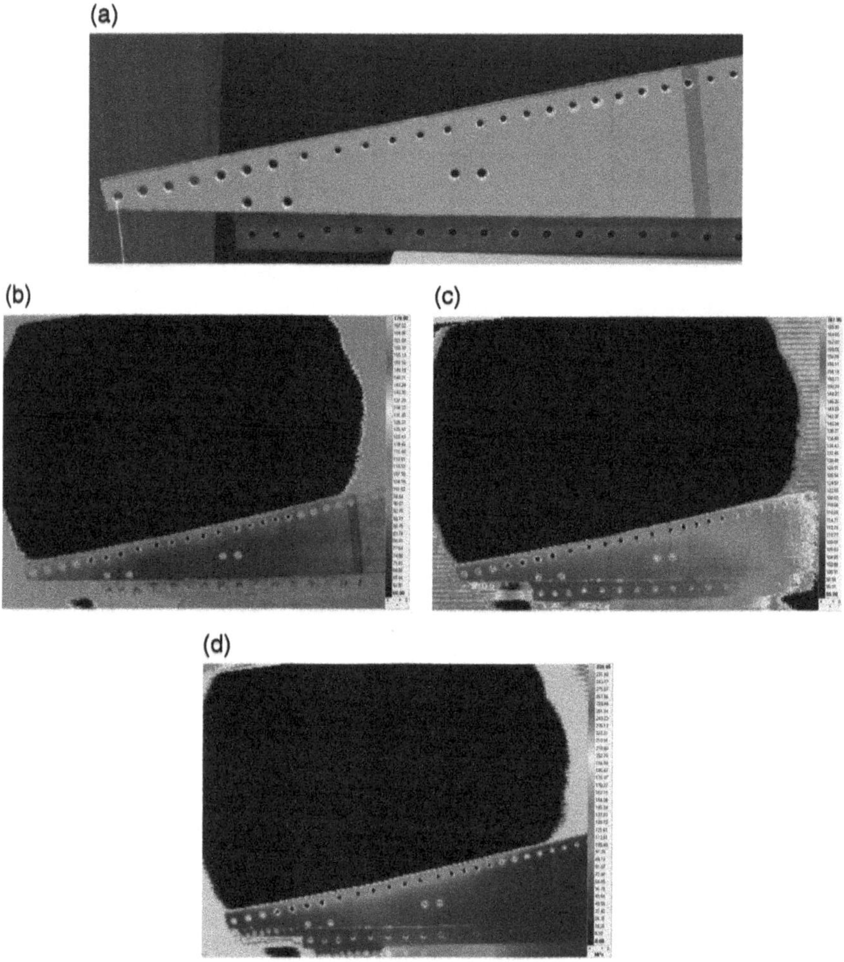

Figure 5.41 Location 2 on RHW Upper Panel #09 at 0.5 Hz with six lamps. (a) Digital camera image. (b) and (c) First and second harmonic phase images, respectively. (d) An amplitude image.

issues difficult. As such the amplitude images were less effective. Nevertheless, in most cases, the amplitude images were useful in confirming the results obtained by the phase images as well as by visual inspections.

5.5 Conclusion

IR thermography for detecting corrosion in both military and civil aircraft [33–35] is currently in its infancy. This paper has shown that by using a combination of the first and second harmonics of the measured temperature field, lock-in IR thermography NDI can be used to detect incipient corrosion and damage to paint associated with operational aircraft without the need to enhance (paint) the surface of the wing. Although the present study has focused on a RAAF AP-3C aircraft, the findings should be applicable to a wide range of metal aircraft structures.

Acknowledgment

The authors would like to thank Dr. K. Krishna Pillai and Dr. B. Hinton for their assistance in this work and DSTO for allowing access to the RAAF AP-3C wing.

Disclaimer

The opinions and thoughts expressed in this report are solely by the authors. The RAAF DGTA as an establishment/agency assumes no liability for the contents or use thereof. Trade or manufacturer's names appear herein solely because they are considered essential to the objective of this report. This document does not constitute DGTA certification policy. Consult your local aircraft certification office to its use.

References

1 Wu, D., Zenzinger, G., Karpen, W., and Busse, G. (1996). Nondestructive inspection of turbine blades with lock-in thermography. *Materials Science Forum* 210–213: 289.
2 Wu, D., Salerno, A., Malter, U. et al. (1997). Inspection of aircraft structural components using lockin-thermography. In: *Quantitative Infrared Thermography, QIRT 96* (ed. G. Busse, D. Balageas and G.M. Carlomagno), 251–256. Pisa: Edizione ETS.
3 Avdelidis, N.P. and Almond, D.P. (2004). Through skin sensing assessment of aircraft structures using pulsed thermography. *NDT&E International* 37: 353–359.
4 Del Grande, N.K. (1993). Dual-band infrared imaging for quantitative corrosion detection in aging aircraft. *American Society for Nondestructive Testing Fall Conference*, Long Beach, California (9–11 November 1993).
5 Marinetti, S. and Vavilov, V. (2010). IR thermographic detection and characterization of hidden corrosion in metals: general analysis. *Corrosion Science* 52: 865–872.
6 Forsyth, D.S. (2011). Non-destructive testing for corrosion, AGARD AG-AVT-140. In: *Corrosion Fatigue and Environmentally Assisted Cracking in Aging Military Vehicles, AC/323(AVT-140)TP/346*, 21-1–21-11.

7 Schonberger, A., Virtanen, S., Giese, V. et al. (2010). Non-destructive detection of corrosion applied to steel and galvanized steel coated with organic paints by the pulsed phase thermography. *Materials and Corrosion* 61: 1–5.
8 Rajic, N. (2000). Quantitative approach to active thermographic inspection for material loss evaluation in metallic structures. *Research in Nondestructive Evaluation* 12: 119.
9 Vavilov, V.P. (2003). Non-contact one-sided evaluation of hidden corrosion in metallic constructions by using transient infrared thermography. *Revista de Metalurgia (Madrid)* 235.
10 Thomas, R.L., Favro, L.D., and Kuo, P.K. (1985). Thermal-wave imaging for nondestructive evaluation. *Canadian Journal of Physics* 64: 1234.
11 Rosencwaig, A. (1982). Nondestructive evaluation with thermal waves. *Journal of Photoacoustics* 1: 371–386.
12 Busse, G. (1980). Photothermal transmission probing of a metal. *Infrared Physics* 20: 419–422.
13 Luukkala, M., Lehto, A., Jaarinen, J., and Jokinen, M. (1982). *Photothermal Imaging and Thermal Surface Waves as a NDT Tool for Coatings*, 591–594. San Diego, CA: IEEE.
14 Almond, D.P., Patel, P.M., Pickup, I.M., and Reiter, H. (1985). Evaluation of the suitability of thermal wave interferometry for the testing of plasma sprayed coatings. *NDT International* 18: 17–24.
15 Almond, D.P., Patel, P.M., Morris, J.D., and Reiter, H. (1988). Thermal wave inspection employing the infrared detection technique. *Proceedings of the SPIE – The International Society for Optical Engineering* 917: 84–90.
16 Miller, R.M. (1986). Thermal wave imaging. *Proceedings of the SPIE – The International Society for Optical Engineering* 685: 16–22.
17 Busse, G., Wu, D., and Karpen, W. (1992). Thermal wave imaging with phase sensitive modulated thermography. *Journal of Applied Physics* 71: 3962–3965.
18 Meola, C., Carlomagno, G.M., Squillace, A., and Giorleo, G. (2002). Non-destructive control of industrial materials by means of lock-in thermography. *Measurement Science and Technology* 13: 1583.
19 Meola, C., Carlomagno, G.M., and Giorleo, L. (2004). Geometrical limitations to detection of defects in composites by means of infrared thermography. *Journal of Nondestructive Evaluation* 23: 125.
20 Maldague, X. (2002). Introduction to NDT by active infrared thermography. *Materials Evaluation* 6: 1060–1073.
21 Bai, W. and Wong, B.S. (2001). Photothermal models for lock-in thermographic evaluation of plates with finite thickness under convection conditions. *Journal of Applied Physics* 89: 3275–3282.
22 Cramer, K.E., Howell, P.A., and Syed, H.I. (1995). Quantitative thermal imaging of aircraft structures. *Proceedings SPIE, Thermosense XVII* 2473: 226–232.

23 Huth, S., Breitenstein, O., Huber, A. et al. (2002). Lock-in IR-thermography – a novel tool for material and device characterization. *Solid State Phenomena* 82–84: 741–746.
24 Jones, R., Krishnapillai, M., Cairns, K., and Matthews, N. (2010). Application of infrared thermography to study crack growth and fatigue life extension procedures. *Fatigue and Fracture of Engineering Materials and Structures* 33 (12): 871–884.
25 Jönsson, M., Rendahl, B., and Annergren, I. (2010). The use of infrared thermography in the corrosion science area. *Materials and Corrosion* 61: 961–965. https://doi.org/10.1002/maco.200905525.
26 Jones, R. and Pitt, S. (2006). An experimental evaluation of crack face energy dissipation. *International Journal of Fatigue* 28: 1716–1724.
27 Krishnapillai, M., Jones, R., Marshall, I.H. et al. (2005). Thermography as a tool for damage assessment. *Composite Structures* 67 (2): 149.
28 Wallbrink, C., Wade, S.A., and Jones, R. (2007). The effect of size on the quantitative estimation of defect depth in steel structures using lock-in thermography. *Journal of Applied Physics* 101 (10): 104907.
29 Zalameda, J.N. (2001). Synchronized electronic shutter system (SESS) for thermal nondestructive evaluation. *Proceedings SPIE, Thermosense XXIII* 4360: 616. https://doi.org/10.1117/12.421044.
30 Krishnapillai, M., Jones, R., Marshall, I.H. et al. (2006). NDTE using pulse thermography: numerical modeling of composite subsurface defects. *Composite Structures* 75: 241–249.
31 Paynter, R.J.H. and Dutton, A.G. (2003). The use of a second harmonic correlation to detect damage in composite structures using thermoelastic stress measurements. *Strain* 39: 73–78.
32 Matzkanin, G.A. and Yolken, T.H. Nondestructive evaluation techniques for detecting hidden corrosion. *The AMMTIAC Quarterly* 2 (2): 2–6.
33 John, D., Weir, J.D., and DiMarzio, D. (2004). Non-destructive testing of corrosion under coatings, strategic environmental research & development program. *Final Report, SERDP Pollution Prevention Project 1137* (September 2004). https://apps.dtic.mil/dtic/tr/fulltext/u2/a439080.pdf.
34 Han, X., Favro, L.D., Li, L. et al. (2001). Quantitative thermal wave corrosion measurements on a DC-9 belly skin in the presence of irregular paint thickness variations. *Review of Progress in Quantitative Nondestructive Evaluation* 20A: 483–486.
35 Corrosion Detection Technologies. *Final Report, Prepared for the North American Technology and Industrial Base Organization.* https://apps.dtic.mil/dtic/tr/fulltext/u2/a370959.pdf.
36 Wong, A., Jones, R., and Sparrow, J.G. (1988). Thermoelastic constant or thermoelastic parameter. *Journal of Physics and Chemistry of Solids* 48: 749–753.

6

Electrochemical Impedance Spectroscopy for Nondestructive Evaluation of Corrosion Processes

V.S. Raja

Department of Metallurgical Engineering and Materials Science, Indian Institute of Technology Bombay, Mumbai, India

6.1 Introduction

Nondestructive testing of corrosion of metals/alloys is of interest for two types of studies: (i) estimating corrosion rate/damage of metals, determination of coating life, and the corrosiveness of the environment and (ii) the elucidation of corrosion mechanism(s). In many cases, both these aspects of corrosion are time dependent in nature. So, a chosen technique must not interfere with the normal course of change as imposed by external variables such as environmental conditions – including temperature, pressure, and chemistry changes – in an industrial process or due to the changes that might occur within a structural component, such as changing corrosion tendency of a metal and aging of coating. For example, one might wish to determine, the corrosion rate of a pipeline carrying crude oil online real time, when the process variables such as crude chemistry, pumping rate, temperature, and inhibitor chemistry are expected to change intentionally or otherwise in order to take immediate remedial action.

Several simple nondestructive techniques (NDT) such as electrical resistance (ER), linear polarization resistance (LPR), and ultrasound (US) are conveniently employed to estimate the corrosion rate of pipeline to evaluate the process conditions, including the corrosion control measures. Such techniques, however, lack the ability to resolve such questions as why the system under study exhibits a particular corrosion behavior; to answer this question it is necessary to know the corrosion mechanism(s) responsible for such a change. Electrochemical impedance spectroscopy (EIS) can provide answers to many such questions. More often, EIS is employed as a laboratory technique to study diverse corrosion problems: coatings, inhibitors, passivity, and microbial

Non-Destructive Evaluation of Corrosion and Corrosion-assisted Cracking, First Edition.
Edited by Raman Singh, Baldev Raj, U. Kamachi Mudali and Prabhakar Singh.

corrosion, to name a few. This chapter outlines the application of EIS to determine corrosion rate as well as to investigate corrosion mechanisms from the perspective of it being employed as an NDT. Theoretical considerations of AC response of various systems and formalisms are not discussed in this review. For detailed study of theory of EIS, readers can refer to other publications [1–3].

The application of AC impedance is much older than its use in corrosion science and engineering, and the technique was originally meant to analyze passive electrical circuits. When it was applied to study electrochemical corrosion, it was rechristened as electrochemical impedance spectroscopy. In many ways, EIS can be a more powerful tool in the analysis of corrosion behavior of metals and alloys than conventional DC polarization techniques. However, unlike DC methods such as Tafel extrapolation, some prior knowledge of corroding systems is essential for meaningful analysis of EIS test data. Therefore, some of the basic aspects of electrochemical interface representing some typical corroding systems are briefly outlined before discussing the application of EIS as an NDT.

6.2 Structure of Electrochemical Interfaces

When a metallic phase comes into contact with a corrosive medium, charge separation occurs at the interface region of the metal and the medium, leading to the formation of a capacitor, termed double layer capacitor (Figure 6.1a). Since corrosion is a DC process occurring as a result of charge transfer across this double layer, it would not be possible should the double layer behave like a pure capacitor. In order to account for the flow of charges across the so-called double layer, the double layer is considered to consist of a capacitor and a resistor arranged parallel to each other. Accordingly, an electrical equivalent of corrosion reaction can be given by the circuit shown in Figure 6.1b. Such an arrangement of electrical elements satisfactorily explains electrochemical phenomena such as corrosion and electrodeposition.

In general, overall corrosion reaction involves one more step, in addition to the charge transfer occurring at the electrode/electrolyte interface. That is, migration of reactants (corrosive species) from the bulk (of the environment) to the interface and reaction (corrosion) products away from the interface to the bulk is essential for a sustained corrosion process. As the charge transfer and migration (of reactants and products) occur in series and are coupled, the kinetics of the slower of the two steps controls the overall reaction kinetics. In order to include the transport of the species in the overall corrosion process, another component called Warburg impedance, named after the scientist who formulated a mathematical relation, needs to be included. The modified equivalent circuit considering the diffusion phenomena is shown in Figure 6.1c.

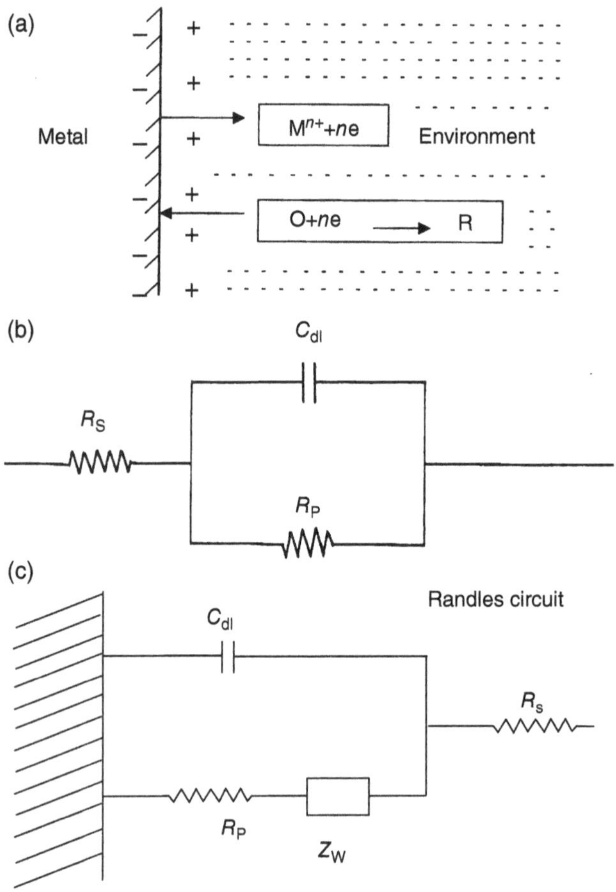

Figure 6.1 (a) Schematic of electrochemical interface showing charge separation at the metal/solution interface and the corrosion reactions occurring along the interface. (b) Electrical equivalent circuit representing (a) and (c) Polarizable interface with mass transport limitations. (M^{n+}, O, R, C_{dl}, R_p, Z_W, R_s, respectively, represent metallic ions having n valancy, oxidised species, reduced species, double layer capacitance, polarization resistance, Warburg impedance, and solution resistance. Note that R_s corresponds to resistance between the electrochemical interface and the tip of a reference electrode used to measure the potential, not shown in the cartoon of (a)).

A variation from the above description of electrical equivalent of a corroding metal occurs when a corroding metal surface is covered with oxides or passive films or protective organic coatings (Figure 6.2a). In such cases species have to surmount such barriers before they attack metals. Considering the fact that current must flow through the barrier films and the barrier films too can exhibit resistive (through which DC current flows) and capacitive components, an electrical circuit of the kind as shown in Figure 6.2b can be used to represent metals exhibiting passivity or coated with organic coating. Nyquist and Bode

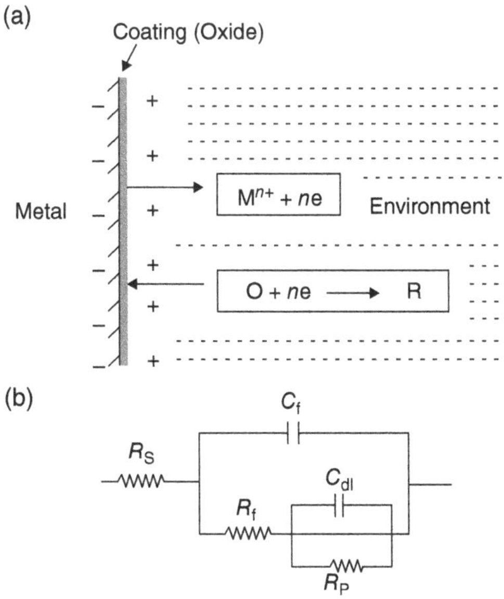

Figure 6.2 (a) Schematic of electrochemical interface showing metal covered with coating/oxide, charge separation at the coating (oxide)/solution interface, and the corrosion reactions occurring across them. (b) Equivalent circuit for coated metal/environment system; R_f, C_f, C_{dl}, and R_p, respectively, represent coating/oxide resistance, coating capacitance, double layer capacitance, and polarization resistance.

representation of the data can be seen in other publications [4, 5]. Needless to emphasize the fact that the above equivalent circuit corrosponding to the coating/oxide requires modification should the physical structure of the oxide(s)/coatings change. For example, there can be multilayered oxides, each having different properties (capacitor and resistor values) requiring different equivalent circuits. A few such variations will be shown later, but the readers need to get familiarized with such modifications for an effective analysis of EIS data. The idea behind constructing interface models truly representing the physical structure of the corroding surface is to analyze the role of each of the constituents of the surface either in impeding or aiding metallic corrosion. In reality it may prove to be difficult to satisfactorily fit EIS data to these models due to several limitations of the EIS technique.

6.3 Data Analysis

The main objective of data analysis is to extract elements of equivalent circuits such as those described in Figures 6.1 and 6.2. Analysis of EIS test data is done either through graphical analysis [6] or fitting to some preassigned electrical equivalent circuits. In fitting EIS data to various electrical equivalent circuits, caution must be exercised of the fact that a given data can be satisfactorily fit to several different equivalent circuits, and it is imperative that the chosen electrical equivalent circuit must truly represent physical phenomena of the concerned electrochemical corrosion interface to avoid frivolous circuits. In employing graphical method of analysis, it is possible that the spectra is not adequately resolved to extract various elements of an equivalent circuit. This could happen when the time constants (τ) given by RC of the two processes do not differ by a factor of 100. In such cases, data should be fit using nonlinear least square method for which several commercial software programs are available. Detailed description of the procedures for data analysis can be obtained in the published literature [1, 6]. It should be pointed out that the elements of equivalent circuits are the expressions of the characters of the electrochemical corroding system.

6.4 Estimation of Corrosion Rate

Corrosion rate of metals is estimated through determination of R_p ($R_p = |Z|_{\text{at } \omega = 0}$) and equating R_p to corrosion current density, i_{corr}, through Stern–Geary relation:

$$R_p = \left(\frac{\beta_a \beta_c}{\beta_a + \beta_c}\right)\frac{1}{2.303 i_{corr}} \tag{6.1}$$

The first term as shown in the parenthesis of the right side of the equation is constant and is called Stern–Geary coefficient (B), which depends on the values of anodic (β_a) and cathodic (β_c) Tafel slopes. Accordingly, the Eq. (6.1) becomes

$$i_{corr} = \frac{B}{2.303 R_p} \tag{6.2}$$

Though R_p can be obtained by a much simpler way by another nondestructive technique, namely, LPR, the advantage of using EIS over LPR lies in the fact that the measured R_p using LPR contains the solution resistance (R_s) (a resistance between the working and reference electrodes, which is not connected to the corrosivity of metals/alloys) (see Figure 6.1c), while the measured R_p value using EIS is free from that. Because the measured R_p value of LPR technique contains R_s, the estimated corrosion rate by LPR technique can become considerably lower than the actual value (see Eq. 6.2) especially when the R_s is significant. This factor becomes important when corrosion rates need to be measured in poorly conducting corrosive environments as the poorly conducting media exhibit high R_s.

Another factor that needs consideration in relating R_p to i_{corr} is the mechanism of corrosion. Depending on whether the rate of corrosion reaction is anodically (example passive metal) or cathodically controlled (diffusion controlled), the following relations are applied to determine i_{corr}:

$$i_{corr} = \frac{\beta_c}{R_p}(\text{anodically controlled}) \tag{6.3}$$

$$i_{corr} = \frac{\beta_a}{R_p}(\text{cathodically controlled}) \tag{6.4}$$

Usually Tafel slopes are not determined, and a value of 120 mV decade^{-1} is assumed for both the anodic and cathodic Tafel slopes resulting in a value of 20 for Stern–Geary coefficient (B), which is often used for estimating i_{corr}. However, in practice this is not true, and deviation from the assumed values does occur, causing error in the determined i_{corr}.

The uncertainty in the interpretation of EIS data is yet another problem in proper evaluation of i_{corr}. Extraction of R_p from electrochemical impedance spectra exhibiting two semicircles, as well as inductive loops, to estimate corrosion rate has been met with problems. Epelboin et al. [7] have shown that the use of charge transfer resistance (R_{ct}) rather than R_p is more appropriate to determine corrosion rate. However, later, Lorenz and Mansfeld [8] have shown that it is not a correct method to determine i_{corr}. They showed that the use of R_p better reflects true i_{corr} of the system, namely, Fe in 0.5 M H_2SO_4 containing triphenyl benzyl phosphonium chloride, than its R_{ct}. This leads to the conclusion that some verification of the kind such as weight loss measurement is essential for unambiguous use of R_p or R_{ct} in i_{corr} determination.

6.5 Evaluation of Organic Coating

Organic coatings can be evaluated for their:

1) Overall ability to protect substrates against corrosion.
2) Water uptake.
3) Fraction of delaminated area due to corrosion.
4) Kinetic parameters, such as diffusion coefficients of species responsible for corrosive attack.

Coating capacitance, pore resistance, double layer capacitance, and polarization resistance are some of the most important parameters that are useful to evaluate the above factors affecting coating performance.

Coating resistance (R_f) also called as pore resistance (R_{po}) is a measure of the capacity of a coating to protect metal against corrosion. This is a useful

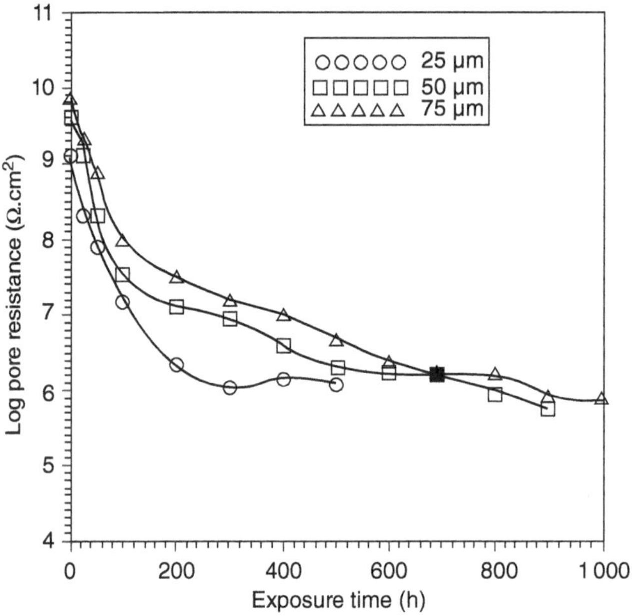

Figure 6.3 Variation in pore resistance with time of exposure to salt spray for pigmented alkyd system on a phosphated mild steel. *Source*: Raja et al. [9]. Reproduced with permission of Elsevier.

parameter determined for assessing performance of organic coatings. Figure 6.3 is an illustration of this fact for an alkyd coating [9]. The sum of film resistance and polarization resistance is a measure of the durability of coating/metal system against corrosion. Though R_p should ideally depend on the metal/electrolyte interface, meaning it is a reflection of inherent tendency of a corroding metal, the poor performance of a coating, in the sense that the migration of the environment through the coating, can substantially affect R_p.

The fundamental property of a coating, namely, dielectric constant, can be significantly altered due to absorption of constituents of corrosive environments such as water and various ions:

$$C_f = \frac{\varepsilon_f \varepsilon_o A_f}{d_f} \tag{6.5}$$

where ε_f, A_f, and d_f, respectively, represent dielectric constant, area, and thickness of the exposed coating, ε_o the permittivity of free space ($8.84 \times 10^{-12}\,\mathrm{F\,m^{-1}}$), and C_f coating capacitance.

Accordingly, the volume fraction of water, V_w, absorbed by the coating with time can be estimated *in situ* by measuring the coating capacitance nondestructively through the following equation:

$$V_w = \log\left[C_{(t)} / C_{(o)}\right] / \log 80 \tag{6.6}$$

where $C_{(t)}$ and $C_{(o)}$, respectively, represent coating capacitance measured after exposure of time period, t, and at the beginning of exposure. Figure 6.4 shows variation of water content with time of thermally sprayed low density polyethylene coating [9–13]. Correlation in water content determined through EIS has been found to correlate well with that determined using gravimetric studies [14].

De Rosa et al. [15] used changes in capacitance, during early stages of coating exposure, as a parameter to evaluate water absorption by organic coatings. The authors determined the diffusion coefficients based on the following equation:

$$\frac{\ln\left(C_{(t)}\right) - \ln\left(C_d\right)}{\ln\left(C_\infty\right) - \ln\left(C_d\right)} = \frac{4}{\sqrt{\pi}} \frac{\sqrt{tD}}{2d_f} \tag{6.7}$$

where C_d, $C_{(t)}$, and C_∞ represent capacitance of the coating in dry state and after exposure for time t and ∞, respectively. D and d_f represent diffusion coefficient for water permeation and coating thickness, respectively. For a

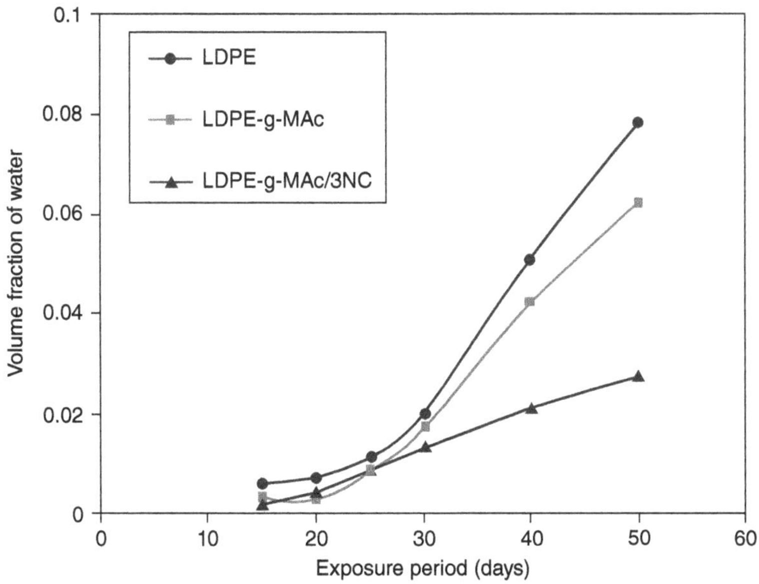

Figure 6.4 Volume fraction of water absorbed by different coatings. *Source:* David et al. [10]. Reproduced with permission of Elsevier.

large enough porosity in the coating, these authors also derived an equation (Eq. 6.8) to correlate the porosity of the coating to differential capacitance:

$$\frac{d\left(\ln C_{(t)}\right)}{d\sqrt{t}} = \frac{4}{\sqrt{\pi}}\sqrt{\ln\frac{C_{\infty}}{C_{d}}}\frac{\sqrt{D_{w}\theta}}{2d_{f}} \tag{6.8}$$

Some examples of applications of EIS to study the performance of the coatings can be found in the literature [16–18]. In the above equation, θ represents the coating porosity defined as $\theta = V_p/V_c$, where V_p and V_c are the volume of pores and the nominal volume of the coating, respectively, and D_w is the diffusion coefficient of water vapor in gases ($D_w = 10^{-1}$ cm^2 s^{-1} at room temperature). Plots of capacitance logarithm, ln(C) versus $\sqrt{t}$, and phase angle, ϕ, against $\sqrt{t}$ for the early stages of exposure (<400 s) provide information about (i) Fickian-like sorption, (ii) localized corrosion behavior, and (iii) multistage sorption of coating. The schematic diagram (Figure 6.5) showing a raise in ln(C) during early stages of exposure is an illustration of a Fickian-like behavior. In such a case, the phase angle remains close to 90°.

Coating properties are also evaluated by subjecting them to different relative humidity (RH) values [19, 20]. These authors use a parameter called comparative diffusion coefficient, D^*, based on the following relationship:

$$D^* = \frac{d(\text{capacitance})}{d(\text{time})}\left[RH_2(\%) - RH_1(\%)\right] \tag{6.9}$$

where RH_2 and RH_1 refer to two different humidity values to which a coating is exposed. These authors have found that D^* of the aged sample is higher than

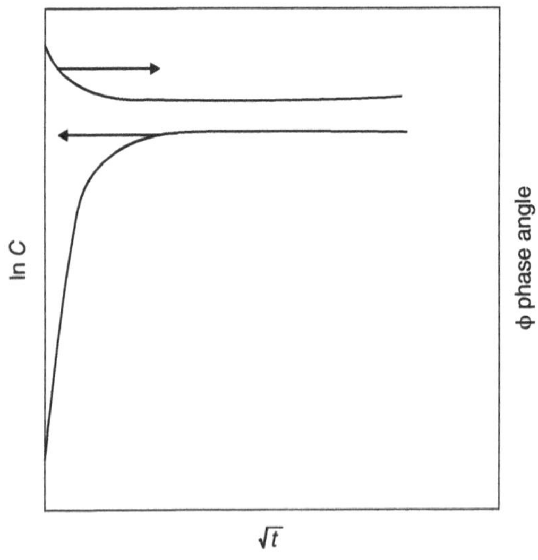

Figure 6.5 A schematic of capacitance logarithm, ln(C), and impedance phase angle, ϕ, as a function of $\sqrt{t}$ for a paint coating exposed to a corrosive liquid.

the fresh sample, and they attributed this difference to the plasticizing effect of water and polymer. The permeability coefficient can be further calculated by multiplying water solubility (S in kg m^{-3}) with D^*.

Yet another method of evaluating coating performance is through break-point frequency (f_b) [21–23]. The break-point frequency (the frequency corresponding to the phase angle of 45°) unlike R_f and C_f, as shown by the following equation, is independent of the coating thickness.

$$f_b = \frac{1}{2\pi\varepsilon\varepsilon_o\rho} \tag{6.10}$$

where ε and ρ represent dielectric constant of the coating and specific resistance of the coating, respectively. As shown in Figure 6.6 the f_b increases with time of exposure [9]. It needs to be pointed out that only the high frequency region of the Bode plot that should be employed to measure break-point frequency and not the one that appears at low frequency that corresponds to dissolution process. Scully [24] used the break-point frequency as a parameter to provide numerical rating for the protection behavior of the coating, so that the same can be compared with the ASTM visual rating system:

$$\text{Numerical rating} = 10 - 2\log f_{45} \tag{6.11}$$

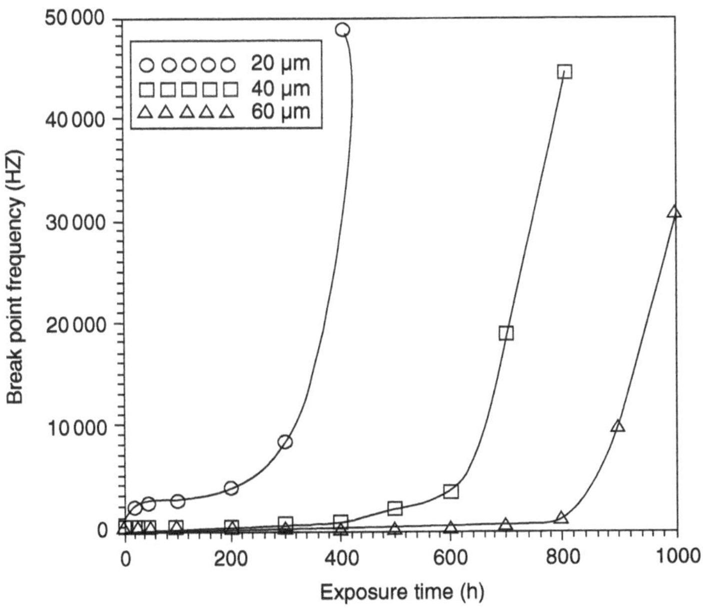

Figure 6.6 Variation of break-point frequency with exposure time to salt spray for unpigmented alkyd system on mild steel phosphated with phosphating (I) formulation. *Source:* Raja et al. [9]. Reproduced with permission of Elsevier.

Thus, break-point frequency has been found to be an important parameter in examining performance of organic coating toward corrosion protection.

Delaminated area (A_d) of coating has also been related to various electrical circuit elements through the following equations:

$$R_p = \frac{R_p^o}{A_d} \tag{6.12}$$

$$R_f = \frac{R_f^o}{A_d} \tag{6.13}$$

$$C_{dl} = C_{dl}^o A_d \tag{6.14}$$

$$\sigma_t = \frac{\sigma_t^o}{A_d} \tag{6.15}$$

The degree of delamination can also be related to break-point frequency through the following equation, when the delaminated area is small.

$$f_b = K\frac{A_d}{A_f} \tag{6.16}$$

where K is considered as the material property which is $1/2\pi\varepsilon\varepsilon_o\rho$.

It should be noted that, the material that is the coating, property does not change with time indicating ε is constant which is not the case as we noticed from Eq. (6.6)

Hack and Scully [21] indeed found the A_d calculated from EIS data to correlate with those determined using ASTM D610. One of the ways of estimating the protective behavior of the coating is to determine R_p and R_f values with exposure time.

A better measure of coating property that truly excludes possible changes in coating resistivity and electric constant has been suggested by Mansfeld [25]. This he did by normalizing f_b with f_{min} (minimum of frequency) as shown below

$$\frac{f_b}{f_{min}} = \left(\frac{C_{dl}}{C_f}\right)^{1/2} = \left(\frac{C_{dl}^o}{C_f^o}\right) D^{1/2} \tag{6.17}$$

can increase electrolyte intake and thereby lower R_p. It should be noted that $R_f + R_p$ (=R_t) represents *faradaic* process of corrosion phenomena.

EIS can be used to study complex coatings. Raja et al. [26] studied the role of graphite, added as dispersoid to promote electrical conductivity of an acrylic coating, on possible galvanic corrosion with AZ31 alloy, the substrate. A physical model for the composite coatings was first formulated to construct an equivalent circuit. The physical model, electrical equivalent circuit, and the variation of resistances of the circuit are shown in Figure 6.7. A steep decline in the resistance

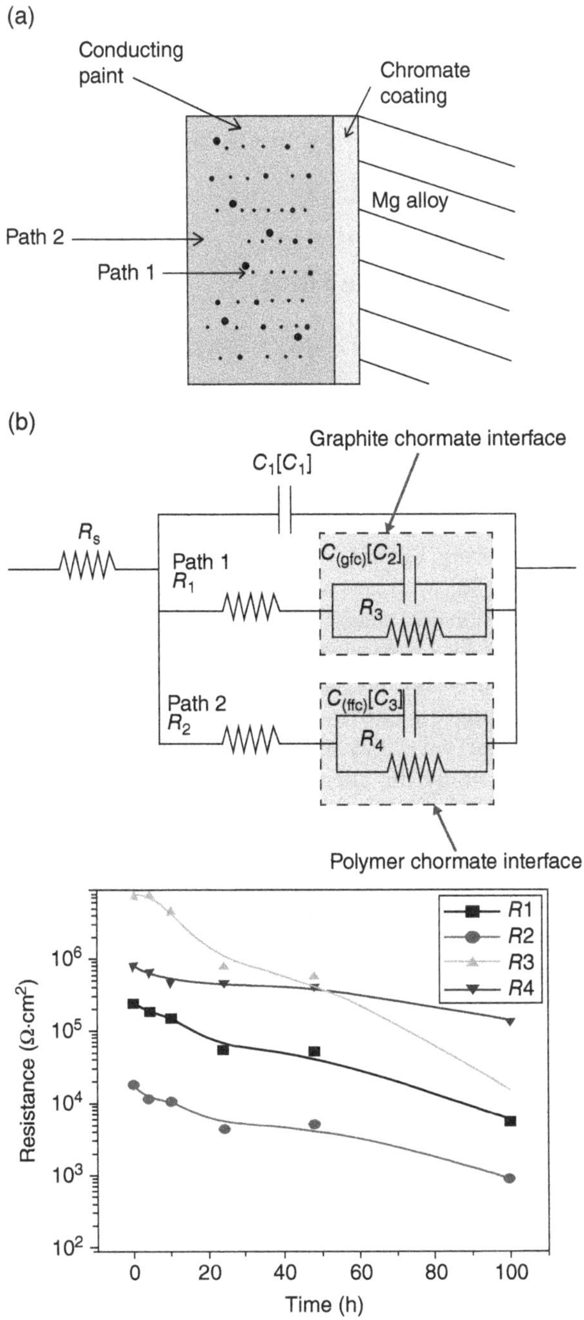

Figure 6.7 (a) A physical model for environmental/conducting coating/AZ31 magnesium alloy system. Paths 1 and 2 represent graphite particle-assisted current conduction and conduction by the polymer matrix, respectively. (b) Equivalent circuit model to fit the experimental data of acrylic coated AZ31 magnesium alloy exposed to 3.5 wt.% sodium chloride solution. Paths 1 and 2 represent graphite particle-assisted current conduction and conduction by the polymer matrix, respectively. R_s, R_3, and R_4 represent solution resistance, faradaic resistance across graphite/chromate, and coating/chromate interfaces, respectively. C_f, $C_{g/C}$, and $C_{f/C}$ correspond to capacitance of coating, graphite/chromate interface, and coating/chromate interface, respectively. (c) Variation of the coating resistances with exposure time is shown. *Source:* Raja et al. [26]. Reproduced with permission of Elsevier.

of electrochemical corrosion reaction associated with graphite-AZ31 galvanic coupling with time is shown in Figure 6.7c.

6.6 Monitoring Passivity and Oxide Films

6.6.1 Passivity

Unlike conventional organic coatings, passivation of metals/alloys is a dynamic process in which metal/alloys can self-repair the passive film, whenever the latter suffers damage. This happens through continuous supply of passiving element(s) by the alloy to the surface. Passivity is developed over a time period. Destruction of passive films can lead to localized corrosion such as pitting corrosion, crevice corrosion, and stress corrosion cracking. EIS has been employed to study mechanism of passivation of alloys [27–30]. In a review, Raja [4] has discussed the application of EIS for alloy dissolution and passivity.

For a given metal, localized attack on passive films could depend on the nature of the environment, exposure time, etc., all of which can be effectively monitored, without unduly affecting the corrosion tendency of metals using EIS. Time-resolved EIS has been carried out by several researchers. Saravanan et al. [31] used this technique to study the effects of plasma immersion ion implantation (PIII) of nitrogen on the passivation and pitting tendency of austenitic stainless steels in order to optimize implantation parameters. Figure 6.8a is shown to bring out the fact that phase angle and modulus of impedance of Bode plots can be used to distinguish the modifications occurred on the alloy surface due to various treatment parameters. The difference between the surface attack on the untreated and PIII-treated 304L alloy is shown in Figure 6.8b. The change in R_p value gives a direct indication of the corrosion resistance due to PIII treatment (Figure 6.8b).

EIS technique is also powerful enough to determine charge carriers in passive films and therefore it is able to examine the role of alloying elements and the environment on passivity. The relationship between capacitance and applied potential is given by the Mott–Schottky equations [32]:

$$\frac{1}{C^2} = \frac{2}{\varepsilon\varepsilon_0 e N_D}\left(E - E_{fb} - \frac{kT}{e}\right) \quad \text{for } n\text{-type semiconductor} \tag{6.18}$$

$$\frac{1}{C^2} = -\frac{2}{\varepsilon\varepsilon_0 e N_A}\left(E - E_{fb} - \frac{kT}{e}\right) \quad \text{for } p\text{-type semiconductor} \tag{6.19}$$

where C is space charge capacitance (F cm^{-2}), ε is dielectric constant of passive film, ε_o is the permittivity of free space (8.84×10^{-12} F m^{-1}), e is elementary electron charge, 1.602×10^{-19} coulomb ($-e$ for holes), N_D is charge donor density

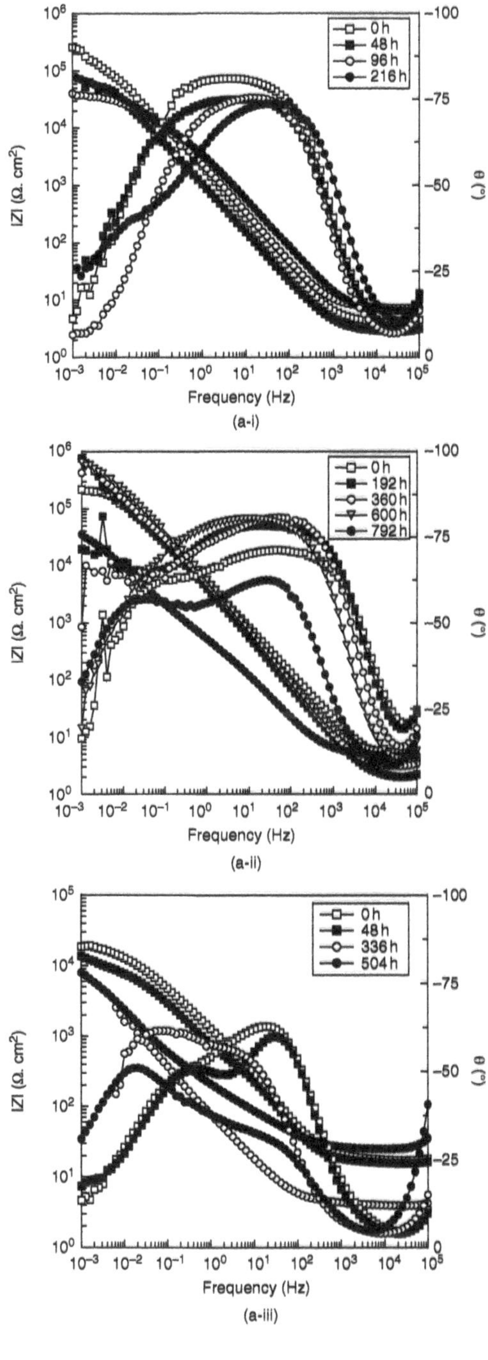

Figure 6.8 (a) Bode plots of (a-i) untreated, (a-ii) 380 °C, and (a-iii) 500 °C PIII-treated Type 304L SS for different time of immersion in 3.5 wt.% NaCl. (b) Optical micrographs of untreated and PIII-treated Type 304L SS specimen after polarization studies. (b-i) Untreated, (b-ii) 380 °C, and (b-iii) 500 °C treated, Type 304L SS after subjecting them to potentiodynamic polarization test in 3.5 wt.% NaCl. Note that severe attack on 500 °C treated alloy suffers localized attack attributable to sensitization. (c) Variation of R_p with immersion time for the PIII-treated alloys in 3.5 wt.% NaCl: (a) Type 304L SS. The plot shows that 380 °C treatment gives the highest corrosion resistance and 500 °C treatment is indeed detrimental to corrosion resistance of the alloy as reflected by R_p values. *Source:* (a–c) Saravanan et al. [31]. Reproduced with permission of Elsevier.

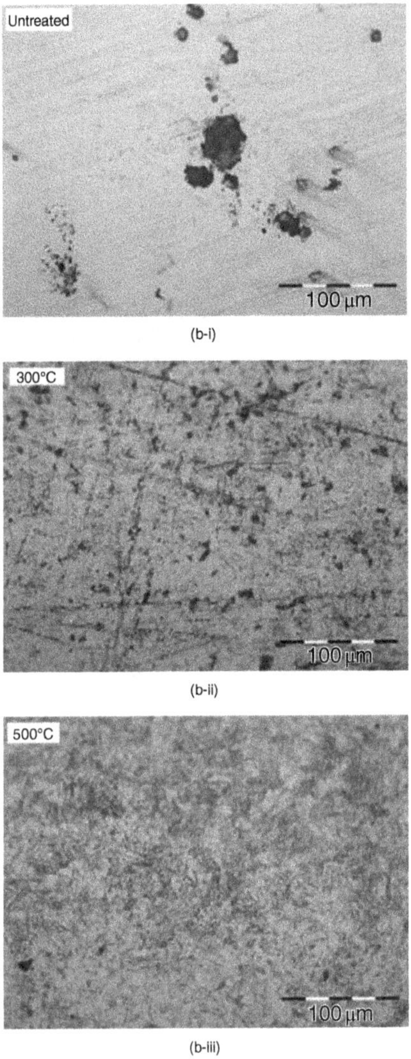

Figure 6.8 (Continued)

(cm^{-3}), N_A is charge acceptor density (cm^{-3}), E is applied potential (V_{SCE}), E_{fb} is flat band potential (V), k is the Boltzmann constant (1.38×10^{-23} J K^{-1}), and T is absolute temperature (K).

Arya et al. [33] have used Mott–Schottky plots to determine the charge carriers to explain the role of the alloying elements and the pH of the environment

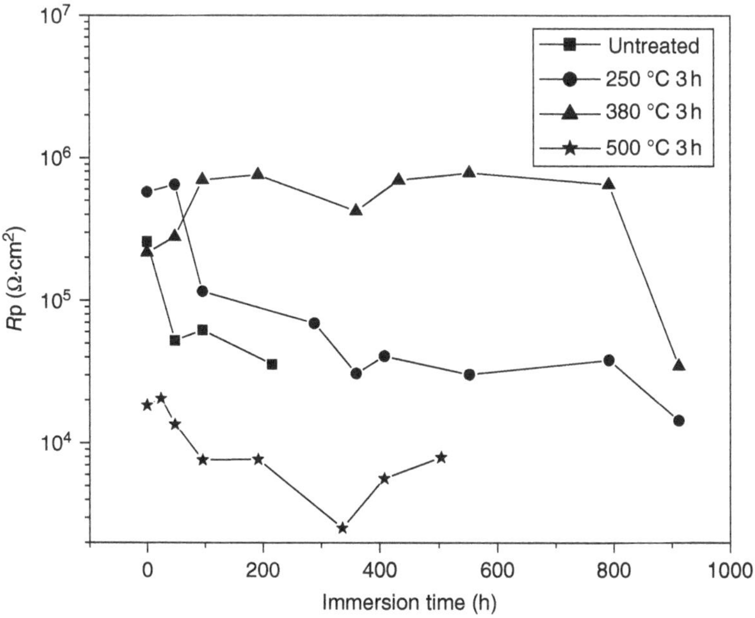

Figure 6.8 (Continued)

on the passivation tendency of a variety of stainless steels including high nitrogen stainless steels. Donor density was calculated from the positive slope of experimental plots of $1/C^2$ versus E. Figure 6.9 shows the variation of charge density with solution pH.

6.6.2 Oxide Films

Aluminum and magnesium alloys can be protected through thick oxide films, formed *in situ* through processes known as anodization and plasma electrolyte oxidation (PEO). EIS has been found to be a powerful tool to assess the performance of these coatings against corrosion with respect to aluminum alloys [34] and magnesium alloys [35, 36] and has been employed to evaluate such coatings. Published literature shows that several electrical equivalent circuits are employed depending on the proposed oxide film structures and the nature of corrosives.

A criterion named damage function (DF) was first introduced by Mansfeld and Kendig [37] from the point of view of monitoring coating performance on continuous exposure to corrosive environment. In this method, the modulus of

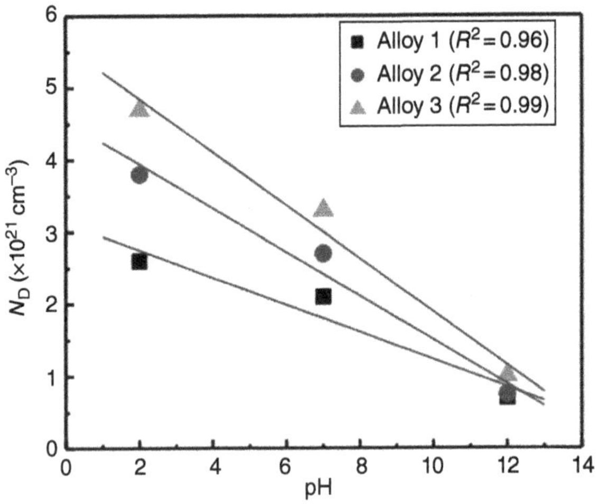

Figure 6.9 Dependence of N_D on pH and alloy composition: Alloy 1(0.070C–18.59Cr–2.00Mo–1.00N–0.011S–0.010P–0.30Si), Alloy 2(0.29C–17.50Cr–3.00Mo–0.50N–0.007S–0.014P–0.42Si), and Alloy 3(0.25C–17.13Cr–2.10Mo–0.010S–0.026P–0.49Si). Source: Arya et al. [33]. Reproduced with permission of Trans Tech Publications Ltd.

impedance at low frequency (usually at 0.1 Hz) at various time intervals is obtained, and the DF is estimated as per the following equation:

$$DF = \log\left|\frac{Z_d}{Z_t}\right| \quad \text{at low frequency} \tag{6.20}$$

where $|Z_d|$ and $|Z_t|$ represent initial impedance value and impedance value for $t = t_{corr}$ at low frequency (usually at 0.1 Hz), respectively. For a perfect coating DF value corresponds to zero as its $|Z|$ value is expected to remain unchanged. Based on this criterion, Kumar et al. [34] showed that both molybdate and vanadate additions to sulfuric acid bath, employed in anodization of the alloy, give rise to DF values of 0.456 and 0.442, respectively, as compared with 0.60, a value obtained for AA-2024 alloy anodized in sulfuric acid without these additives. Further, they employed an equivalent circuit as shown in Figure 6.10a to fit the impedance data of Figure 6.10b and c to analyze the influence of molybdate and vanadate additions on the stability of the oxide films. The study also clearly established the beneficial effect of hot-water sealing of the anodized oxide layer.

Ghasemi et al. [35, 36] using EIS examined the role of structure and thickness of PEO coatings produced through different electrolytes. The authors found good correlation among EIS data (Figure 6.11), microscopy (Figure 6.12), and polarization behavior. Their studies also bring out the fact that equivalent circuit can significantly vary depending on the structure of the oxide films.

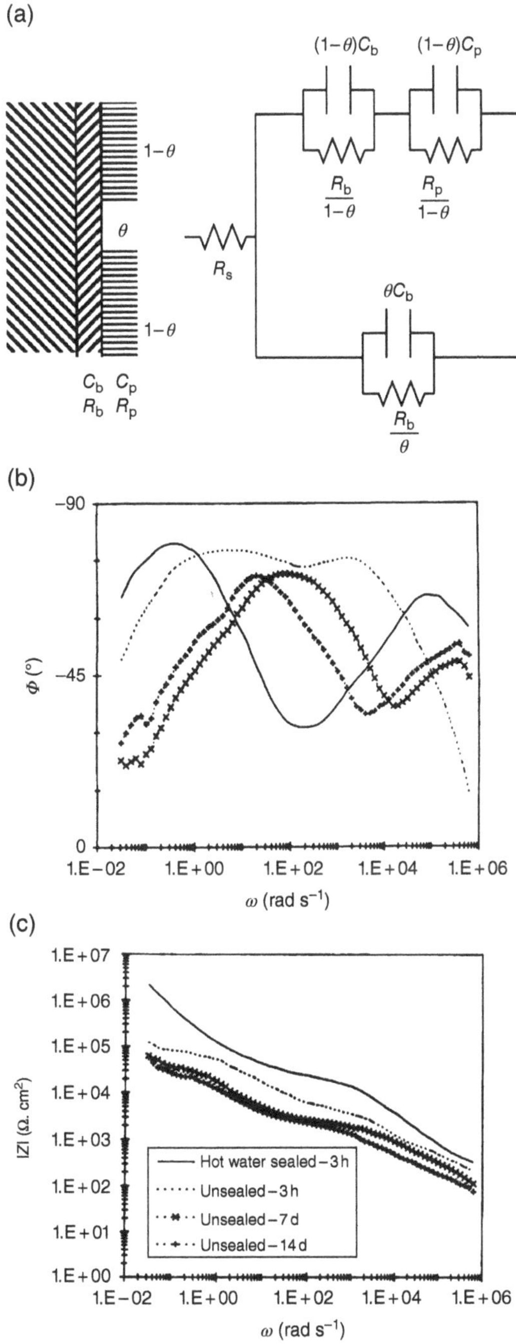

Figure 6.10 (a) Equivalent circuit employed to evaluate the coating (θ and $[1-\theta]$ denote the fraction of pitted surface and fraction of unaltered surface; R_p and R_b denote the resistance of porous and barrier layers; C_p and C_b denote the capacitance of porous and barrier layers; and R_s denotes solution resistance). EIS data: (b) Bode phase angle and (c) Bode modulus of impedance for sealed and unsealed (V) treated samples for different exposure time intervals. *Source:* Kumar et al. [34]. Reproduced with permission of Elsevier.

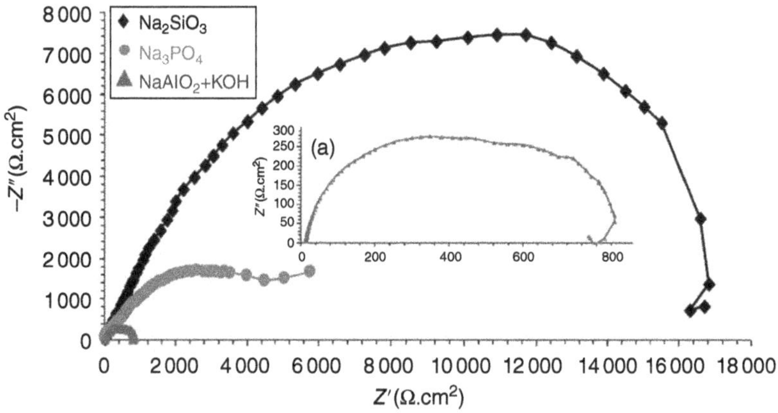

Figure 6.11 Impedance curves of samples immersed in various salt solutions. (a) $NaAlO_2$ curve in large scale. *Source:* Ghasemi et al. [36]. Reproduced with permission of Elsevier.

6.7 Localized Corrosion

Localized corrosion such as pitting corrosion, crevice corrosion, intergranular corrosion (IGC), and stress corrosion cracking is one of the most insidious forms of corrosive failures of metallic components, and i_{corr} cannot be a direct measure of damage due to any of these forms of localized corrosion. Therefore, the study of mechanisms and monitoring of localization corrosion requires different kinds of analysis. Oltra and Keddam [39] have brought out the complexities in the application EIS to pitting and stress corrosion cracking, generalized in terms of planar and of deeper defects (deep pits and cracks) suffering active corrosion. Mansfeld and his coworkers [38] have done extensive work on pitting corrosion of Al alloys and Al-based metal matrix composites. They have used an equivalent circuit as shown in Figure 6.13 to study pitting of AA 6061 alloy and its composites with SiC and graphite and AA7075 alloy in T6 and T73 conditions in chloride medium as well as the influence of $CeCl_3$ on pitting corrosion. Notably, they have shown that even pitting over as small as 5×10^{-4} area fraction of the surface could significantly alter EIS data. Using the EIS data, they are able to estimate the initiation time for pitting and crevice corrosion. However, these authors have reported a raise in R_p values on the commencement of pitting, which has been attributed due to the cathodic protection of the passive surface by the anodic dissolution occurring within the pits (Figure 6.14).

Intergranular corrosion of Al alloys has been studied by a few authors. Venugopal et al. [40] have used R_p as a parameter to determine IGC behavior Al–Zn–In sacrificial anodes. They have shown that the alloys (sacrificial anodes) exhibiting a continuous decrease in R_p value suffer little or no IGC,

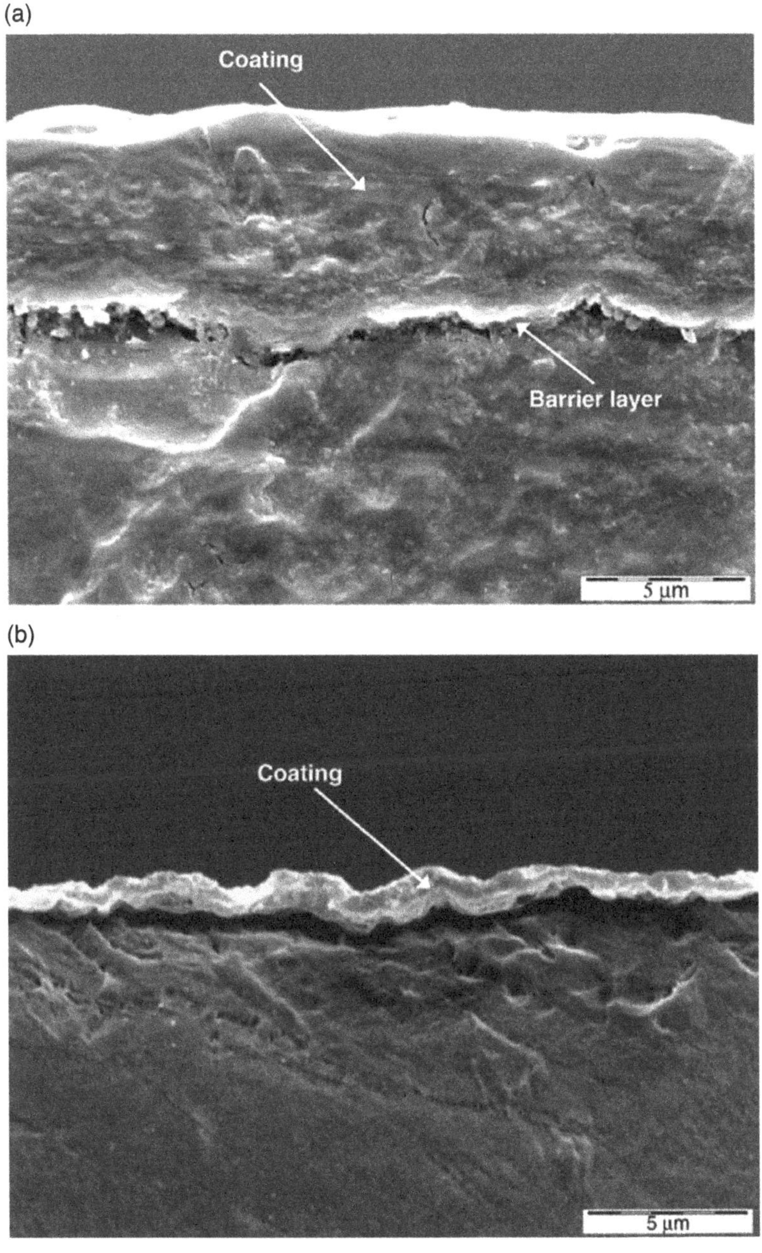

Figure 6.12 Cross-sectional images of coatings obtained using (a) silicate and (b) $NaAlO_2$ additives. The corresponding Nyquist plots are shown in Figure 6.11. *Source:* Ghasemi et al. [36]. Reproduced with permission of Elsevier.

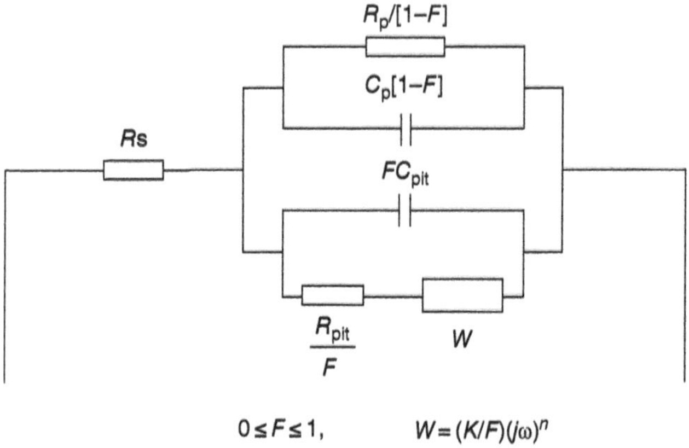

Figure 6.13 Model for the impedance of the pitting process on Al-based materials. R_p, polarization resistance of the passive surface; C_p, capacitance of the passive area; R_{pit}, polarization resistance of the pitted area; C_{pit}, pit capacitance; *F*, area fraction of the pit; and *W*, transmission line behavior. *Source:* Mansfeld et al. [38]. Reproduced with permission of Electrochemical Society, Inc.

due to the surface enrichment of activators during uniform dissolution that destroys passivation. In contrast, the alloys showing fluctuation in R_p values, with exposure time, suffered IGC. When grain falling occurs, as a result of IGC, fresh alloy surface having low activator is generated exhibiting good passivity and high R_p. These authors have further correlated the variation of R_p with time to the current efficiency of the sacrificial anode.

EIS has been explored to study IGC/exfoliation corrosion by a few authors [41, 42]. Keddam et al. [42] have employed an equivalent circuit having two time constants as brought out in Figure 6.15. The low and high frequency resistances and the low frequency capacitance have been shown to follow closely the exfoliation corrosion tendency of the alloy in EXCO solution prepared as per the ASTM G34-90 (Figure 6.16). Further, the EIS study could distinguish the pitting corrosion that occurred in AA 2219-T87 from that of the exfoliation suffered by AA 6013-T6 and 2024-T351. Although de Damborenea and Conde [41] employed the same equivalent circuit as that was employed by Keddam et al. [42] to study IGC of AA8090 (Al–Li) alloy, they had also employed an equivalent circuit with three time constants to fit the observed EIS data of the long-term exposure conditions. In this case, the third time constant has been attributed to the adsorption of anions.

Application of EIS as a technique to study stress corrosion cracking behavior of metallic alloys has been explored by a few researchers [39, 43, 44]. Bosch et al. [44] and Bosch [43] have proposed an equivalent circuit consisting of two broad elements: transmission line model for stress corrosion crack wall and tip

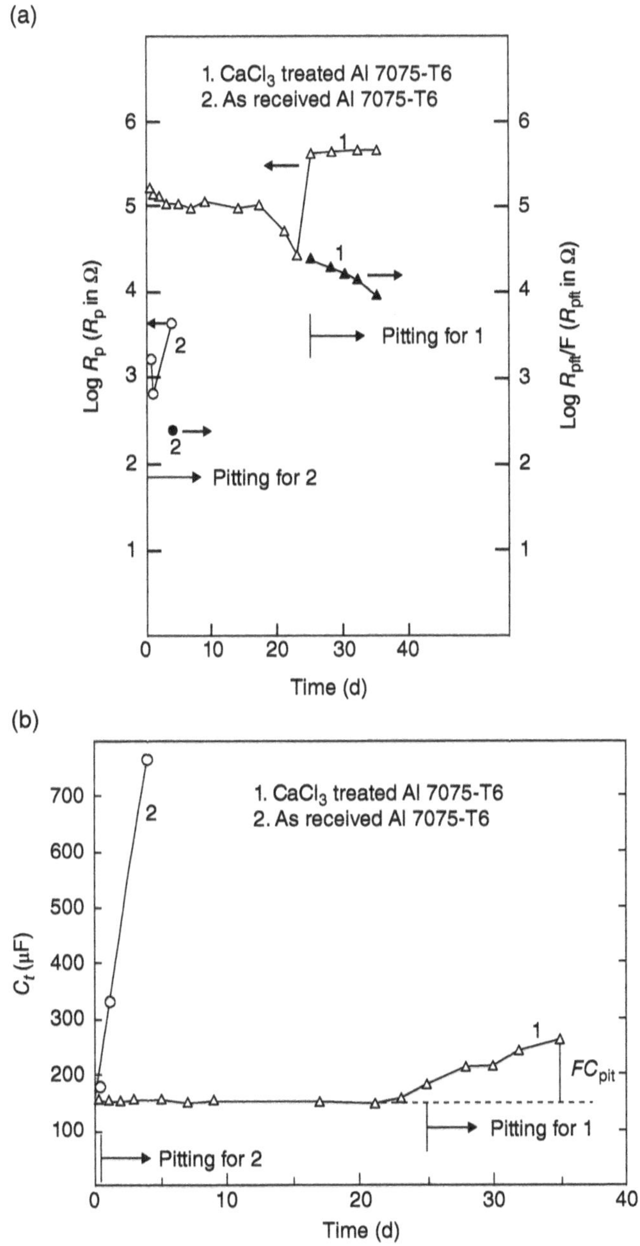

Figure 6.14 Time dependence of R_p and R_p/F (a), and C_t [$=C_p(1-F)+FC_{pit}$], (b) for Al 7075-T6 (as received vs. $CeCl_3$ treatment) during exposure to NaCl. *Source:* Mansfeld et al. [38]. Reproduced with permission of Electrochemical Society, Inc.

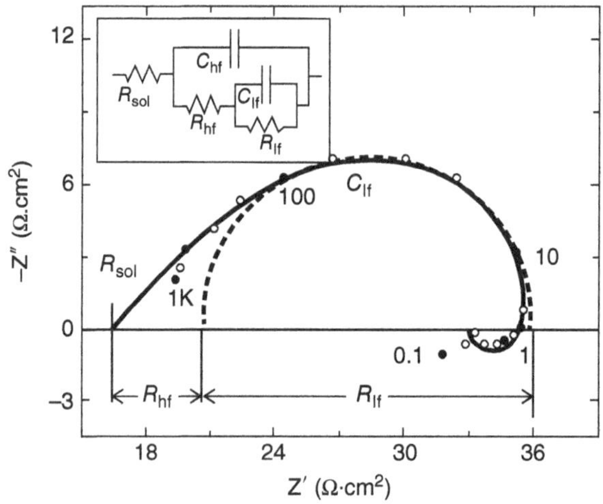

Figure 6.15 Typical feature of impedance diagram obtained for 2024T351 and 6013T6 samples in EXCO solution at 2 h immersion. The insert depicts the equivalent electrical circuit used. R_{sol}, solution resistance; R_{hf}, high frequency resistance; R_{lf}, low frequency resistance; C_{lf}, interface capacitance; and C_{hf}, high frequency capacitance. *Source:* Keddam et al. [42]. Reproduced with permission of Elsevier.

and a simple Randle circuit for uncracked specimen surface. Both these circuit elements are parallel to each other, meaning current will flow through crack as well as uncracked surface of specimens. Bosch [43] is rather extensive in studying the impedance behavior of Type 304L SS under the following three conditions:

1) SSRT of sensitized specimens in a 5 N H_2SO_4 + 0.1 M NaCl solution at room temperature.
2) Constant load testing of Type 304 SS in a boiling acidified sodium chloride solution.
3) SSRT testing at 300 °C in an autoclave having 0.01 M Na_2SO_4 solution.

Bosch [43] had showed that the obtained data obey Kramers–Kronig transformations (KKT) validating the application of EIS for SCC studies. The above study showed that phase shift ($\phi_{\text{stressed specimen}} - \phi_{\text{unstressed specimen}}$) can be used as a parameter to monitor crack initiation and growth (Figure 6.17). On the other hand, the modulus of impedance does not seem to depend significantly on the crack length. Furthermore, the study brings out the following noteworthy information:

> The phase shift depends on frequency as well as the crack length. There is an optimum frequency at which the phase shift becomes the highest for any given crack length; and as the crack length is increased, the phase shift also becomes high at the optimized frequency.

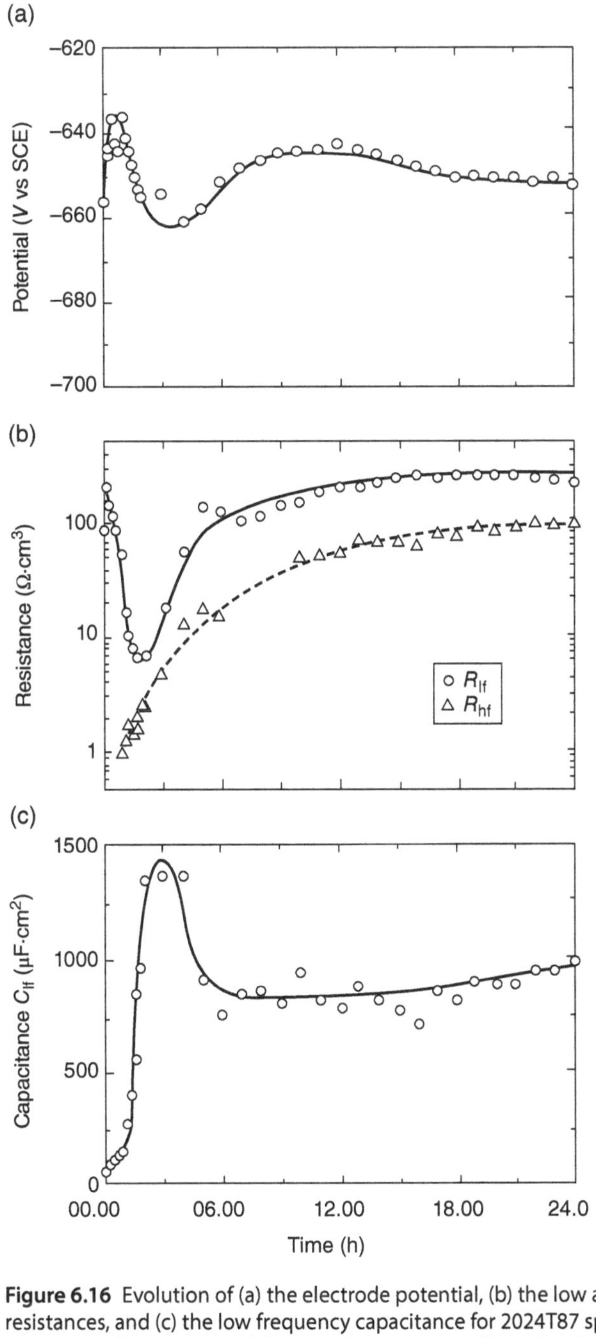

Figure 6.16 Evolution of (a) the electrode potential, (b) the low and high frequency resistances, and (c) the low frequency capacitance for 2024T87 specimen in EXCO solution. Figure 6.15 illustrates the definition of these terms. (Note that R_{hf} may correspond to R_{lf}, a faradaic resistance not seen in 2024 or 6013 alloy specimens.) *Source:* Keddam et al. [42]. Reproduced with permission of Elsevier.

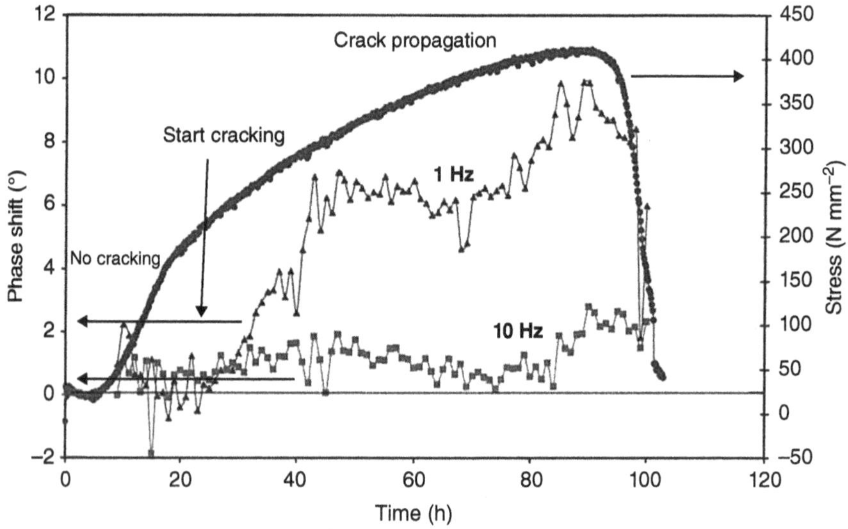

Figure 6.17 Crack initiation and propagation is related to the evolution in phase shift during an SSRT test of solution annealed type 304 SS in 0.01 M Na_2SO_4 at 300 °C. Source: Bosch [43]. Reproduced with permission of Elsevier.

The application of low frequency during slow strain rate testing is constrained by higher rate of film damage, leading to noncompliance of the stability criteria set by KKT. However, this is overcome in constant load test as the film damage phenomenon is much slower than that occurs in SSRT. For these reasons, the measured phase shift in constant load testing is higher than that seen in SSRT.

6.8 Atmospheric Corrosion

Application of EIS to monitor atmospheric corrosion of bare metals and coatings has been found to be promising [45–51]. While the electronics/instrumentation for EIS studies employed in the above studies are similar to those employed for normal (bulk electrolyte) studies, significant variation in the so-called electrochemical cell has been incorporated to study thin film electrolyte/metal interface, representing atmospheric corrosion. This is because of the fact that in the cases involving thin film electrolyte/metal interface corrosion studies, the use of remote reference/counter electrode is not possible. Hence, simple two-electrode arrangement has been employed. Simpson et al. [51] developed a grid-like electrode by sputter depositing gold on an alkyd-coated steel panel to monitor the coating degradation using EIS (Figure 6.18a).

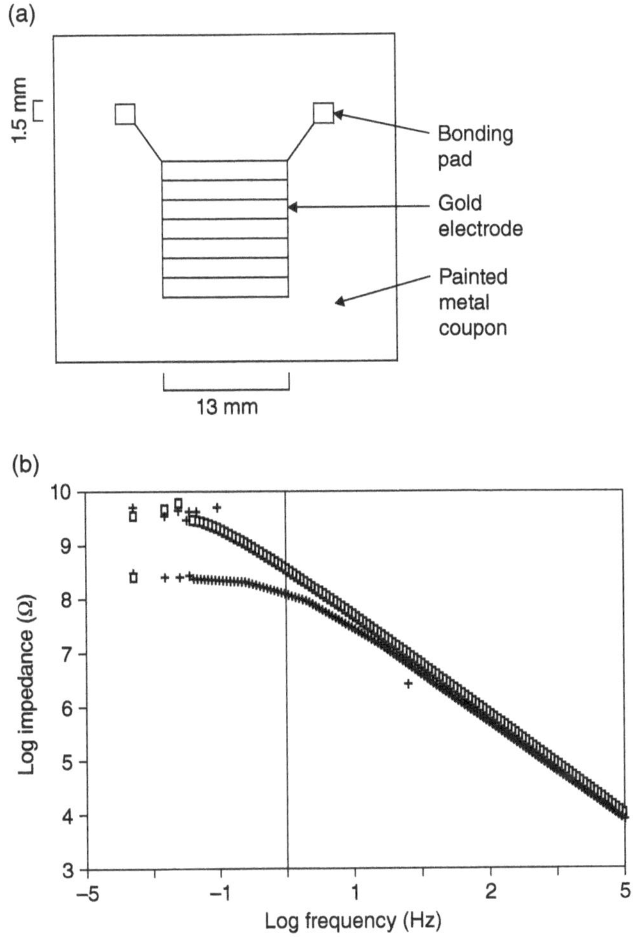

Figure 6.18 (a) Schematic of the prototype atmospheric corrosion probe used for monitoring paint coatings. The gold grid serves a counter/reference and the substrate working electrode. Source: Simpson et al. [51]. Reproduced with permission of Electrochemical Society, Inc. (b) EIS data after 1 day and 2 weeks exposure of the ATMEIS (impedance probe) in pH 3 H_2SO_4 monitored using remote electrodes (RR/RC) or the gold electrode in air (AIR). (□) RR/RC, 1 day; (+) AIR, 1 day; (Δ) RR/RC, 2 weeks; and (◊) AIR, 2 weeks.

This was a bi-electrode arrangement, with the deposited gold electrode acting as reference/counter and the steel substrate as working electrode. Using sulfuric acid as bulk corrosive medium, they compared the data obtained using such a probe (two-electrode arrangement) with that obtained using three-electrode arrangement. They found both the data are similar in nature. They also studied the EIS behavior of the probe in air-exposed conditions, after removing the electrolyte. Regardless of the electrode arrangement, the data was found to be very similar (Figure 6.18b).

In the above study, the impedance across the gold-(counter)-electrode/electrolyte interface has been considered to be negligible, especially when used in acidic solution, and the total impedance measured by the system was attributed to the coating. However, this can only be true when the coating under study has much higher impedance than that of the gold-(counter)-electrode/electrolyte interface. However, when the coating degrades, as usually happens on prolonged atmospheric exposure, its impedance can drop to smaller values, and the contribution of the impedance of the gold/electrolyte interface becomes significant enough to be ignored for assessing the coating performance. Amirudin and Thierry [45, 46], therefore, modified equivalent circuit to incorporate the contribution of the counter electrode impedance to fit the EIS data. They also showed that the coating impedance also depends significantly on the solution conductivity that is used for the study.

Atmospheric corrosion behavior of weathering steel of composition Fe–0.09C–0.005S–0.081P–0.35Mn–30Si–0.48Cr–0.027Ni–0.28Cu–0.37Al has been studied by Li et al. [49] by employing EIS. In their online study, the authors have employed comb-like electrodes that consist of alternate layers of steel strips and epoxy films (Figure 6.19). Each of the alternate steel strips was electrically connected, thus forming two-electrode configuration. By exposing them to controlled relative humidity (RA) in a chamber, they were able to show that the steel suffered corrosion only at 70 RH and above within the test duration, with the application of EIS. Though the trend in R_p with time followed that of the corrosion rates measured by weight loss method,

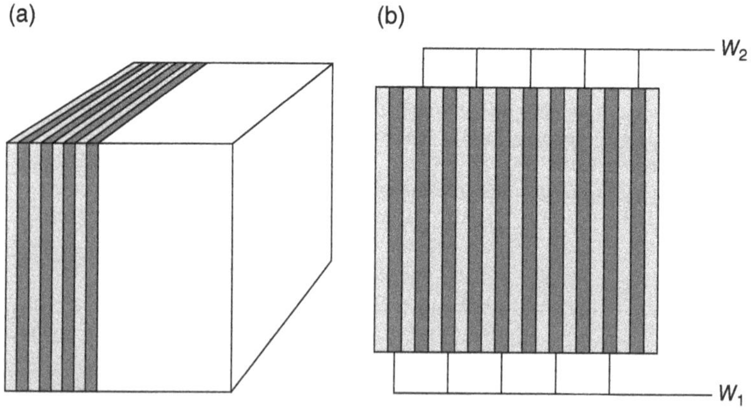

Figure 6.19 The design of monitoring electrode: (a) schematic of assembly method and (b) schematic of lead connection. *Source:* Li et al. [49]. Reproduced with permission of Elsevier.

the corrosion rates of the alloy could not be established from the EIS data (R_p) as Stern–Geary coefficient was unavailable and that is one of the limitations of EIS.

A small variation in the above configuration was carried out by Thee et al. [52] to study thin film corrosion. These authors first made two comb-like electrodes and embedded them in an epoxy resin in such a way that both of them were electrically isolated. These authors have studied the role of electrolyte thickness on the corrosion rate and the corrosion evolution of steel with wet–dry cyclic exposures.

A simple bi-electrode of Zn separated by a polymer membrane and all of them embedded in an epoxy handle formed a sensor cell assembly for monitoring thin layer corrosion of Zn [50].

In another study, EIS has been employed to obtain patina properties of bronze containing Sn. The impedance was mainly employed to determine film thickness and its porosity [48].

6.9 Fouling

Fouling by corrosive environments of metallic structures causes several effects. Scale deposition in heat exchanger tubes and pipelines carrying water and calcareous deposition by seawater on submerged structures cause reduction in corrosion rates. On the other hand, microbial fouling can either induce localized corrosion or prevent corrosion depending on the prevailing environmental conditions and the nature of microbes. Hence, studying fouling phenomena is very valuable from industrial standpoint.

Calcareous deposition has been found to significantly increase R_p values, though it does not significantly affect the capacitance [53]. Gabrielli et al. [54] studied both the mechanism and the nature such as compactness and thickness of calcium carbonate scales formed from water, in relation to scale formation in boilers and cooling water systems. The high frequency resistance has been related to scale thickness and has been found to increase significantly with time. The capacitance, however, did not vary with time (Figure 6.20). Hence, it looks that R_p can be used as a parameter to determine the scaling tendency of boiler as well as cooling water. Such results indicate large potential for EIS as nondestructive technique for monitoring water treatment for boilers and cooling water systems.

In investigating the role of microbes on corrosion, the parameters such as R_p and R_{pit} have been found to be valuable [55]. For those systems such as Al alloys that are passivating, EIS data showed a remarkable variation in these parameters due to the onset of microbial corrosion enabling the technique to be very useful (Figure 6.21) [55].

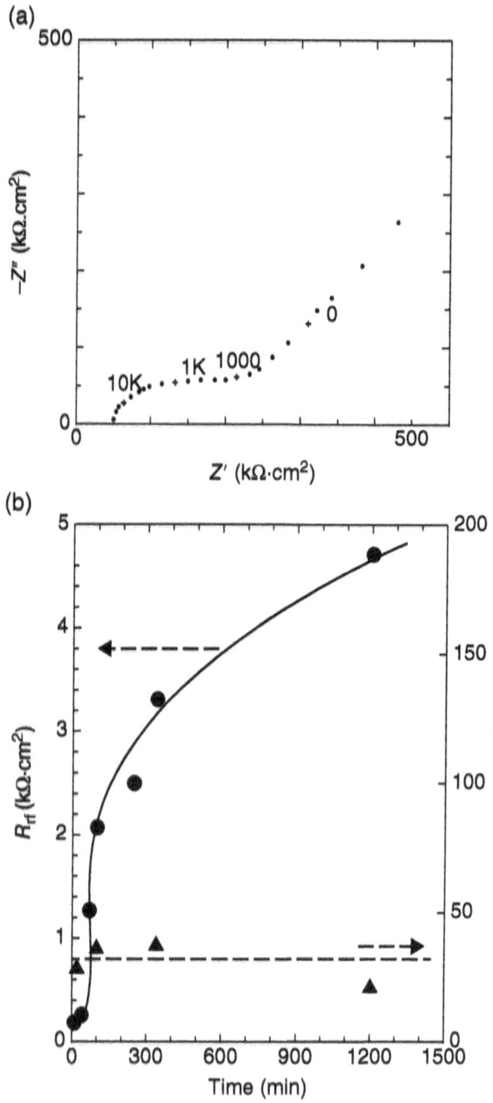

Figure 6.20 Impedance measured at a potential where oxygen reduction was very low ($I \approx 0$) on an electrode totally covered by scale. (a) Impedance measured after 180 min of scaling at 2000 rpm. (b) Change of R_{hf} and C_{hf} with respect to time. *Source:* Gabrielli et al. [54]. Reproduced with permission of Elsevier.

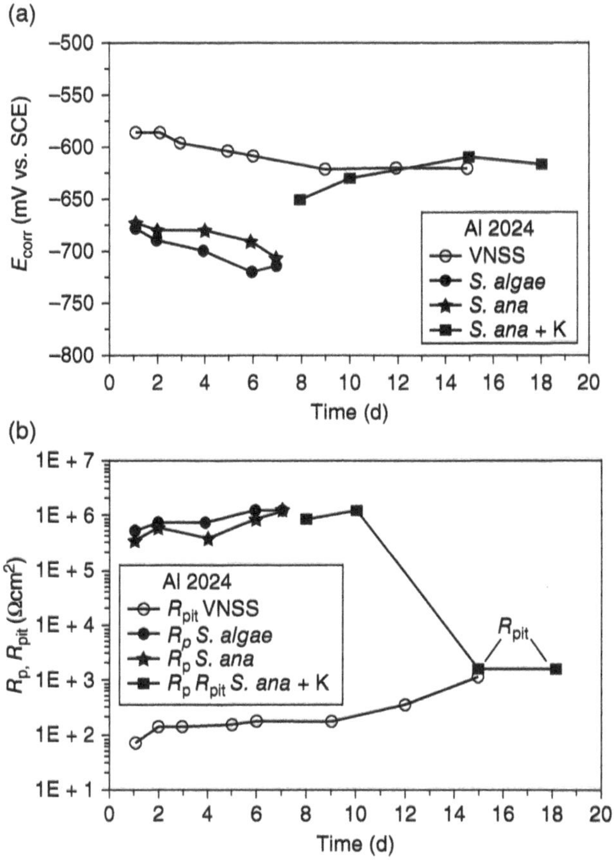

Figure 6.21 Time dependence of E_{corr} (a) and R_p and R_{pit} (b) for Al 2024 exposed to VNSS containing *S. algae* or *S. ana*. *Source:* Nagiub and Mansfeld [55]. Reproduced with permission of Elsevier.

6.10 Concrete Structure

Concrete structures exposed to corrosive atmospheres fail, as steel reinforcement suffers corrosion. The ever-growing infrastructures in civil society necessitate development of reliable ND techniques to monitor corrosion. Unfortunately, so far no NDT is available that can accurately estimate corrosion of steel reinforcements in aging concrete structures. Field and laboratory experiences in using several electrochemical techniques have been reviewed by Videm [56]. One of the most versatile parameters to assess corrosion of steel in concrete has been corrosion potential of steel. The numerical value of corrosion potentials has been used to predict whether the steel is in passive or active condition. Generally, the former implies a protected state and the later corrosive state of steels. Thus, corrosion potential remained a qualitative measure to assess corrosion.

Several studies were undertaken to quantify corrosion. Linear polarization, potentiostatic polarization, and galvano static pulse techniques, which are relatively simple to implement, have been explored but were met with some limitations. Despite the complexities of instrumentation and data analysis, studies were undertaken to explore the use of EIS to estimate corrosion of steel in concrete. One of the most practical studies, employing EIS for evaluating the performance steel reinforcement in typical bridge decks, is by Brown et al. [57]. They measured the modulus of impedance $|Z|$ and phase angle (ϕ) at three frequencies, viz. 5000 Hz, 1 Hz, and 1 mHz. It was presumed that at 5000 Hz the impedance values corresponded to solution resistance (R_s) and at 1 mHz the cumulative effects of corrosion, concrete, passive film, and coating, if the latter is present. Though EIS could not be used to quantify corrosion rate, the study could reveal the occurrence of corrosion, which was in agreement with the corrosion potential data (Figure 6.22).

Feasibility of EIS as a monitoring technique to assess efficacy of cathodic protection to control corrosion of steel rebar in concrete structures was examined by Thompson et al. [58] and Pruckner et al. [59]. The change in frequency at maximum phase angle ($\omega_{\phi m}$) and the maximum phase shift with time, obtained from wide frequency experiments, were used as criteria to evaluate the efficacy of cathodic protection, since the polarization resistance obtained from EIS studies could not be correlated to the cathodic protection efficiency [58]. On the contrary, Pruckner et al. [59] were able to correlate R_{ct} to the cathodic protection efficacy. They followed Haruyama and Tsuru [60] methodology of employing two frequency data to obtain R_{ct} values. They have also pointed out the limitation of this methodology to estimate R_{ct} in those cases where Nyquist plot results in depressed semicircles. The published literature thus shows the real difficulties in measuring the active corrosion rate of steel in concrete through EIS technique, though it has been found useful in predicting the initiation time for corrosive attack.

6.11 Instrumentation and Testing

Typical electrochemical impedance measurement systems consist of a potentiostat of a reasonable wide frequency bandwidth, interfaced with a frequency response analyzer or a lock-in amplifier, and a corrosion cell similar to those employed in DC polarization studies. Two types of configurations, namely, two (working and counter) or three (working, counter, and reference) electrodes cells, can be used, although the latter is extensively used as it enables better control of working electrode potential. In the application of EIS to study

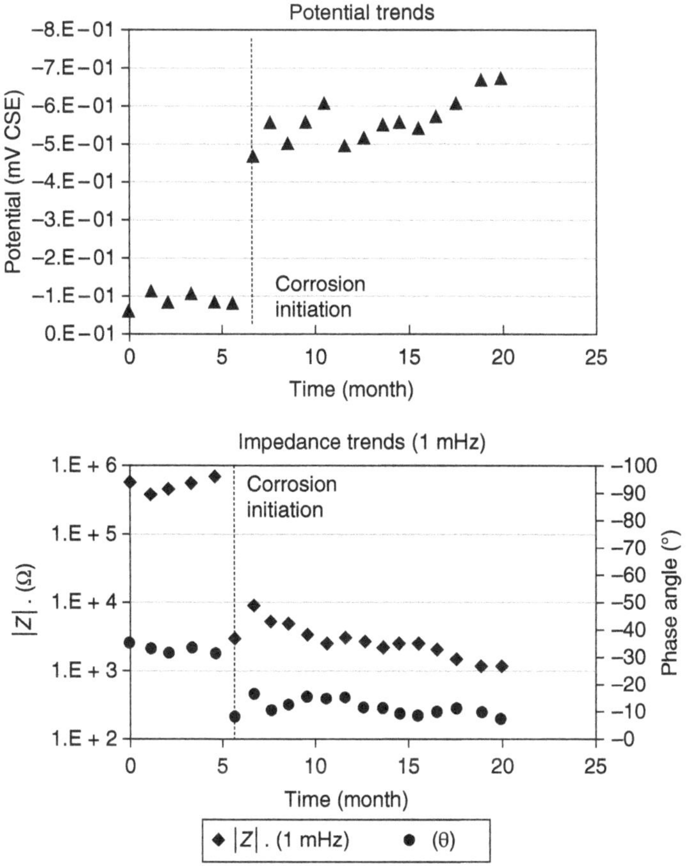

Figure 6.22 Sample potential and impedance trends for ECR. *Source:* Brown et al. [57]. Reproduced with permission of ASTM International.

atmospheric corrosion, where bulk electrolytes cannot be employed, electrodes and cell configuration is significantly varied as discussed in the section on atmospheric corrosion.

In EIS study AC volt of small magnitude, the value of which is dictated by the need to maintain linearity between the applied voltage and the resultant current, is applied. This usually lies in the range ±5 to ±10 mV, depending on the Tafel slope [1], though higher amplitudes are also employed in the published literature.

Despite its ability to provide more comprehensive information of corroding systems in comparison with other widely employed nondestructive techniques for corrosion monitoring, the sophisticated and expensive instrumentation combined with complex/tedious data analysis, a procedure requiring high level of competence, make the field application of EIS as a tool to monitor corrosion a less preferred choice. Possibilities of employing limited number of frequencies corresponding to various corrosion phenomena (marked by differing time constants) can be explored. Mansfeld and Kendig [37] measured protective behavior of anodized oxide layer of aluminum alloys at a single frequency using DF as a parameter. Kendig and Scully [61] use limited-frequency studies to determine break-point frequency.

6.12 Concluding Remarks

Among the several electrochemical and non-electrochemical techniques employed for studying corrosion, EIS has been found to be one of the most powerful techniques. With EIS, it is possible to resolve changes in corrosion mechanism of corroding systems, with time, without affecting their normal course ("steady state"). Using this technique, capacitance of double layer and oxide or polymer film and polarization resistance (or change transfer resistance) of corrosion processes, resistance of oxide or polymer films can be obtained, which can be used to estimate corrosion rates and to reveal the corrosion mechanisms. More importantly, the technique is also capable of studying diffusion phenomena in all corrosion situations. LPR technique, being a relatively much simpler and more versatile, competes strongly with EIS. However, in poorly conducting environments the i_{corr} estimated through LPR will be lower than the actual value, as R_s is added to the measured R_p value, whereas the EIS technique is able to isolate R_s from R_p and thus reflecting true i_{corr} of metals. The biggest advantage of EIS that merits application over the other electrochemical techniques is its ability to study protective behavior of thick coatings such as paints and oxides as the capacitive components of these coatings do not permit the application of DC techniques to probe such informations.

A normal EIS study requires applying a small AC voltage (±5 to ±10 mV) over a wide frequency range (1 mHz to 100 kHz), and so it is quite a time-consuming technique. However, the measurement duration can be reduced significantly, either by simultaneous application of multiple frequencies or by carrying out measurements at a select few frequencies. Though the test period can be significantly reduced, as suggested above, they are limited by (i) complicated electronics and low signal-to-noise ratio in the former and (ii) the requirement to know the time constants of single- or multiple-step corrosion processes of the corroding system, in the second case. The biggest challenge in the application of EIS technique for online and real-time corrosion monitoring is that a much higher level of technical competence is required to analyze the data to resolve complex corrosion phenomena than several other competing techniques. Further, for a rapidly changing corroding system, requiring low frequency data to monitor an industrial process, this technique is not applicable.

Acknowledgment

Meticulous reading of the proof by Dr. Srinivas Palli is acknowledged.

References

1 Cottis, R. and Turgoose, S. (1999). *Electrochemical Impedance and Noise.* Houston, TX: NACE International.
2 Macdonald, J.R. (1987). *Impedance Spectroscopy: Emphasizing Solid Materials and Systems.* New York: Wiley-Interscience.
3 Mansfeld, F. and Lorenz, W.J. (1991). Electrochemical impedance spectroscopy (EIS): application in corrosion science and technology. in: *Techniques for Characterization of Electrodes and Electrochemical Process* (ed. R. Varma and J.R. Selman), 581. New York: Wiley.
4 Raja, V.S. (1998). Study of alloy dissolution, passivity and protective coatings by electrochemical impedance spectroscopy. *Transactions of the Indian Institute of Metals* 51: 351.
5 Raja, V.S. (2002). Electrochemical impedance spectroscopy for the development of coatings for corrosion protection. *Transactions-Metal Finishers Association of India* 11: 181.
6 Mansfeld, F., Tsai, C.H., and Shih, H. (1992). Software for simulation and analysis of electrochemical impedance spectroscopy (EIS) data. *ASTM International* 1154: 186.
7 Epelboin, I., Gabrielli, C., Keddam, M., and Takenouti, H. (1981). Alternating-current impedance measurements applied to corrosion studies and corrosion-rate determination. in: *Electrochemical Corrosion Testing,* ASTM STP 727 (ed. F. Mansfeld and U. Bertocci), 150. West Conshohocken, PA: American Society for Testing and Materials.
8 Lorenz, W.J. and Mansfeld, F. (1981). Determination of corrosion rates by electrochemical DC and AC methods. *Corrosion Science* 21: 647.
9 Raja, V.S., Gayathridevi, R., Venugopal, A. et al. (1998). Evaluation of blistering performance of pigmented and unpigmented alkyd coatings using electrochemical impedance spectroscopy. *Surface & Coatings Technology* 107: 1.
10 David, R., Tambe, S.P., Singh, S.K. et al. (2011). Thermally sprayable grafted LDPE/nanoclay composite coating for corrosion protection. *Surface & Coatings Technology* 205: 5470.
11 Singh, S.K., Tambe, S.P., Gunasekaran, G. et al. (2009). Electrochemical impedance study of thermally sprayable polyethylene coatings. *Corrosion Science* 51: 595.
12 Singh, S.K., Tambe, S.P., Raja, V.S., and Kumar, D. (2007). Thermally sprayable polyethylene coatings for marine environment. *Progress in Organic Coatings* 60: 186.

13 Singh, S.K., Tambe, S.P., Samui, A.B. et al. (2006). Maleic acid grafted low density polyethylene for thermally sprayable anticorrosive coatings. *Progress in Organic Coatings* 55: 20.
14 Liedheiser, H. and Kendig, M. (1976). The electrical properties of protective polymer coatings as related to corrosion of the substrate. *The Journal of the Electrochemical Society* 123: 982.
15 De Rosa, L., Monetta, T., and Bellucci, F. (1998). Moisture uptake in organic coatings monitored with EIS. *Materials Science Forum* 289–292: 315.
16 Kale, S.S., Raja, V.S., and Bakare, A.K. (2014). Interface electrochemical corrosion behavior of alclad Al–Zn–Mg–Cu–Cr Al alloy adhesively bonded with carbon fiber reinforced plastic stiffener. *Corrosion Engineering, Science and Technology* 49: 116.
17 Phanasgaonkar, A. and Raja, V.S. (2009). Influence of curing temperature silica nanoparticles and cerium on surface morphology and corrosion behavior of hybrid silane coatings on mild steel. *Surface & Coatings Technology* 203: 2260.
18 Raja, V.S., Panday, C.K., Saji, V.S. et al. (2006). An electrochemical study on deformed galvanneal steel sheets. *Surface & Coatings Technology* 201: 2296.
19 de Wit, J.H.W., Lenderink, H.J.W., Vander Weijde, D.H., and Van Westing, E.P.M. (1995). Possibilities for life time prediction of organic coatings with electrochemical impedance measurements. *Materials Science Forum* 192–194: 253.
20 van der Weijde, D.H., van Westing, E.P.M., and de Wit, J.H.W. (1998). Monitoring the effect of environmental changes on coating properties with EIS. *Materials Science Forum* 289–292: 237.
21 Hack, T.P. and Scully, J.R. (1991). Defect area determination of organic coated steels in seawater using the breakpoint frequency method. *The Journal of the Electrochemical Society* 138: 33.
22 Haruyama, S., Asari, M., and Tsuru, T. (1987). Mass transport in blistering of coating steel. In: *Proceedings of the Symposium on Corrosion Protection by Organic Coatings*, vol. 87-2 (ed. M.M. Kendig and H. Leidheiser), 197. Pennigton, NJ: The Electrochemical Society.
23 Hiriyama, R. and Haruyama, S. (1991). Electrochemical impedance for degraded coated steel having pores. *Corrosion* 47: 952.
24 Scully, J.R. (1989). Electrochemical impedance of organic-coated steel: correlation of impedance parameters with long-term coating deterioration. *Journal of the Electrochemical Society* 136: 979.
25 Mansfeld, F. (1995). Use of electrochemical impedance spectroscopy for the study of corrosion protection by polymer coatings. *The Journal of Applied Electrochemistry* 25: 187.
26 Raja, V.S., Saji, V.S., Venugopal, A. et al. (2010). Electrochemical impedance behavior of graphite dispersed electrically conducting acrylic coating on AZ31 magnesium alloy in 3.5 wt.% NaCl solution. *Progress in Organic Coating* 67: 12.

27 Annergren, I., Keddam, M., Takenouti, H., and Dominique Thierry, D. (1996). Modeling of the passivation mechanism of Fe–Cr binary alloys from ac impedance and frequency resolved *rrde* – I. Behaviour of Fe–Cr alloys in 0.5M H_2SO_4. *Electrochimica Acta* 41: 1121.

28 Annergren, I., Keddam, M., Takenouti, H., and Dominique Thierry, D. (1997). Modeling of the passivation mechanism of Fe–Cr binary alloys from ac impedance and frequency resolved *rrde* – II. Behaviour of Fe–Cr alloys in 0.5M H_2SO_4 with an addition of chloride. *Electrochimica Acta* 42: 1595.

29 Falkenberg, F., Raja, V.S., and Ahlberg, E. (2001). An alternative explanation for the apparent passivation of molybdenum in 1M hydrochloric acid. *The Journal of the Electrochemical Society* 148: 132.

30 Venugopal, A. and Raja, V.S. (1997). AC impedance study on the activation mechanism of aluminium by indium and zinc in 3.5% NaCl solution. *Corrosion Science* 39: 2053.

31 Saravanan, P., Raja, V.S., and Mukherjee, S. (2013). Effect of alloyed molybdenum on corrosion behavior of plasma immersion nitrogen ion implanted austenitic stainless steel. *Corrosion Science* 74: 106.

32 Morrison, S.R. (1980). *Electrochemistry at Semiconductor and Oxidized Metal Electrodes*, 64. New York: Plenum Press.

33 Arya, S.B., Raja, V.S., and Tiwari, A.N. (2013). A comparative study of semiconducting behavior of passive films of high nitrogen and Ni and Mn free stainless steels. In: *A Century of Stainless Steels*, Advanced Materials Research, vol. 794 (ed. V. Kain, L.K. Singhal, K.P. Dwivedi, et al.), 626. India: Trans Tech Publications.

34 Kumar, C.S., Rao, V.S., Raja, V.S. et al. (2002). Corrosion behaviour of solar reflector coatings on AA 2024 – an impedance study. *Corrosion Science* 44: 387.

35 Ghasemi, A., Raja, V.S., Blawert, C. et al. (2008). Study of the structure and corrosion behavior of PEO coatings on AM 50 magnesium alloy by electrochemical impedance spectroscopy. *Surface & Coatings Technology* 202: 3513.

36 Ghasemi, A.R., Raja, V.S., Blawert, C. et al. (2010). The role of anions in the formation and corrosion resistance of the plasma electrolytic oxidation coatings. *Surface Coating Technology* 204: 1469.

37 Mansfeld, F. and Kendig, M. (1985). Technical note: impedance spectroscopy as quality control and corrosion test for anodized Al alloys. *Corrosion* 41: 490.

38 Mansfeld, F., Lin, S., Kim, S., and Shih, H. (1990). Pitting and passivation of Al alloys and Al-based metal matrix composites. *Journal of the Electrochemical Society* 137: 78.

39 Oltra, R. and Keddam, M. (1990). Application of EIS to localized corrosion. *Electrochimica Acta* 35: 1619.

40 Venugopal, A., Angal, R.D., and Raja, V.S. (1996). Effect of grain boundary corrosion on the impedance behaviour of Al-Zn-In alloy in 3.5% NaCl solution. *Corrosion* 52: 138.

41 de Damborenea, J.J. and Conde, A. (2000). Intergranular corrosion of 8090 Al-Li: interpretation by electrochemical impedance spectroscopy. *British Corrosion Journal* 35: 48.

42 Keddam, M., Kuntz, C., Takenouti, H. et al. (1997). Exfoliation corrosion of aluminum alloys examined by electrode impedance. *Electrochimica Acta* 42: 87.

43 Bosch, R.W. (2005). Electrochemical impedance spectroscopy for the detection of stress corrosion cracks in aqueous corrosion systems at ambient and high temperature. *Corrosion Science* 47: 125.

44 Bosch, R.W., Moons, F., Zheng, J.H., and Bogaerts, W.F. (2001). Application of electrochemical impedance spectroscopy for monitoring stress corrosion cracking. *Corrosion* 57: 532.

45 Amirudin, A. and Thierry, D. (1995). Application of electrochemical impedance spectroscopy to study the atmospheric corrosion of painted metals. *Materials Science Forum* 192–194: 317.

46 Amirudin, A. and Thierry, D. (1995). Electrochemical sensor for monitoring atmospheric corrosion of polymer coated metal. *British Corrosion Journal* 30: 214.

47 Amirudin, A., Barreau, C., Massinon, D., and Thierry, D. (1992). Application of impedance spectroscopy to study the atmospheric corrosion of galvanized steel coated with epoxy paint. *Materials Science Forum* 111–112: 291.

48 Gianni, L., Cavallini, M., and Adriaens, A. (2013). Corrosion behavior of bronze in urban environments: influence of tin content. *Corrosion Engineering, Science and Technology* 48: 334.

49 Li, C., Ma, Y., and Wang, F. (2010). EIS monitoring study of atmospheric corrosion under variable relative humidity. *Corrosion Science* 52: 3677.

50 Nazarov, A.P. and Thierry, D. (2006). Probing the atmospheric corrosion of metals zinc. *Protection of Metals* 42: 478.

51 Simpson, T.C., Moran, P.J., Moshier, W.C. et al. (1989). An electrochemical monitor for the detection of coating degradation in atmosphere. *Journal of the Electrochemical Society* 137: 2761.

52 Thee, C., Hao, L., Dong, J. et al. (2014). Atmospheric corrosion monitoring of a weathering steel under an electrolyte film in cyclic wet-dry condition. *Corrosion Science* 78: 130.

53 Deslouis, C., Festy, D., Gil, O. et al. (2000). Characterization of calcareous deposits in artificial sea water by impedances techniques: 2-mg$(OH)_2$ deposit of mg(OH), without $CaCO_3$. *Electrochemical Acta* 45: 1837.

54 Gabrielli, C., Keddam, M., Khalil, A. et al. (1997). Study of calcium carbonate scales by electrochemical impedance spectroscopy. *Electrochimica Acta* 42: 1207.

55 Nagiub, A. and Mansfeld, F. (2002). Evaluation of microbiologically influenced corrosion inhibition (MICI) with EIS and ENA. *Electrochimica Acta* 47: 2319.

56 Videm, K. (1998). Field and laboratory experience with electrochemical methods for assessing corrosion of steel in concrete. *Material Science Forum* 289–292: 3.

57 Brown, M.C., Weyers, R.E., and Sprinkel, M.M. (2006). Service life extension of Virginia bridge decks afforded by epoxy-coated reinforcement. *Journal of ASTM International* 3: 1.

58 Thompson, N.G., Lawson, K.M., and Beavers, J.A. (1988). Monitoring cathodically protected steel in concrete structures with electrochemical impedance techniques. *Corrosion Engineering* 44: 581.

59 Pruckner, F., Theiner, J., Eri, J., and Nauer, G.E. (1996). In-situ monitoring of the reinforced concrete by electrochemical impedance spectroscopy. *Electrochimica Acta* 41: 1233.

60 Haruyama, S. and Tsuru, T. (1981). A corrosion monitor based on impedance method. in: *Electrochemical Corrosion Testing*, ASTM STP 727 (ed. F. Mansfeld and U. Bertocci), 167. West Conshohocken, PA: American Society for Testing and Materials.

61 Kendig, M. and Scully, J.R. (1990). Basic aspects of electrochemical impedance application for the life prediction of organic coatings on metals. *Corrosion* 46: 22.

7

Electrochemical Noise as Nondestructive Evaluation Technique for Understanding and Monitoring Corrosion

Girija Suresh[1], U. Kamachi Mudali[1] and Baldev Raj[2]

[1] *Corrosion Science and Technology Group, Indira Gandhi Centre for Atomic Research, Kalpakkam, India*
[2] *National Institute of Advanced Studies, Bengaluru, India*

7.1 Introduction

Corrosion is a deleterious degradation process affecting plant equipment, causing reduced plant performance and unexpected failures that can be costly in terms of repair, environmental damage, and even harm to human life in plant and even outside the plant. The several manifestations of corrosion range from uniform to localized corrosion. Understanding the various forms and their combinations is essential in determining their importance and to finding appropriate technology for detection, characterization, and monitoring. Localized corrosion such as pitting, environmental cracking, corrosion fatigue, and microbiologically influenced corrosion (MIC) is most common occurrence in chemical processing, petrochemical, water treatment, and nuclear industries. This calls for periodic inspection and corrosion monitoring to combat corrosion. Among the various corrosion inspection and monitoring techniques, the nondestructive evaluation (NDE) methods are the most effective and commonly used in industries [1]. NDE plays a vital role at various stages of manufacturing – in material property measurements, process design for materials manufacturing, online process control, and quality control [2]. Corrosion detection is a subset of the larger fields of NDE and nondestructive inspection. Corrosion itself is an intriguing multidisciplinary domain. Corrosion engineering combines several disciplines, including mechanics, structures, materials science, chemistry, physics, and numerous sensor technologies. The various NDE methods commonly employed for corrosion detection include visual/optical, ultrasonic testing, eddy current testing, thermography, nuclear radiation-based techniques, etc. In terms of corrosion, visual/optical

Non-Destructive Evaluation of Corrosion and Corrosion-assisted Cracking, First Edition.
Edited by Raman Singh, Baldev Raj, U. Kamachi Mudali and Prabhakar Singh.

nondestructive methods are generally used to detect and measure the topographic changes caused by pitting corrosion on the exposed surface or subsurface corrosion damage in built-up structure. Ultrasonic testing can be used for detection of stress corrosion cracking (SCC), pitting, and intergranular corrosion morphology [3]. One of the typical applications of ultrasound testing is the in-service inspection of the coolant circuit pipes and the bypass lines in boiling water reactor (BWR), where the pipings are made of AISI type 316 stainless steel (SS) [3] and failures are due to intergranular stress corrosion cracking (IGSCC) at weld heat-affected zones. Eddy current testing is one of the techniques for accurate detection and sizing of defects related to corrosion damage such as pits, cracks, and thickness loss [3]. Both amplitude and phase analyses are used in corrosion pit detection and sizing with the assumption that amplitude of eddy current is proportional to the depth of the pit. Thermography is a technique used for detecting coating defects such as blisters and corrosion spots [4] wherein the differences in thermal properties of degraded and nondegraded areas of a coated panel under the influence of a temperature gradient through the thickness of the coated panel is imaged. The resulting thermographic image after image processing identifies the size, location, and extent of degradation. Sonic IR [5] or vibro-thermography has been attracting attention as a nondestructive crack detection method especially for small cracks initiated at the surface, like that for SCC. When a material with a surface crack is fluctuated by acoustic energy, friction between closed crack surfaces causes localized temperature rise, which is detected by Sonic IR. Radiography is extensively used to measure material loss in pipelines with insulation. The digital electronic radiographic image indicates material loss at specific points that can be gauged and graphically plotted. The ground-penetrating radar (GPR) is a potential method for periodic inspection and maintenance of concrete structures [6, 7]. Most of these NDE techniques so far discussed form a part of corrosion inspection, which are performed at periodic intervals to check corrosion. Acoustic emission (AE) has been widely used as an NDE technique for structural integrity monitoring and for leakage detection in pressure systems. It is also a viable tool for online corrosion monitoring. The rapid release of energy within a material due to changes in local stress and strain fields generates transient elastic waves that propagate over a wide range of frequencies ranging from audible frequencies to frequencies in MHz range and can be detected by piezoelectric sensors. AE can occur due to atomistic processes like dislocation motion or due to macroscopic processes like deformation and crack growth. It is also generated during film cracking, gas evolution, hydrogen migration, hydrogen cracking, plastic zone formation, and discontinuous cracking during SCC, as all these processes strain the lattice. However, a few processes during corrosion such as dissolution of metal and film formation do not emit AE, since the lattice is not strained. Migration of atoms during film growth will stress the lattice and produces AE, but when the

Table 7.1 Comparison of detection by AE and EN for some corrosion process.

Process	Detection by AE	Detection by EN
Dissolution of metals	No	Medium to high amplitude
Film formation	No	Low amplitude
Evolution of gas	Yes, low amplitude	Low amplitude
Hydrogen migration	Low amplitude	Low amplitude
Hydrogen cracking	High amplitude	High amplitude
SCC/discontinuous cracking	High amplitude	High amplitude
Motion of electrons	No	Yes

film growth is slow, the energy released will be too low to be detected. Detection of IGSCC by AE is not reliable due to very small amount of energy that is relaxed during deformation. Table 7.1 gives an overview of various corrosion processes detected by AE and electrochemical noise (EN). EN is a nondestructive technique for real-time online corrosion monitoring that requires no external perturbation to decode useful information on the nature and rate of general and localized corrosion from naturally corroding surfaces. The EN technique for corrosion monitoring that emerged in 1968 [8] has evolved into a sophisticated field of technology, particularly because of its credentials over other monitoring techniques and the simple instruments used for the measurement. EN gives an instantaneous online indication of the corrosion events as these occurs. The "signatures" from EN sensors can be used to derive mechanistic information about the corrosion type; especially localized corrosion gives strong EN signals.

7.2 Electrochemical Noise: Definition

The term electrochemical noise [8–15] is a generic term used to describe the low level, low frequency (<1 Hz) spontaneous fluctuations of potentials or current that occurs during an electrochemical process.

7.2.1 EN Measurement

Electrochemical potential noise (EPN) (in the range of mV) is measured between a working electrode or a working electrode pair (coupled through zero resistance ammeter [ZRA]) and a noise-free reference electrode or a pseudoreference electrode (made of the same material as that of the working electrodes). Electrochemical current noise (in the range of μA) is measured between a pair of identical working electrodes using a ZRA to ensure that the two electrodes are at the same electrode potential. Simultaneous measurements of current and potential noise are made at open circuit potential, whereas potentiostatic and galvanostatic [10, 12, 15] methods are used to measure current and potential noise individually.

7.2.2 EN Data Collection

Data collection involves sampling of the measured parameter. If sampling is done above the Nyquist frequency ($f_s/2$), where f_s is the sampling frequency, the results are same as the sampling done at frequencies below Nyquist frequency. This phenomenon is called aliasing [12, 15]. To avoid aliasing, components greater than Nyquist frequency have to be eliminated before analog-to-digital (A/D) conversion using low-pass filter [12, 15–17].

7.2.3 EN Data Analysis

The DC drift or trend is defined as the variation of the mean current or potential divided by time and is not useful for the corrosion phenomenon. Any DC drift can introduce new, false frequency components and hence needs to be removed before data analysis. Some of the procedures [18] to remove drift include moving average removal (MAR), linear regression, polynomial fitting, analog or digital high-pass filtering, wavelets [19, 20], and artificial neural networks [21]. EN analysis involves visual examination of the time record as the first part of a data analysis procedure. The EN time records are further analyzed in the time domain (statistical), frequency domain (spectral), shot noise, artificial neural networks, and chaos–wavelet analysis methods. A few of the analysis methods are briefed in this chapter.

7.2.3.1 Visual Examination of the Time Record

The potential or current time record gives typical fingerprints for various corrosion processes like uniform corrosion, passivation, pitting, crevice, IGSCC, and SCC [8, 12, 22–27].

7.2.3.2 Statistical Analysis

A number of statistical parameters can be derived from the EN time record. Some of the important parameters are discussed below:

1. **Localization index (LI):** It is defined as the rms current (i_{rms}) divided by the standard deviation (σ_i). A range of values of LI (Table 7.2) has been reported for different corrosion mechanisms under the name-pitting index (PI) [28].

Table 7.2 Range of LI for various corrosion processes.

Type of corrosion	LI
Uniform	$0.001 < LI < 0.01$
Mixed	$0.01 < LI < 0.1$
Localized	$0.1 < LI < 1$

Source: From Eden et al. [28].

2. **Standard deviation:** The standard deviation of current (σ_i) and potential noise (σ_v) represent the corrosion status and permit the identification of the nature of attack.
3. **EN resistance (R_n):** EN resistance (R_n), a parameter first proposed by Eden [29], is defined as the ratio of the standard deviation of potential noise (σ_v) to the standard deviation of current noise (σ_i) [29, 30]. Noise resistance is represented as

$$R_n = \frac{\sigma_v}{\sigma_i} \tag{7.1}$$

EN resistance (R_n) has been found to be equivalent to polarization resistance (R_p) [31–36].
4. **Skewness and kurtosis:** Skewness and kurtosis indicates the shape character of distribution of the EN data [12, 37, 38]. The skew is a dimensionless description of the extent to which the distributions of values are skewed about the mean, or in simple terms skewness characterizes the asymmetry of distribution of the data set around the mean value of the data set. Kurtosis gives a measure if the data is peaked or flat compared to Gaussian distribution. Kurtosis of a normal (Gaussian) distribution is 3, and higher values indicate more sharply peaked distribution than a normal distribution, and kurtosis <3 indicates flat-topped distribution. Skewness and kurtosis of EN data is evaluated from the EN data as

$$\text{Kurtosis} = \frac{m_4}{m_2^2} \quad \text{and} \quad \text{skewness} = \frac{m_3}{m_2^{2/3}} \tag{7.2}$$

where m_2, m_3, m_4 are second, third, and fourth moments, respectively,

$$m_2 = \frac{1}{N}\sum_{i=1}^{N}\left(i-\overline{i}\right)^2, \quad m_3 = \frac{1}{N}\sum_{i=1}^{N}\left(i-\overline{i}\right)^3, \quad m_4 = \frac{1}{N}\sum_{i=1}^{N}\left(i-\overline{i}\right)^4$$

and i represents potential or current noise.

Table 7.3 Skewness and kurtosis values for various corrosion processes.

	Potential		Current	
Mechanism	**Skewness**	**Kurtosis**	**Skewness**	**Kurtosis**
General	< ±1	<3	< ±1	<3
Pitting	< −2	≫ 3	> ±2	≫3
Transgranular SCC	+4	20	−4	20
Intergranular SCC ≠ 1	−6.6	18 to 114	1.5 to 3.2	6.4 to 15.6
Intergranular SCC ≠ 2	−2 to −6	5 to 45	3 to 6	10 to 60

Source: From Reid et al. [40].

Skewness and kurtosis have been used to identify corrosion mechanism [12, 16, 39]. Reid et al. [40] have patented values of kurtosis and skewness parameters to identify corrosion mechanism (Table 7.3).

7.2.3.3 Spectral Analysis

Fast Fourier transform (FFT) and mathematical entropy method (MEM) are the two mathematical tools to convert time domain to frequency domain to obtain noise spectra. Noise spectra are represented either as amplitude ($V/\sqrt{Hz}$ or $A/\sqrt{Hz}$) vs. frequency or as noise power (V^2/Hz or A^2/Hz) vs. frequency [10, 15]. Few parameters derived from spectral analysis are defined below:

1. **Roll-off slope and frequency:** In a power spectral density (PSD) plot, the white noise for uniform corrosion appears in the lower frequency region, and the $1/f^n$ noise for localized corrosion appears in the higher frequency range. The slope of the $1/f^n$ noise is called the roll-off, and the frequency at the cross point between the white noise and the $1/f^n$ noise is defined as the roll-off frequency [22, 27]. Roll-off slope of −2 to −4 has been reported for localized corrosion, while values close to zero represent uniform corrosion [27]. Although roll-off slope has been used by many investigators, there are controversial views in using this parameter for determining corrosion mechanism.
2. **Spectral noise resistance (R_{sn}):** The ratio of PSD of potential to that of current, at every frequency, is termed as the spectral noise resistance [41].

$$R_{sn}(f) = \left| \frac{V_{PSD}(f)}{I_{PSD}(f)} \right|^{1/2} \tag{7.3}$$

where $V_{PSD}(f)$, $I_{PSD}(f)$ are the PSD functions of voltage and current signals at a particular frequency, f. At very low frequencies, the spectral noise resistance R_{sn}^0 is determined as the DC limit of these plots as

$$R_{sn}^0 = \lim_{f \to 0} \{R_{sn}(f)\} \tag{7.4}$$

For some systems R_{sn}^0 has been found to be equivalent to R_n and R_p [16, 42, 43].

7.2.3.4 Shot Noise Analysis

According to shot noise theory [12, 44–46], signals are composed of packets of data departing from the baseline. In corrosion systems, current noise signals could be considered as a series of statistically independent packets of charge having a short duration and shot noise is produced by current being carried by discrete charge carriers. The theory assumes some restrictions [12, 46–49] on the noise generating process when applied to EN analysis. Based on the assumptions, it is possible to determine the charge in each event (q) and the frequency of appearance of each event (f_n):

$$q = \frac{\sqrt{PSD_v \times PSD_i}}{B} \tag{7.5}$$

$$f_n = \frac{B^2}{(A \times PSD_v)} \tag{7.6}$$

where PSD_v and PSD_i are the PSD values at low frequency, B is the Stern–Geary coefficient, and A is the surface area of specimen exposed to corrosion. The charge in individual transient is related to the mass of metal lost in the corrosion events, and f_n gives information about the rate at which these events are happening or frequency of the events. Short noise parameters can be related to corrosion process. Uniform corrosion is associated with high q and high f_n, while localized corrosion gives high value of q and low f_n. For passive systems, q is small and f_n depends on processes that takes place in the passive layer. The cumulative probability, $F(f_n)$, at each f_n is derived from the set of f_n values for a particular experiment. The f_n values are initially arranged in the ascending order for N data sets. The cumulative probability is evaluated as $M/(N+1)$, where M is the rank in the data set. The Weibull distribution function [44, 46] is a cumulative probability function that is used to predict life and failure rate. The mean time to failure, t_n, which corresponds to $1/f_n$, is expressed as

$$\ln\left(\ln \frac{1}{1 - F(t_n)}\right) = m \ln t_n - \ln n \tag{7.7}$$

where m and n are scale and shape parameters.

7.2.3.5 Wavelet Analysis

Wavelet transforms are used to study stationary and nonstationary EN signals [50, 51]. The wavelets are small waves that grow and decay in a limited time period and have an average value of zero. A set of wavelets of varying amplitude, duration, and location can then be constructed such that their sum reproduces the signal of interest. An EN signal (time record), X_n ($n = 1, ..., N$), can be represented as a linear combination of basis functions $\Phi_{j,k}$ and $\psi_{j,k}$, which are formed from the mother wavelet $\psi(t)$ and father wavelet $\Phi(t)$ by translating in time and dilating in scale [50–54]:

$$\varphi_{j,k}(t) = 2^{-j/2}\psi\left(2^{-j}t - k\right) \tag{7.8}$$

$$\phi_{j,k}(t) = 2^{-j/2}\phi\left(2^{-j}t - k\right), \tag{7.9}$$

j, $k \in Z$ where $k = 1, 2, ..., N/2$, N is the number of data record, Z is set of integers, and $j = 1, 2, ..., j$, j is a small natural number that depends on N. For full decomposition of a finite length signal, $2^j = N$.

The time record $x(t)$ can be represented as

$$x(t) = \sum_k s_{j,k}\phi_{j,k}(t) + \sum_k d_{j,k}\psi_{j,k}(t) + \sum_k d_{j-1,k}\psi_{j-1,k}(t) + \cdots + \sum_k d_{1,k}\psi_{1,k}(t) \tag{7.10}$$

where $s_{j,k}$, $d_{j,k}$, ..., $d_{l,k}$ are the wavelet coefficients, j is a small natural number that depends on N and the basis function, k takes a value from 1 to the number of coefficient in the specified component, and s and d are the approximation or smooth coefficient and detailed coefficient, respectively. There are three operations involved in the fast wavelet transform (FWT) algorithm, which includes a low-pass filter, a high-pass filter, and a downsampler. Each filter has a downsampler after it [50–54]. At the first stage, the low frequencies of the signal pass through the low-pass filter, and the high frequencies pass through high-pass filter, producing the smooth coefficient set $S_1 = (s_{1,1}, s_{1,2}, ..., s_{1,N/2})$ and detailed coefficients, $D_1 = (d_{1,1}, d_{1,2}, ..., d_{1,N/2})$, respectively. The detail is kept aside for analysis, and the smooth signal is further filtered in the second stage to produce S_2 and D_2. This process is iterated successively j times to decompose the signal into $D_1, D_2, ..., D_j$ and S_j coefficients. Each set of coefficients $D_1, D_2, ..., D_j$ is called a crystal. The set of approximation coefficients, S, carries information about the general trend of the signal. The set of detailed coefficients, D, carries information about the local fluctuations in the signal. Each new set of crystal has half the resolution and contains half the number of coefficients and is designated with a successive higher number.

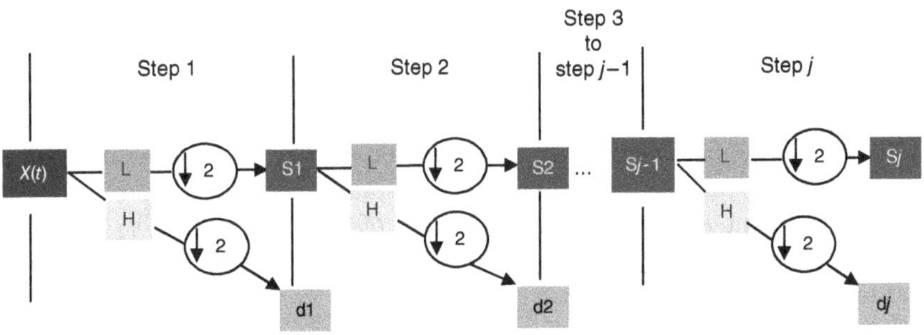

Figure 7.1 General scheme of fast wavelet transform (FWT) algorithm. *Source:* Abelle et al. [51]. Reproduced with permission of Elsevier.

Each of the $j+1$ signals is called the partial signal, which resembles the fluctuations of the original signal at a particular interval of frequency. A typical schematic of the FWT algorithm [52] is shown in Figure 7.1. The plot of standard deviation of partial signal (SDPS) as a function of crystal name is called SDPS plots [54]. It could be used to indicate the extent of corrosion. The plot of the energy of the partial signals against crystal numbers called as energy distribution plots (EDP) [50–54] gives information about the corrosion mechanism.

7.3 Applications of Electrochemical Noise

EN covers a wide spectrum of applications that includes identifying and distinguishing various corrosion types [27, 55], monitoring crack propagation in SCC [56, 57], evaluation of corrosion inhibitors [58], coating evaluation [59], monitoring MIC [55], and many others. Some of the applications are discussed in the forthcoming section.

7.3.1 Identification of Corrosion Types

7.3.1.1 Passivation and Uniform Corrosion

Figure 7.2a and b shows EN time record for passivation and uniform corrosion of 304L SS. Passivation process is characterized by low amplitude (nA range), highly repetitive current noise (white noise feature) and almost nil potential fluctuations, as shown in Figure 7.2a. Higher amplitude and medium repetitive rate of current and potential noise are indicative of uniform corrosion (Figure 7.2b). The noise amplitude for uniform corrosion would be lower than that of localized corrosion and higher than passivation process.

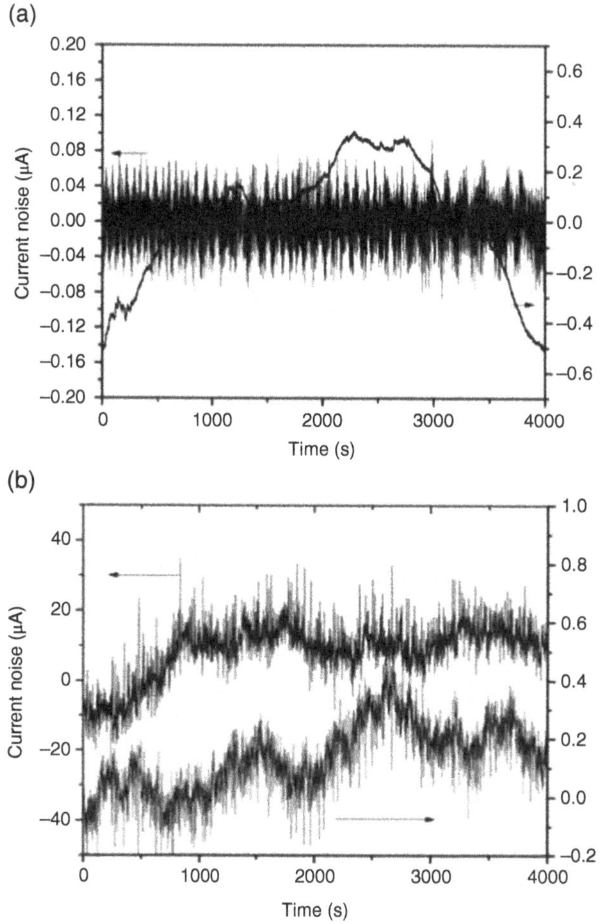

Figure 7.2 EN time record for (a) passivation of 304L SS in 0.1% NaOH and (b) uniform corrosion of 304L SS in 5% H_2SO_4. *Source:* Girija et al. [27]. Reproduced with permission of Elsevier.

7.3.1.2 Localized Corrosion

EN generated from localized corrosion is composed of high amplitude, distinct potential and current transients.

7.3.1.2.1 Pitting Corrosion

The shape of the transient during pitting corrosion depicts a sudden drop in potential with an exponential recovery till the initial value is reached. When a small, high rate dissolution site such as a pit is produced, the remainder of the surface acts as the cathode. Because oxygen reduction is sluggish on the oxide-covered surface and is diffusion limited, the potential of the pit and the passive surface falls rapidly [60]. The potential drop corresponds to pit initiation and the local minimum corresponds to the start of repassivation. The slow recovery of the potential does not result from repassivation, as no current flows to the pitted area during potential recovery. When a pit initiates on the metal surface, part of the charge required for the growth of the pit is drawn out from the passive area nearby. Once the pit ceases, the passive surface of metal is recharged by the cathodic reaction. The potential increase after the local minimum reflects recharging of the capacitance of the passive surface by reduction of oxygen [11, 61]. Metastable pits have a limited lifetime of few seconds and immediately repassivate [34] (Figure 7.3a), whereas stable pits continue to grow [62] (Figure 7.3b).

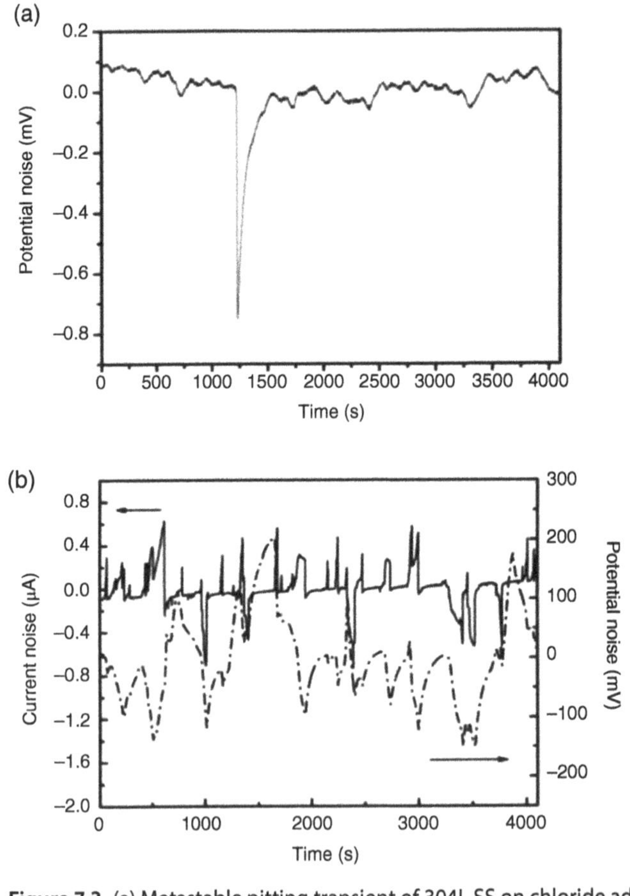

Figure 7.3 (a) Metastable pitting transient of 304L SS on chloride addition to 0.1% NaOH. *Source:* Girija et al. [27]. Reproduced with permission of Elsevier. (b) Stable pitting of 304L SS in 0.05 m $FeCl_3$. *Source:* Girija and Kamachi Mudali [62]. Reproduced with permission of NACE International.

The EDP obtained from wavelet analysis are useful for tracing the mechanism of corrosion. EDP of the EN time records taken after 1 and 25 h of immersion of 304L SS [62] in 0.05 M $FeCl_3$ are shown in Figure 7.4.

According to literature reports [52, 53], the position of the maximum relative energy distribution of a crystal (defined in Section 7.2.3.5) corresponds to dominant process contributing to the signal. For an eight-level decomposition, EDP can be divided into three zones. The relative energy distribution concentrated on smaller time scale crystals D1–D3 gives information about rapid events such as metastable pitting, the medium time scale D4–D6 crystal is

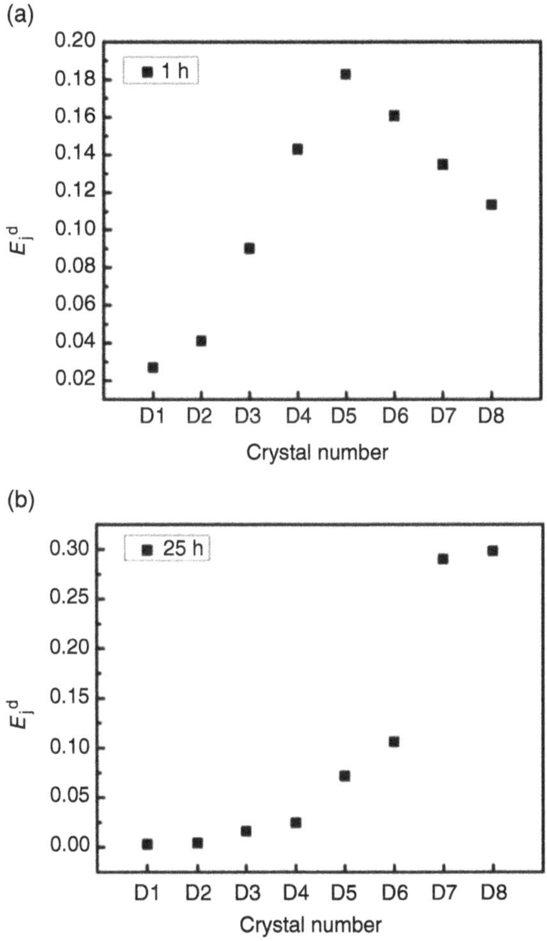

Figure 7.4 EDP of current noise after (a) 1 h of immersion and (b) 25 h of immersion. *Source:* Girija and Kamachi Mudali [62]. Reproduced with permission of NACE International.

associated with repassivation/propagation of pits, and the large time scale D7, D8 crystal gives information on diffusion or growth of pits [63–66]. From the EDP after 1 h immersion, it is observed that the relative energy is defined on crystals D2, D3, D4, D5, D6, D7, and D8 and the maximum contribution is on D5. It is inferred that during the initial immersion period, three processes are occurring on the specimen surface, namely, (i) metastable pitting, (ii) propagation and repassivation of pits, and (iii) stable pitting, and the predominant process is propagation/repassivation of pits. From the EDP after 25 h of immersion, it is found that the energy concentration on D1–D3 is negligible and that on D4, D5 has decreased. Most of the energy is concentrated on D6–D8 with maximum deposition on D8 crystal, which indicates that the dominant process occurring on the specimen surface is stable pitting.

7.3.1.2.2 Crevice Corrosion

Noise transients during crevice corrosion of SS are similar to that produced by a stable pit, but with an extended life. The film breakdown occurs within the crevice, and the metal dissolution continues as a large current transient. The onset of crevice attack is accompanied by a rapid fall in potential noise [12, 67–70] as a result of the active crevice polarizing the passive external surface.

The noise output of the electrode undergoing crevice corrosion [12, 70] is characterized by relatively long bursts of regular peaks occurring at regular interval, as shown in Figures 7.5 and 7.6. According to Cottis et al. [12], in cases where the passive surface cannot sustain the current needed to maintain the crevice in the active state, potential oscillations with chaotic nature will be observed.

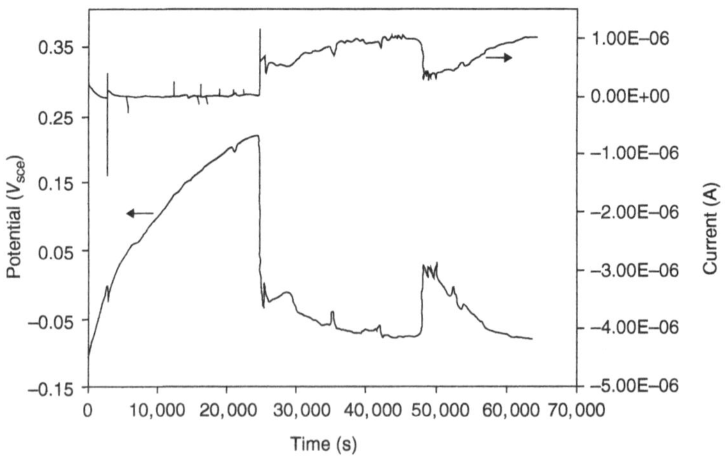

Figure 7.5 Typical EN time record during crevice corrosion. *Source:* Cottis [12]. Reproduced with permission of NACE International.

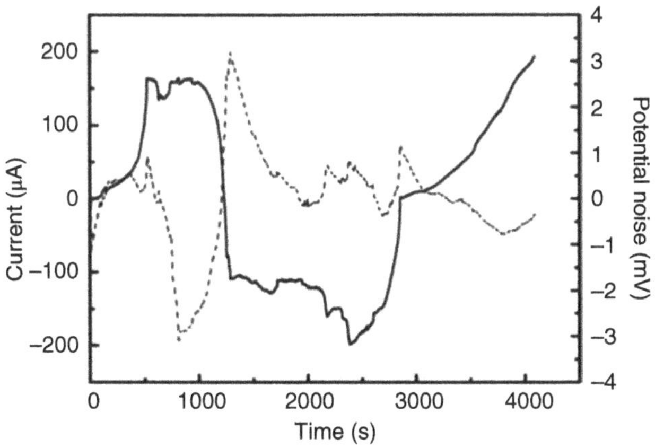

Figure 7.6 EN time record during crevice corrosion of 304L SS in 0.05 M $FeCl_3$. *Source:* Suresh et al. [70]. Reproduced with permission of *Journal of the Electrochemical Society of India.*

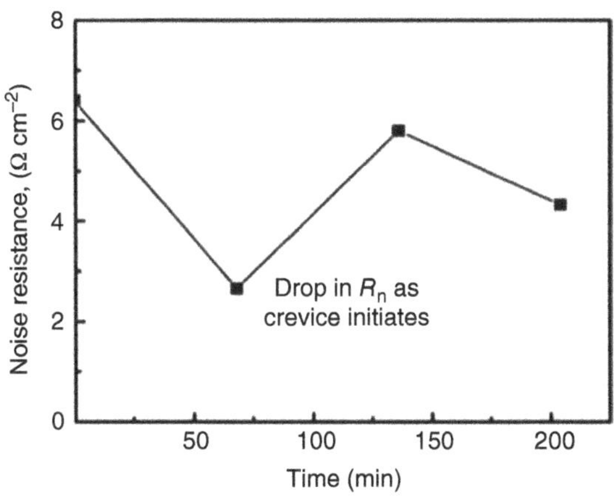

Figure 7.7 R_n–time plot obtained for the immersion of 304L SS in 0.05 M. *Source:* Suresh et al. [70]. Reproduced with permission of *Journal of the Electrochemical Society of India.*

In other cases, the potential will fall and remain low. Crevice initiation can be observed by a profound drop in noise resistance [70] (Figure 7.7).

7.3.1.2.3 Stress Corrosion Cracking

Anodic transients that occur during SCC are the result of metal dissolution at the crack wall and hence are similar to pit initiation. The current transients during SCC are of greater amplitude and repeat at longer time intervals than those that occur during pitting. Anita et al. [71] have obtained widely spaced

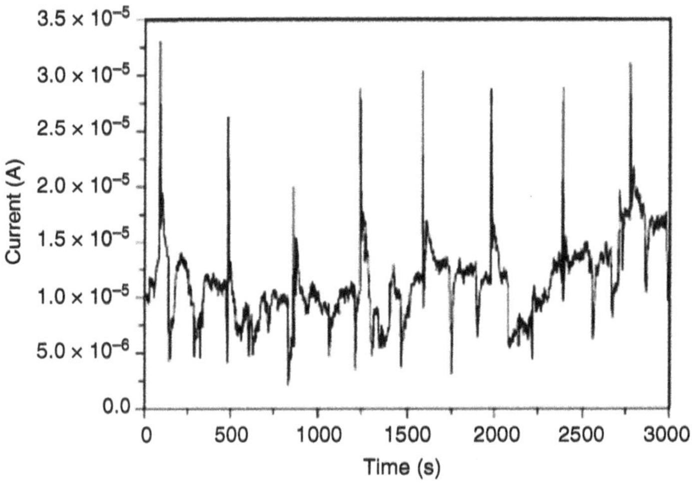

Figure 7.8 Current noise during propagation of SCC, 32 h after exposure of 316 SS U-bend specimen in boiling acidified NaCl. *Source:* Anita et al. [71]. Reproduced with permission of Elsevier.

and high amplitude current transients during propagation of SCC of 316 SS U-bend specimens in boiling acidified NaCl solution (Figure 7.8). Similar observations of widely spaced current transients have been reported by Kim and Kim [72] in their study on Pb-assisted SCC of alloy 600 in simulated environment of a steam generator sludge pile at high temperatures under potentiostatic conditions. It is seen in Figure 7.9 that the current transients between points "B" and "C" are characterized by repetitive occurrences of sudden rise and steep decay with short time intervals during initiation of SCC. The current increase followed by slower decay with longer time intervals in the time record at points D and E in Figure 7.10 are attributable to its propagation.

In order to evaluate the stochastic characteristics of initiation and propagation of SCC, Kim and Kim [72] performed the same test under open circuit conditions to measure both potential and current transients simultaneously. Tests were carried out for 278 h in the leaded caustic solution (40 wt% NaOH with 0.01 wt% Pb), followed by addition of 500 ppm CuO, and the test was continued for another 122 h. CuO is an oxidizing species that will ennoble the corrosion potential and promote Pb-assisted SCC to occur under open circuit conditions. EN time records obtained during the test are shown in Figure 7.10. The points corresponding to A, C, D, and F in Figure 7.10a and b are characterized by current rise accompanied by potential drops, which occurs when oxide film breaks and metal dissolves. The potential recovery corresponds to repair of surface oxide film. The repetitive and faster current transients at points A and C are attributed to initiation of SCC, while the more localized transients at D and F are attributed to propagation of SCC.

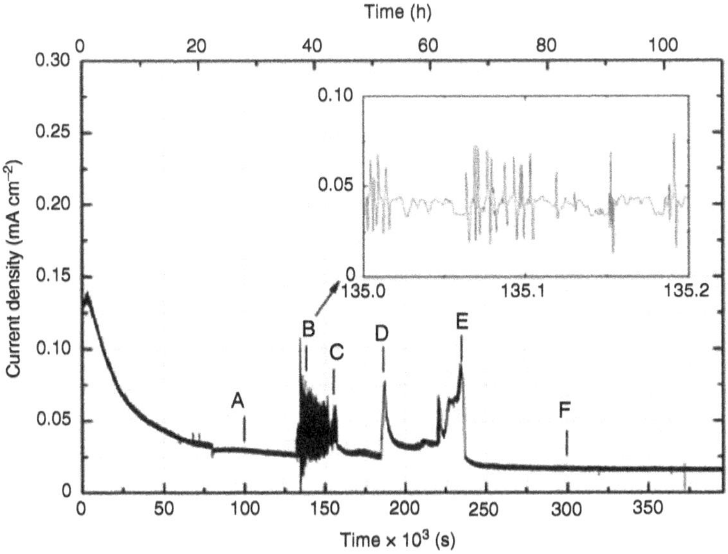

Figure 7.9 Current noise from the stressed C-ring specimen in 40 wt% NaOH solution containing 0.01 wt% PbO at 315 °C after 110 h exposure. *Source:* Kim and Kim [72]. Reproduced with permission of Elsevier.

Stochastic analysis of the time records was done by applying Weibull distribution function using Eq. (7.7). Figure 7.11 represents the plots of $\ln\{\ln[1/(1-F(t_n))]\}$ vs. $\ln t_n$ calculated from time records. The mean time to failure, t_n, obtained from the EN measurement of two identical unstressed alloy 600 specimens in 40 wt% NaOH solution without impurities at 290 °C is also plotted for comparison. Only general corrosion occurred on the unstressed specimen surface without any cracking. The mean time to failure for the initiation and the propagation of SCC shifted to a higher value, analogous to the shift of f_n to a lower frequency, when compared with that for the general corrosion. The shape parameter "m" of the Weibull distribution function is an indicator of types of corrosion. The value of "m" was determined from the figure by a linear curve fitting method. A value of 1.46 was obtained for the unstressed specimen that underwent general corrosion. Similarly, a value of 0.47 was evaluated for the initiation of SCC on the stressed specimen in the leaded caustic solution, and 0.21 for its propagation in the leaded caustic solution plus CuO. The shape parameter "m" was used to distinguish between initiation and propagation of SCC.

7.3.1.2.4 Microbiologically Influenced Corrosion

Microbiologically influenced corrosion occurs when a metal is exposed to aquatic environment containing microorganisms that tend to adhere to the metal surface, multiply, grow, and produce biofilm. The microorganisms and

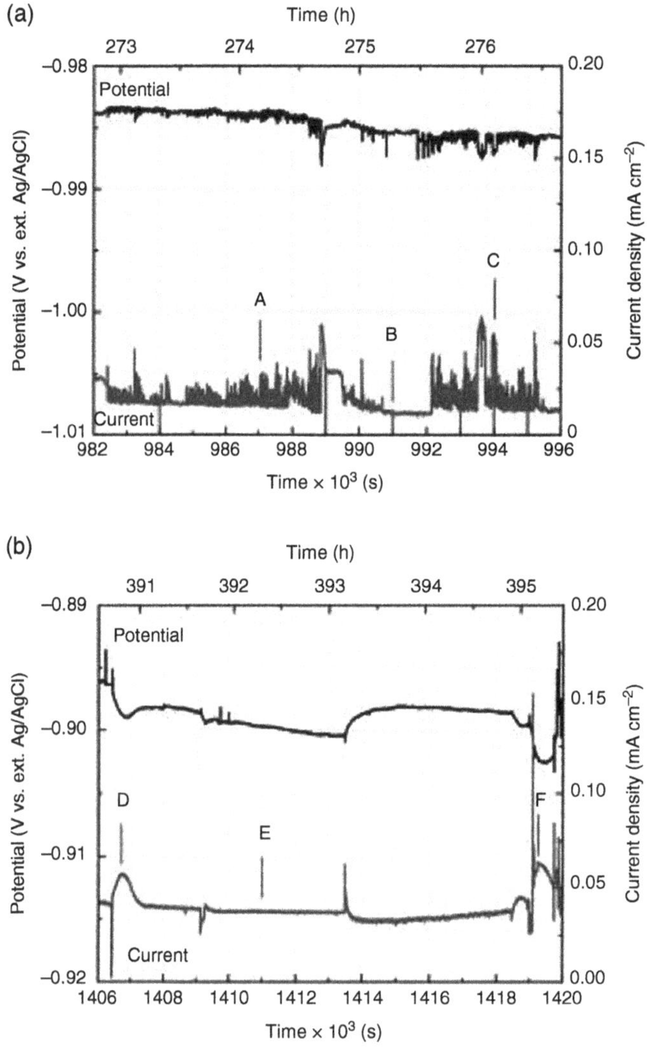

Figure 7.10 Potential and current noise–time records for (a) leaded caustic solution and (b) leaded caustic solution containing CuO at 290 °C. *Source:* Kim and Kim [72]. Reproduced with permission of Elsevier.

their products such as enzymes, organic and inorganic acids, exopolymers, and volatile compounds like ammonia and hydrogen sulfide can alter the equilibrium of anodic and cathodic reactions at the metal–biofilm interface, leading to MIC. EN is one of the electrochemical techniques to detect MIC. Pujar et al. [55] carried out EN studies on 316LN SS in sterile and biofilm-forming medium

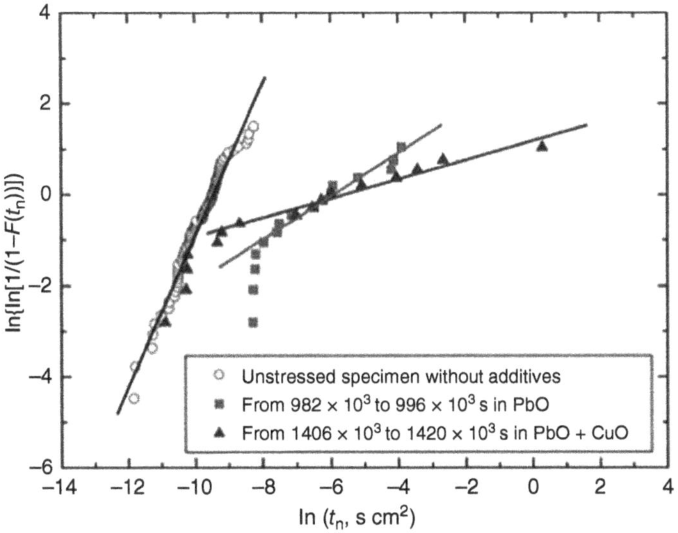

Figure 7.11 Plot of $\ln\{\ln[1/(1-F(t_n))]\}$ vs. $\ln t_n$. *Source:* Kim and Kim [72]. Reproduced with permission of Elsevier.

containing chloride ions. The current and potential time records after 8 days of immersion in biofilm-forming medium is shown in Figure 7.12a. The time record shows a drop in potential and a corresponding rise in current noise, which is indicative of crevice attack. However, in the sterile medium (Figure 7.12b), very low amplitude fluctuations are seen, indicating stable passive film on the metal surface. Shot noise analysis of the EN time records indicated significant changes in shot noise parameters during MIC initiation.

The characteristic charge and characteristic frequency evaluated from shot noise analysis are plotted against time of exposure to microbial and sterile medium in Figure 7.13a. It is observed that the characteristic frequency (f_n) is low and characteristic charge (q) is high in microbial medium and the reverse trend is observed for sterile medium. A low f_n and high q are associated with localized corrosion, and it is evident that the SS is undergoing localized attack in the microbial medium for the exposure period. Post-experimental scanning electron microscopy (SEM) image (Figure 7.13b) showed crevice attack of the steel in the microbial medium.

7.3.2 Evaluation of Corrosion Inhibitors

The efficiency of organic inhibitors is associated with the adsorption of the inhibitor on metal surface forming a protective layer; as a result, the metal surface is protected from the environment. EN is a potential technique for the

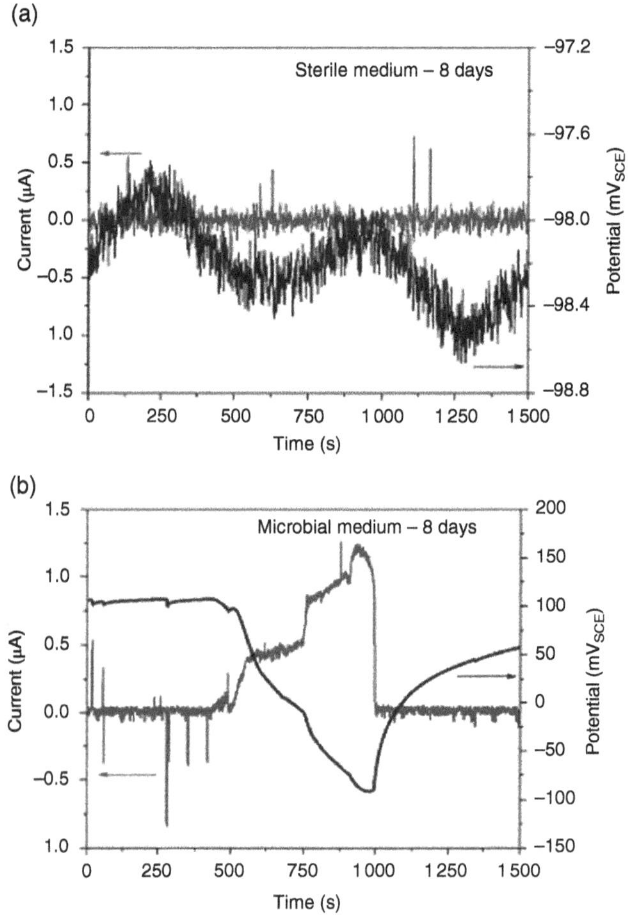

Figure 7.12 EN time record after 8 days of immersion in (a) sterile and (b) microbial medium. *Source:* Pujar et al. [55]. Reproduced with permission of NACE International.

evaluation of inhibitors used for corrosion protection. Markhali et al. [73] used EN to evaluate corrosion inhibition of azole compounds (benzotriazole [BTR], benzothiazole [BNS]) on SS in acidic media. Figure 7.14 shows the electrochemical current noise taken at different durations of exposure of austenitic SS in hydrochloric acid solution in the presence and absence of inhibitor. During the initial stages of immersion (Figure 7.14a), the amplitude of current fluctuations is found to be higher in the acid medium without inhibitor. With time of immersion, metastable pitting transients appeared in the time record of the SS immersed in hydrochloric acid without inhibitor (Figure 7.14b), while transients are negligible in the presence of inhibitor. Characteristic charge (q)

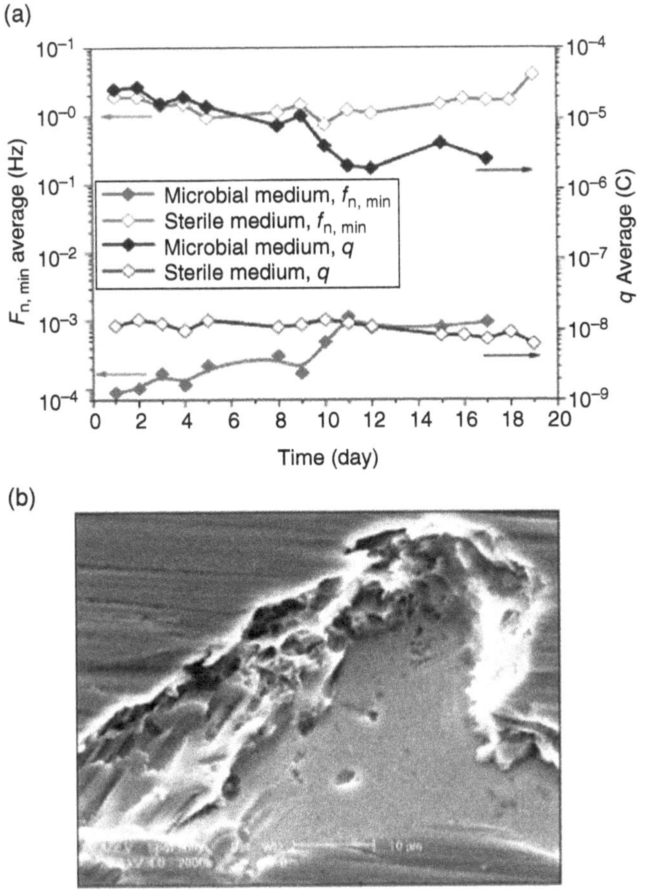

Figure 7.13 (a) Frequency of events (F_n) and charge (q) as a function of time for 316LN SS in sterile and microbial medium. (b) SEM image of crevice attack of the steel in microbial medium. *Source:* Pujar et al. [55]. Reproduced with permission of NACE International.

evaluated from shot noise analysis of the EN signals (Table 7.4) showed an increase in acid and a decrease in inhibited acid solution, indicating that corrosion resistance is enhanced in the inhibited solution.

EN studies indicated that SS that are susceptible to pitting attack in hydrochloric acid could be protected from corrosion attack in the presence of azole inhibitors.

7.3.3 Evaluation of Organic Coatings in Marine Atmosphere

Metikis-Hukovic et al. [74] studied the protective qualities of phenol-formaldehyde/coumarone-indene resin-based coatings on low alloy steel under stimulated

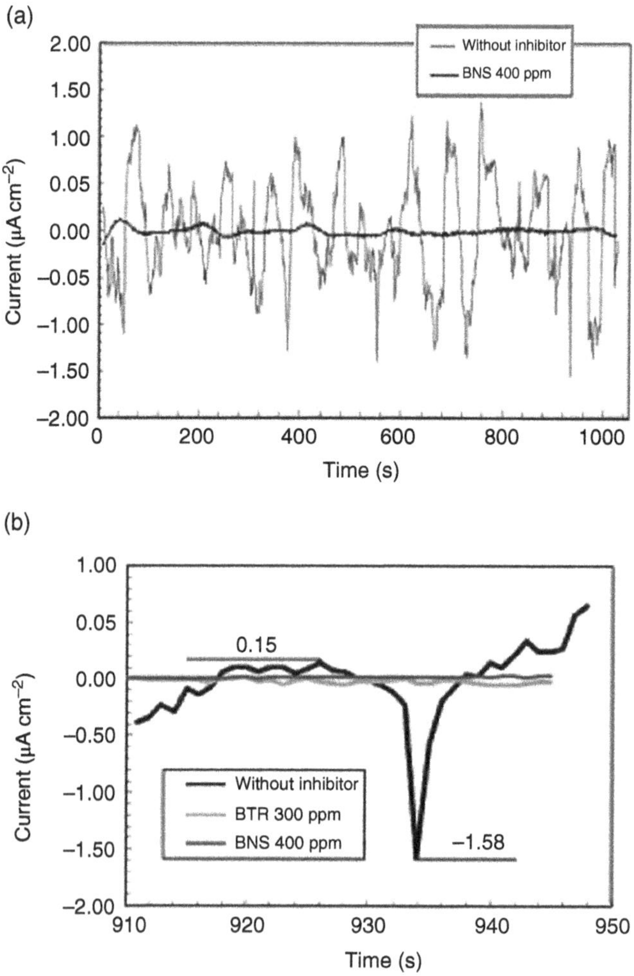

Figure 7.14 EN time record (with and without inhibitor) during (a) initial stages of immersion of stainless steel in hydrochloric acid (b) after 4 h of immersion. *Source:* Markhali et al. [73]. Reproduced with permission of Elsevier.

Table 7.4 Characteristic charge for the stainless steel in 1 M HCl, with and without inhibitor.

Inhibitor	Conc. (ppm)	*B* (mV/decade)	*q* (C)
No inhibitor	—	46.2	6.2×10^{-5}
BTR	50	20.9	3.8×10^{-5}
	100	17.6	3.3×10^{-5}
	300	19.6	2.6×10^{-5}
BNS	200	24.5	3.9×10^{-5}
	300	26.0	1.6×10^{-5}
	400	26.1	5.8×10^{-6}

Source: Markhali et al. [73]. Reproduced with permission of Elsevier.

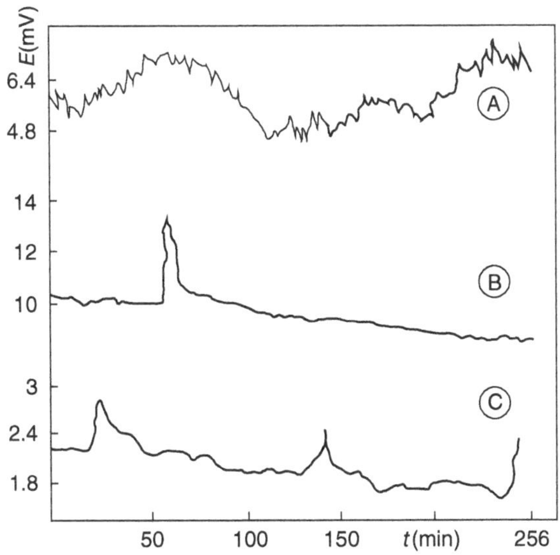

Figure 7.15 Potential noise–time plot for EN sensor coated with (A) phenol-formaldehyde resin, (B) coumarone-indene resin, and (C) uncoated surface after 30 h. *Source:* From Metikis-Hukovic et al. [74].

conditions of marine atmospheric environment using potential noise measurements. Figure 7.15 shows potential noise–time plots obtained under simulated marine (20% NaCl, $T = 308\,K$) atmospheric environments for EN sensor coated with phenol-formaldehyde-based coating (curve A), coumarone-indene-based coatings (curve B), and uncoated mild steel (curve C) after 30 h immersion. There are many similarities between potential time records for the sensor coated with coumarone-indene-based resin and for the unprotected steel under the same conditions. Figure 7.16 shows the potential noise spectra for 6 and 30 h exposure, respectively. The slopes of the power spectral plots were evaluated. It was observed that for an exposure of 6 h, the slopes appeared larger on surfaces coated with coumarone-indene polymer and unprotected steel surfaces, which is attributed to an increase in the onset of pitting corrosion. After 30 h of exposure, the difference in slope decreased, indicating that the surface of the steel coated with phenol-formaldehyde polymer has been attacked in the meantime by pitting corrosion.

EN analysis indicated that phenol-formaldehyde-based coating offered good corrosion protection for mild steel in marine atmospheric environments. In another study [75] related to on EN application for coating performance, EN resistance and slope of PSD were found be useful in evaluating coating performance of thick polyurethane film coated on cold-rolled steel samples. The corrosion test was carried out in 0.5 M NaCl environment. It was

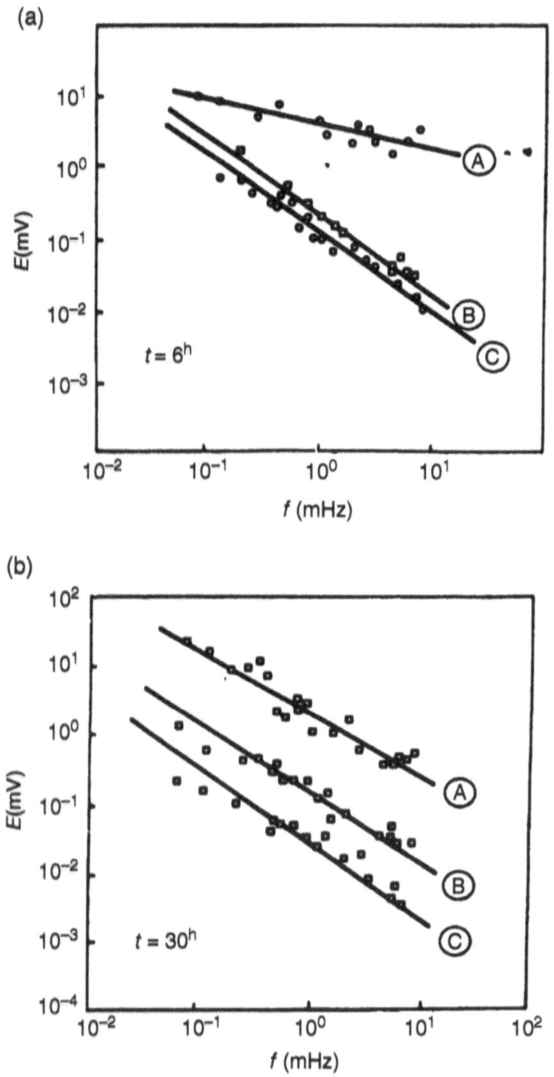

Figure 7.16 Potential noise spectra under simulated marine atmospheric environments for EN sensor coated with (A) phenol-formaldehyde resin, (B) coumarone-indene polymer, (C) and uncoated surface after (a) 6 h and (b) 30 h exposure. *Source:* From Metikis-Hukovic et al. [74].

found that the PSD plots (Figure 7.17) transformed from white noise feature to 1/*f* trend after 2000 h of immersion. The noise resistance plot (Figure 7.18) showed a decrease after 2000 h of exposure, followed by a rise and a drop again after 3000 h of exposure. The coating failure was marked by increase in slope of the PSD plot and drop in resistance.

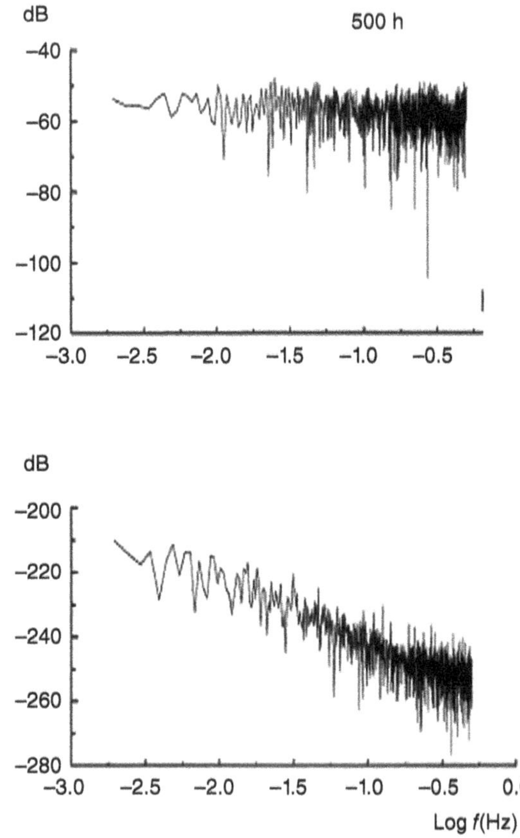

Figure 7.17 Current PSD plots for polyurethane film-coated cold-rolled steel after 500 and 2000 h of exposure in 0.5 M NaCl. *Source:* Mojika et al. [75]. Reproduced with permission of Elsevier.

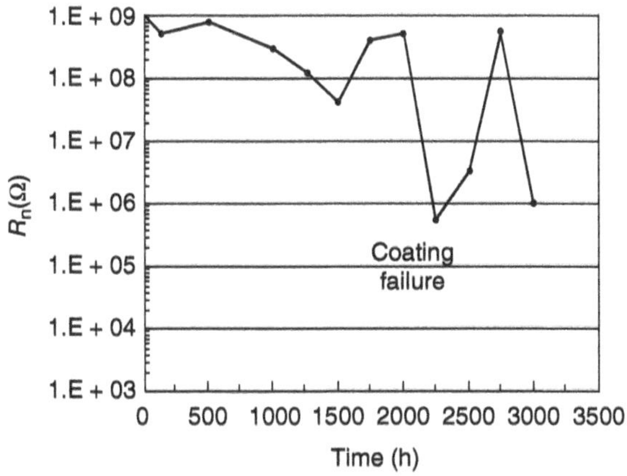

Figure 7.18 R_n–time plots for polyurethane-coated cold-rolled steel in 0.5 M NaCl. *Source:* Mojika et al. [75]. Reproduced with permission of Elsevier.

7.3.4 Online Monitoring of Delamination in TiN Coatings on 316L SS Subjected to Wear

Corrosive wear or tribocorrosion is the degradation of a material in the presence of both wear and corrosion. The material loss during corrosive wear is much more than that when only either of the two exists. Fretting is a special corrosion wear process that occurs at the contact area between two materials under load and subject to minute relative motion by vibration or some other force. For a coating system, the corrosion wear depends on various factors such as the coating thickness, coating and substrate property, adhesion, etc. EN technique is sensitive to changes occurring during corrosion wear and hence could be used as a tool to monitor fretting corrosion and coating delamination due to fretting. Quan et al. [76] carried out online EN monitoring in a fretting test performed on TiN-coated 316L SS sliding against corundum in 0.5 M H_2SO_4 medium under a load of 5 N, 10 Hz frequency. Physical vapor-deposited (PVD) TiN coatings have a columnar structure with pinholes that permit seepage of corrosive media to reach the substrate and result in localized corrosion. The electrochemical behavior of TiN coatings depends on the substrate properties. If the substrate is passive, the electrochemical response describes the coating properties; if the substrate is active, the electrochemical response is dominated by dissolution of the substrate at the pores. Figure 7.19a shows a steady-state phase of fretting corrosion, during which there is successive removal and reformation of passive film in the tribo-activated wear scar on the TiN coatings. The low current noise during the steady-state phase suggested that TiN coatings are resistant during the steady state of corrosion wear. As exposure time increased, a sudden and sharp rise in current and simultaneously a sharp drop in potential occurred in the time record corresponding to the delamination of the coating. The delamination was confirmed by SEM (Figure 7.19b).

7.3.5 Corrosion Monitoring Using EN for Industrial Applications

Mild steel suffers from corrosion damage due to the presence of sulfuric acid and hydrochloric acid generated in the flue gas due to moisture. A multi-element sandwich type of sensor [77] that monitors the corrosion of steel inside the pipeline carrying the flue gas is shown in Figure 7.20. The multi-element probe is useful in making simultaneous measurements of EN, electrochemical impedance spectroscopy (EIS), ZRA, and electrical resistance (ER). This is a controlled temperature sensor and is useful in detecting the effects of temperature variations in the plant. The response of the sensor reflecting the corrosion behavior of mild steel in the flue duct during warm-up from cold state is shown in Figure 7.21.

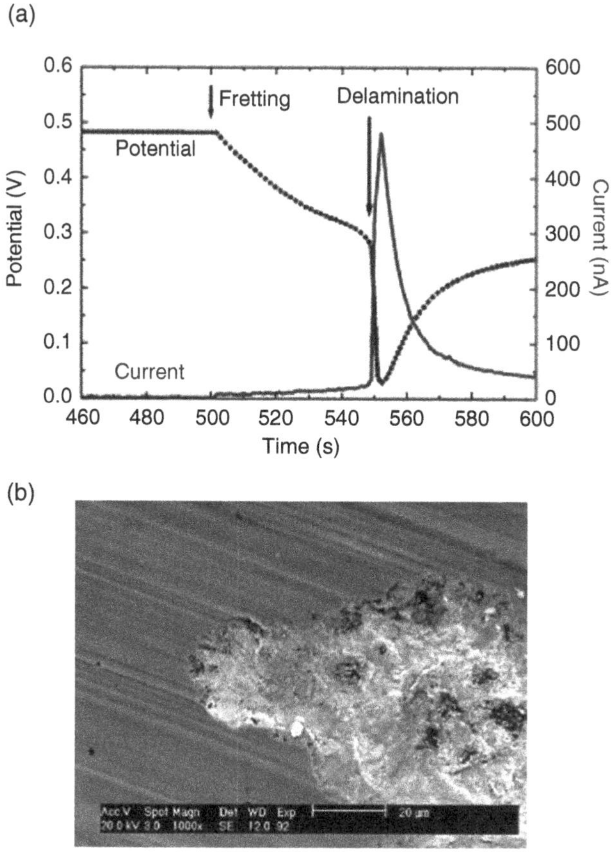

Figure 7.19 (a) EN time record and (b) SEM micrograph of wear scar on TiN-coated AISI 316 stainless steel sliding against corundum in 0.5 M H_2SO_4 (5 N, 10 Hz, 200 μm). *Source:* Quan et al. [76]. Reproduced with permission of Elsevier.

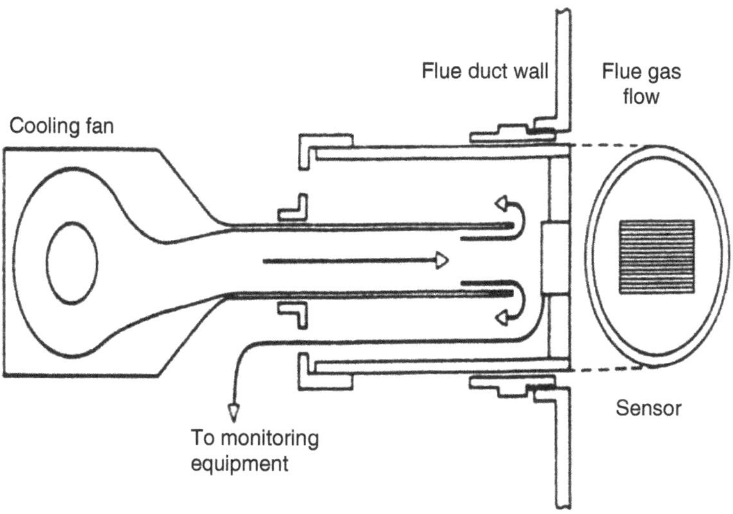

Figure 7.20 Multi-element sandwich type of sensor that monitors the corrosion of steel inside the pipeline carrying the flue gas. *Source:* Cox et al. [77]. Reproduced with permission of Elsevier.

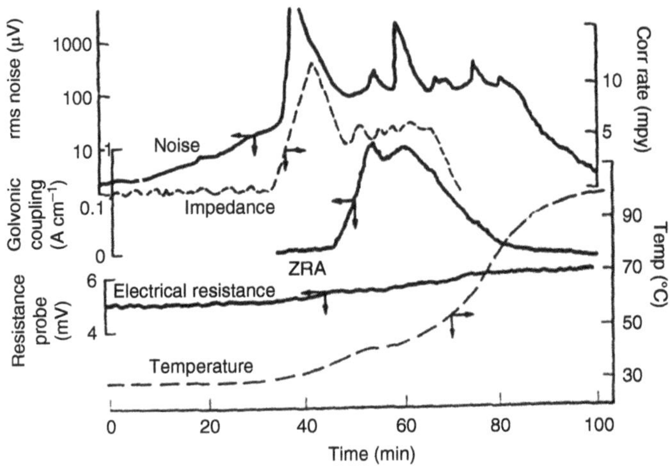

Figure 7.21 Corrosion behavior of mild steel in the flue duct during warm-up from cold state. *Source:* Cox et al. [77]. Reproduced with permission of Elsevier.

From the figure, it is seen that the EN and EIS traces rose immediately with increase in temperature, indicating the onset of corrosion attack. EN trace increased more rapidly due to its sensitivity to localized phenomenon. The ER probe response was almost negligible to the corrosion process. It is noted that the EN response is the most sensitive, compared with ER, impedance, and ZRA. The noise traces established that the corrosion activity in the flue duct increased only during warm-up period of the plant.

7.3.6 Corrosion Monitoring Relating to Nuclear High Level Waste Storage Application

The Department of Energy (DOE) sites in the United States uses underground storage tanks made of carbon or SS to store radioactive liquid waste generated from weapon production and nuclear fuel reprocessing sites [78–80]. Historically, corrosion monitoring and control in nuclear waste tanks of DOE's site were accomplished through waste chemistry sampling and analysis, coupon measurement, linear polarization, and ER, but with limited degrees of success. Corrosion monitoring using EN probes has been established since 1995 in carbon and SS waste tanks holding caustic and acidic waste produced from nuclear installations of Hanford Site, Savannah River Site (SRS), Oak Ridge Reservation (ORR), and Idaho National Engineering and Environmental Laboratory (INEEL).

7.3.6.1 Corrosion Monitoring at Hanford Site

A prototype EN probe was deployed in 1966 in one of the waste tanks, after which a full-scale system was deployed in the successive years, from 1967

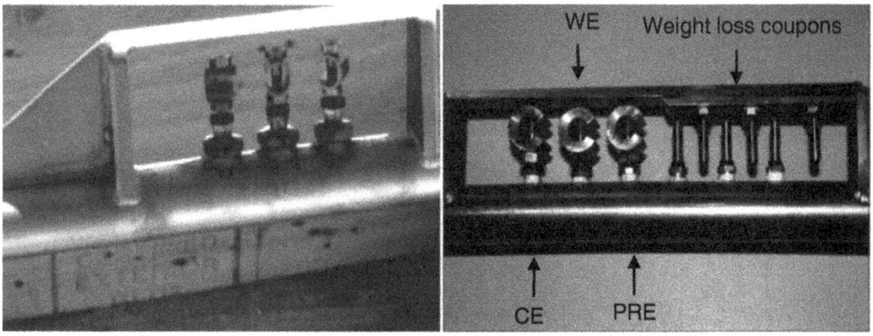

Figure 7.22 Hanford probe (waste tank 241-AN-107) for monitoring in vapor and liquid waste: along the length are 4 two-channel electrode arrays. CE, counter electrode; PRE, pseudoreference electrode; WE, working electrode. *Source:* Edgemon [80]. Reproduced with permission of NACE International.

to 2001, with improvements made each time based on prior experience. Figure 7.22 shows the photograph of the Hanford EN probe [80] deployed in one of the waste tanks.

The three identical electrode probe is made of SS immersed in the waste that is connected to automated data collection system. Four channels comprise C-ring (stressed, notched, pre-cracked) and four bullet electrodes for the detection of SCC and pitting/uniform corrosion, respectively. To assist SCC, one electrode of each C-ring array was pre-cracked and strained prior to immersion in waste. Individual radiation-resistant insulated conductors are attached to each electrode and lead up through the SS probe body to the probe top, from where cables are used to transmit signal from the probe electrodes to the data collection hardware. Figure 7.23 shows actual EN data collected from one of the tanks of Hanford (241-AZ-101) [81].

EN time records showed pitting transients that were observed after five separate water additions (7000–26 000 gal). As the tank was inhibited, pitting corrosion decreased with time, and the tank was re-equilibrated.

7.3.7 Laboratory-scale Corrosion Monitoring in Nitric Acid and Simulated High Level Waste Using 304L EN Probes

Laboratory-scale EN monitoring experiments using a three identical electrode probe (Figure 7.24) made of the 304L structural material were established. Monitoring experiments [82] performed on 304L SS under solution-annealed and sensitized microstructure in nitric acid (4, 8, 12 M) at 298 and 323 K solution temperature indicated passivation of the SS owing to the formation of a stable chromium-rich passive film [82]. Typical EN time record showing

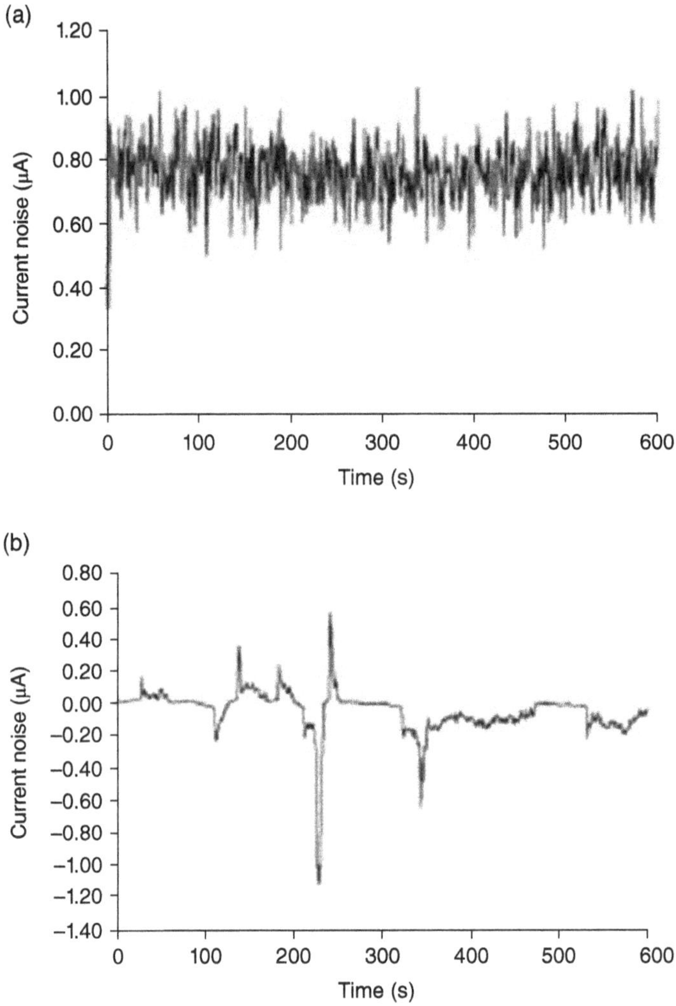

Figure 7.23 EN record (a) prior to water addition in November 1997 and (b) after water addition in November 1997. *Source:* Edgemon et al. [81]. Reproduced with permission of NACE International.

features of passivation of 304L SS in 4 M nitric acid is shown in Figure 7.25a. The EN probe detected localized attack in sensitized 304L SS in 4 M nitric acid at 323 K acid temperature (Figure 7.25b). This is well depicted by distinct current and potential transients that appeared as a result of corrosion attack along the grain boundaries in the sensitized regions of the specimen, where chromium content falls below 12%.

Figure 7.24 EN probe made of structural material of Indian nuclear reprocessing and high level waste storage medium tested in laboratory. *Source:* Girija and Kamachi Mudali [82]. Reproduced with permission of Taylor & Francis.

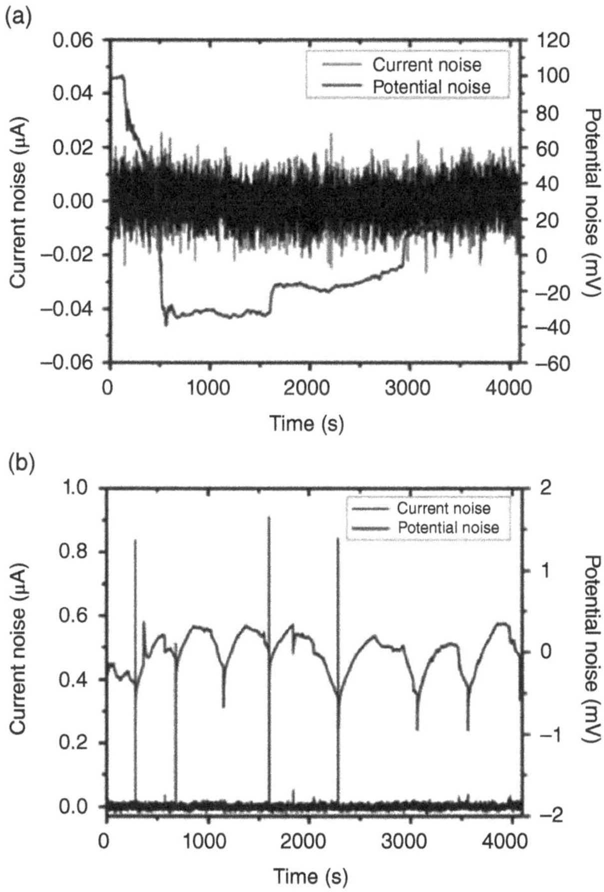

Figure 7.25 EN time record for 304L SS taken after 85 h of immersion of (a) solution-annealed 304L SS in 4 M nitric acid (298 K) and (b) sensitized 304L SS in 4 M nitric acid (323 K). *Source:* Girija and Kamachi Mudali [82]. Reproduced with permission of Taylor & Francis.

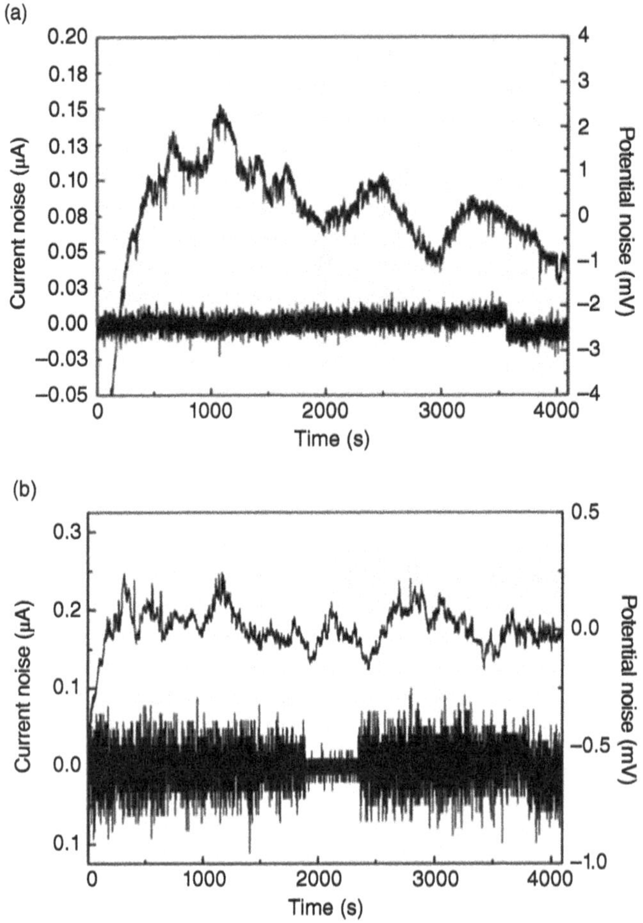

Figure 7.26 EN time record for solution-annealed 304L SS in (a) 3 M nitric acid and (b) simulated HLW. *Source:* Girija et al. [83]. Reproduced with permission of Springer.

Figure 7.26 compares the EN time record for solution-annealed 304L SS in 3 M nitric acid and simulated high level waste (HLW) [83]. Both the time records indicate features of passivation of the SS. But in simulated HLW (prepared in 3 M nitric acid), the time record showed burst of current noise occurring at intermittent intervals. This is attributed to the destabilization of passive film by the ions present (Figure 7.26b) in the waste medium on long-term exposure [83].

This observation was further confirmed from noise resistance evaluation, which showed a decrease in simulated HLW [83], compared with the plain 3 M nitric acid (Figure 7.27).

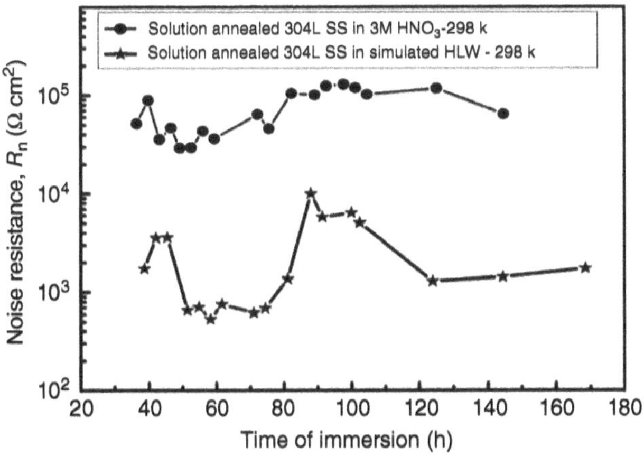

Figure 7.27 R_n–time plots for solution-annealed 304L SS in 3 M nitric acid and simulated HLW at 298 K. *Source:* Girija et al. [83]. Reproduced with permission of Springer.

7.3.8 Electrochemical Noise to Detect IASCC in Boiling Water Reactor Conditions

BWR core components generally suffer from irradiation-assisted stress corrosion cracking (IASCC). As neutron dose accumulates, radiation damage is manifested as change in mechanical properties, phase transformations, and irradiation-induced segregation (RIS), which plays an important role in IASCC. Wang et al. [84] have studied the effect of RIS on IASCC through a series of stress corrosion tests in simulated BWR environment. The authors have studied the SCC characteristics of irradiated samples using EN technique. Tests were performed on sensitized (SEN) AISI type 304 SS C-ring specimens. A notch of 1 mm depth was cut at the central region of the inner radius. The irradiated samples were prepared by bombarding the notched areas with 5 MeV protons at 723 K using a tandem accelerator to simulate neutron irradiation damage in a commercial BWR. All SCC specimens were subjected to an environment of high purity oxygen-saturated water at 561 K. An EN setup as shown in Figure 7.28a, with three electrodes, was used to monitor the corrosion/cracking behavior.

Figure 7.28b shows the coupling current and potential of sensitized 304 SS (unirradiated and irradiated). The current increased only during the initial test period and subsequently leveled off at a later stage. Since the working and counter electrode surface areas are roughly equal, the general oxidation current was ruled out, and the current rise was attributed to the dissolution current generated by IASCC at the crack tip. For the sensitized 304 SS, the level of current increase was significantly lower, suggesting the absence of cracking. After test, hairline cracks were observed on the notch surface of the irradiated sensitized type 304 SS, and no cracking occurred on the 304 SS specimen.

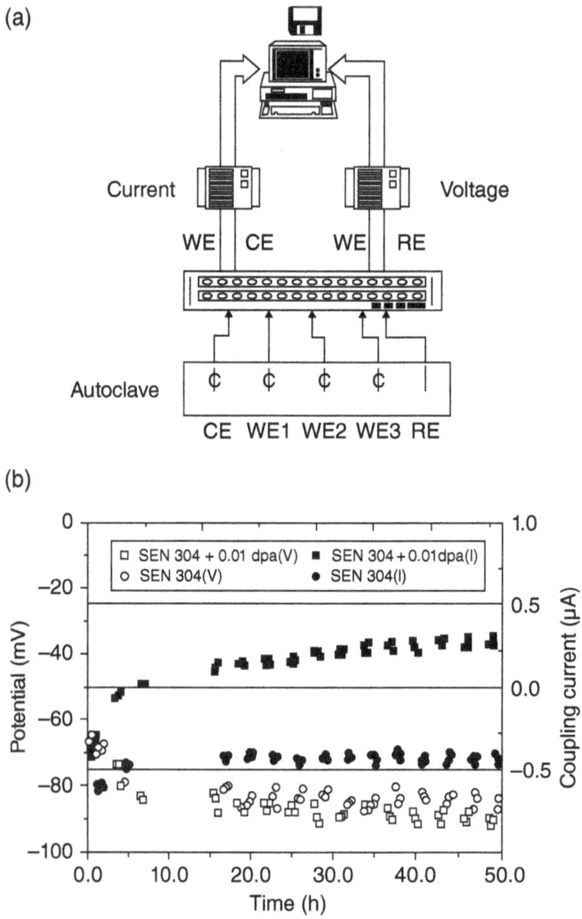

Figure 7.28 (a) EN setup. (b) Coupling current and potentials of sensitized 304 SS and irradiated sensitized 304 SS. *Source:* Wang et al. [84]. Reproduced with permission of Elsevier.

7.3.9 EN to Detect IASCC for Pressurized Water Reactor Conditions

Van Nieuwenhove [85] carried out potential noise measurements on sensitized SS pressure tubes in a recirculating test loop by simulating pressurized water reactor (PWR) conditions except for the radiation. A schematic of the pressure tube with EN sensor is shown in Figure 7.29. Cracks were initiated by gradually increasing the stress in the pressure tube wall up to very high values.

The pressure in the tube was increased gradually from 24 MPa (stress level 0.61) to 42.4 MPa (stress level of 1.88) at which time the tube bursts. In Figure 7.30a, very some small (0.1 mV) positive and negative short-duration spikes are seen at stress level between 0.9 and 1 ks. The rise time and decay time are very short (<1 s), whereas for the current transitions under BWR

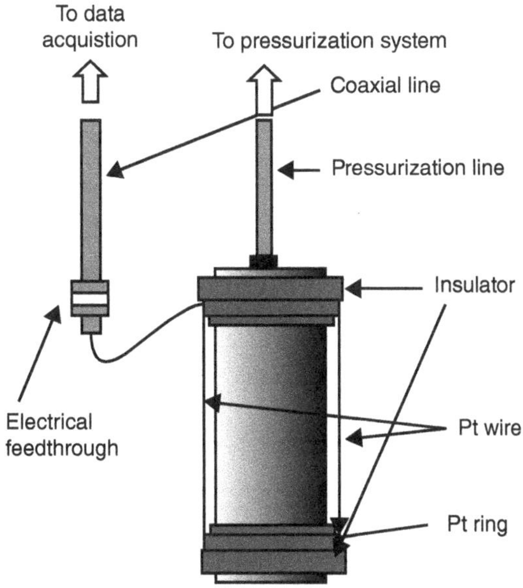

Figure 7.29 Schematic of pressure tube with EN sensor. *Source:* Van Nieuwenhove [85]. Reproduced with permission of NACE International.

conditions, the rise time is very short (1 s) while the decay time is considerably longer (100–1000 s). In Figure 7.30b, corresponding to a stress level of 1.5, the amplitude of the sharp spikes (positive and negative) increased up to ~0.25 mV and showed a very erratic behavior. In Figure 7.30c, corresponding to a stress level between 1.71 and 1.78, some very distinct high amplitude (up to 1 mV) spikes were observed, relative to a less noisy background noise. There was a strong reduction of noise level at 5600 s when the pressure in the tube was raised from 38.8 to 39.1 MPa (stress level of 1.65), and subsequently the noise level increased at higher stress levels.

The amplitude and frequency of the potential spikes were correlated to applied stress, and the spikes corresponded to crack initiation process. Microscopic investigations using SEM revealed microcracks.

7.3.10 Combined EN Technique and DCPD Measurements in Simulated Boiling Water Reactor Conditions

Seifer et al. [86] combined the direct current potential drop method (DCPD) method and EN technique to monitor the initiation of SCC in CT specimens of sensitized SS in simulated BWR conditions. A schematic of the experimental setup is shown in Figure 7.31. The DCPD method has been used as an online monitoring tool for crack growth measurements since then. Initial experiments were carried out at 298 K under constant straining. After 8 h of constant straining, DCPD showed onset of crack advance, and at the same time, a drop occurred in the EPN measured by EN technique (Figure 7.32a). Fractography by SEM as shown in Figure 7.32b confirmed IGSCC initiation

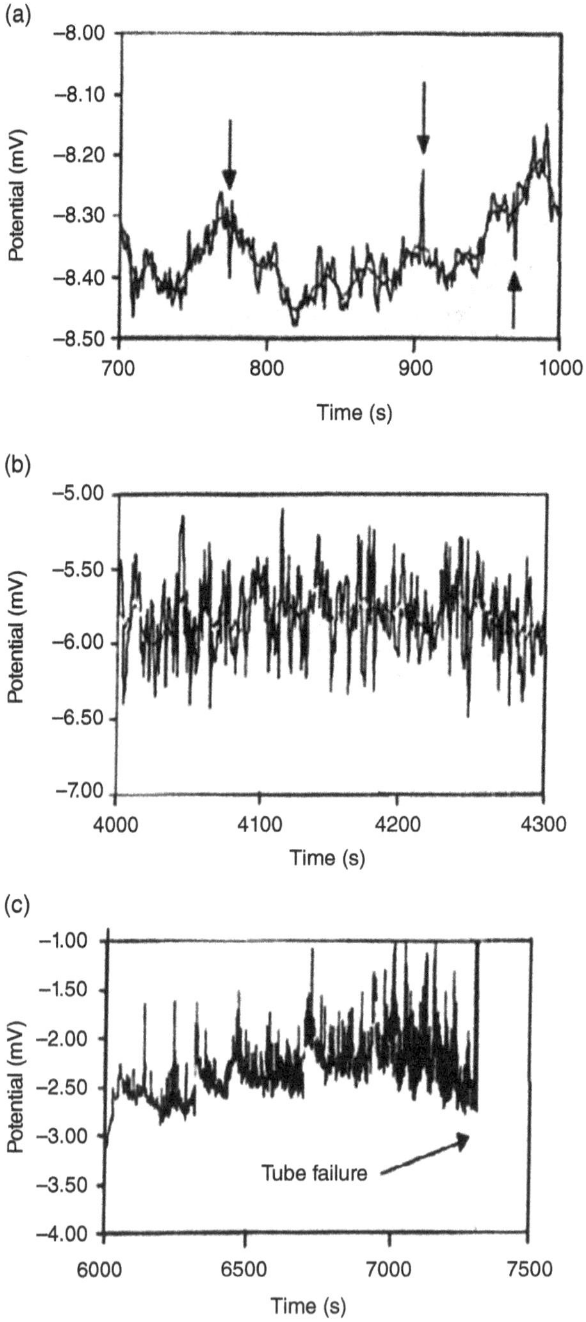

Figure 7.30 Potential noise behavior at different stress levels ks: (a) 0.9 < ks < 1, (b) ks = 1.5, and (c) 1.71 < ks < 1.88. *Source:* Van Nieuwenhove [85]. Reproduced with permission of NACE International.

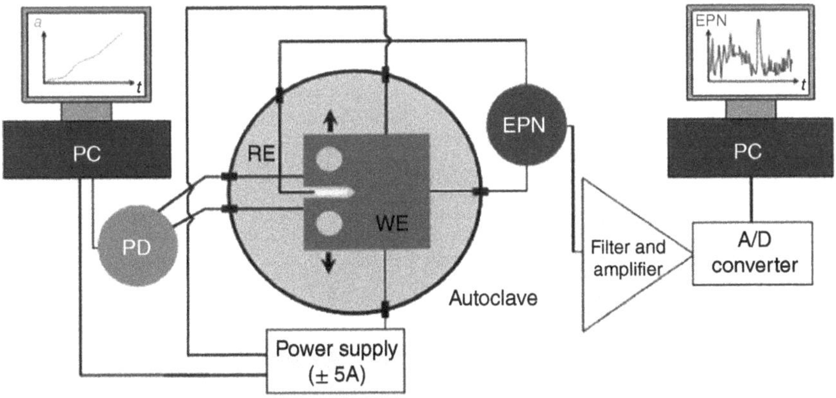

Figure 7.31 Schematic of the experimental setup for EPN and DCPD measurements with C(T) specimens. *Source:* Seifert and Ritter [86]. Reproduced with permission of International Atomic Energy Agency (IAEA).

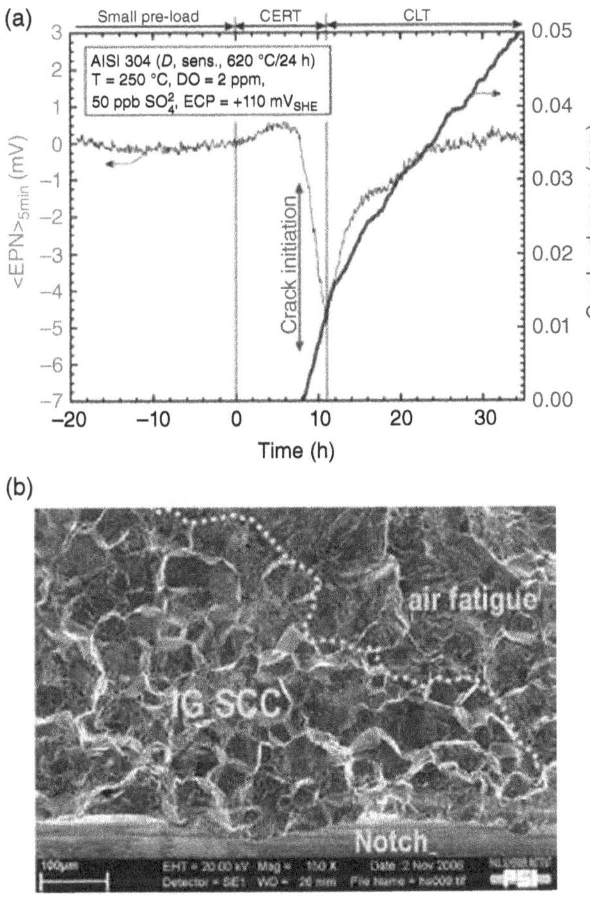

Figure 7.32 (a) Mean values of EPN and crack advance measured by DCPD in a combined constant extension rate–constant load test on a sharply notched sensitized stainless steel CT specimen at 298 K. (b) Post-experimental fractography by SEM showing IGSCC. *Source:* Seifert and Ritter [86]. Reproduced with permission of International Atomic Energy Agency (IAEA).

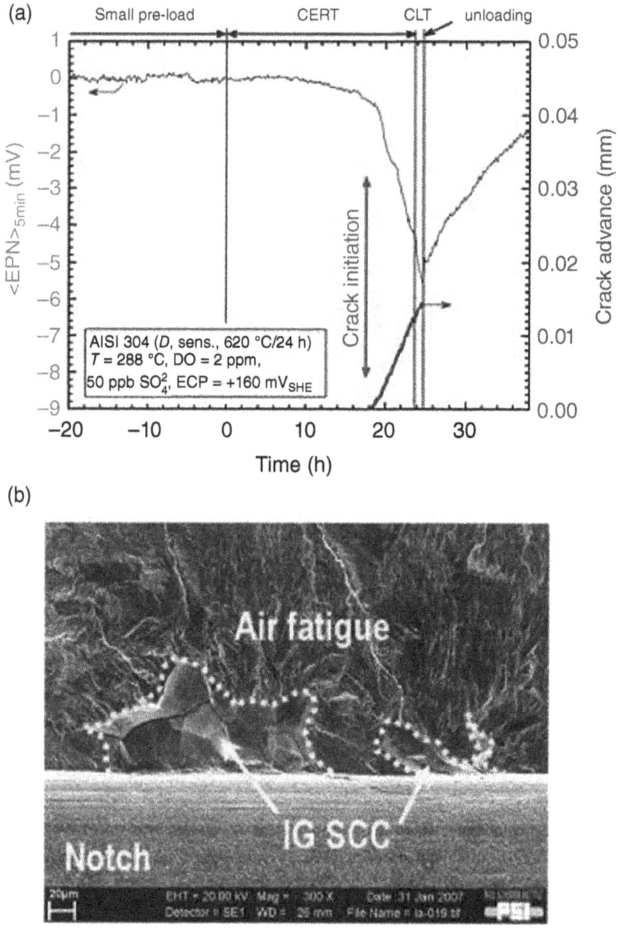

Figure 7.33 Mean values of EPN and crack advance measured by DCPD in a combined constant extension rate–constant load test at 313 K on a sharply notched stainless steel C(T) specimen (a) sensitized. (b) Post-experimental fractography by SEM showing IGSCC. (c) Solution annealed. (d) Post-experimental fractography by SEM showing TGSCC. *Source:* Seifert and Ritter [86]. Reproduced with permission of International Atomic Energy Agency (IAEA).

and growth along the whole notch root. Three hours after SCC initiation, the loading was shifted to constant load. The DCPD showed crack growth, while EPN resumed to original value.

This behavior was attributed to the absence of any further surface crack growth along the notch root and low conductivity of high purity water at 298 K. The crack grew along the depth and moved away from the surface and the EN

(c)

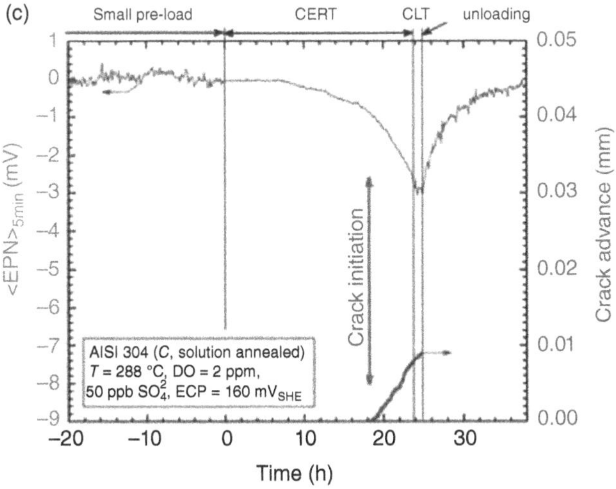

(d)

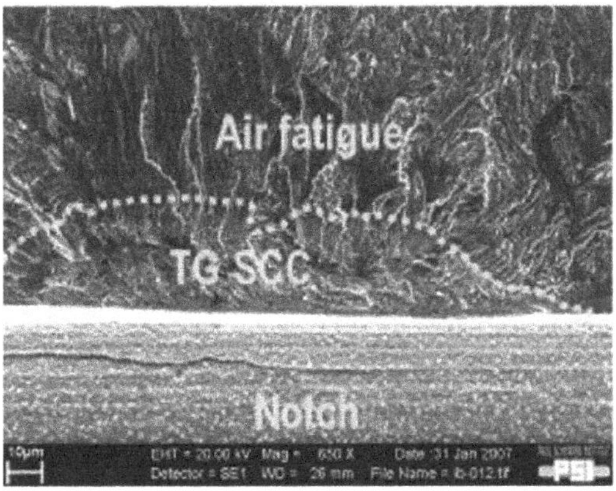

Figure 7.33 (Continued)

reference electrode that was placed at the surface of the specimen near the notch. Only SCC initiation and surface crack growth was thus detectable by EN technique. Results of similar experiments carried out with solution-annealed and sensitized CT specimens at 313 K with a slower extension rate of the pull rod are shown in Figure 7.33. SCC initiation occurred 18 h after start of the strain, which was sensed by a drop in EPN and crack advance by DCPD technique. In the sensitized specimen, an average crack advance of ca. 15 μm and a drop of 15 mV potential were detected by DCPD and EN technique, respectively. But an average crack advance of ca. 9 μm and a drop of 3 mV occurred in the solution-annealed specimen.

7.3.11 Online Electrochemical Noise Monitoring Under Flow Conditions

An online electrochemical probe made of three identical ring electrodes of 70Cu30Ni alloy, embedded in Teflon tube, installed in an absorption heat pump/transformer, is shown in Figure 7.34. An absorption heat pump uses a mixture solution containing the working fluid and absorbent. The absorbent transports the working fluid from the absorption zone to generation zone, where the vapor is separated from the absorbent for heat recovery and further processing. LiBr/water solution, which is highly corrosive, is commonly used as the mixture where LiBr serves as absorbent and water as the working fluid.

The probe is used to monitor the corrosion of the copper alloy in LiBr solution under real operating conditions by EN technique and LPR. Urzúa et al. [87] carried out corrosion monitoring of 70Cu30Ni alloys in LiBr solution to assess its performance as an absorption heat pump/transformer. Localization index and noise resistance were used for monitoring corrosion behavior of the copper alloy. LI, which was monitored as a function of temperature during operation, is represented in Figure 7.35. From the plot it is seen that localization index lies in the regime of localized attack (0.1–1 LI) at temperatures between 40 and 60 °C, beyond which mixed corrosion (0.01–0.1 LI) prevailed. The authors established from EN technique that corrosion mechanism of the copper alloy in LiBr was controlled by passive film at lower temperatures, where bromide-promoted localized attack occurred. Conditions of erosion corrosion existed due to enhanced solubility of the corrosion products at higher temperatures. Figure 7.36 shows the response of noise resistance from

Figure 7.34 Online EN probe in an adsorption heat pump/transformer. *Source:* Urzúa et al. [87]. Reproduced with permission of Elsevier.

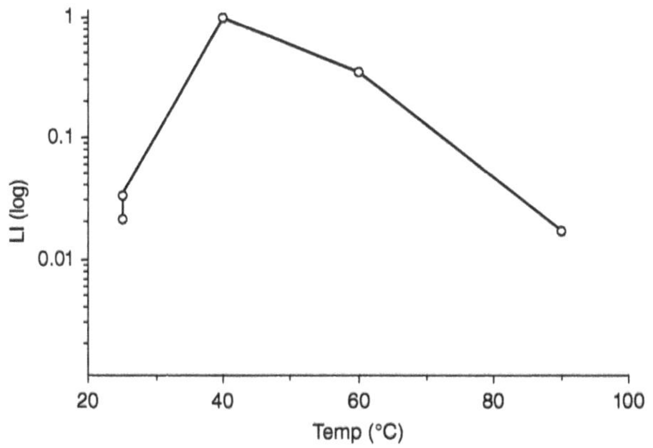

Figure 7.35 Localization index (LI) monitored as a function of temperature. *Source:* Urzúa et al. [87]. Reproduced with permission of Elsevier.

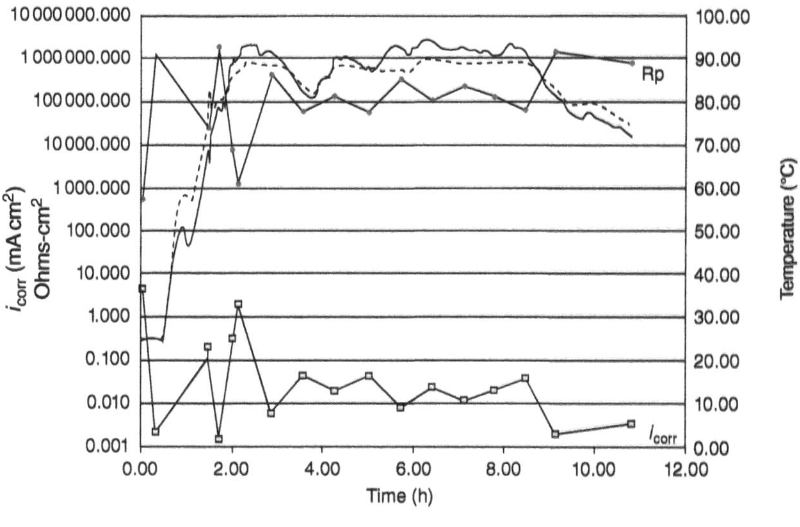

Figure 7.36 Variation of electrochemical parameters (EN and LPR) as a function of time during monitoring. *Source:* Urzúa et al. [87]. Reproduced with permission of Elsevier.

EN and corrosion current density from LPR. The inlet and outlet temperature increased during start-up, reaching a steady state of 90 °C, and decreased during shutdown. During start-up, the noise resistance oscillated between 0.9 kΩ cm^2 and 1 MΩ cm^2, reaching a steady state of 100 kΩ cm^2. During shutdown, the noise resistance again increased to 1 MΩ cm^2. Similarly, the i_{corr} obtained from LPR decreased from a steady-state value of 0.05 mA cm^{-2} during start-up to 0.002 mA cm^{-2} during shutdown.

The monitoring experiments showed a direct correlation between temperature and electrochemical parameters.

Limitations of EN EN as a corrosion monitoring technique has several credentials over other monitoring techniques, but there are challenges that stand a barrier to its wide application. A few of the limitations are listed below:

1) Though EN measurement is simple, the data acquired is enormous and analysis is redundant.
2) Caution to prevent DC drifting either by using high-pass filters or by digital methods is necessary before analysis in order to avoid misinterpretations.
3) Aliasing (discussed in Section 7.2.2) and quantization noise are other challenges to be addressed before data analysis. This calls for experts to handle the data and interpretations in plant applications. Converting the continuous analog voltage or current to a digital form will introduce a limit on the resolution of the digital signal, which will only be able to represent a discrete set of values. This effect is known as quantization and introduces a noise into the signal known as quantization noise.
4) There is divergence over the globe in the measurement method, analysis, and interpretation of EN data. There are controversial views over using various parameters for data analysis.
5) Some limitations also exist for specific systems; to quote an example, EN signals are relatively small for redox systems, and hence no specific advantage exists over other deterministic methods such as impedance for such systems. Quantitative estimation of uniform corrosion rates by EN resistance (R_n) method requires confirming that R_n follows polarization resistance (R_p) for that particular system. R_n is applicable only for determining uniform corrosion rates.
6) If multiple corrosion events are happening on the material surface, data handling becomes complicated and signal overlapping can lead to wrong interpretations.

References

1 Bossi, R.H., Iddings, F.A., Wheeler, G.C., and Moore, P.O. (2002). *Radiographic Testing, Nondestructive Testing Handbook*, 3e, vol. 4. Columbus, OH: American Society for Nondestructive Testing.
2 Raj, B., Jayakumar, T., and Thavasimuthu, M. (2012). *Practical Non Destructive Testing*, 3e. New Delhi: Narosa Publishing House.
3 Raj, B., Kalyanasundaram, P., and Devangan, S.K. (2002). Corrosion detection and monitoring in austenitic stainless steels using nondestructive evaluation and testing and evaluation techniques. In: *Corrosion of Austenitic Stainless*

Steels: Mechanism, Mitigation and Monitoring (ed. H.S. Khatak and B. Raj), 287. New Delhi: Narosa Publishing House.
4 McKnight, M.E. and Martin, J.W. (1989). Detection and quantitative characterization of blistering and corrosion of coatings on steel using infrared thermography. *Journal of Coating Technology* 61: 57.
5 Favro, L.D., Thomas, R.L., and Han, X.Y. (2001). Sonic IR imaging of cracks and delaminations. *Analytical Sciences* 17: s451–s453.
6 Laurens, S., Balayssac, J.P., Rhazi, J. et al. (2005). Non-destructive evaluation of concrete moisture by GPR: experimental study and direct modeling. *Materials and Structures/Materiaux et Constructions* 38 (283): 827.
7 Hugenschmidt, J. and Mastrangelo, R. (2006). GPR inspection of concrete bridges. *Cement and Concrete Composites* 28 (4): 384.
8 Iverson, W.P. (1968). Transient voltage changes produced in corroding metals and alloys. *Journal of Electrochemical Society* 115: 617.
9 Kearns, J.R., Scully, J.R., Roberge, P.R. et al. (eds.) (1996). *Electrochemical Noise Measurements for Corrosion Applications*, ASTM STP 1277. Philadelphia: ASTM.
10 Dawson, J.L. (1996). Electrochemical noise measurement: the definitive in-situ technique for corrosion applications. In: *Electrochemical Noise Measurement for Corrosion Applications*, ASTM STP 1277 (ed. J.R. Kearns, J.R. Scully, P.R. Roberge, et al.), 3. Philadelphia: ASTM.
11 Macdonald, D.D. (1992). The point defect model for the passive state. *Journal of Electrochemical Society* 139: 3434.
12 Cottis, R.A. (2001). Interpretation of electrochemical noise data. *Corrosion* 57: 265.
13 Bertocci, U. and Kruger, J. (1980). Studies of passive film breakdown by detection and analysis of electrochemical noise. *Surface Science* 101: 608.
14 Pujar, M.G., Suresh, G., and Kamachi Mudali, U. (2000). Electrochemical impedance and noise analysis for localised corrosion assessment and monitoring. *Workshop on Advances in Corrosion Testing*, p. 4.1. Kalpakkam, India: Indian Institute of Metals.
15 Girija, S., Kamachi Mudali, U., Raju, V.R., and Raj, B. (2005). Electrochemical noise technique for corrosion assessment – a review. *Corrosion Reviews* 23: 107.
16 Cottis, R.A. and Loto, C.A. (1990). Electrochemical noise generation during SCC of a high-strength carbon steel. *Corrosion* 46: 12.
17 Bastos, N., Huet, F., Nogueira, R.P., and Rousseau, P. (2000). Influence of aliasing in time and frequency electrochemical noise measurements. *Journal of Electrochemical Society* 147 (2): 671.
18 Bertocci, U., Huet, F., Nogueira, R.P., and Rousseau, P. (2002). Drift removal procedures in the analysis of electrochemical noise. *Corrosion* 58: 337.
19 Homborg, A.M., Tinga, T., Zhang, X. et al. (2012). Time-frequency methods for trend removal in electrochemical noise data. *Electrochimica Acta* 70: 199.

20 Jian, L., Weikang, K., Jiangbo, S. et al. (2013). Determination of corrosion types from electrochemical noise by artificial neural networks. *International Journal of Electrochemical Science* 8: 2365.
21 Luaces, M.M., Luaces, V.M., and Ohanian, M. (2006). Trend-removal in corrosion processes using neural networks. *Proceedings of the 5th WSEAS International Conference on Artificial Intelligence*, Knowledge Engineering and Data Bases, Madrid, Spain (15–17 February 2006), p. 248
22 Cao, C.H., Chang, X., and Lin, H. (1989). Features of electrochemical noises during pitting corrosion. *Journal of Chinese Society for Corrosion and Protection* 9: 21.
23 Hladky, K. and Dawson, J.L. (1981). The measurement of localized corrosion using electrochemical noise. *Corrosion Science* 21 (4): 317.
24 Okada, T. (1991). A theoretical analysis of the electrochemical noise during the induction period of pitting corrosion in passive metals. *Journal of the Electrochemical Society of India* 297: 349.
25 Gabrielli, C. and Keddam, M. (1992). Review of applications of impedance and noise analysis to uniform and localized corrosion. *Corrosion* 48: 794.
26 Cheng, Y., Luo, J., and Wilmott, M. (2000). Spectral analysis of electrochemical noise with different transient shapes. *Electrochimica Acta* 45: 1763.
27 Girija, S., Kamachi Mudali, U., Raju, V.R. et al. (2005). Determination of corrosion types for AISI type 304L stainless steel using electrochemical noise method. *Material Science and Engineering A* 407: 188.
28 Eden, D.A., John, D.G., and Dawson, J.L. (1987). Corrosion monitoring. International Patent WO 87/07022. World Intellectual Property Organization.
29 Eden, D.A., Hladky, K., John, D.G., and Dawson, J.L. (1986). Electrochemical noise – simultaneous monitoring of potential and current noise signals from corroding electrodes. In: *Corrosion/86*, Paper no. 274. Houston, TX: NACE International.
30 Dawson, J.L., Farrell, D.M., Aylott, P.J., and Hladky, K. (1989). Corrosion monitoring using electrochemical noise measurements. In: *Corrosion/89*, Paper no. 31. Houston, TX: NACE International.
31 Tan, Y.-J. (1999). Interpreting electrochemical noise resistance as a statistical linear polarisation resistance. *Journal of Corrosion Science and Engineering* 1 (11).
32 Bertocci, U., Gabrielli, C., Huet, F. et al. (1997). Noise resistance applied to corrosion measurements. II. Experimental tests. *Journal of Electrochemical Society* 144: 37.
33 Bierwagen, G.P. (1994). Calculation of noise resistance from simultaneous electrochemical voltage and current noise data. *Journal of Electrochemical Society* 141: L155.
34 Chen, J.F. and Bogaerts, W.F. (1995). The physical meaning of noise resistance. *Corrosion Science* 37: 1839.

35 Brusamarello, V., Lago, A., and Franco, C.V. (2000). Analysis of different methods to calculate electrochemical noise resistance using a three-electrode cell. *Corrosion* 56 (3): 273.

36 Eden, D.A. and Rothwell, A.N. (1992). Electrochemical noise data: analysis, interpretation and presentation. In: *Corrosion/92*, Paper no. 292. Houston, TX: NACE International.

37 Eden, D. (1998). Electrochemical noise – the first two octaves. In: *Corrosion/98*, Paper no. 386. Houston, TX: NACE International.

38 Montespereelli, G. and Gusmano, G. (2004). The use of electrochemical noise analysis on corroding systems. *Fluctuations and Noise Letters* 4 (3): R39.

39 Shi, Y.Y., Zhang, Z., Cao, F.H., and Zhang, J.Q. (2008). Dimensional analysis applied to pitting corrosion measurements. *Electrochimica Acta* 53: 2688.

40 Reid, S., Bell, G.E.C., and Edgemon, G.L. (1998). The use of skewness, kurtosis, and neural networks for determining corrosion mechanism from electrochemical noise data. In: *Corrosion/98*, Paper no. 176. Houston, TX: NACE International.

41 Mansfeld, F., Han, L.T., and Lee, C.C. (1996). Analysis of electrochemical noise data for polymer coated steel in the time and frequency domains. *Journal of Electrochemical Society* 143 (12): L286.

42 Xiao, H. and Mansfeld, F. (1994). Evaluation of coating degradation with electrochemical impedance spectroscopy and electrochemical noise analysis. *Journal of Electrochemical Society* 141: 2332.

43 Girija, S., Kamachi Mudali, U., Khatak, H.S., and Raj, B. (2007). The application of electrochemical noise resistance to evaluate the corrosion resistance of AISI type 304 SS in nitric acid. *Corrosion Science* 49 (11): 4051.

44 Cottis, R.A., Al-Awadhi, M.A.A., Al-Mazeedi, H., and Turgoose, S. (2001). Measures for the detection of localized corrosion with electrochemical noise. *Electrochimica Acta* 46: 3665.

45 Al-Mazeedi, H. and Cottis, R.A. (2004). A practical evaluation of electrochemical noise parameters as indicators of corrosion type. *Electrochimica Acta* 49: 2787.

46 Sanchez-Amaya, J.M., Cottis, R.A., and Botana, F.J. (2005). Shot noise and statistical parameters for the estimation of corrosion mechanisms. *Corrosion Science* 47: 3280.

47 Cottis, R.A. and Turgoose, S. (1999). Corrosion testing made easy-electrochemical impedance and noise. In: *Corrosion/99* (ed. B.C. Syrett). Houston, TX: NACE International.

48 Cottis, R.A. and Turgoose, S. (1995). Electrochemical noise measurements – a theoretical basis. *Material Science Forum* 192–194: 663.

49 Bertocci, U. and Huet, F. (1995). Noise analysis applied to electrochemical systems. *Corrosion* 51 (2): 131.

50 Wharton, A., Wood, R.J.K., and Mellor, B.G. (2003). Wavelet analysis of electrochemical noise measurements during corrosion of austenitic and superduplex stainless steels in chloride media. *Corrosion Science* 45: 97.

51 Abelle, A., Bethencourt, M., Botana, F.J. et al. (2001). Use of wavelets to study electrochemical noise transients. *Electrochimica Acta* 46: 2353.

52 Aballe, A., Bethencourt, M., Botana, F.J., and Marcos, M. (1999). Using wavelets transform in the analysis of electrochemical noise data. *Electrochimica Acta* 44: 4805.

53 Aballe, A., Bethencourt, M., Botana, F.J., and Marcos, M. (1999). Wavelet transform-based analysis for electrochemical noise. *Electrochemical Communication* 1: 266.

54 Shahidi, M., Hosseini, S.M.A., and Jafari, A.H. (2011). Comparison between ED and SDPS plots as the results of wavelet transform for analyzing electrochemical noise data. *Electrochimica Acta* 56: 9986.

55 Pujar, M.G., George, R.P., Muralidharan, P., and Kamachi Mudali, U. (2011). Microbiologically influenced corrosion in UNS S31653: detection and analysis using electrochemical noise technique. *Corrosion* 67 (11): 115004-1–115004-11.

56 Loto, C.A. and Cottis, R.A. (1989). Electrochemical noise generation during stress corrosion cracking of the high-strength aluminum AA 7075-T6 alloy. *Corrosion* 45 (2): 136.

57 Anita, T., Pujar, M.G., Mallika, C. et al. (2012). Evaluation of the effect of nitrogen on stress corrosion cracking of austenitic stainless steels using electrochemical noise technique. *Proceedings, of the Eleventh International Conference on High Nitrogen Steels and Interstitial Alloys*, Chennai, India (27–29 September 2012), pp. 213–220.

58 Monticelli, C., Brunoro, G., Frignani, A., and Trabanelli, G. (1992). Evaluation of corrosion inhibitors by electrochemical noise analysis. *Journal of the Electrochemical Society of India* 139: 706.

59 Eden, D.A., Hoffman, M., and Skerry, B.S. (1985). Application of electrochemical noise measurements to coated systems. *Polymer Material Science Engineering* 53: 388.

60 Kelly, R.G., Inman, M.E., and Hudson, J.L. (1966). Analysis of electrochemical noise for type 410 stainless steel in chloride solutions. In: *Electrochemical Noise Measurements for Corrosion Applications*, ASTM STP 1277 (ed. J.R. Kearns, J.R. Scully, P.R. Roberge, et al.), 101–113. Philadelphia: ASTM.

61 Pistorious, P.C. (1966). The effect of some fundamental aspects of the pitting corrosion of stainless steel on electrochemical noise measurements. In: *Electrochemical Noise Measurements for Corrosion Applications*, ASTM STP 1277 (ed. J.R.|. Kearns, J.R. Scully, P.R. Roberge, et al.), 343. Philadelphia: ASTM.

62 Girija, S. and Kamachi Mudali, U. (2014). Electrochemical noise analysis of pitting corrosion of type 304L stainless steel. *Corrosion* 17: 283.

63 Zhao, B., Li, J.-H., Hu, R.-G. et al. (2007). Study on the corrosion behavior of reinforcing steel in cement mortar by electrochemical noise measurements. *Electrochimica Acta* 52: 3976.
64 Davoodi, A., Pan, J., Leygraf, C., and Norgren, S. (2007). Integrated AFM and SECM for in situ studies of localized corrosion of Al alloys. *Electrochimica Acta* 52: 7697.
65 Li, L., Li, Y., and Wang, F. (2008). Pitting mechanism on an austenite stainless steel nanocrystalline coating investigated by electrochemical noise and *in-situ* AFM analysis. *Electrochimica Acta* 54: 768.
66 Cai, C., Zhang, Z., Cao, F. et al. (2005). Analysis of pitting corrosion behavior of pure Al in sodium chloride solution with the wavelet technique. *Journal Electroanalytical Chemistry* 578 (1): 357.
67 Conde, A. and Williams, D.E. (1999). Crevice corrosion and pitting detection on 304 stainless steel using electrochemical noise. *Materials and Corrosion* 50: 585.
68 Simoes, A.M.P. and Ferreira, M.G.S. (1987). Crevice corrosion studies on stainless steel using electrochemical noise measurements. *British Corrosion Journal* 2 (1): 21.
69 Oldfield, J.W. and Sutton, W.H. (1978). Crevice corrosion of stainless steels: II. Experimental studies. *British Corrosion Journal* 13 (3): 104.
70 Suresh, G., Kamachi Mudali, U., and Raj, B. (2008). Electrochemical parameters for crevice corrosion of 304LN SS in highly aggressive chloride medium. *Journal of the Electrochemical Society of India* 57: 77–82.
71 Anita, T., Pujar, M.G., Shaikh, H. et al. (2006). Assessment of stress corrosion crack initiation and propagation in AISI type 316 stainless steel by electrochemical noise technique. *Corrosion Science* 48: 2689–2710.
72 Kim, S.-W. and Kim, H.-P. (2009). Electrochemical noise analysis of PbSCC of Alloy 600 SG tube in caustic environments at high temperature. *Corrosion Science* 51: 191–196.
73 Markhali, B.P., Naderi, R., Mahdavian, M. et al. (2013). Electrochemical impedance spectroscopy and electrochemical noise measurements as tools to evaluate corrosion inhibition of azole compounds on stainless steel in acidic media. *Corrosion Science* 75: 269–279.
74 Metikis-Hukovic, M., Stupnisek Lisac, E., and Loncar, M. (1991). Electrochemical noise measurement for the evaluation of the protective properties of organic coatings in marine atmospheric environments. *J. Electrochemistry* 7 (3): 128.
75 Mojika, J., Garcia, E., Rodriguez, F.J., and Genasca, J. (2001). Evaluation of the protection against corrosion of a thick polyurethane film by electrochemical noise. *Progress in Organic Coatings* 42: 218–225.
76 Quan, Z., Wu, P.Q., Tang, L., and Celis, J.P. (2006). Corrosion-wear monitoring of TiN coated AISI 316 stainless steel by electrochemical noise measurements. *Applied Surface Science* 253: 1194–1197.

77 Cox, W.M., Geary, D., and Dawson, J.L. (1984). Corrosion, industrial problems and techniques. In: *Proceedings of the International Arabian Conference on Corrosion, Kuwait (1984)* (ed. V. Ashworth), 83–89. Oxford, UK: Pergamon Press.
78 Michael, T.T., Edgemon, G.L., and Ronald, E.M. (2002). Development and deployment of advanced corrosion monitoring systems for high-level waste tanks. *WM'02 Conference*, Tucson, AZ (24–28 February 2002), p. 1.
79 Patrick, J.P., Lister, T.E., and Mizia, R.E. (2003). Development of an electrochemical noise surveillance system for determining compliance of INEEL high level waste tanks to corrosion standards. In: *Corrosion/2003, San Diego CA (March 2003)*, Paper no. 03406. Houston, TX: NACE International.
80 Edgemon, G.L. (2001). Electrochemical noise based corrosion monitoring at the Hanford site: third generation system development, design and data. In: *Corrosion/2001*, Paper no. 01282. Houston, TX: NACE International.
81 Edgemon, G.L., Nelson, J.L., Ohl, P.C., and Bell, G.E.C. (1997). Hanford prototype corrosion probe operational experience. In: *Corrosion/97*, Paper no. 97124. Houston, TX: NACE International.
82 Girija, S. and Kamachi Mudali, U. (2014). Electrochemical noise resistance evaluation of 304L SS in nitric acid and simulated nuclear high level waste. *Corrosion Engineering Science and Technology* 49 (5): 335.
83 Girija, S., Kamachi Mudali, U., and Raj, B. (2011). Corrosion monitoring of type 304L stainless steel in nuclear near-high level waste by electrochemical noise. *Journal of Applied Electrochemistry* 41: 973–981.
84 Wang, L.H., Kai, J.J., Tsai, C.H., and Fong, C. (1998). Comparison of stress corrosion cracking susceptibility of thermally-sensitized and proton-irradiated 304 stainless steel using electrochemical noise techniques. *Journal of Nuclear Materials* 258: 2046.
85 Van Nieuwenhove, R. (2000). Electrochemical noise measurements under pressurized water reactor conditions. *Corrosion* 56 (2): 161.
86 Seifert, H.-P. and Ritter, S. (2009). *Environmentally-Assisted Cracking in Austenitic Light Water Reactor Structural Materials*. Final Report of the KORA-1 Project, Nuclear Energy and Safety Research Department, Laboratory for Nuclear Materials, Paul Scherrer Institute, Switzerland, PSI Bericht Nr. 09-03 (March 2009).
87 Urzúa, R., Siqueiros, J., Morales, L. et al. (2009). On-line corrosion monitoring of 70 Cu 30 Ni alloy in a LiBr solution under absorption heat pump flow conditions. *Portugaliae Electrochimica Acta* 27 (2): 127–142.

8

Evaluation of Cracking and Spallation of Oxide Scales by Acoustic Emission

M.B. Venkataraman[1], Prabhakar Singh[2] and R.K. Singh Raman[3]

[1] *Research School of Engineering, The Australian National University, ACT, Australia*
[2] *Center for Clean Energy Engineering, University of Connecticut, Storrs, CT, USA*
[3] *Department of Mechanical & Aerospace Engineering, Department of Chemical Engineering, Monash University, Clayton Campus (Melbourne), Victoria, Australia*

8.1 Introduction

Corrosion and high temperature oxidation damage are critical considerations in various industries because of their economic impacts and human safety factors. Losses sustained by corrosion in industry and in the military amount to billions of dollars annually [1]. In the event of loss of health or life through explosions, unpredictable failures of chemical equipment, airplane components, trains, or automobiles due to corrosion, the indirect losses are generally difficult to assess and are beyond monetary interpretations. Thus, it becomes additionally necessary to develop techniques and strategies for detecting and monitoring corrosive/oxidative degradations. Research efforts in the past several decades have resulted in the development of new methods and techniques and evaluation of their field use.

While the standard techniques employing in-service coupon in hostile environments are still valuable, they may be inconvenient, labor intensive, too slow, and cost prohibitive in cases where teardown inspection and shutdowns are involved. Such disadvantages can be circumvented by use of online nondestructive monitoring techniques from which data can be acquired/analyzed remotely. These techniques can be broadly characterized into electrochemical methods, eddy current testing, electromagnetic, sonic, and other reflected wave techniques, acoustic emission (AE), and optical/fiber-optic sensing. Each of these techniques has certain advantages over the others depending on the target application. Among the techniques

Non-Destructive Evaluation of Corrosion and Corrosion-assisted Cracking, First Edition.
Edited by Raman Singh, Baldev Raj, U. Kamachi Mudali and Prabhakar Singh.

listed, AE has proved to be invaluable for early-stage detection of high temperature corrosion damage. Several studies have shown that AE (coupled with other characterization techniques) can be used not only for *in situ* monitoring of large areas of structure and locating the source of subcritical cracks but also to understand the damage mechanism. Hence, this chapter will elaborate on the potential of AE for identifying oxidation-related processes such as breakaway oxidation, spallation, and internal cracking. The state of the art in AE technology and current research trends in this area will also be discussed.

8.2 Acoustic Emission

The conventional technique of evaluating field-tested coupons for corrosion prediction is quite inconvenient and retrospective and is not suitable for preventive maintenance. Several techniques have been developed in the better half of the last century for online detection of corrosion processes and lifetime prediction of various components, especially in harsh industrial environments. Recent research has been focused on the nondestructive evaluation (NDE) techniques for high temperature corrosion monitoring, such as electrochemical, electromagnetic, ultrasonic, guided wave, fiber-optic and reflected light, thermogravimetric (TG), thermography, fluorescence, hydrogen test probes, radiographic testing using X-rays and gamma rays, molecular imprinted polymer, AE, and triboluminescent sensing [2]. Among these AE is arguably one of the most versatile nondestructive monitoring techniques prevalent in the industry.

AE comprises of transient mechanical waves spontaneously generated by sudden localized variation in strain within a body. These strain changes can be caused various metallurgical/materials phenomena, such as dislocation movement and crack initiation and growth or disruptions in oxide scale. The resultant surface elastic waves are detected by sensors/transducers and converted into a voltage–time waveform. Practical challenges lie in the ability to discriminate between the signal and background noise and their accurate quantitative interpretation. The signals may be small (low amplitude) but frequent to the extent of being approximately continuous, such as those arising from submicroscopic phenomena, e.g. dislocation movement. These low amplitude quasi-continuous signals cannot be differentiated from the background noise. On the other hand, signals can also be large amplitude and low frequency, appearing as bursts. Application of AE to nondestructive testing (NDT) has been restricted to the monitoring of these burst emissions. Burst emissions are thought to result from large-scale (intense) phenomena such as discontinuous crack propagation. In the last few decades, AE monitoring has been used to detect corrosion phenomena such as spallation/cracking in oxide scales during isothermal and cyclic oxidation [3–5] and discontinuous propagation of stress corrosion cracking. The fundamental principles of AE and its utility to NDT and online monitoring of high temperature oxidation processes is discussed here.

Several mechanisms have been proposed in literature for generation of acoustic/stress waves due to structural alteration in a solid material. Some of the established sources of AE include movement of dislocations and grain boundaries [6, 7], formation and growth of twins [8], generation and propagation of cracks [8], fracture of brittle inclusions and surface films [9], fiber breakage and delamination [10], phase transformations [11], and microseismic/seismic activity in geological materials [12]. A suitable transducer can be employed to detect these transient elastic waves propagating through the material. A complete discussion of the fundamentals of AE is beyond the scope of this chapter. The interested readers can refer to the recent compilation by Grosse and Ohtsu [13] that covers most aspects of AE fundamentals including basics for beginners to sophisticated AE algorithms and its applications to real large-scale structures. Table 8.1 summarizes the parameters of interest from the captured AE signals. Kostopoulos et al. [14] have explained their statistical

Table 8.1 Captured AE parameters of interest.

AE parameters	Symbol	Units
Duration	DUR	μs
Counts	CNT	Number
Amplitude	AMP	dB
Counts to peak	CNP	Number
Rise time	RT	s
Fall time	FT	μs
Absolute energy	ABE	J
Signal strength	SS	pVs
Rise angle	RAN	deg
Decay angle	DAN	deg
Average frequency	AF	kHz
Initiation frequency	INF	kHz
Reverberation frequency	REVF	kHz
FFT amplitude	FTAMP	–
FFT peak frequency	FFTPF	kHz

Source: Kostopoulos et al. [14]. Reproduced with permission of Elsevier.

pattern recognition analysis including data preprocessing, noise reduction methods, and pattern recognition algorithms for identification of failure mechanisms in oxide composites using AE.

A few parameters that are essential for correlation of AE signals with corrosion/oxidation processes have been discussed here. The most common parameters are based on the number of AE peaks and assessment of energy. However, the analyses of amplitude distribution and frequency content can also yield useful information. Figure 8.1 presents the primary AE parameters associated with AE signals. For larger energy phenomena like spallation of oxide layer, a single event contains information about amplitude of the signal and the ring down counts (RDC). RDC increases with amplitude but depends on the threshold selected for noise isolation. The amplitude of signal is affected by the sample geometry, surface roughness, thickness, etc. Even while analyzing two samples of same composition, these extraneous effects can lead to different peak amplitude/energy associated with the same event. Hence, while comparing across different samples, multiple parameters like RDC, rise time, fall time, and event duration are used in conjunction with peak amplitude to analyze the event characteristics [15].

Event energy is another important parameter for oxide scale investigation. Event energy is proportional to the square of the signal amplitude and provides more weightage to higher energy events. The root mean square (RMS) voltage,

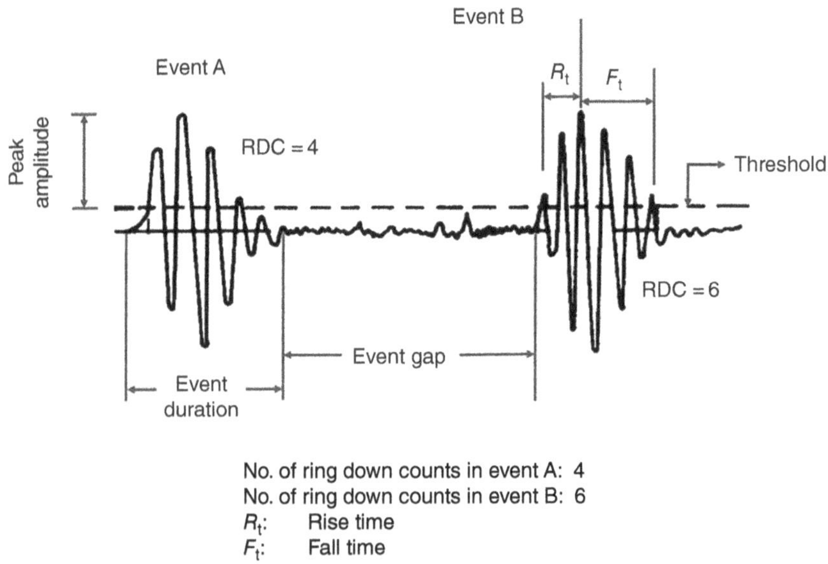

Figure 8.1 Various parameters of acoustic emission signals. *Source:* Raj et al. [15]. Reproduced with permission of Elsevier.

on the other hand, is associated with the energy rate [16]. Peak amplitude distribution curve for a resonant transducer can be empirically correlated using a power law relationship of the form [17]:

$$F(V)=\left(\frac{V}{V_0}\right)^{-b}$$

where $F(V)$ is the number of AE events having an amplitude more than V and V_0 is the lowest detectable amplitude. The exponent b in the equation (also called the "b-parameter" and determined using the slope of log $F(V)$ versus log (V/V_0) plot) contains characteristic information about the source of AE events. For example, a large value of b indicates predominantly low energy, high frequency emissions, whereas low b signifies dominance of high energy, low frequency events [17]. Thus, event energy and amplitude distribution plots can be used to differentiate between events having significantly different energy signatures such as breakaway oxidation, spallation, and scale cracking during cooling [15].

The analysis techniques discussed above pertain to time domain analysis. Sometimes, frequency domain analysis of AE data can reveal invaluable information. Standard fast Fourier transformation (FFT) technique can be used to convert time domain AE data into frequency domain. However, while dealing with AE signals, frequency content is time dependent and hence is usually difficult to be handled by conventional FFT. A possible solution is using small divisions of time series and sequential application of Fourier transform techniques. This procedure is called short-time/windowed Fourier transformation. Like all mathematical techniques, this procedure has certain limitations. Broader window lengths provide better frequency resolution, whereas smaller windows provide better time resolution [18, 19].

Frequency spectrum analysis essentially determines frequency content by measuring the power spectrum of AE signals. The predominant frequency present in the power spectrum is inversely proportional to the time duration of a dynamic event, hence providing useful information about the time scale of oxidation-related processes [15]. This in turn can be correlated to the source of the dynamic event. Figure 8.2 shows the frequency spectrum of AE signal obtained during oxidation of 2.25Cr-1Mo steel at 1173 K [20]. Typical frequency spectrum during oxidation can be divided into three stages. As seen from the figure, the frequency spectrum consists of three regions: pre-breakaway region, where the AE activity is minimal; a breakaway region, where a sudden rise in the AE activity is detected; and an internal cracking region that shows enormous high energy AE activity. Consequently, AE activities at different frequencies can be correlated to different physiological events like microcracking, internal cracking, spallation, and delamination based on the difference in the transient nature of the events.

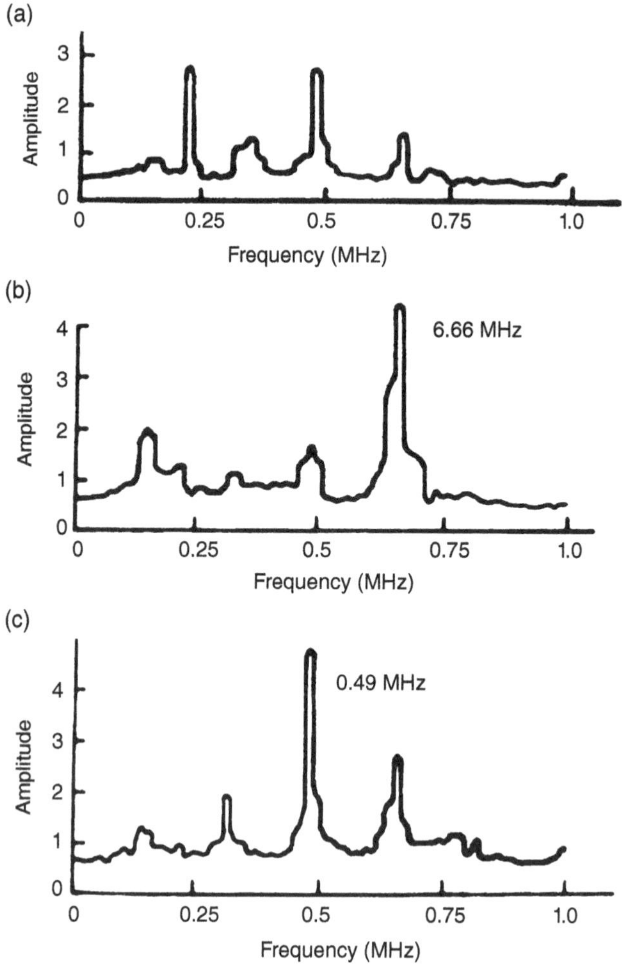

Figure 8.2 Frequency spectrum of signal during (a) pre-breakaway, (b) breakaway, and (c) cracking of oxide layer during cooling for a 2.25Cr–1Mo steel oxidized at 1173 K. *Source:* Raj et al. [15]. Reproduced with permission of Elsevier.

8.3 Research Trends in Use of AE for Oxidation Monitoring

AE as a nondestructive technique for monitoring structural health has been prevalent for more than three decades. Commercial AE monitoring systems enable early detection and pinpoint the location of flaws/cracks that could otherwise lead to catastrophic failure of structures. AE provides an indispensable tool for investigation of dynamic behavior of materials. Extensive studies on AE from deformation of materials have been reported in *Journal of Acoustic Emission*, conference proceedings of IAES (Japan), ICAE (US), EWGAE (Europe), AECM (US/Europe), and several books [13, 21]. These studies have enhanced our knowledge on the origins of AE signals and the methodology to utilize it effectively in NDE applications.

The use of AE for monitoring and inferring oxidation-related physiological changes including breakaway oxidation and spallation of oxide layer is relatively recent. Scale cracking during oxidation was first detected using AE by Coddet et al. [22]. Based on microscopic and TG data, the authors were able to decipher the mechanistic evolution of oxide scale and the associated disruption phenomena during oxidation of titanium at 800 °C. The experimental findings were supplemented by the observation of intense AE events during oxide layer cracking, hence, corroborating the hypothesis of oxide layer fracture during high temperature oxidation of titanium. Furthermore, Coddet et al. [22] reported strong AE signals during the cooling of sample that were associated with thermal stresses. The authors were able to correlate the AE signals to the structural changes in the oxide layer. Parameters such as sample thickness, surface area of sample, composition of oxidizing atmosphere, etc. are known to affect the breakaway oxidation behavior in a high temperature. AE can be used to understand the effect of these parameters on the shift in breakaway instant. For example, Coddet and coworkers [23] reported the parametric study on breakaway oxidation of zirconium using AE.

AE is most powerful when it is coupled with other characterization techniques such as thermogravimetric analysis (TGA) or creep testing [24, 25]. Bennett et al. [26] presented an experimental setup capable of simultaneously measuring mass change and AE (shown in Figure 8.3). An experimental setup for simultaneous measurement of AE and TG data is shown in Figure 8.3. The authors studied the oxidation behavior of Fe-20Cr-25Ni-Nb alloy under isothermal conditions. The major finding of this study was that no AE events were detected during the isothermal oxidation process. However, there was an enormous amount of AE activity during cooling of sample that preceded the weight loss in TG. The AE signals were explained in terms of strain-energy model, in which the difference in thermal expansion of oxide and metal acts as the source of stress, leading to initiation of internal cracks. The study concluded that crack initiation precedes spallation and was able to determine the time scale for propagation of internal cracks in the oxide scale. Similarly, thermocyclic oxidation damage on Ti-6Al-4V alloy between 600 and 900 °C was correlated with AE amplitude distribution analysis by Baram et al. [27]. The study demonstrated that AE trends closely follow the material damage mechanisms identified by other techniques. Plastic deformation of the scale and partial stress relief due to microcracking were reported as the source of AE during isothermal conditions, whereas higher intensity AE signals were attributed to scale spallation.

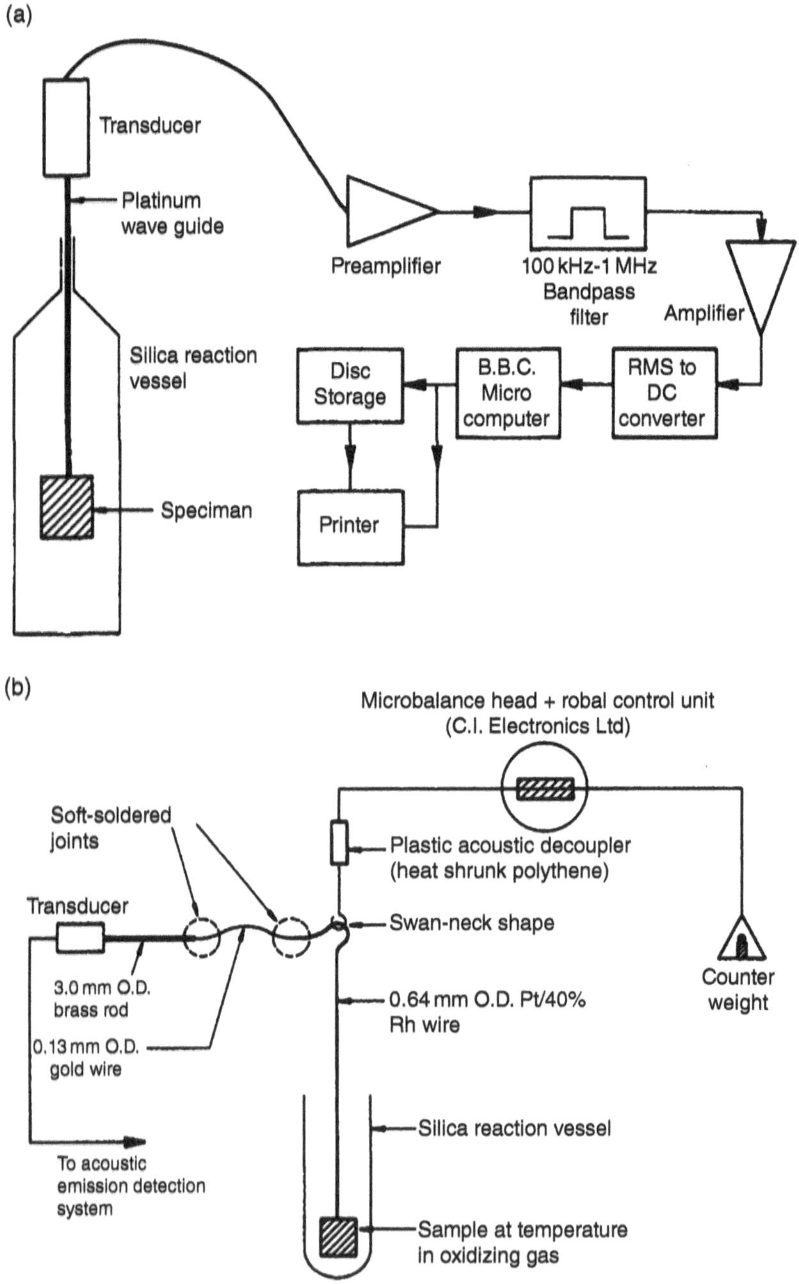

Figure 8.3 Schematic diagram of (a) AE apparatus and (b) AE detector coupled to a controlled atmosphere microbalance. *Source:* Bennett et al. [26]. Reproduced with permission of Elsevier.

AE monitoring has been used to confirm the results of TG tests suggesting difference in spallation behavior (their initiation and intensity) of 9Cr–1Mo steel with different grain sizes and Si contents [28]. Results of TG tests of steel (with 0.26% Si) in air at 700 °C (Figure 8.4) suggested that oxide scale on fine grain size (90 μm) suffered a pronounced weight loss after 60 min, whereas such weight loss was delayed until 240 min for the scale on coarse grain size (360 μm) steel. AE monitoring during oxidation under identical conditions with two different grain sizes suggested sudden and pronounced increase in AE activity after similar durations for fine- and coarse-grained steels, respectively (Figure 8.5), corroborating the TG results. Hence, the authors have ascribed the weight losses to spallation of oxide layer. As suggested by both AE and TG results, onset of pronounced scale spallation is significantly delayed in the case of steel with higher Si content (0.75%). Thus, the study is able to establish the strength of AE technique in monitoring the spallation of oxide scales.

Raj et al. [15] have used AE to investigate the oxidation behavior in Cr–Mo steels, particularly, the spallation and internal cracking. 9Cr–1Mo and 2.25Cr–1Mo steels were subjected to isothermal oxidation at different temperatures between 500 and 950 °C. The study was able to correlate the severity of scale damage during cooling to the intensity of AE signals. The major finding of the study is the identification of "*b*-parameter" during different stages of oxidation, namely, isothermal oxidation, breakaway, and cooling. Figure 8.6 shows the cumulative amplitude distribution plots at 900 and 950 °C, where three different regions can be clearly identified. The slope of these regions (*b*-parameter) provides information about the nature of the events they represent. The first region (the low amplitude regime) is characterized by high "*b*" value that is indicative of low energy events that occur almost continuously, such as the isothermal scale growth. Second region is representative of less frequent events with relatively higher energy, such as the breakaway oxidation. The third region provides information about the burst events that are very high energy but quite infrequent, such as scale cracking during cooling. A comparison between the *b*-parameter at different temperatures can be used to understand the scale disruption and degradation mechanism. Frequency spectrum analysis is instrumental in establishing the frequency and energy signatures of different processes, such as microcracking and internal cracking, and their occurrence frequencies. Such a methodology is extremely valuable from an industry perspective as AE signals from similar alloys under service can be interpreted conveniently.

Breakaway oxidation and spallation behavior in 2.25Cr–1Mo steel [29] and copper [30] have also been investigated using AE technique. Analysis of the time domain parameters obtained during oxidation can successfully determine the breakaway transition point. In the initial studies, Khanna et al. [29, 30] had used the same material for test specimen and the wave guide. However, this can lead to discrepancy related to the thermal gradient in the sample that

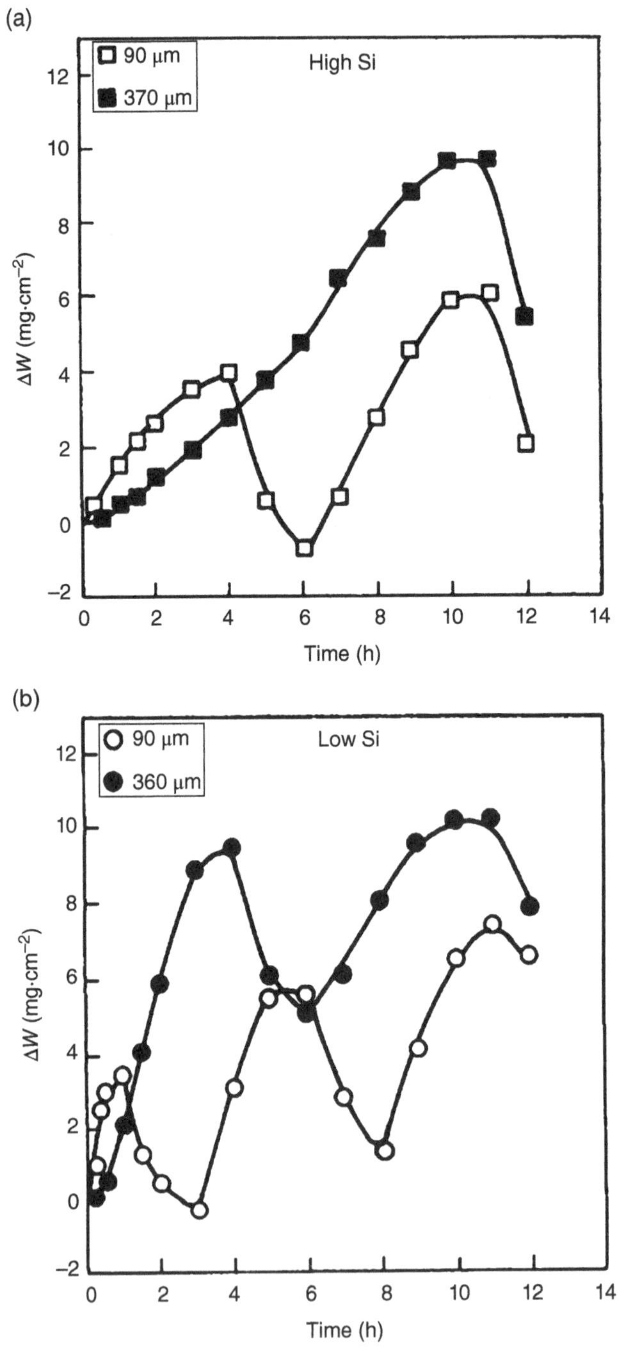

Figure 8.4 Weight change vs. time plots during air oxidation (700 °C) of low and high Si steels with different grain sizes. *Source:* Raman et al. [28]. Reproduced with permission of Springer.

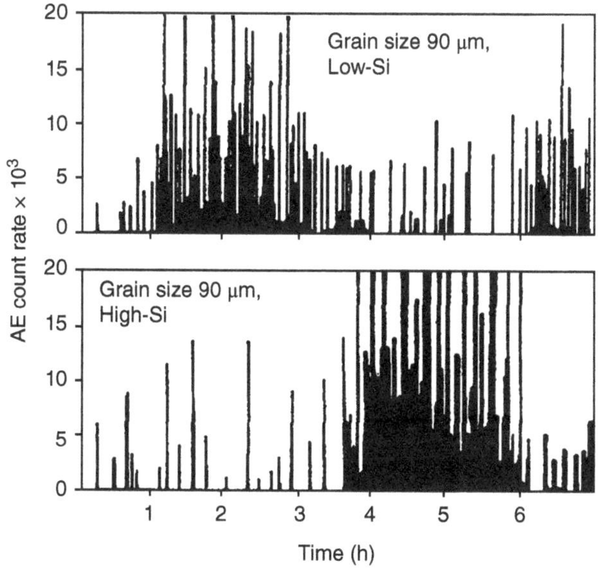

Figure 8.5 AE count rate vs. time tracings during air oxidation (at 700 °C) of low and high Si steels with different grain sizes. *Source:* Raman et al. [28]. Reproduced with permission of Springer.

cannot be accounted for in the analysis. Subsequent modifications in the setup were implemented to eliminate such discrepancy by using Pt-wire as a wave guide [31]. In a similar study, Christl et al. [3] were able to corroborate the evidences presented by Khanna et al. [29] on 2.25Cr–1Mo steels by identifying the different failure modes and assessing the extent of damage in the oxide layer. The authors were able to explain the oxidation mechanism in different environments (pure O_2, air, and Ar-50% H_2O) using AE. The scales were primarily composed of hematite or magnetite depending on the oxygen partial pressure of the oxidizing environment. The *b*-parameter for buckling and delamination of scales with different composition was correlated with the scale damage.

It has been quite well established that addition of small amounts of reactive elements such as Ce, Y, Zr, Hf, etc. improves the high temperature oxidation resistance of certain alloys [32]. Improved adhesion of the oxide layer to the substrate due to the presence of these reactive elements has been suggested as a mechanism for this improvement. Ashary et al. [33] have used AE testing to validate the improved oxidation resistance in the presence of Ce, Y, and Hf in several alumina- and chromia-forming alloys. The authors have reported a difference in scale cracking and spallation behavior during cyclic oxidation for CoCrAl alloy with and without Y addition. It was observed that the AE activity during cooling was orders of magnitude higher in case of CoCrAl as compared

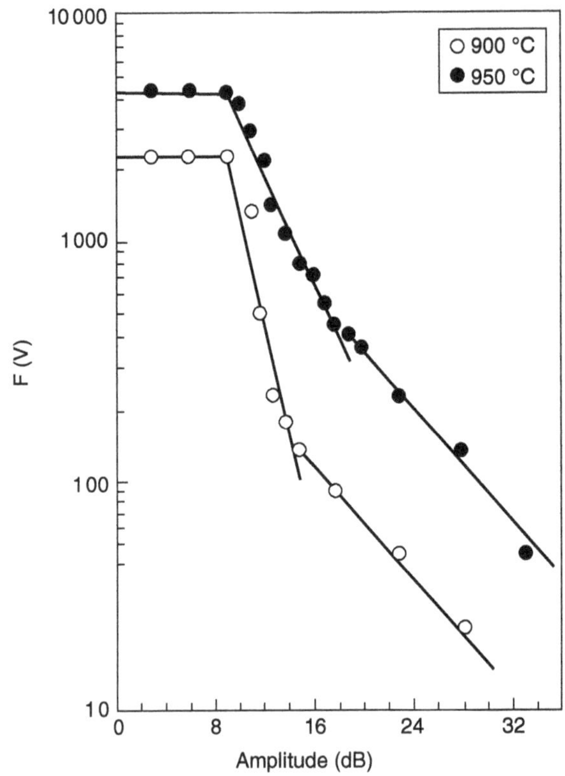

Figure 8.6 Cumulative amplitude distribution plot for 9Cr–1Mo steel subjected to isothermal oxidation at 900 and 950 °C. *Source:* Raj et al. [15]. Reproduced with permission of Elsevier.

with CoCrAlY. Similar observations have been reported for NiCrAl alloys as well. These results are important in terms of conclusively validating the role of reactive elements in improving scale adherence, especially under cyclic loading conditions. Khanna et al. [34] and Huang et al. [35] have, respectively, shown improved adherence of oxide scales when S impurity is present in Ni-based and FeCrAl alloys. Subdued AE activity was observed in case of alloys containing higher S impurity indicating enhanced resistance to scale spallation.

Certain AE parameters are particularly useful in investigation of oxidation-related phenomena. Jonas et al. [36–38] measured the RDC, amplitude distribution, and cumulative energy of AE signals to investigate cyclic oxidation in Hastelloy X and Incoloy 800. In conjunction with hydrogen permeation measurements, it was possible to detect even miniscule damages to the oxide surface coating, including microcracks that were not detectable using metallographic techniques. Similar results have been reported for AE monitoring of plasma-sprayed stabilized zirconia coatings during thermal cycling [39, 40].

Although AE technology has been subject of significant scientific investigations in the past, there exists enormous scope for methodical future research. Quantitative prediction of structural stability and lifetimes, and correlating the AE data with established degradation mechanisms is still elusive. Advanced signal processing/analysis concepts such as pattern recognition and signature analysis of different oxide phases and oxidation-related processes are necessary for exploiting the full potential of AE.

8.4 Conclusions

Online nondestructive corrosion monitoring techniques are invaluable to the industry for convenient preventive maintenance. This chapter presents a comprehensive review of the AE-based NDT and its suitability to specifically address the online monitoring and early-stage detection of deterioration due to high temperature oxidation. A few highlights of the research trends in AE are:

1) AE is a powerful, versatile, and easy-to-implement technique for early-stage detection of corrosion/high temperature oxidation damage in metallic components.
2) AE has been shown to be an effective characterization tool for identifying spallation, breakaway oxidation, internal cracking, scale buckling, delamination, etc. at both laboratory and industrial scales.
3) Considerable research during the past few decades in advancing the AE technology has led to improvement in design, data interpretation, and accuracy of AE signals.
4) Advances in AE signal processing/interpretation are necessary to fully exploit the immense potential of AE, hence enabling quantitative prediction of structural stability and lifetimes in the near future.

References

1 Koch, G.H., Brongers, M.P.H., Thompson, N.G. et al. (2002). *Corrosion Cost and Preventive Strategies in the United States*. NACE Report No. FHWA-RD-01-156. Houston, TX: NACE International.

2 Agarwala, V.S. (2003). *ASM Handbook: Volume 13A, Corrosion: Fundamentals, Testing and Protection* (ed. S.D. Cramer and B.S. Covino Jr.). Materials Park, OH: ASM International.

3 Christl, W., Rahmel, A., and Schütze, M. (1987). Application of the acoustic emission technique for the detection of oxide scale cracking during thermal cycling. *Materials Science and Engineering* 87: 289–293.

4 Evans, H.E. and Lobb, R.C. (1984). Conditions for the initiation of oxide-scale cracking and spallation. *Corrosion Science* 24 (3): 209–222.
5 Singh Raman, R.K., Gnanamoorthy, J.B., and Roy, S.K. (1992). Influence of alloy grain size on the high temperature oxidation behaviour of some steels with different Cr contents. *Transactions of the Indian Institute of Metals* 45 (6): 391–398.
6 Heiple, C.R. and Carpenter, S.H. (1987). Acoustic emission produced by deformation of metals and alloys. *Journal of Acoustic Emission* 6: 177–204.
7 Yudin, A.A. and Ivanov, V.I. (1985). Acoustic emission in plastic deformation of metals (review). Report 1. *Strength of Materials* 17 (6): 842–851.
8 Dunegan, H.L., Harris, D.O., and Tatro, C.A. (1968). Fracture analysis by use of acoustic emission. *Engineering Fracture Mechanics* 1 (1): 105–122.
9 Jaffery, D. (1980). The failures of MnS inclusions in steel under uniaxial stress and their relevance to AE studies. In: *Fifth International Acoustic Emission Symposium*. Tokyo, Japan: Japanese Society for Non-destructive Inspection (NDI).
10 Harris, D.O., Tetelman, A.S., and Darwish, F.A.I. (1972). Detection of fiber cracking by acoustic emission. In: *Acoustic Emission ASTM STP 505*. Philadelphia, PA: American Society for Testing and Materials.
11 Speich, G.R. and Fisher, R.M. (1972). Acoustic emission during martensite formation. In: *Acoustic Emission ASTM STP 505*. Philadelphia, PA: American Society for Testing and Materials.
12 Lord, A.E. Jr. and Koerer, R.M. (1979). Acoustic emission in geological materials. In: *Fundamentals of Acoustic Emission* (ed. K. Ono). Los Angeles, CA: Materials Department, School of Engineering and Applied Sciences, UCLA.
13 Grosse, C.U. and Ohtsu, M. (2008). *Acoustic Emission Testing, Basics for Research – Applications in Civil Engineering*, 414. Berlin: Springer.
14 Kostopoulos, V., Loutas, T.H., Kontsos, A. et al. (2003). On the identification of the failure mechanisms in oxide/oxide composites using acoustic emission. *NDT & E International* 36 (8): 571–580.
15 Raj, B., Jha, B.B., Khanna, A.S., and Bhattarcharya, D.K. (1991). Acoustic emission studies towards better understanding of high temperature oxidation in Co-Mo steels. *International Journal of Pressure Vessels and Piping* 45: 301–326.
16 Wadley, H.N.G., Scruby, C.B., and Speake, J.H. (1980). Acoustic emission for physical examination of metals. *International Metal Reviews* 3: 41–64.
17 Pollock, A.A. (1973). Acoustic emission amplitudes. *NDT International* 6: 264–269.
18 Gabor, D. (1991). Theory of communication. *Journal of the Institute of Electrical Engineering, London* 93 (3): 891–906.
19 Hubbard, B.B. (1996). *The World According to Wavelets: The Story of Mathematical Technique in the Making*. Wellesley, MA: AK Peters Ltd.

20 Jha, B.B., Mishra, B.K., and Ojha, S.N. (2012). Oxidation behaviour of 2.25Cr–1Mo steel and its characterisation using acoustic emission. *Canadian Metallurgical Quarterly* 51 (2): 221–227.
21 Kishi, T., Ohtsu, M., and Yuyama, S. (2000). *Acoustic Emission – Beyond the Millennium*. Amsterdam: Elsevier.
22 Coddet, C., Beranger, G., and Chretien, J.F. (1978). Application of acoustic emission technique to the detection of oxide layer cracking during oxidation. In: *Materials and Coatings to Resist High Temperature Corrosion* (ed. D. Holmes and A. Rahmel), 175–183. London: Applied Science Publishers.
23 Coddet, C., de Barros, G., and Beranger, G. (1984). Application of the acoustic emission technique to the understanding of the high-temperature oxidation behavior of zirconium. *Proceedings of the International Congress on Metallic Corrosion,* Toronto, Vol. 4, pp. 66–72.
24 Grgic, D. and Amitrano, D. (2009). Creep of a porous rock and associated acoustic emission under different hydrous conditions. *Journal of Geophysical Research: Solid Earth* 114 (B10): B10201.
25 Schütze, M. (1986). The healing behavior of protective oxide scales on heat-resistant steels after cracking under tensile strain. *Oxidation of Metals* 25 (5–6): 409–421.
26 Bennett, M.J., Buttle, D.J., Colledge, P.D. et al. (1989). Spallation of oxide scales from 20%Cr-25%Ni-Nb stainless steel. *Materials Science and Engineering: A* 120–121 (Part 1): 199–206.
27 Baram, J., Itzhak, D., and Rosen, M. (1979). Acoustic emission monitoring of high-temperature cyclic oxidation damage in a titanium-aluminum-vanadium alloy. *Journal of Testing and Evaluation* 7 (3): 172–176.
28 Raman, R.K.S., Gnanamoorthy, J.B., and Roy, S.K. (1994). Synergistic influence of alloy grain size and Si content on the oxidation behavior of 9Cr–1Mo steel. *Oxidation of Metals* 42 (5–6): 335–355.
29 Khanna, A.S., Jha, B.B., and Raj, B. (1985). Detection of breakaway oxidation and spalling in the oxide scales of 2 1/4Cr-1Mo steel using acoustic emission technique. *Oxidation of Metals* 23 (3–4): 159–176.
30 Khanna, A.S., Jha, B.B., and Raj, B. (1987). Acoustic emission technique: an alternative method to study the brittle oxide formed on copper. *Oxidation of Metals* 27 (1–2): 95–102.
31 Jha, B.B., Khanna, A.S., and Raj, B. (1987). Study of the oxidation behaviour of 9 Cr-1 Mo steel using the acoustic emission technique. *Journal of Materials Science* 22 (8): 2823–2827.
32 Whittle, D.P. and Stringer, J. (1980). Improvements in high temperature oxidation resistance by additions of reactive elements or oxide dispersions. *Philosophical Transactions of the Royal Society of London. Series A, Mathematical and Physical Sciences* 295 (1413): 309–329.
33 Ashary, A., Meier, G.H., and Pettit, F.S. (1983). Acoustic emission study of oxide scale cracking during alloy oxidation. In: *Proceedings of Conference on*

High Temperature Protective Coatings, Atlanta, GA (7–8 March 1983). Warrendale, PA: Metallurgical Society.

34 Khanna, A.S., Jonas, H., and Quadakkers, W.J. (1989). Use of acoustic emission technique to study the spalling behaviour of oxide scales on Ni-10Cr-8Al containing sulphur and/or yttrium impurity. *Materials and Corrosion* 40 (9): 552–558.

35 Huang, T.T., Lin, Y.C., Shores, D.A., and Pfender, E. (1984). Corrosion of FeCrAl, FeCrAlY, and FeCrAlHf alloys in high temperature H_2–H_2O–H_2S environments. *Journal of the Electrochemical Society* 131 (9): 2191–2196.

36 Jonas, H. and Golczewski, J.A. (1984). Measurements of acoustic emission due to thermocycling of oxidized high temperature alloys. *Journal of Nuclear Materials* 120 (2–3): 272–277.

37 Jonas, H., Stöver, D., and Hecker, R. (1985). On the application of acoustic emission analysis to evaluate the integrity of protective coatings on high-temperature alloys. In: *Progress in Materials Analysis* (ed. M. Grasserbauer and W. Wegscheider), 263–273. Vienna: Springer.

38 Jonas, H., Stöver, D., and Hecker, R. (1985). On the application of acoustic emission analysis to evaluate the integrity of protective oxide coating. *Proceedings of the Second World Conference of Acoustic Emission*, Las Vegas, NV (28 October–1 November 1985), Vol. 4, Issue 2–3, pp. S78–S81.

39 Berndt, C.C. (1985). Acoustic emission evaluation of plasma-sprayed thermal barrier coatings. *Journal of Engineering for Gas Turbines and Power* 107 (1): 142–146.

40 Ravi Shankar, N. and Berndt, C.C. (1983). Acoustic emission from thermally cycled plasma-sprayed oxides. *American Ceramics Society Bulletin* 62: 614–619.

9

Nondestructive Testing and Corrosion Monitoring

Alec Groysman

Faculty of Chemical Engineering, Technion, Israel Institute of Technology, Haifa, Israel

9.1 Introduction

All our knowledge begins from sensations.
Leonardo da Vinci (1452–1519), the Italian polymath

People from the ancient times used *nondestructive testing* for detection of different defects in materials (including corrosion phenomena and their consequences). In order to realize that, people used their organs of sense, usually *sight, hearing, touch,* and *smell.* Chemists also used *taste.* Senses of *sight* and *touch* are used for visual inspection up to now in many industries. *Smell* of different substances (e.g. acetone, gasoline, ammonia, hydrogen sulfide, etc.) in the environment can indicate their leakage as a result of corrosion holes. I was called many times to look at the surface of heat exchanger tube bundles, pipes, tanks, reactors, other equipment, and structures for the presence of corrosion products or various defects (pits and cracks) formed as a result of corrosion. Even the color of corrosion products (*sight*) may provide an apparent idea about what substances caused corrosion [1].

I remember when I was a student and returned from Moscow to my native town Yalta in the Crimea (the Black Sea) by train, at some station, a special person (a worker) went along the whole train and hit with a metallic hammer on the metallic springs and heard audible *sounds.* According to such *sounds,* a worker could detect cracks and other defects in springs of the train and thus prevent catastrophic failure. Even today, mechanical engineers *hear* how pumps work, and according to specific *sound* of stones' hits, they can detect

Non-Destructive Evaluation of Corrosion and Corrosion-assisted Cracking, First Edition.
Edited by Raman Singh, Baldev Raj, U. Kamachi Mudali and Prabhakar Singh.

cavitation in centrifugal pumps. *Sonic* inspection with hammers allows determining quality of concrete according to *soundness* of concrete.

We should provide a clear determination of what we call *nondestructive testing* (*NDT*) and *corrosion monitoring* (*CM*).

Nondestructive testing, also called nondestructive examination (NDE) or nondestructive evaluation (NDE) or nondestructive inspection (NDI), is a wide group of analysis techniques used in industry for evaluating the properties of materials without causing damage.

The term *NDT* historically encompasses *visual examination, ultrasonic testing* (*US* or *UT*), *eddy current, acoustic emission* (*AE*), *radiography examination, infrared (IR) thermography, penetrant testing,* and *magnetic flux* [2, 3]. All these methods are based on physical phenomena (*electromagnetic radiation* or *sound*) taking place inside a metal or on a metal surface. The history of development of NDT is described in literature [4]. It is important to emphasize two points. *On-site metallography* (*replicas' method*) includes use of optical or electron microscope that needs surface polishing, and thus it is strictly speaking "destructive." In this sense, surface for US measurements must also be smooth and needs polishing. Earlier *chemical spot tests* also needed small destruction of outer surface of alloys for determination of individual alloying elements. Nowadays, these *chemical tests* are not used, and they are replaced by quick *spectroscopic analysis* of alloys, which is really NDT (see Section 9.4.1.12). Some physicochemical methods (for instance, *scanning electron microscopy* [SEM], *energy-dispersive spectroscopy* [EDS], *scanning tunneling microscopy* [STM], *atomic force microscopy* [AFM], etc.; see Section 9.3.3) used for the assessment of corrosion surface, corrosion products, and biodeterioration and the microcharacterization of different materials are also NDT. Most electrochemical methods are NDT too (Section 9.6). Some methods, for instance, *hydrogen monitoring,* may be NDT (using hydrogen probes) and destructive (using analytical method of inert gas fusion principle) (Section 9.5.3).

The term *corrosion monitoring means the measurement of a metal loss rate, crack, or pit growth, erosion, and deposition rate of scale and fouling for a specific piece of a metallic equipment or structure exposed to the environment* [1, 5–7]. Shortly, *CM is the regular measurement of corrosion effect or corrosiveness of the environment.* CM methods are based on physical, chemical, and physicochemical properties of a metal and the environment.

The goal of both NDT and CM is to provide an indication of the potential for degradation of the equipment and structures before significant damage occurs and to allow reduction in the rate of degradation to an acceptable level. Many of the CM methods such as hydrogen probes and chemical analytical and some physicochemical methods are nondestructive in the sense of this term.

NDT and CM occupy separate important niches in corrosion and materials science and technology. There are nearly no industries or plants that do not use

NDT and CM methods. We will concentrate on using *NDT for assessment of corrosion* in equipment and structures.

Information "explosion" occurs in any kind of human activity; NDT and CM are not exclusion. For instance, when I searched for "corrosion monitoring" on Google in 2010, I obtained 473 000 references; "NDT and corrosion," 146 000 references; "acoustic emission and corrosion," 66 500; "electrochemical corrosion monitoring," 49 700; "electrochemical noise measurements," 1 130 000; etc. Thus, nowadays it is impossible to encompass all information (literature) concerning NDT and CM, which is published all over the world (books, encyclopedias, monographs, journals, proceedings, reports, etc.). We should restrict our study only to literature mostly published during the last 10 years. We will list main principles, concepts, and examples of use of NDT and CM in different industries. It was important in some cases to provide historical notes on our subject.

NDT and CM are very versatile and developing interdisciplinary topics that found important application in corrosion science and technology. We will try to unite theory and practice of important subjects included in this chapter, emphasizing their achievements of use in different branches of the industry. We are eyewitnesses to fast implementation of theoretical achievements of NDT and CM in different industrial applications.

Books on corrosion science and engineering published during the last 10 years contain chapters about *CM* [1, 8–10], but there are nearly no textbooks on the same corrosion engineering with sections on *NDT* in spite of its importance in risk-based inspection [11]. Unfortunately these two important *fields*, NDT and CM, are developing separately, and our task is to connect them – to find how achievements of both fields may be used for decisions on practical tasks of detection, control, and prevention of corrosion. Corrosion specialists know well how NDT helps in CM. *Therefore, we consider NDT as a vital part of CM technique.* Few studies focus on a combination of NDT and CM [2, 12], for instance, use of NDT for CM under thermal insulation [13] and use of NDT and electrochemical techniques of corrosion assessment of reinforced concrete structures (bridges, buildings, etc.) [12].

Many standards related to NDT have been published by ASTM since 2000 [14]. Special monographs [15–19], reviews [2, 5–7, 12, 13, 20–23], technical committee reports and standards of NACE [24–43], ASTM [44–54], API [3, 55–57], ASME [58], ASNT [59], ANSI [60], and EN [61], and patents [6, 62–67] on *CM* have been published since 2000. Textbooks on corrosion and corrosion control published during the last 10 years contain description of CM methods [1, 68]. Some papers [69, 70], even published 24 years ago, remain classical and are recommended for reading. Many corrosion conferences include a session on *CM*. The European Federation of Corrosion (EFC) Working Party 4 on Nuclear Corrosion and the European Cooperative Group on Corrosion Monitoring of Nuclear Materials published CM methods in nuclear power plant environments (essentially in high temperature water at

300 °C) and the long-term storage of radioactive waste [71]. "*Corrosion monitoring* in district heating systems" project was conducted in Nordic countries [72]. Many conferences on NDT take place around the world [73, 74], and corrosion specialists usually are not informed about them. Monographs [4, 75] and reviews [76] on NDT have also been published during the last 10 years. Modern curricula courses on *corrosion science and technology* include *CM* [77]. Most information about NDT and CM is dispersed in periodical literature and special publications on manufacture of NDT and CM equipment and companies producing corrosion-resistant materials, chemicals, paints, etc.

This chapter describes the principles and industrial achievements in NDT and CM during the last 10 years (2001–2010). Some studies that have an essential role in the developing of NDT and CM methods, but published before 2001, are cited in this chapter.

The NDT and CM methods have advantages as well as limitations. There is no "ideal" NDT and CM technique, but many of them help to detect defects in technological processes and to optimize the injection of chemicals (demulsifiers, neutralizers, corrosion inhibitors, dispersants, biocides, oxygen scavengers, etc.). The NDT and CM have become an important aspect of the design and operation of many industrial plants because they enable engineering and management personnel to be aware of the damage caused by corrosion and the rate of deterioration [70, 78, 79]. This chapter will help to select NDT and CM methods correctly in any industrial plant for any structure and equipment.

Monitoring of protective properties of anticorrosion coatings and cathodic protection is beyond the scope of this chapter and can be found in literature [15, 18, 34, 35, 38–40, 42, 43, 60–62, 80–105].

In any case, we have enumerated briefly some monitoring methods using in cathodic protection in Section 9.12.

9.2 Why Should We Monitor *Corrosion*?

The most incomprehensible in the world
That it is comprehensible.

Albert Einstein (1889–1956)

It is impossible to prevent corrosion of many metals and alloys because they are thermodynamically unstable in the environment. If corrosion of metals is inevitable, we want to know how long equipment or structure will serve under particular conditions safely and efficiently. Various electronic devices are used in industries. Their reliable functioning depends on the quality of electric contacts made of different metals and alloys. If corrosive species are present in the environment, electric contacts may corrode, reliability of the data received by these devices diminishes, and corrosion may result in malfunctioning and unpredicted disasters. Therefore, we should monitor corrosive aggressiveness of the environment where electronic or other devices are in contact with it. The third reason for CM requirement is related to preventive anticorrosion measures. Various methods are used for corrosion control. Such chemicals as corrosion inhibitors, neutralizers, anti-scaling dispersants, biocides, demulsifiers, and scavengers of oxygen and hydrogen sulfide are used in order to prevent or diminish corrosion. We must know the efficiency of these "anticorrosive" chemicals. When cathodic protection is used for corrosion control of the underground and submerged equipment and structures, we should know its efficiency. If coatings are used for corrosion control, we should know their efficiency and protective properties. In other words, CM helps us to define the efficiency of the corrosion control methods. To sum up, there are three important questions for industry-related CM. The first question is a "qualitative" one. Does corrosion occur? Is there any equipment or structure that suffers from corrosion? Many side issues exist. What is the corrosion form? Is the corrosion uniform or localized? Here is a wide spectrum of corrosion phenomena: pitting, crevice, stress corrosion cracking (SCC), galvanic corrosion, dealloying, intergranular corrosion, erosion, cavitation, microbiologically induced corrosion (MIC), underdeposit corrosion, fretting corrosion, and corrosion fatigue. CM can help define corrosion form in order to use correct preventive measures. The second question is related to the previous one. What are the reasons and factors influencing corrosion? Certainly this question is connected to the selection of anticorrosion measures, type of corrosion-resistant materials, coating systems, inhibitors, and other chemicals used for corrosion control. The third question is a "quantitative" one. What is the corrosion intensity? What is the corrosion rate value? *The corrosion rate is the corrosion of material occurring in the unit of time.* This is not always the corrosion product quantity. Corrosion may change with time and with distance on the metal surface. Even if we know the corrosion intensity, that is, for corrosion rate of 0.01 mm yr^{-1}, or 0.1 mm yr^{-1}, or 1 mm yr^{-1}, the following questions remain. Is this value small or large? Is this situation dangerous or can we exist with such corrosion rate? Can we continue to use the equipment? The replies are beyond the scope of this chapter and you may find answers partly in literature [1].

We will describe CM and NDT methods for corrosion detection and its monitoring.

9.3 Corrosion Monitoring and NDT

An engineer is a person who can explain how a mechanism works,
But cannot explain why it does not work.

M. Hazernov (*A Large Book of Aphorisms*, EKSMO-Press, Moscow, 1999, p. 842)

Three main factors influence corrosion: type of metal, environment, and conditions at a boundary between metal and environment. Therefore, we can use the properties of a metal, of an environment, and of a metal–environment boundary for CM and NDT. In the light of this, CM methods can be divided into three groups: control of a metal condition, of the environment, and of a metal–environment border. Control of a metal condition is based on the physical properties of metals: mass, thickness, and electrical resistance (ER) of a metal sample. Control of the environment is based on the chemical, physical, physicochemical, and microbiological properties and conditions of the environment. Control of the metal–environment boundary is based on the detection of its physicochemical properties.

The NDT is used as a rule for examination and control of a metal condition; only *replication* (*on-site metallography*) and *spectrochemical analysis* of alloys relate to examination of the metal–environment boundary. There is no generally accepted classification of CM and NDT methods. Some authors [19] classify according to corrosion phenomenon type, for example, methods of detection of pitting corrosion, MIC, hydrogen damage, etc. Most authors [1, 2, 5–7] classify CM and NDT methods according to their main principle of measuring, for example, physical, physicochemical (e.g. electrochemical), and chemical methods. We agree with the last authors.

9.3.1 Control of Physical Properties of a Metal (Physical Methods)

Historically, physical methods were the first: *man's eyes* (sometimes with magnification glass) or *optical devices* (video cameras, microscope, stereo video microscope, and borescope) were used for visual control of properties of corroded metal surface [2, 106]. Visual methods are possible in most cases during shutdown. Bell-hole excavations are carried out for evaluating corrosion on external surfaces of pipelines [107].

Video cameras are used for inspection of inner surfaces (including coating assessment) of tanks and voids on ships [108] and in radioactive environments that are not suitable for personnel.

A *borescope* (or *boroscope*) is the general name of an optical device consisting of a rigid or flexible tube with an eyepiece on one end and an objective lens at the other linked together by a relay optical system in between. The optical system is usually surrounded by optical fibers used for illumination of the remote object. An internal image of the illuminated object is formed by the objective lens and magnified by the eyepiece that presents it to the viewer's eye. *Rigid or flexible borescopes* may be fitted with a video or charge-coupled device (CCD) camera [109]. Sometimes *borescopes* are divided into *fiberscopes* (*flexible borescopes*), *videoscopes* (*video borescopes*), and *rigid borescopes*. Criteria for selecting a *borescope* are usually image clarity and access. Devices for use inside the human body are referred to as endoscopes.

Remote visual inspection (*RVI*) is used for real-time views and images from inside voids such as tubes, pipes, heat exchangers, tray towers, refractory-lined vessels, engines, rotating machinery, and any enclosed structures [110]. RVI includes *video borescopes* and *pan–tilt–zoom cameras. Optical microscope and profilometry* are used for analyzing and measuring depth of pits [111]. *Optical imaging techniques* eliminate the influences of human subjectivity by digitally capturing the sample images under enhanced illumination conditions and then subjecting them to image analysis managed by computer software [112].

Optical monitoring was used for obtaining information of the kinetics of film formation inside pipelines in district heating systems [72] and measuring the depth of pits [113]. *Vertical scanning interferometry* was used for evaluating pitting corrosion: surface topography was measured with high precision of 0.1 nm [113]. A *laser profilometer* with a resolution of 1 μm in height was used for estimation of pit depth and density on carbon steel surface [114].

Sentry (sentinel) holes are used for periodic visual examination of the equipment and indicate when corrosion allowance has been consumed [9]. *Corrosion allowance* is the thickness in millimeter that an engineer should add for design thickness of equipment because of corrosion. Sentry holes are small holes drilled from the outside surface of the equipment at appropriate positions most susceptible to corrosion attack, to a depth corresponding to the design wall thickness minus the corrosion allowance [21]. When corrosion has consumed the corrosion allowance thickness, a small leak develops, providing a timely warning of the need for replacement. For instance, *sentinel holes* are useful in preventing catastrophic failure due to erosion at pipe bends [9].

In many cases (e.g. in the progress of reinforcement corrosion in concrete), *visual inspection* will not observe anything until a significant damage has developed, which causes a high cost of repair in the short term [115].

The following NDT based on physical phenomena are used for CM: *ultrasonic (US), AE, eddy current, penetrant testing, radiographic examination, magnetic particle inspection, magnetic induction, and IR thermography* [2, 15, 23, 116]. Visual inspection and US measurements are used for determining the lifetime of metallic structures [117]. *On-site metallography* and *spectrochemical analysis* of alloys also belong to NDT. *Weight loss* (WL) and ER measurements are based on the loss of corroded metal samples and belong to CM. All these NDT and CM physical methods will be described in Section 9.4.

9.3.2 Control of the Environment (Chemical Analytical, Physicochemical, Physical, and Microbiological Methods)

These methods are based on chemical, physical, physicochemical, and microbiological analyses of media that are in contact with metallic equipment. It is impossible to use the WL method in the case of CM of copper alloys or stainless steels after immersion in cold water because of very low mass changes of coupons made of these alloys. One can use various sensitive physicochemical analytical methods for the determination of concentration of metallic ions (Fe^{2+}, Fe^{3+}, Cu^{2+}, Zn^{2+}, Cr^{3+}, Ni^{2+}) in media as a result of corrosion of metallic equipment. On the other hand, chemical analytical methods are used for the determination of aggressive components in media – such as pH, Cl^-, SO_4^{2-}, HCO_3^-, CO_3^{2-}, O_2, CO_2, H_2S, NH_3, CN^-, Fe^{3+}, Cu^{2+}, hardness (Ca^{2+} and Mg^{2+}), alkalinity, other dissolved (silicates, phosphates, etc.) and suspended solids, microorganisms influencing corrosion, contaminants inducing erosion and cavitation in water and steam, and some physicochemical properties of liquid media (electrical conductivity and redox potential) – and for the determination of aluminum, chromium, iron, copper, lead, and tin in engine oils. For organic liquids (fuels, solvents, etc.), it is important to measure water content, dissolved oxygen concentration, and electrical conductivity. For crude oil, it is important to measure total acid number (inorganic and organic acids including naphthenic acids), sulfur, nitrogen, water, and salt content. For boiler feed water (BFW), it is important to define the concentration of dissolved oxygen, copper and iron cations, silica, and total dissolved solids. For steam and high temperature gases, it is important to measure temperature in order to eliminate water condensation and, as a result, dew point corrosion and cavitation. For the determination of the aggressiveness of atmosphere, we should measure the relative humidity, the temperature and its changes, the period during which a metal surface is covered by a film of electrolyte that is capable of causing atmospheric corrosion (time of wetness), sulfur oxides (SO_x), nitrogen oxides (NO_x), ammonia and its derivatives, ozone, hydrogen sulfide, and dust in the atmosphere. The determination of hydrogen gas in some processing (hydrogen monitoring) is related to the analytical methods. For the determination of the corrosiveness of soil, we should measure its electrical conductivity. In some cases, it is preferable to identify the microorganisms, humidity, and pH of soil [82]. Microbiological analysis of water, soil, atmosphere, crude oil, fuels, and sludge is very important for control of potential MIC [118].

Control of process (technological) or environmental parameters includes the measuring of flow rate, flow regime, temperature, pressure, dew point temperature, heat transfer resistance, the presence of solid particles in streams, humidity and temperature at surface, surface wetness, ambient humidity, etc.

The analysis of residue concentrations of corrosion inhibitors, neutralizers, oxygen and hydrogen sulfide scavengers, and biocides in technological streams is important for control of efficiency of anticorrosion treatment, as well as for the environment.

9.3.3 Control of the Metal–Environment Boundary (Physicochemical Methods)

These methods are based on physicochemical properties of the metal–environment border and may be divided into *electrochemical* methods, *radioactive* methods, *identification of corrosion products and deposits forming on metal surface*, and *examination of morphology of metal surface underdeposits*.

Among *electrochemical* methods are *redox* and *corrosion potential measurements*, *polarization measurements* (*potentiodynamic polarization* or *Evans diagrams* and *Tafel extrapolation*), *linear polarization resistance* (*LPR*), *galvanic probes* (*zero resistance ammeter* [*ZRA*]), *electrochemical noise measurements* (*ENM*), *electrochemical impedance spectroscopy* (*EIS*), and *harmonic distortion analysis* (*HDA*).

Radioactive methods such as *thin layer activation* (TLA) and *neutron activation* (NA) are used for the determination of corrosion resistance (corrosion rate) of some alloys under specific conditions [15, 21].

Microbiological analysis of deposits (*sessile* bacteria) on metal surface is also very important for assessing MIC.

SEM, STM, and AFM are used for the examination of metal surface morphology [119–121]. EDS, X-ray photoelectron spectroscopy (XPS) (also known as electron spectroscopy for chemical analysis [ESCA]), Auger electron spectroscopy (AES), focused ion beam SEM, Mössbauer spectroscopy, secondary ion mass spectrometry (SIMS), grazing X-ray diffraction (grazing XRD), time-of-flight secondary ion mass spectrometry (ToF-SIMS), Fourier transform infrared spectroscopy (FTIR), thermogravimetric analysis–FTIR (TGA-FTIR), inductively coupled plasma mass spectrometry (ICP-MS), headspace gas chromatography–mass spectrometry (GS-MS), X-ray diffraction analysis, X-ray fluorescence (XRF), Raman spectroscopy, ultraviolet (UV)–visible reflectance studies, ultraviolet photoelectron spectroscopy (UPS), electron probe microanalysis (EPMA), spectroscopic ellipsometry (SE), reflection electron energy loss spectroscopy (REELS), synchrotron radiation (SR), and synchrotron radiation–grazing incidence X-ray diffraction (SR-GIXRD) are used for the identification of chemical composition and phase identity of corrosion products and properties of oxidized metallic surface (passive layers) and even organic coatings [106, 119–131]. SEM and EDS was used for the evaluation of pitting corrosion [114]. Confocal laser microscopy (CFLM), electron microscopy (EM), and environmental scanning electron microscopy (ESEM) are used for monitoring biofilms and possible MIC [9, 132–136]. Software tools are developed for quantitative mapping (chemical microanalysis) by means of SEM [137].

The mass spectrometric methods such as *thermal desorption mass spectrometry* (*TDMS*), *chromato-mass spectrometry* (*CMS*), *matrix-assisted laser desorption/ionization* (*MALDI*), and *surface-assisted laser desorption/ionization* (*SALDI*) are used in research of corrosion processes [138]. *TDMS* is used for studying accumulation of gases in metals and alloys, formation of corrosion products, and transformation of inhibitors. *CMS* is used for rapid analysis of organic compounds (for instance, inhibitors and products of their transformation). *MALDI* and *SALDI* as modern methods of mass spectrometry analysis are used for analysis of high molecular and unstable compounds (corrosion products, organic and inorganic inhibitors). An important advantage of these methods is a possibility for direct surface analysis without preliminary sample preparation.

Useful information can be obtained from the color of corrosion products forming on the metal equipment [1, 139]. Slippery surface of brown, orange, black, green, or intermediate colors usually recognizes the presence of biofouling.

9.4 Physical Methods of Corrosion Monitoring

Science and technology have had yet an obvious limit,
for instance, smell of Vicki is not transformed by phone.
Romain Gary (1914–1980), the French novelist

We will describe following physical methods: different *NDT*, WL, and ER measurements.

9.4.1 NDT for Corrosion Monitoring

The following NDT techniques based on physical phenomena are used for CM: *ultrasonic (US), AE, eddy current, penetrant testing, radiographic examination, magnetic particle inspection, magnetic induction, IR thermography, on-site metallography (replication), and spectroscopic chemical analysis* of alloys [2, 15, 23].

9.4.1.1 Ultrasonic Technique

Ultrasound is a cyclic *sound* pressure with a frequency greater than the upper limit of human hearing. Although this limit varies from person to person, it is ~20 kHz in healthy young adults, and thus, 20 kHz serves as a useful lower limit in describing *ultrasound*. *Ultrasonic* is an adjective referring to *ultrasound (US)*. Approximate frequency ranges corresponding to US, with some applications, are shown in Figure 9.1. Two regions, *acoustic* and *ultrasound*, are used for NDT.

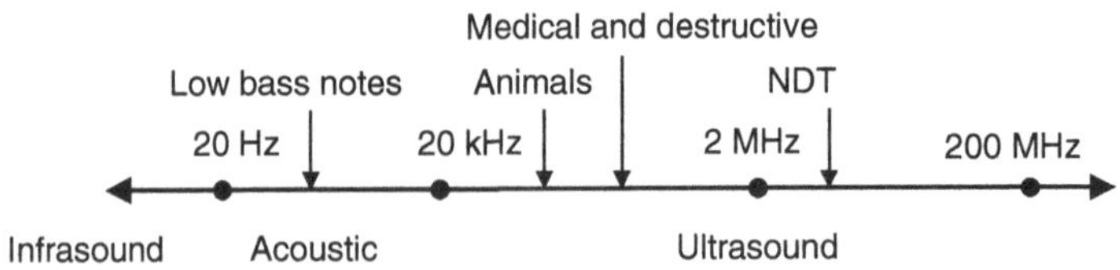

Figure 9.1 Ultrasound range diagram.

US testing is a type of NDT *commonly used to find flaws in materials and to measure the thickness of objects* and thus to *monitor corrosion*. Frequencies of 2–50 MHz are commonly used. Thus, US uses high frequency sound waves transmitted through the metal. The sound is reflected from other surfaces (the opposite metallic wall, anomalies in metal). The time taken for the sound to transverse the thickness of metal and return to the probe is displayed as metal thickness. US devices are portable or fixed and measure thicknesses from 1 to 300 mm to an accuracy of 0.1 mm [23], sometimes to 1% of the wall thickness [27]. Usually the maximum operating temperature for the solid coupled probes in US is 250 °C [27], but in some cases, 350 °C can be reached, and new probes are being developed for measuring up to 500 °C [140]. The deviation of US from WL data is ~2% [141]. US technique has been used for ~50 years.

US waves in solids can roughly be divided into two groups, the *bulk waves* and *guided waves*. The majority of US testing is carried out with *bulk waves*. In this case, the wavelength of the waves is small compared with the dimensions of the object under inspection. Relatively localized information is obtained with this type of waves. The disadvantage of use of *bulk waves* is overcome by using of *guided waves*, which allow obtaining a *wall thickness map*.

Guided waves, named also *guided Lamb waves* in honor of the English mathematician *Horace Lamb* (1849–1934) who analyzed and described acoustic waves of this type in 1917, are *elastic waves* whose particle motion lies in the plane defined by the plate normal and the direction of wave propagation. In this case, the wavelength is typically in the order of wall thickness or even larger. The theory of this method is described in literature [142, 143]. The *guided waves* may propagate over very long distances and hence may provide global information. A variety of *guided wave* modes exist in plates and pipes, and majority of these wave modes are dispersive. This means that their phase velocity depends on the frequency and wall thickness. This property is used for instantaneous monitoring of the wall thickness of plates and pipes. This method is capable of determining the wall thickness of large areas with only a limited number of US transducers. The wall thickness is determined between two transducer arrays. These transducers excite specific *guided wave* modes that are highly dispersive. Noncontact *electromagnetic acoustic transducers* (EMAT) in thin metal plates are also widely used [144]. *Ultrasonic guided Lamb wave tomography* is based on propagation and measuring *guided Lamb waves* in metals and is used for instantaneous CM of airframe skins, fuel storage tanks, pressure vessels, and pipelines [145–147] and rapid inspection of large sections of aircraft structures for flaws such as disbonds, delaminations and corrosion [148], and thinning changes of pipelines at gas plant and refineries [142, 149].

US thickness measuring instruments usually are the most accurate means for obtaining thickness measurements on installed pipes larger than 33 mm and in the case of uniform corrosion [3]. *Radiographic* profile techniques are preferred for pipe diameters of 33 mm and smaller. When corrosion is nonuniform, *radiography, ultrasonic scanning,* and *eddy current* devices may be used [3]. When *US thickness* measurements are carried out above 65 °C, instruments, couplants, and procedures should be used that will result in accurate measurements at higher temperatures. Measurements should be adjusted with an appropriate temperature correction factor [3]. Each of the NDT techniques has practical limits with respect to accuracy.

The energy company General Electric (GE) developed two *US* sensor types for online wall thickness monitoring (automated thickness measurement system): low temperature "Rightrax Flex 120 °C" (−40 up to 120 °C) and high temperature "Rightrax HT 350 °C" (−20 up to 350/500 °C) [140]. These sensors are intrinsically safe, simply bond, or clamped onto the inspection area (usually at selected critical points) and thus allow receiving online corrosion data excluding erecting scaffoldings, removing insulation, shutdown plants, eliminating excavation and thus decrease inspection cost. These sensors can measure metal thicknesses of 3–19 mm, suitable for pipe diameters of 3″ up to 30″, and even more. The developed systems have data acquisition and data analysis software with the following functionalities: temperature correction, minimum/maximum/average thickness values, alarms for allowable thickness, and short-term, long-term, and maximum corrosion rates. These *US* sensors are self-calibrating, and the data obtained by the sensors are integrated with process variables such as temperature, pressure, flow rate, pH, and anticorrosive chemical injections [140].

Often US measurements do not show localized thinning and localized dangerous corrosion. Thus, perforation due to local corrosion often happens in places far from US measurement points [116].

Traditional manual NDT has disadvantages such as very large number of points measuring (several ten thousands), time consuming, low effectiveness, and relatively large cost.

Automated US scanners (crawlers) are used to rapidly inspect the thickness of large metallic structures at small measurement intervals, providing a high detailed thickness map of the scanned surface (tank floors, courses and roofs, vessels, and towers). Scan velocities are >0.5 $m\,s^{-1}$ [150]. Most of the data in oil refineries' units is received from ultrasonic measurements carried out on shutdown (every 4–5 years) basis.

Phased array ultrasonic technology is an advanced method of US testing [151–154]. NDT is typically performed in industry using either radiography or US. Radiography has the disadvantages of being a safety hazard and poor at detecting cracks. Manual US is much better at detecting planar discontinuities, but it is slow and the results are highly dependent on the operator. Automated US testing typically involves large, expensive, and inflexible systems, though the results are reproducible. Portable *phased array US* equipment is highly computerized and fast and can be operated in manual, semiautomated, or fully automated modes. From a practical viewpoint, *ultrasonic phased arrays* are merely a technique for generating and obtaining ultrasound; once the ultrasound is in the material, it is independent of the generating technique (piezoelectric, electromagnetic, laser, or phased arrays). Consequently, many of the details of US testing remain unchanged. Applications include detection and sizing of SCC [155, 156], hydrogen-induced cracking (HIC), crevice corrosion (detection corrosion under a gasket seat without removing the bolts) [155], and MIC [157]. According to a few studies, US was not effective for detection of SCC and HIC [150]. Another NDT method, *AE*, overcomes this disadvantage.

9.4.1.2 Acoustic Emission

We have listened many times to the sound of breaking glass, plate, or ceramic vase, falling tree, and cracking ice. Primitive men heard cracking stones and fracturing bones or wood. They used their experience and intentionally "listened" to different structures in order to detect if they were in danger. These are several examples how we may hear fractured sounds from different objects subjected to stress. A "tin cry" is the characteristic sound heard when a bar of tin is bent. The bar of tin will "cry" repeatedly when bent until it breaks. Variously described as a "screaming" or "crackling" sound, the effect is caused by the shearing of crystals in the metal. The sound is not particularly loud, despite terms such as "crying" and "screaming." Niobium, indium, and gallium exhibit a similar effect.

AE is a phenomenon of sound and ultrasound wave generation (see Figure 9.1) by materials that undergo deformation and fracture [158]. *AE is the generation of transient elastic waves during the rapid release of energy from localized sources within a material* [159]. In other words, AEs are the stress waves produced by the sudden internal stress redistribution of the materials caused by changes in the internal structure. When a structure is subjected to an external stimulus (change in pressure, load, or temperature), localized sources trigger the release of energy, in the form of stress waves, which propagate to the surface and are recorded by sensors. With the right equipment and setup, motions of the order of picometers (10^{-12} m) can be identified. Sources of AE vary from natural events such as earthquakes and rock bursts to the initiation and growth of cracks, slip and dislocation movements, melting, twinning, and phase transformations in metals and fiber breakage and fiber–matrix debonding in composites [160]. Most of the sources of AEs are damage related; thus, the detection and monitoring of these emissions are commonly used to predict material (metals, alloys, polymers, ceramics, composites, such as wood, concrete, etc.) failure. AE is unlike most other NDT techniques in two ways. The first difference pertains to the origin of the signal. Instead of supplying energy to the object under examination, AE technique simply listens for the energy released by the object. AE tests are often performed on structures while in operation, as this provides adequate loading for propagating defects and triggering AEs.

The second difference is that AE technique deals with dynamic processes, or changes, in a material. This is particularly meaningful because only active features (e.g. crack or pit growth) are highlighted.

Sources generating AE in different materials are unique. In metals, primary macroscopic sources are crack jumps, processes related to plastic *deformation* (or when a material is loaded at or near its yield stress), *fracturing*, and debonding of inclusions. At the microscopic level, as plastic deformation occurs, atomic planes slip past each other through the movement of dislocations. These atomic-scale deformations release energy in the form of elastic waves, which "can be thought of as naturally generated ultrasound" traveling through the object. When cracks exist in a metal, the stress levels present in front of the crack tip can be several times higher than the surrounding area. Leaks, frictions, knocks, and chemical reactions belong to secondary class of AE. Quantitative and qualitative characteristics of AE waves generated by sources of different nature depend on material properties and environmental factors. Both sources are related to corrosion phenomena. Certain corrosion reactions generate elastic waves (sounds), which may be detected by sufficiently sensitive instrumentation to provide an identifiable acoustic signature. Correlation of these signatures with known standards provides an NDT for monitoring corrosion. Thus, electrochemical corrosion reactions (anodic and cathodic) are accompanied by the radiation of signals having a low amplitude such as white noise [161]; the destruction of oxide films on the metal surface followed by localized corrosion (pitting, crevice, erosion, cavitation, SCC) produces AE of an explosive type with relatively small amplitude [161]. Oxygen reduction on stainless steel surfaces during corrosion processes also contributes to AE [162]. Different corrosion phenomena, especially localized corrosion such as SCC, HIC, pitting corrosion, crevice corrosion, and corrosion fatigue, are accompanied by evolution of energy that allows defining by AE method [75]. AE was proposed for use in practical industrial NDT of different structures during the 1960s. Today, AE NDT is standardized [163–173] and is used practically in all industries for different types of structures and materials [75]. Theory and practice of using AE is described in literature [75]. The AE NDT is based on the detection and conversion of high frequency elastic waves to electrical signals. AE testing can be performed in the field with portable instruments or in a stationary laboratory setting. Typically, systems contain a sensor, preamplifier, filter, and amplifier, along with measurement, display, and storage equipment (e.g. oscilloscopes, voltmeters, and personal computers). AE sensors respond to dynamic motion that is caused by an AE event. This is achieved through transducers, which convert mechanical movement into an electrical voltage signal. The transducer element in an AE sensor is almost always a piezoelectric crystal, which is commonly made from a ceramic such as lead zirconate titanate. Sensors are coupled to the structure by means of a fluid couplant (grease or oil) and are secured with tape, adhesive bonds, or magnetic hold-downs.

The output of each piezoelectric sensor (during structure loading) is amplified through a low-noise preamplifier, filtered to remove any extraneous noise, and furthered processed by suitable electronic equipment. Transducers are selected on basis of operating frequency, sensitivity, and environmental characteristics and are grouped into two classes: resonant and broadband. The signal that is detected by a sensor is a combination of many parts of the waveform initially emitted. AE source motion is completed in a few millionths of a second. As the AE leaves the source, the waveform travels in a spherically spreading pattern and is reflected off the boundaries of the object. The sensitivity of an AE system is often limited by the amount of background noise nearby. Noise in AE testing refers to any undesirable signals detected by the sensors. Examples of these signals include frictional sources (e.g. loose bolts or movable connectors that shift when exposed to wind loads) and impact sources (e.g. rain, flying objects, or wind-driven dust) in bridges. To compensate for the effects of background noise, various procedures can be implemented. Some possible approaches involve fabricating special sensors with electronic gates for noise blocking, taking precautions to place sensors as far away as possible from noise sources, and electronic filtering.

AE technology is a nondestructive real-time monitoring used in almost all industries.

We will describe only the use of AE for detection of corrosion in industrial applications.

The first wide description and analysis of use of AE for *corrosion detection* was done probably in 1986 [174] and then reviewed [161] in 1990. Since the 1990s, AE was spread as a nondestructive CM for detecting localized corrosion of different equipment and structures: pressure vessels (containing liquefied petroleum gas [LPG], etc.), reactors, storage tanks, piping systems, steam generators, aerospace structures, wings, bulkhead, fuel tanks, rocket engines, offshore and onshore platforms, bridges, and bucket trucks. AE is used for detection of destruction (we may use also term *corrosion*) of a variety of composite and ceramic components. A useful book on AE was published in 2000 [75]. Based on these and other sources published during the last 10 years, we can summarize the use of AE for detection corrosion. AE was used for detection and monitoring *general corrosion* [162, 175–184], *pitting corrosion* [183, 185–189], *crevice corrosion* [182, 190, 191], *SCC* [183], *chloride SCC* [192], and *hydrogen damage* [193]. AE monitoring was used for studying the influence of laser peening and shot peening (surface treatment) on the pitting corrosion of 316L stainless steel [194] and early detection of *HIC* [195–198] and *sulfide stress cracking* (*SSC*) in wet hydrogen sulfide media [199], and it was also possible to differentiate between stress-oriented hydrogen-induced cracking (SOHIC) and SSC [200], corrosion damages of buried pipes in refinery [201], and damage assessment in fiber-reinforced polymer-matrix composites, in particular glass-fiber-reinforced parts or structures (e.g. fan blades). AE was used for detection of SCC during early stages in carbon steel [202, 203], aluminum [204], and zircaloy-4 alloy [205] structures and corrosion fatigue crack growth in metallic constructions [206–211], early detection of exfoliation corrosion of aluminum alloys [212], detection of fatigue damage in reinforced concrete slabs [213] and other materials [160], and corrosion detection in storage tanks, reinforced concrete structures, process equipment [214, 215], and corrosion under insulation [216]. In cases of insulation, only small holes in insulation are required for sensor mounting. In high temperature processes, waveguides are used to guide the AE waves from the hot surface to the edge where the sensor is mounted. In large cryogenic vessels, permanent sensors are mounted under insulation for periodic control.

One of the *disadvantages* of AE is that commercial AE systems can only estimate *qualitatively* the damage in the material. *Quantitative AE nondestructive inspection* was developed and used for detecting corrosion in pressure vessels and piping [75, 150, 217, 218] or deaerators [75]; detecting flaw initiation and growth in weld joints of LPG pressure vessels that has undergone *SCC* [219]; diagnosing the condition of operating high energy equipment and piping of nuclear and fossil power plants and chemical and refinery reactors [220, 221]; revealing SCC, HIC, and caustic embrittlement in steels used in nuclear, chemical, and refinery plants, metal bridges, and cranes [75, 150, 222–224]; evaluating the remaining lifetime of tubes that have undergone SCC [221, 225], corrosion of reinforced concrete beams [226], and water cooling system of boiling water reactor (BWR) [227]; and identifying cracks in low density polyethylene (LDPE) reactor tubes [228]. AE technique was used to monitor and predict the rate of corrosion of mild steel [179, 180, 229]; it was shown that corrosion films as thin as 10 μm or as light as 3.5 mg could be detected [178].

The corrosion of 1 mm of steel results in up to 12 mm of the hydrated iron oxide [214]. AE is caused by microfractures and delaminations during expansion of iron oxides. The emission from corrosion usually releases much less energy than emission from crack growth and so is more difficult to detect in the field environment. The sensor frequencies used to detect these signals are determined more on the basis of the environmental noise under test conditions than on the frequency spectrum of the emission. In low-noise environments, the signals from corrosion may be detected at distances of tens of meters by using monitoring frequencies down to 10 kHz. However, in a live process plant, it may be necessary to use several hundreds of kHz to obtain the noise. In this case, detection distance may be limited to less than half a meter [214].

AE is used for online tank floor (the bottoms of aboveground storage tanks) inspection including annular rings (detection of corrosion, leaks, and remaining lifetime) [75, 230, 231]. This method need not be used to empty or clean the tank. AE method allows inspection of one tank of diameter 50 m during 8 h. Highly sensitive AE sensors are attached to the outside of the tank wall and the tank monitored, following a period of conditioning during which valves are closed and heaters/agitators turned off. Noise sources are eliminated by sophisticated pattern recognition methods. The corrosion of the floor is detected together with leaks that are active during the actual monitoring period.

The coupling of AE and thermogravimetric techniques are used for high temperature degradation and receiving information of the growth of the scales and mechanical stresses of alloys in oil refining and petrochemical industries [232].

AE was used for detection of failure of a coating during testing and influence of temperature on protective properties of coatings [233] and for monitoring damage of rubber coating on metallic surfaces [234]. The AE stress source with respect to coatings is a result of the penetration of water, gases, and ions through the coating and their interaction with the coating/substrate interface. This interaction at the interface can result in coating disbondment, blistering, and/or ruptured blisters, which are AE sources. A second source of AE is the corrosion process on the steel surface under the coating. AE for monitoring protective properties of coatings did not find further use and development.

The advantages of AE technique are as follows:

1) AE is a unique noninvasive method as a test monitoring technique in that it originates in the material and does not require an external source of energy as ultrasonic or electrochemical techniques do.
2) AE is of high sensitivity and fast diagnosis. Sensors are mounted on external surfaces of equipment and structures (up to 170 °C).
3) In-service (*in situ*) online real-time monitoring for corrosion behavior of metals, alloys, polymers, ceramics, and composite materials. AE detects the activities inside the materials.
4) AE inspection does not require access to the whole examination area. For example, to cover a total area of a 16 m diameter sphere, 30–40 sensors are needed [150].
5) AE is the only NDT that can detect the onset and continuation of pitting, crevice corrosion, SCC, and HIC, in service and online.
6) Early and rapid detection of cracks and flaws.

The disadvantages of AE technique are as follows:

1) Complicated devices, skillful and knowledgeable personnel, and expertise are needed.
2) Environments are generally very noisy, and the AE signals are usually very weak. Thus, signal discrimination and noise reduction are very difficult, yet extremely important for successful AE applications.
3) In spite of developing *quantitative* AE technique, in most cases, commercial AE systems can only estimate *qualitatively* the damage in the material and approximately how long the components will serve. Thus, other NDT methods are still needed to perform more thorough examinations and provide quantitative results.

9.4.1.3 Eddy Current Technique Uses an Electromagnetic Field Generated by an Electrical Coil Mounted in a Probe

The alternated electromagnetic field (e.g. tube wall) induces eddy currents in a conducting material. The induced eddy currents in return generate an alternating electromagnetic field opposing the original electromagnetic field and result in a change in coil impedance, which is measured by the eddy current instrument. The change in coil impedance depends on the amount of metal loss or the depth of defects (cracks and pits) [107]. By calibrating the eddy current instrument, the defects are measured with high accuracy. Eddy current technique is a noncontact one used for measuring general corrosion, for example, detection of corrosion under insulation [235] and detection and evaluation of sensitization and intergranular corrosion in austenitic stainless steels [236]. The inductive scanning system (based on eddy current technique) using a heterodyning sensor in combination with real-time digital signal processing was used for detection of corrosion in reinforced steel bars embedded within the concrete [237].

Some physical methods cannot be used in particular conditions, for instance, under insulation at high temperatures (above 100 °C) or inside heat exchanger tubes. In such cases, other physical methods are used [235]: *ultrasonic guided wave* (see Section 9.4.1.1), *pulsed eddy current (PEC)*, *radiography* (see Section 9.4.1.9), and *IR thermography* (see Section 9.4.1.11).

PEC is a nonintrusive and noncontact NDT method. Therefore, it can be applied for wall thickness monitoring at high temperatures, up to 540 °C [238], and CM naphthenic acid corrosion in refineries [239]. PEC employs a pulsed magnetic field to generate eddy currents in the steel. Since carbon and low-alloy steels are ferromagnetic, only the top layer of the steel is magnetized [240, 241]. The eddy currents diffuse into the test specimen until they eventually reach the far surface. Then, they induce a voltage signal in the receiver coils of the PEC probe. As long as the eddy currents experience free expansion in the steel, their strength decreases relatively slowly. Upon reaching the far surface, their strength decreases rapidly. The moment in time when the eddy currents first reach the far surface is indicated by a sharp decrease in the PEC signal. The onset of the sharp decrease point is a measure of wall thickness. An earlier onset of this sharp decay of one PEC signal compared with a reference signal indicates wall loss. PEC has a much better reproducibility than ultrasonic wall thickness measurements and has been applied to monitor wall thickness in piping of refineries and oil production platforms. Physical methods (NDT techniques) used for CM under thermal insulation are summarized in literature [235].

9.4.1.4 Electromagnetic Methods

Electromagnetic methods (sensors) were suggested for detection of corrosion beneath (including material loss) [242] and moisture within paint coatings on aircraft aluminum alloys [243] and pipeline CM [244]. *Remote field electromagnetic technique* (*RFET*) is used as computerized NDT in pipeline and tank inspection [245].

Electromagnetic acoustic transducers based on physical effects such as Lorentz force and magnetostriction are developed as an in-line inspection (ILI) tool for the detection of SCC and coating disbondment [246].

9.4.1.5 Other Physical NDT Methods

Acoustic pulse reflectometry (APR) is the technique when a wideband acoustic pulse is shot into the tube and any reflections that are created by changes in the cross section of the tube are recorded by a microphone. This method allows detecting any fault in tube systems: pits, holes, wall thinning, and deposits [247].

Acoustic vibro-modulation technique, based on nonlinear interaction of ultrasound and vibrations in the presence of defects, was developed for nondestructive detection of SCC, corrosion-induced delamination of structural elements, fatigue cracks in various materials (metals, polymers, and composites), debonding, and crevice corrosion [248]. One of the unique features of this technique is its ability to differentiate cracks, delaminations, and debonding from notches, voids, and other heterogeneities because of specific nonlinear interaction that occurs only at the contact-type interface.

Nonlinear elastic wave spectroscopy is being developed for detection of stress corrosion damage in aircraft structures [249]. In the presence of stress corrosion damage, the material starts to behave nonlinearly around the damage location. This behavior manifests itself in the bi-harmonic excited signal spectrum as sidebands and harmonics of the excited frequencies. The magnitude and number of these effects are related to the size of damage.

Laser scattering technique has been explored for assessment of pitting corrosion, intergranular corrosion, and fatigue damage [2].

The following physical techniques are used to monitor *reinforcement corrosion*: visual inspection, audio, IR thermography, magnetic methods, stress wave transmission, ground-penetrating radar, tomography, petrography, rebound hammer, ultrasonic pulse velocity, and radiography [12, 250] (see Section 9.8). Various models are reviewed for determining the corrosion rate of reinforcement and prediction of the time to cracking [250].

Most physical methods are *NDT techniques* that allow monitoring uniform and localized corrosion, including SCC, corrosion fatigue, hydrogen damage, intergranular corrosion, exfoliation corrosion, and MIC [2]. Advanced NDT such as *multi-array automated ultrasonic testing* (AUT) and *swept low frequency eddy current* (SLOFEC) are used in oil refineries for detailed mapping of the inspected surfaces that resolve small pits and thus allow to detect localized corrosion [251].

9.4.1.6 Penetrant Testing

Penetrant testing is based on the properties of some liquids containing a contrasting (generally red-colored) or fluorescent compound to penetrate into small defects in solid materials. After removing the redundant penetrant liquid, a developer is applied. The developer is a white powder with absorbing properties. The remaining penetrant in the defect is sucked into the developer and indicates the failure (cracks) in metals, ceramics, and polymers.

9.4.1.7 Magnetic Particle Inspection

Magnetic particle inspection allows detection of defects (both open to the surface and subsurface, for instance, SCC) in ferromagnetic materials. The magnetic lines of force are running from the south to the north pole in a magnetized material. These lines are interrupted and leave the metal at locations of defects. At these locations, compounds that can be magnetized such as iron particles (generally with fluorescent agents) are attracted. The iron particles concentrate at locations of leaving lines of forces and indicate defects.

9.4.1.8 Magnetic Induction

Magnetic induction (DUALSCOPE and PERMASCOPE) is based on magnetic induction principle of ferromagnetic materials and is used for measuring thicknesses of nonmagnetic layers on ferromagnetic materials (e.g. stainless steel, copper or zinc coating thickness on carbon steel).

9.4.1.9 Radiographic Examination

Radiographic examination (*radiography*), using X-rays (are emitted by electrons outside the nucleus) or gamma rays (are emitted by the nucleus), can detect general and localized corrosion. *Flash radiography*, using short pulses of X-rays, allows detection of corrosion under insulation. X-ray radiographic methods are used for online wall thickness measurements of insulated pipes [21, 252] and detection of intergranular corrosion and flow-accelerated corrosion. Nondestructive measurement by *X-ray computed tomography* of thickness of corrosion layer formed on carbon steel archaeological samples (excavated and removed) from the soil after 400 years allows to predict corrosion penetration for the overpack material in high level radioactive waste disposal [253].

9.4.1.10 Thin Layer (Proton) Activation and Neutron Activation

Thin layer (proton) activation (TLA) and *neutron activation* (NA) methods are related to *radioactive methods* and sometimes are called *surface-active techniques* [27]. TLA method involves irradiation of a surface by a nuclear accelerator, which provides an accurate path profile of the surface with a low level of radioactivity producing gamma radiation, or introduction of a small quantity of radioisotope tracer into the surface of a coupon or a surface of equipment [15, 21, 23]. Loss of material due to corrosion–erosion is proportional to the decrease of radioactivity (after correction to natural decay) monitoring with a suitable detector. In the TLA method, a depth of 100–300 μm is activated, while the NA method allows activation of a significant thickness of most materials [27]. TLA has unique advantages in the measurement of erosion and wear and can be monitored remotely and with high sensitivity (~0.1–1% of the thickness of the activated material). These methods measure only metal loss. Corrosion products not removed from the activated surface influence the results.

9.4.1.11 Infrared Thermography

All objects emit IR radiation and the amount of radiation increases with temperature. The IR radiation is a part of the electromagnetic spectrum with a wavelength between 9 and 14 μm, not visible to the naked eye, but can be viewed with IR camera. When viewed by a thermographic camera, warm objects stand out well against cooler backgrounds. As a result, thermography makes it possible to "see" an object, without visible illumination [254]. Thermography is a type of IR imaging. IR cameras detect radiation in the IR electromagnetic spectrum and produce images of the surface with information about the surface temperatures or temperature differences. IR is used as a visual technique for the identification of real or potential corrosion. IR technology is used for detection of temperature deviations, namely, hot or cold spots at equipment or pipes, and detection of wall thinning in stainless steel pipes [255]. IR thermography is used in detecting corrosion in refinery units [254] and for monitoring blisters under organic coatings and filiform corrosion [256].

The *limitations* of IR techniques are as follows:

1) Surface conditions (reflection, dirt, rust, and other compounds formed on the surface) influence results.
2) Weather conditions (e.g. strong wind) can completely obscure thermal data.

9.4.1.12 On-site Metallurgical Analysis

The usual method of metallographic investigation that involves cutting of large metallic pieces from a construction, so that laboratory preparation and examination can be performed, generally renders the component of the construction unfit for service or necessitates a costly repair. As a result, metallographic investigations are avoided, and important microstructural information is not available to evaluate the component for satisfactory performance. Therefore, an *in situ on-site field metallurgical examination* is needed to aid in the proper determination of a construction life [257, 258].

On-site metallurgical analysis of fully operational or of stationary or nonremovable metallic parts of different equipment and structures is widely used for detection of corrosion and materials' properties such as crack type (SCC, HIC, fatigue crack, and creep), other defect types, microstructure, heat treatment, hardness (for tensile strength estimation), and type of alloy (chemical content of alloys). These *in situ* measurements of microstructure, hardness, and chemical content of alloys of metallic equipment also represent NDT methods. *On-site metallurgical analysis* unlike traditional NDT methods can predict such dangerous failures as SCC and HIC. *On-site metallurgical testing* can reveal which metallic components are actually damaged and must be replaced and which components can be put back into operation without risk. *On-site metallurgical methods* are routine at many power, oil refining, petrochemical, and chemical plants. We will describe three methods of *on-site metallurgical analysis* that found wide use in industry: *field metallographic replicas, on-site hardness testing,* and *on-site chemical analysis of alloys.*

Field Metallographic Replicas

Field metallographic replica is the *replication technique* that produces an exact copy of the surface that can be peeled away and examined microscopically in the laboratory. This method also is named *in situ metallography,* or *on-site metallography,* or *nondestructive metallography,* or *nondestructive analysis of surface topography,* or *replication microscopy technique,* or *surface replication* [258, 259]. This method is "half on-site" as replicas are made on real constructions and final determination is carried out in laboratory, but principally, we can use optical microscope also in field. *Replication* is used as a tool for evaluating microstructures and other surface features in lieu of laboratory evaluations of real parts of metallic structures. Producing a replica requires abrasive preparation using successively finer grits, followed by a final diamond-impregnated cloth polish. The last polishing step will leave the surface with a mirror finish from which an etchant can develop the characteristics of the microstructure. A copy of the microstructure is taken by either a softened polymer film or by molding with polymer material. Metallographic replicas from aluminum foil by means of knurling were also investigated [260], but they did not find further application. The cellulose acetate and two-component silicone rubbers are used as replicating materials, and the former found wide application in practice. The replicating materials (cellulose acetate) soften in acetone and are applied to a surface that has been wet with acetone. They are stripped off when dry. Thicknesses of 22 or 35 (up to 100) μm are used for finer detail replication. The topography of the replica is a negative copy of the relief on the surface, but because of simplicity of preparation, such replicas are usually used for observations. The replicas are examined by optical microscope or SEM.

Typical examples of applications of metallurgical replication are shown in Figure 9.2.

Replication technique is vital for metallic constructions after a fire [262] and determination of remaining life of metallic constructions and equipment [257]. Standard API 570 [3] obligates that only qualified NDT personnel and proper procedures must be used, points out needed inspection for specific types of corrosion and cracking, and even shows specific equipment and

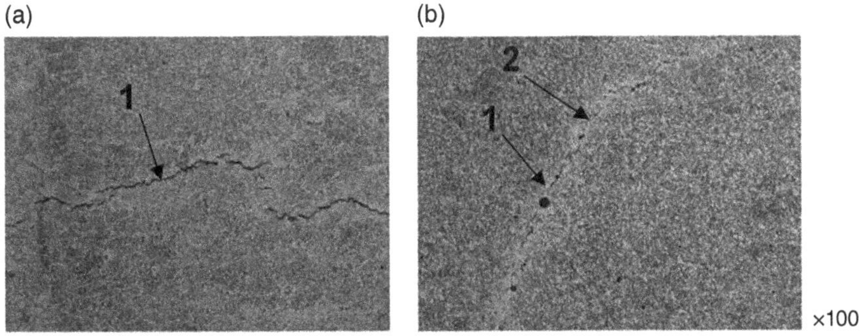

Figure 9.2 Replica images by optical microscope (×100). Carbon steel heat exchanger worked under hydrogen service (heat-affected zone). (a) Surface. (b) After grinding 3 mm of surface. (1) Cracks. (2) Characteristic structure for decarburization. *Source:* Adapted from Groysman et al. [261].

points of inspection. This type of NDT has been in use for several decades and is standardized [263]. It can confirm observations made by other NDT methods such as AE tests, penetrant examination, and ultrasonic flaw detection and go a step further to identify the nature of the flaw, for instance, to determine whether a crack is intergranular corrosion or transgranular fatigue.

The advantages of replication technique are as follows:

1) Efficient preparation and collection of copies can be made on-site by skilled personnel, and then the evaluation can be done in the laboratory with the best microscopes by qualified metallurgists.
2) Replicas can be archived for comparison with later retesting.
3) The replica technique can be used for verification of the resulting microstructure of a welding or heat treatment procedure.
4) It is possible to estimate if fires or other overheating events have caused unacceptable degradation.

On-site Hardness Testing

Principally, mechanical strength can only be determined by destructive testing. On the other hand, one obtains an estimation of the strength based on a relation between hardness and ultimate tensile strength. Several types of portable hardness testers have been developed for on-site NDT [258]. Some of these testers follow the standard principles for hardness testing, that is, Vickers, Brinell, and Rockwell (the names of hardness tests). Other testers use a related principle from which the results can be transformed into standardized scales through empirical tables [258]. Vickers measurements made by portable equipment follow the standards known from stationary equipment [264]. Measurements of hardness were used for detecting SCC of gas pipelines [265].

On-site Chemical Analysis of Alloys

Often, we need to know the type of a metal/alloy, concentration of alloying elements, and even concentration of carbon, sulfur, phosphorus, silica, and some other elements in alloys of equipment or construction. Transportable devices for fast *on-site nondestructive chemical analysis* of alloys were developed and came into wide use based on *XRF spectroscopy* and *optical emission spectrometry* (OES) [258].

XRF spectroscopy is the emission of characteristic "secondary" (or fluorescent) X-rays from a material that has been excited by bombarding with high energy X-rays or gamma rays. This technology analyzes the composition of alloys by measuring the spectrum of the fluorescent X-rays. Each of the elements present in a sample produces a unique set of characteristic X-rays that is a "fingerprint" for that specific element [266]. Based on this physical principle, various handheld (portable) XRF alloy analyzers were developed and are widely used in industry [267]. The advancements in handheld XRF analyzer technology gave possibility to measure even low concentrations (ppm) of trace elements in alloys of metallic constructions as well as light elements with atomic numbers as low as 12 (magnesium) [266]. All that is necessary for analysis is to expose the sample (surface of structure/equipment) to the instrument for a few seconds and then read the final identification from the display or printout.

Optical emission spectrometry, also called *atomic emission spectroscopy* (*AES*) or *arc spark emission spectroscopy*, is based on the *emission spectrum* of a chemical element, which is the spectrum of frequencies of electromagnetic radiation emitted by the element's atoms when they are returned to a lower energy state. Each element's emission spectrum is unique [268]. The spectra of samples (liquid or solid) containing many elements can be very congested, and spectral separation of nearby atomic transitions requires a high-resolution spectrometer. Since all atoms in a sample are excited simultaneously, they can be detected concurrently using a polychromator with multiple detectors. This ability to simultaneously measure multiple elements is a major advantage of *OES* compared with atomic absorption (AA) spectroscopy. Solid materials (metals and alloys among them) can be directly vaporized and excited by a spark between electrodes. Spark and arc excitation sources use a current pulse (spark) or a continuous electrical discharge (arc) between two electrodes to vaporize and excite atoms in solids. The electrodes are either metal (for instance, tungsten) or graphite. If the sample to be analyzed is a metal or alloy, it can be used as one electrode. By measuring the intensity of the peaks in this spectrum, *OES* analyzers can produce qualitative and quantitative analysis of the material composition with uncompromising accuracy. Thus *OES* is an important tool for fast and accurate elemental analysis of metals and alloys. Portable *OES* alloy analyzers can provide carbon content analysis with precision of 0.001% at 0.04% and identification of steel grades based on 0.02% difference in carbon concentration.

We will describe physical methods based on WL of metallic samples: WL and ER.

9.4.2 Weight Loss Method

WL method is based on the insertion of metallic specimen called *coupons* of known mass and area in a process stream, immersion during some period (30–300 days), removing, cleaning from corrosion products and other deposits, and weighing. Knowing the difference in mass of the coupon of known area and immersion period, we can calculate the corrosion rate of this metal/alloy in particular media. This method is described and analyzed in literature [1, 5, 6]. Coupons are manufactured in different dimensions and forms and are inserted into a process stream by means of retractable or retrievable holders (Figure 9.3). Special spools are recommended to install in the pipelines [269]. The immersion time of coupons must be determined for every particular case.

(a)

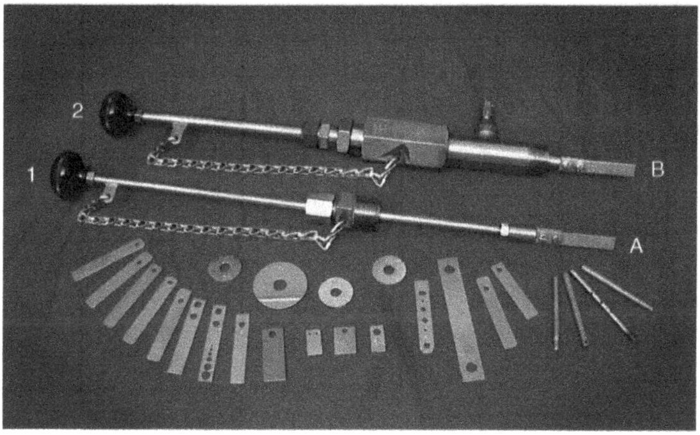

(b)

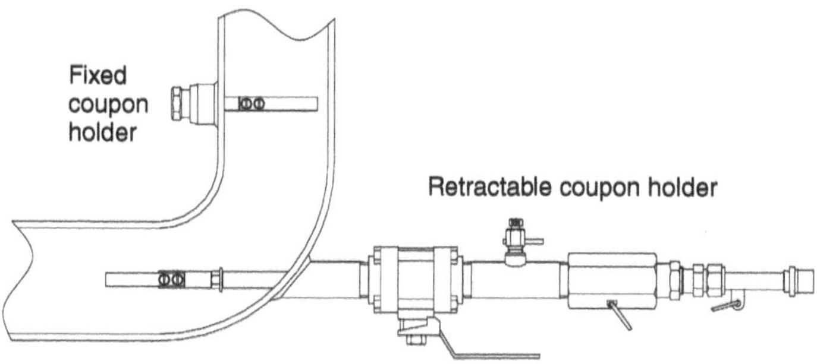

Figure 9.3 Various coupons with retractable holders (a) for corrosion examination by WL method and their installation in a process stream (b). (1) Retrievable coupon holder (up to 8 atm). (2) High pressure or packing gland holder (up to 105 atm). (A and B) Coupons. *Source:* Courtesy of Metal Samples.

Usually 30–40 days are recommended for the determination of corrosion rate of coupons in water systems. This period is correct for carbon steel. Copper alloys, stainless steel, and titanium are resistant to water. Therefore, immersion period for these alloys must be increased to 180 days and even more. Sometimes, a period of 30–40 days for carbon steel coupons is unacceptable. If water is not treated with chemicals (corrosion inhibitors, anti-scaling agents, and biocides), corrosion rate of carbon steel may be high (~1 mm yr^{-1}), and we should change coupons every 5–10 days. On the other hand, if chemicals are injected into water, corrosion rate of carbon steel is low (about 0.001–0.05 mm yr^{-1}), and we may change coupons every 90–120 days and even more. If corrosion rate of carbon steel is measured in a two-phase system, for example, hydrocarbons–water in the overhead region of the crude oil distillation column, the immersion period of coupons depends on the inhibitor's efficiency and may change from 40 to 180 days and even more. One has to keep in mind that some initiation period (above 60 days) is needed for the determination of pitting, crevice, and underdeposit corrosion. After some period of immersion that we choose according to corrosiveness of a stream and goals defined, the coupons must be taken out. We may take their pictures with, and then without corrosion products (after removing), and compare with the original coupon before installation.

Color identification of corrosion products formed on the coupons' surface may provide preliminary information about the presence of corrosive species in the streams caused by corrosion [1]. If dense films of corrosion products are formed on the coupons' surface (for instance, iron oxides or iron sulfides), their thickness can be measured. Then corrosion products must be removed by means of a knife made of some hard polymeric material (Teflon, polyethylene, etc.) for their subsequent chemical identification. Then, the coupon's (carbon steel and copper alloys) surface must be cleaned mechanically with eraser or chemically with hydrochloric acid (5–8 wt% HCl) inhibited with 0.2% of organic corrosion inhibitor (usually liquid amines or their mixtures) for decrease in loss of clean metal during chemical cleaning.

Here are some recommendations regarding correct installation of coupons into equipment at the plant:

1) A coupon must be installed against the flow, but in such manner that prevents any erosion or impingement.
2) A bypass with three and more places for coupon installation is often used in cooling water systems (Figure 9.4). The bypass should be connected to the return riser at the cooling tower or at the hottest part of the cooling water system. It is very important to install copper alloy coupons after carbon steel and aluminum ones. Copper and galvanized pipes as well as brass valves are not recommended for connection of bypass to the riser. Bypass is usually made of polymeric material (polyvinyl chloride, etc.).
3) The coupons must be mounted closely (but not in contact) to the surface of equipment to examine corrosion rate.

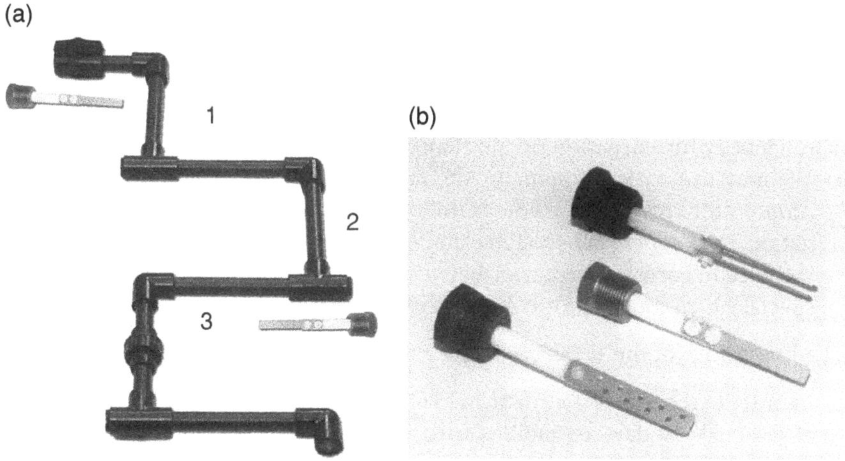

Figure 9.4 Bypass piping assemblies (a) and coupons (b) installed inside water systems. (1–3). Places for insertion of coupons. *Source:* Courtesy of Metal Samples.

4) Flow rates of 1–2 $m\,s^{-1}$ of stream are adequate.
5) We should remember that a coupon gives information about corrosion situation only in the place of its installation. Corrosion situation may change with moving away from the place of a coupon's installation. Therefore, when more coupons are installed, better *corrosion situation map* may be received.

The WL method is standardized [25, 32, 44–46], widely used for corrosion rate determination and rate of deposit formation in industry [13, 31, 239, 270–278], under laboratory conditions [121, 279] and for determination of corrosivity of soils [280, 281]. Coupons have been used since the 1930s for determining the efficiency of corrosion protection [34, 60, 272]. Coupons (including pipe segments and welded coupons with heat-affected zones [HAZ]) are used for CM at high temperatures in industrial boilers [137]. Using a light microscope with digital imaging processing techniques, it is possible to determine the material loss within an accuracy of 4 μm. In this case, a minimal exposure period of 3 weeks should be applied to obtain sensible results [137].

The benefits of the *WL method* are as follows:

1) The corrosion rate calculated by this method is very precise: 3–5% of reproducibility. The reliability of this method is high. It is recommended to compare other CM methods (for instance, electrochemical ones) with the *WL* method.
2) It is possible to measure the corrosion rate in any media (without any relationship to its electrical conductivity), namely, in gases, in liquids, and in solids, and under wide process conditions – up to 450 °C and to 350 Bar in the case of retrievable holders and without restrictions in the case of fixed probes.

3) This is the only method for obtaining quantitative information about pitting corrosion rate. The ENM give qualitative information about pitting tendency (see Section 9.6.5).
4) It is possible to examine a coupon's surface and corrosion products and then to identify the corrosion phenomenon and corrosion mechanism.
5) We may use *scale coupons* for the examination of a scale formation, *disk coupons* for crevice corrosion examination, *stressed coupons* for SCC examination, *rod coupons* for erosion examination, *welded coupons* for studying the variance in corrosion behavior between welded and non-welded metals, and special *gauze* coupons for collection of biofouling (Figures 9.5 and 9.10).

The limitations of the *WL method* are as follows:

1) In order to receive reliable data, in many cases, we need to wait for a long time: at least 30–40 days or more. Corrosion rates can be calculated only after removal of a coupon from the system. If corrosion rate is low (~0.001 mm yr^{-1}),

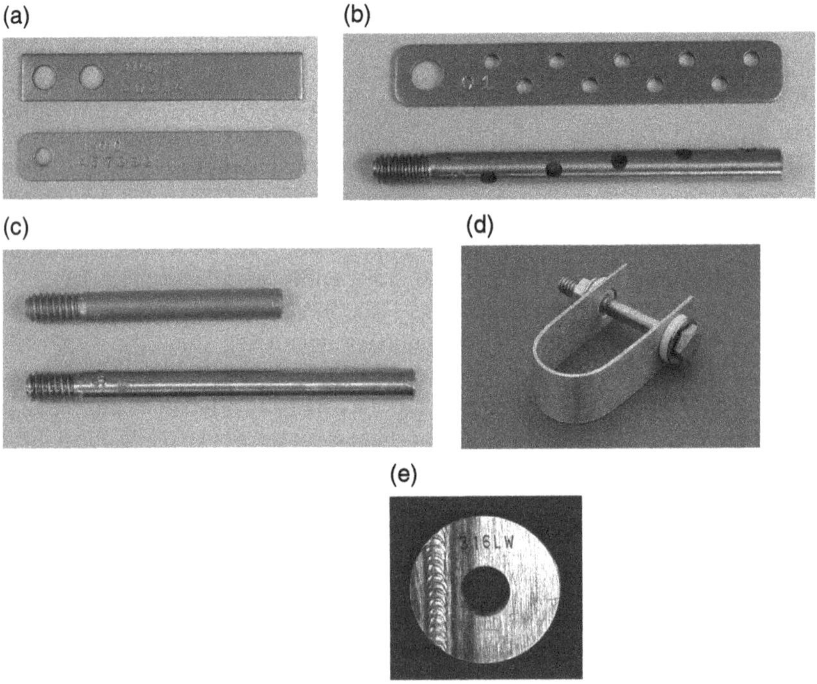

Figure 9.5 Various coupon forms for the examination of different corrosion phenomena. (a) *Flat* coupons for the examination of general and pitting corrosion. (b) *Scale* coupons for examination of scale formation. (c) *Rod* (*cylindrical*) coupons for erosion examination. (d) *Stressed* coupons for stress corrosion cracking examination. (e) *Welded* coupons. *Source:* Courtesy of Metal Samples.

we should wait longer, for about 4–9 months. In order to receive information about pitting corrosion, we need to wait for at least more than 60 days.

2) Coupons provide general and averaged data about corrosion over some period. Environmental conditions may change every moment, and it is impossible to know what happened during such upsets. Thus, continuous monitoring of the WL is not possible.
3) The *WL method* is time consuming. We have to treat the coupons: to clean, to weigh, and to take into account the *cleaning factor*. When we clean coupon's surface with inhibited acid from corrosion products, some WL of clean metal occurs. We can estimate this error in percent (called *cleaning factor*) in the WL definition. In other words, *cleaning factor* shows the influence of chemical cleaning on the accuracy of WL method. In order to define the *cleaning factor*, we may immerse a clean coupon in the same acid solution with inhibitor for similar period that treated coupon and define its WL. The ratio of the WL determined for the clean coupon to the WL determined for the treated coupon in percent is the *cleaning factor*. If the WL of treated coupon is relatively high (10–100 mg), the cleaning factor is not high; it is about 1–2%. If the WL of treated coupon is low (0.5–1 mg), the cleaning factor may be 50–100%.

Sometimes, *WL method* does not yield solution regarding corrosion of nonferrous alloys, for example, copper alloys in water and selection of their corrosion inhibitors. Sensitivity and accuracy of WL in copper coupons are low in water in the presence of copper corrosion inhibitors. In this case, AA and inductively coupled plasma (ICP) methods are used for the determination of concentration of copper ions appearing in water because of the corrosion of copper alloys.

The main drawback of the *WL method* is that it is not always suitable to use it in the industry because of difficulties of installation and change of coupons every month or once in several months. For example, if there are dangerous processing conditions such as high temperature, pressure, and harmful substances (hydrogen sulfide, hydrogen chloride, ammonia, etc.), it is difficult, not safe, or even impossible to reach the points needed for measurement. The *WL method* is used for determination of uniform and localized corrosion rates, but this method of detection provides historical retrospective information and not real-time rates. Therefore, we will discuss the methods that partly overcome the limitations of the *WL method*.

9.4.3 Electrical Resistance Method

If we take a cylindrical metal specimen, a wire, or a strip (Figure 9.6), we may use the following formula known for a conducting material from physics:

$$R = \rho \cdot \frac{L}{A} \tag{9.1}$$

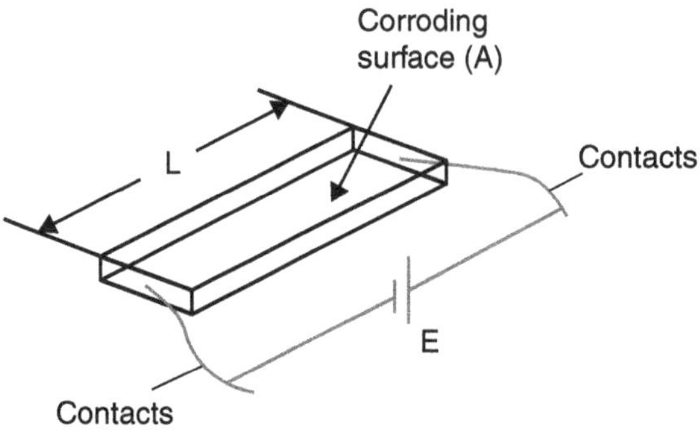

Figure 9.6 Piece of a conducting material (metal) and the operating principle of the electrical resistance (ER) probe. L, distance; E, voltage, volt.

where R is the ER of conducting material, Ω; ρ is the specific electrical resistivity of conducting material, Ω·m; L is the length of a specimen, m; and A is the cross-sectional area of a specimen, m^2.

If a piece of a metal is immersed in some corrosive media, corrosion may occur, and as a result the cross-sectional area A of a specimen will decrease. According to the formula (9.1), the ER of a metal specimen will increase. Thus, a metal specimen (element) shown in Figure 9.6 may be used as a corrosive ER sensor. The temperature influences the ER of a metal. Therefore, an ER probe (a sensor element in contact with a corrosive medium) is mounted together with a reference element for temperature compensation, which is not in contact physically with a medium but being at the same temperature. Thus, a reference element is influenced by temperature in the same way as a sensor, but is not influenced by corrosive media. Sensing elements in the ER probes are made in a variety of geometric configurations (wire, strip, tube, cylindrical, and flush), thicknesses, and alloy materials (Figure 9.7), and they are called *corrosometers*. The selection of a particular element form depends on sensitivity, response time, and element (sensing probe) *lifetime* required. *Span* represents the useful thickness of the ER sensor during the probe *lifetime*, namely, half thickness of the ER sensor. *Response time* is the minimum period in which a measurable change in resistance takes place and governs the speed with which useful results can be obtained. Usually this time changes from tens of minutes to several days [282]. A probe *lifetime*, or a period required for the effective thickness of the exposed element to be consumed, governs the probe replacement schedule.

The lower the element thickness, the higher is the sensitivity, and the lower is its *lifetime*. Since a probe's *lifetime* and response period are directly proportional to each other, the selection of an element is a compromise between sensitivity, data frequency, and probe replacement frequency. Detection levels as low as 1 nm are achieved by elements with thicknesses of

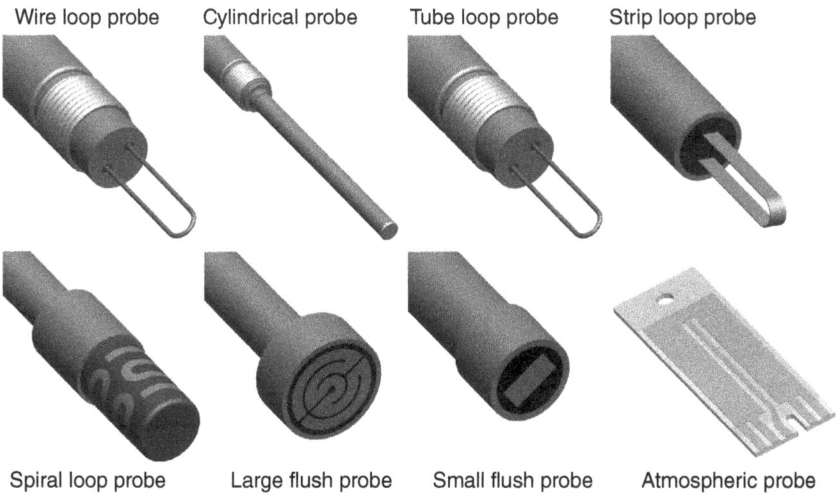

Figure 9.7 Various types of the ER sensing elements. *Source:* Courtesy of Metal Samples.

several millimeters [283, 284]. Flush probes are suited for pipelines where pigging may occur, and for bottom offline monitoring in oil and gas, or multiphase flows where the corrosive water phase exists [285]. Cylindrical elements are suited for more chemically aggressive environments. A portable device can be used for registration of changes of sensor's ER (Figure 9.8a). In this case, we should measure the ER changes of the sensor mounted in a stream periodically. We may continually monitor by means of a transmitter connected to a distributed control system (DCS) (Figure 9.8b). Thus, the advanced ER systems include high-resolution transmitter, data logger, 24 voltage direct current (VDC) power supply to the transmitter, and special recording and retrieval software permitting easy data acquisition and display.

An increase in ER of a sensor is equivalent to decrease in metal thickness. Thus, a slope of a curve "dial reading of ER probe – exposure time" in Figure 9.9 shows that a metal corrodes with time, and as a result, metal thickness and metal quantity diminish. One can calculate corrosion rate according to slope of such curve:

$$k = |\text{slope}| \cdot 0.001 \cdot 365 \left[\text{mm yr}^{-1}\right] = \frac{\Delta L}{\Delta t} \cdot 0.001 \cdot 365 \left[\text{mm yr}^{-1}\right] \qquad (9.2)$$

where $\Delta L/\Delta t$ is a ratio of diminishing thickness of sensor (change of the dial reading ER probe) in μm to the measured period in days, $\mu\text{m d}^{-1}$; $0.001\ \text{mm}\,\mu\text{m}^{-1}$ and 365 days/year are conversion factors. The bigger the slope of a line *dial reading of ER probe versus time*, the bigger is the corrosion rate of a metal.

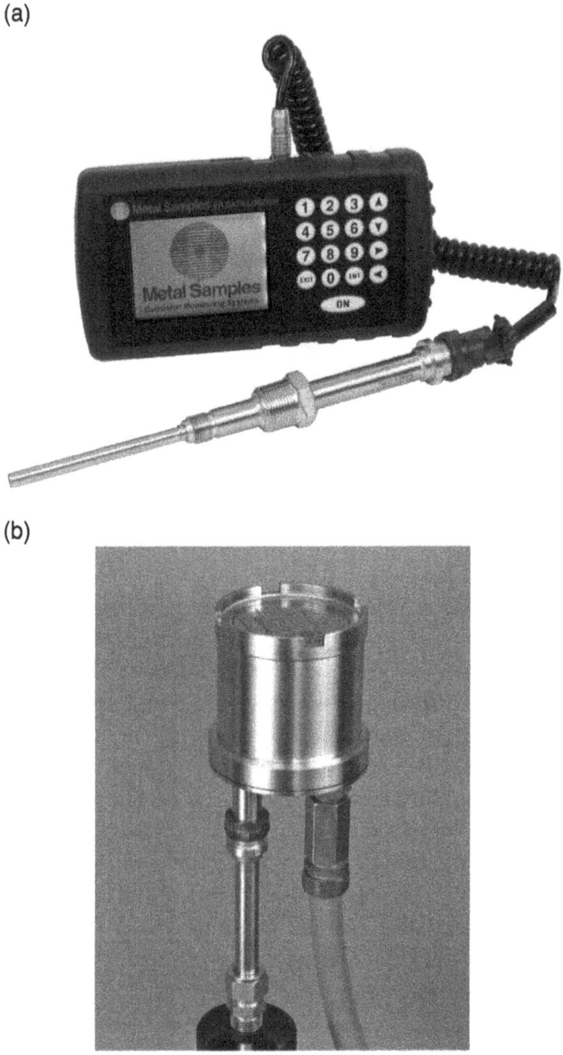

Figure 9.8 Portable device (a) and a transmitter connected to a distributed control system (DCS) (b) for registration of electrical resistance of sensors. *Source:* Courtesy of Metal Samples.

ER method is standardized [47] and described in literature [1, 21, 286] and is used for intrusive erosion control [283, 287], for online monitoring of flow-assisted corrosion in nuclear coolant conditions at 310 °C and 10.1 MPa [288], for CM in district heating systems [72, 289], in the overhead region of crude distillation units in the oil refining industry [1, 270, 273–277, 290], in

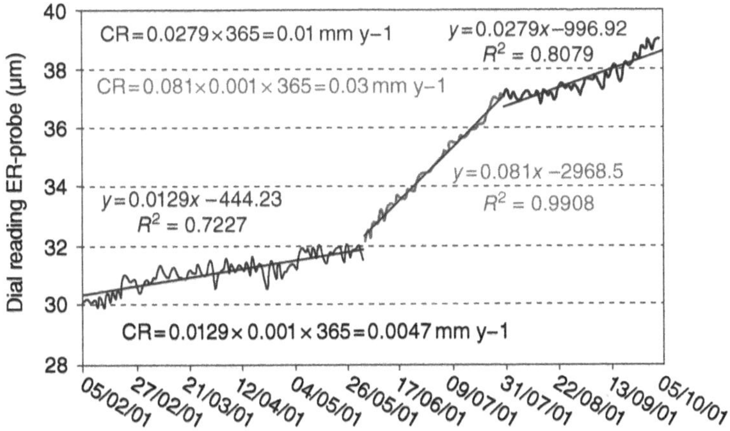

Figure 9.9 ER measured in the corroded thickness of an ER probe versus time. CR, corrosion rates for three various periods.

hydrocarbons [291], for online monitoring of naphthenic acid corrosion in oil refinery units [292], for monitoring of the efficiency of cathodic protection of fuel storage tank bottoms [293] and underground pipelines (as *soil corrosion probes*) [31, 88, 294], for inhibitor selection and optimization [295, 296], for internal CM of subsea production flowlines [297], for online monitoring of fireside corrosion in power plant [298], and for detection of corrosion of grouted and bare steel strands exposed to a high humidity environment [299]. The design of ER probes permits operation up to 537 °C and 700 atm. [283, 300]. Nonintrusive ER technique was also developed for online CM of equipment at plants [301].

The *benefits* of ER method are as follows:

1) One can observe from the curve "ER versus time" that real-time corrosion rate can be measured, permitting detection of any sudden change in corrosion rate and immediate remedial action.
2) Since the measurements in ER method are based on the WL of metallic elements, one can measure the ER and, as a result, the corrosion rate of metals in media both of high electrical conductivity (solutions of electrolytes) and low ones (gases, hydrocarbons, two-phase electrolyte–nonelectrolyte systems, and other nonconducting solvents). ER method is similar to the WL method and is different from electrochemical methods. The latter methods can be used only in media of relatively high electrical conductivity.
3) The lifetime and sensitivity of the ER probes depend on geometrical dimensions of a sensor. Therefore, one can select a sensor of any lifetime and sensitivity required. Very large corrosion rates can be quickly detected (for several hours).

4) It is possible to monitor corrosion without removing the sensor and at any distance. Of course, we should replace a sensor after its full corrosion. We may select a sensor of such configuration that replacement will be needed once a year or in 2 years.

The *limitations* of ER method are as follows:

1) ER method allows the measurement of only uniform corrosion. It is impossible to identify localized corrosion phenomena. Pitting may result in wrong results. Therefore, we should examine the surface of a sensor if sudden changes in ER data occur.
2) Some salts (FeS_2, PbS) and oxides (Fe_3O_4, MnO_2) formed on sensor's surface are electrically conductive and can give rise to wrong results.
3) Some deposits (fouling) formed on the surface of a sensor also give rise to wrong results.
4) If there is no preliminary information about corrosion rate values in a particular medium, it is very difficult to select a correct sensor of a suitable sensitivity and a lifetime. The time response highly depends on the lifetime of ER probe: if you want high sensitivity, the lifetime of the probe decreases correspondingly. Therefore, it is important to define corrosion rate values by means of a WL method in a particular medium at the beginning and then to select an ER probe of a suitable sensitivity and a lifetime.

We should remember to change sensing elements when they reach 80–90% of their nominal life. For instance, if the ER probe with a sensing element thickness of 250 μm is used, we should change it when 225 μm will corrode. However, sensing elements must be immediately changed if localized corrosion has occurred. In order to receive reliable data, we should compare the ER data with the WL method. An example of online CM at the overhead region of the crude distillation unit at the oil refinery by ER and WL methods is described in literature [1].

Conclusion. Physical methods are standardized and used for periodic inspection of equipment and structures [55, 56]. Visual inspection and NDT are standardized in many industries [58, 59]. Different devices based on physical phenomena allow to measure the changes in metal thickness of metallic equipment (*US* and *eddy current* methods) and WL of the metal (*WL coupon* method and *ER* measurements of a corroding metal sensor). The NDT methods (*US scanning, radiographic profile,* and *eddy current*) are used for detection of corrosion–erosion [3]. Surface NDT (*liquid penetrant testing, wet fluorescent magnetic particle testing,* and *US*) are used for detecting hydrogen blistering and HIC [3]. Special sensors were developed for detecting microcracks and the accompanying microstrain in aluminum alloys in aircraft structures [302–304]. The *AE, US, dye penetrant,* and *replica examination* tests are used for detecting cracks; and the *advanced ultrasonic backscatter technique* and *replica examination* are used for detection of high temperature hydrogen attack (HTHA) at the refineries' units [261, 305].

Most physical methods (*US*, *WL*) have limitation in that these methods assume that corrosion rate is constant throughout the entire exposure period. In reality, damage rarely happens at a continuous rate but rather takes place in discrete episodes that can be correlated with specific operational events [15, 306]. Many *physical techniques* are labor intensive and expensive. For instance, it was estimated that 35% of a plant's maintenance budget is spent on the cost of *US* inspection or on physically opening vessels for visual inspection [15].

9.5 Control of the Environment (Chemical Analytical, Physicochemical, and Microbiological Methods)

There is no aspiration more natural
than yearning for knowledge ....

Michel Eyquem de Montaigne (1533–1592), a French writer

Chemical analytical and physicochemical methods probably were the second one after visual analysis in CM methods and include chemical and microbiological analysis of the environment (content of dissolved ions, gases, different organic and inorganic compounds, and microorganisms) and its physicochemical parameters (see Section 9.3.2). These methods allow detecting both corrosion intensity (sometimes also corrosion rate) and corrosiveness of the environment. They must be defined for specific streams at every enterprise: cooling water, boiler feed water (BFW), condensed water, two-phase systems such as water–hydrocarbons, etc. All chemical treatment programs in the cooling water systems require particular water quality, namely, chemicals (corrosion inhibitors, scaling/fouling dispersants, and biocides) work only under specific range of chemical and physical parameters of water [307]. Chemical analytical and physicochemical methods of analyzing different species in water are described in literature [57, 308–312], and their analysis is beyond scope of this chapter.

Photocolorimetric, optical, spectroscopic, or spectrophotometric (IR, UV, and visible light) methods, XRF spectroscopy, AA, ion plasma couple (IPC) or ICP, liquid and gas chromatography, ion-selective electrodes, fiber-optic electrodes, potentiometric methods, mass spectrometry, differential pulse anodic stripping voltammetry, polarography, pH meters, metal oxide electrodes, hydride electrodes, and oxygen and chloride sensors are used in chemical analytical methods [17, 22, 281, 313–317]. Capillary electrophoresis is used for crevice corrosion solution analysis [318]. Organic water-soluble compounds that are present in gasoline are detected by extractive separation and recording nuclear magnetic resonance (NMR) (for ^{1}H and ^{13}C), IR, and mass spectra [315].

Particle-induced X-ray emission (PIXE) spectroscopy is used for qualitative and quantitative analysis of volatile corrosion inhibitor (VCI) concentration in anticorrosion polymer films [319]. Identification of anticorrosion polymer films is investigated by IR spectroscopy and differential scanning calorimetry (DSC) [319]. Vapor pressure measurements are used for determining the efficiency of VCI [320, 321].

A number of dyes such as fluorescein, Columbia blue, and coumarin have been used for a long time as dye penetrants for crack detection [21]. Compounds that change color in the presence of corrosive species are called *chemical sensors* [23]. For instance, ferric ions react with 1,10-phenanthroline and change its color from blue to purple. Colorimetric method is used for detection of ferrous ions and total iron in boiler systems [322]. Online optical particle monitor was developed for determination of iron oxides (corrosion products) in boiler condensate systems [323]. Fluorescent compounds are used as *chemical sensors* for detection of corrosion of aluminum alloys at early stages [324, 325]. Compounds such as fluorescein, morin, Schiff bases, oxines, and hydroxyquinolines were found to fluoresce with aluminum ions and during oxidation in air when incorporated into primer paint coatings for aluminum alloys [326]. Such "smart" coatings utilizing electrochemical properties of color change coatings were developed [21].

Many crude oil refineries combine WL and ER methods with chemical analytical ones of streams (measuring pH, iron, copper, chlorides, sulfides, ammonia, sulfates, and organic acids) [1, 273, 274, 290, 327, 328]. The chemical analysis of iron in oil and gas production is standardized [26]. Monitoring of physicochemical characteristics of crude oils (salt, water, sulfur, nitrogen content, and total acid number) is essential for determining their corrosiveness [329]. Analytical measurements of iron and nickel in crude oil and gas oil are used for evaluation of corrosiveness of these organic liquids in the presence of naphthenic acids [330]. Fiber-optic sensors are used for *in situ* monitoring of concentrations of chlorides in soil and reinforced concrete [331]. Fiber-optic pH sensor was developed for real-time CM, namely, for visualizing corrosion sites and measuring local chemical concentration of H^+ ions [313]. The imaging fiber's distal face containing an immobilized pH-sensitive fluorescent dye is brought into contact with metal surfaces submerged in aqueous solutions, and fluorescence images are acquired as a function of time. The observed changes in fluorescence due to increase in pH at cathodic sites and decrease in pH at anodic sites are indicative of localized corrosion [313]. Chemical sensors were developed to monitor corrosivity of condensate formed in gas-fired appliance heat exchangers [332, 333]. Combinative monitoring of chemical changes (pH, chloride concentration) and electric potential inside the crevice of a metal or at the corroding interfaces using microelectrodes as a function of time has been proposed [21, 334]. Test methods of measuring pH, moisture content of soil (humidity), and soil resistivity are standardized [37, 52–54] and are used in practice [98]. Other physicochemical properties of soil such as soil texture, conductivity of aqueous extract, redox potential, content of chlorides, sulfates, sulfides, nitrates, and calcium and magnesium ions, microbiological analysis of soil, and temperature also are used for determination of corrosivity of soil [280, 281, 335, 336]. Embedded sensors allow following the ingress of aggressive substances (including oxygen) continuously in concrete [115, 337].

Analytical methods concern also the measurement of chemicals used in corrosion control, for example, phosphonates in brine [338], VCI in polymer films [319], and phosphates, nitrites, silicates, and other corrosion inhibitors in water [1].

The question is, what are the dangerous values of analytical parameters (pH, iron, copper, chlorides, microorganisms, etc.) determining corrosion intensity (high or low) in the system? Usually allowable (permittable) values of some analytical parameters in water, fuels, other streams, and soil are recommended, which show the possible danger of corrosion in the system (Table 9.1).

The permittable values depend on a specific system (type of materials and media used). For instance, if equipment is made of iron and copper alloys, total iron and copper content in cooling water systems must not exceed 1 and 0.05 ppm, respectively. For condensed water in overhead region of the crude oil distillation columns at the oil refineries, recommended values are pH = 5.5–6.5, 0.5 ppm total iron, and 20 ppm chlorides. Otherwise, corrosion can occur.

Table 9.1 Effect of chlorides, sulfates, and pH on corrosion of buried steel pipelines.

Parameter	Concentration, ppm (not pH)	Degree of corrosivity
Chlorides	>5 000	Severe
	1 500–5 000	Considerable
	500–1 500	Corrosive
	>500	Threshold
Sulfates	>10 000	Severe
	1 500–10 000	Considerable
	150–1 500	Positive
	0–150	Negligible
pH	<5.5	Severe
	5.5–6.5	Moderate
	6.5–7.5	Neutral
	>7.5	None (alkaline)

Source: From Peabody [82].

Dissolved oxygen plays an important role in corrosion of boilers, and its concentration is strictly restricted ([339], p. 10). The same concerns the content of iron, copper, total hardness, silica, total alkalinity, and conductance of BFW and boiler blowdown ([339], p. 98).

In many cases, it is important to monitor not absolute values, but their trend. Sudden changes of these values can show beginning of corrosion [1]. Sometimes, hydrogen sulfide and ammonia are measured in condensed water at the overhead region of distillation columns in the oil refineries, but these data do not provide useful information about corrosion.

Chemical analytical methods are very sensitive and allow to measure corrosion rates in cases when WL method is not applicable. For instance, copper alloys are resistant to water in the presence of copper corrosion inhibitors, and WL method does not yield reliable results, but AA allows measuring dissolved copper in solution. Any analytical method of determination of metallic cations in water or any other medium after immersion of coupon in this medium allows calculating the corrosion rate of the metal (supposing that the entire area of the coupon dissolves and metallic ions are in the solution) [1].

The benefits of *chemical analytical* methods are as follows:

1) High sensitivity to all metals dissolved in liquids.
2) Estimation of corrosion rate of all metals in liquids under specific conditions.
3) The trend of analytical parameters in the environment can show sudden corrosion or any danger of corrosion occurrence.
4) Determination of concentrations of some metals (Cr, Cd, Pb, and Zn) in the environment can decide the problem of contaminant levels for regulated drinking water suppliers.

The limitations of the *chemical analytical* methods are as follows:

1) *Quantitative* estimation of corrosion rate is applicable for measuring only in liquids under laboratory conditions. The presence of corrosion products on metallic surface can yield wrong results.
2) Chemical content of corrosive species in the environment (water, soil, and atmosphere) provides *qualitative* estimation of corrosion situation.
3) These methods sometimes take time to yield results.

9.5.1 Monitoring of Microbiological Activity Toward Metals

Microorganisms can be responsible for corrosion in many systems: cooling water, drinking water, crude oil, fuel and water storage tanks, heat exchangers, pipes, and other systems. Different methods of microbiological analysis of water, soil, crude oil, fuels, and sludge are described in literature [33, 36, 309, 340–347]. Online corrosion methods for monitoring MIC were reviewed [348]. There is no single objective method of MIC monitoring. The following techniques are used for biofilm monitoring: collection of bacteria on exposed studs (bioprobes, named the Robbins device [349]) or special screen coupons (Figure 9.10) and afterward microbiological characterization of the biofilms, electrochemical devices, and the measurements of the change of heat transfer over a tube array arising from the influence of biofouling, scale, and corrosion products [1]. The *Robbins* and modified *Robbins* devices are used for monitoring *sessile* microorganisms [349, 350]. Periodic examination of *sessile* microorganisms is carried out by indirect microscopic enumeration: removing deposits

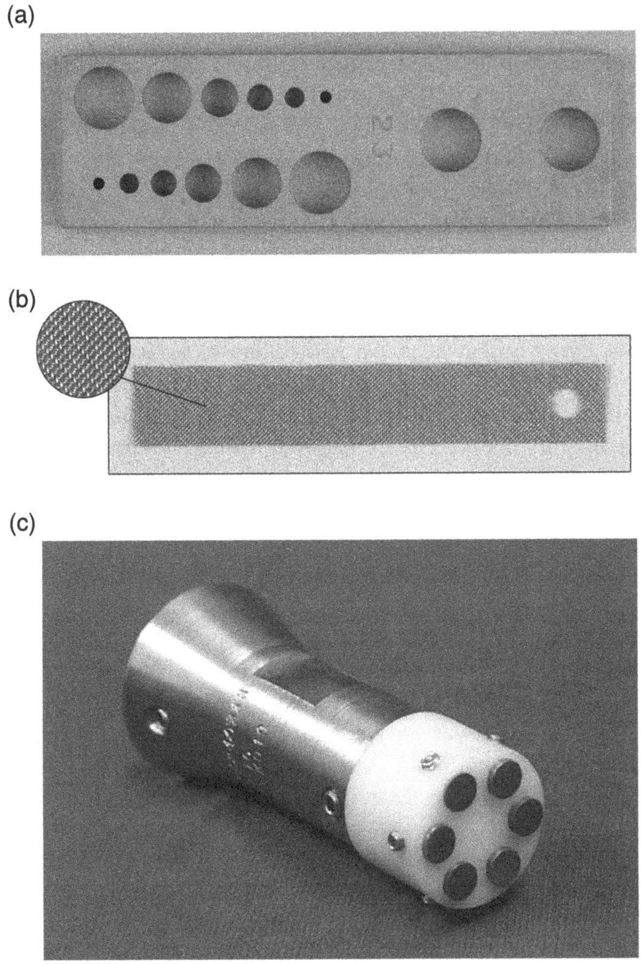

Figure 9.10 Special metallic coupons with holes (a), Ni–Cr screen gauze (b), and *bioprobe*, the *Robbins device* (c), for biofouling collection and afterward examination. *Source:* Bioprobe – Courtesy of Metal Samples.

(swabbing or sonification), suspending in sterilized water, and enumeration using counting (plating) technique by means of *adenosine triphosphate* (ATP), protein, and *deoxyribonucleic acid* (DNA) analysis [350]. Biofilm probes are usually inserted in dead-leg locations where the biofilm formation is most likely to occur. Typically, this uses bottom-of-line access and under-flush insertion.

Bio-traps containing porous polymer beads for trapping biomass were suggested as a rapid means to evaluate biofilms using phospholipid fatty acid and DNA analysis [351]. Biochemical (culturing) methods, genetic (molecular) techniques, microscopic examination (including epifluorescence microscopy [EFM]), surface analysis, and spectrochemical, electrochemical, and piezoelectrical techniques for detection and monitoring of biofilms are described in reviews [132, 133, 351–354]. Combining ENM with EFM was used for control of biofilm growth in seawater [355]. Fiber-optic microprobes [135, 136, 356], turbidimetry [356], and photoacoustic technique [357] were suggested for *in situ* monitoring of biofilms. DNA-based methods with denaturing gradient gel electrophoresis (DGGE) are used for specification of the bacterial and fungal species settled on the concrete surfaces [358] and in aqueous natural gas plants [359]. The total cell number in oil field water was analyzed by 4,6-diamidino-2-phenylindole (DAPI) staining [359]. DAPI is a DNA intercalator that targets both living and dead cells. Molecular biological methods were used for rapid monitoring of *sulfate-reducing prokaryotes* (SRP), which encompass sulfate-reducing bacteria (*SRB*) and sulfate-reducing archaea (*SRA*) [360, 361]. Quantitative polymerase chain reaction (qPCR) technique was developed for detection of *SRP*, *SRA*, and *methanogens* [361]. This technique is based on an enzymatic reaction that copies the specific gene, i.e. the gene for sulfate reduction.

FTIR and *attenuated total reflection (ATR)*, *episcopic differential interference contrast (EDIC) microscopy*, and *piezoelectric technology* based on quartz crystal microbalance (QCM) are used in the analysis of biofilms [362, 363]. The ATR method allows monitoring of the development of a biofilm *in situ* directly at the surface, nondestructively in real time. An IR spectral library of microorganism collection strains was carried out [364]. Chromatographic detection of the C5 to C20 *fatty acid methyl esters* was used for identification of the bacterial consortia formed in asbestos-reinforced concrete pipes carrying drinking water [365].

Microbiological examination includes identification of microbial activity, that is, various kinds of *planktonic* (free-floating bacteria whose movements are controlled by liquid movement – not attached to surfaces) and *sessile* (attached to solid surfaces) microorganisms by means of "kits," biosensors, and conventional methods of microbiological analysis (bacterial growth on special artificial media) [33, 118, 345, 366–369]. Biosensor based on *sulfur oxidase* was used for online monitoring of SRB activity [370]. The response of biosensor to sulfide ions correlated well with offline solution analysis of sulfide ions, but correlation was not linear between the biosensor and LPR data [370]. Direct visualization includes the usage of the fluorescent dyes that cause cells to light up under UV radiation.

Use of commercial "kits" allow quick assay of the presence of enzymes associated with microorganisms suspected to cause corrosion. The most common way to assess microbial populations in various industrial streams is use of artificial growth media for particular microorganisms associated with corrosion. In this case, one should wait for the results longer (~30 days) than in the case of kit [368]. Biochemical assays were developed for the detection of specific microorganisms associated with MIC [369]. Unlike culturing techniques, biochemical assays for detecting and quantifying bacteria do not require growth of bacteria. Instead, these assays measure constitutive properties, including ATP, phospholipids, fatty acids, cell-bound antibodies, and DNA [371]. Thus, microbial activity may be determined by measuring the amount of ATP in industrial streams. This metabolite drives many reactions in the cells. Since ATP is a compound found in all living cells, ATP assay is commonly used to estimate the total number of viable organisms in a sample [368]. The ATP assay is based on light emission of luciferin and the enzyme luciferase. A lysing reagent is added to a water sample to break the cell wall, and the emitted light is measured by a commercially available ATP photometer. The amount of light released during the reaction is directly related to the amount of ATP and the quantity of the microbes in the sample. Detection of other metabolites such as organic acids, some esters, hydrogen sulfide, or methane by analytical methods can indicate biological activity in the biofilm. Identification of proteins (for example, lipopolysaccharide) or nucleic acids may also quickly indicate microbial activity. Online real-time determination of total biological activity with fluorescent bioreporters was described [372]. *Bioscan* device based on the measurement of ATP was developed for monitoring *planktonic* microorganisms [350]. It is recommended to determine the presence of microorganisms in deposits by means of mixing with 2% OsO_4 solution [373]. Microorganisms absorb osmium intensively. Further determination of osmium in the deposits by means of one of physicochemical methods (for instance, EDS) may indicate the presence of microorganisms. *Multisubstrate testing method* based on the microorganisms' isolation from a metal surface and a quantitative estimation of nutrients uptake was suggested for determining seawater biocorrosiveness [374].

Those methods provide information on the types and number of microbes, but usually do not provide any insight into biofilm activity, nor information about corrosion [375]. It was shown that there is no relationship between planktonic counts and sessile counts or MIC [376].

Environmental cell transmission electron microscopy (EC-TEM) and *electron energy loss spectroscopy* (EELS) were used for observation of hydrated bacteria in association with corroding iron and copper and chemical elemental analysis [377].

AFM, *confocal laser scanning microscopy* (CLSM), and *environmental scanning electron microscope* (ESEM) are used for observing surface morphology of carbon steel in the presence of microorganisms in real time [119, 135, 136, 378]. Combination of AFM–SKPM (scanning Kelvin probe microscopy) and fluorescent microscopy allows detection of bacterial attachment to the surface of metals [379]. SEM and EDX are used for determination of the presence of biofilms on stainless steels [380]. The SR technique can be a valuable tool for the assessment of MIC corrosion products and biodeterioration [127]. Synchrotron is a source of intense, bright light of variable wavelength. Synchrotron techniques are particularly suited to nondestructive analyses or to the microcharacterization of different materials including organic and inorganic materials, either amorphous or crystalline.

Some physicochemical methods such as *scanning vibrating probe* allow examining electrochemical conditions at the metal surface when microorganisms are present on it.

The question is, what is the allowable value of microorganisms for use of metallic equipment out of MIC danger? It is recommended that total bacterial count (TBC) in cooling water and fuels must not exceed 10^4 bacteria ml^{-1} and SRB must not exceed 10 bacteria ml^{-1}. Otherwise, MIC occurs or can occur. It is very important to carry out the trend of measurement results. Recommended allowable values of microorganisms depend on a specific system.

For instance, if there are no favorable conditions for the proliferation of microorganisms in some industrial systems, TBC values of 10^4–10^5 bacteria ml^{-1} indicate "out of danger." But if there is leakage of oil into the water system, or there is not enough flow rate of water, or water phase appears in the fuel, even TBC of 100 bacteria ml^{-1} can immediately cause severe growth of microorganisms in water and then give rise to corrosion over several weeks and even months [1]. Chemical and physicochemical analysis of deposits (mixture of corrosion products and other contaminants with microorganisms and their metabolic products) may provide important information about microbiological activity. The presence of iron sulfides in deposits does not always indicate SRB activity. Iron sulfides may be formed as a result of the presence of hydrogen sulfide in the environment, such as in crude oil and fuels, and their leakage into the cooling water. High concentration of chlorides in aqueous solutions shows possible activity of iron and manganese bacteria, especially if high concentrations of iron and manganese cations are detected in solution.

It is very important to know how biofouling and MIC are developing in real time. The most common method that found wide application in industry is the method of measuring deposit accumulation on the metal surface.

9.5.1.1 Deposit Accumulation Test (*Heat Transfer Resistance* Method)

This method is based on the principle that microorganisms attached inside heat exchanger tubes (if water flows inside) decrease in heat transfer. We should measure the resistance to heat transfer from outer surface of the tube to the interior ([342], pp. 137–150; [381]). It is simple to do this, if we measure the inlet and outlet water temperature and flow rate of water in heat exchanger. Some companies manufacture such devices (for instance, corrosion and deposit monitoring system CorrDATS of Rohrback Cosasco Systems [RCS], Inc.) similar to a small heat exchanger that allows carrying out necessary measurements and to calculate the *heat transfer resistance* values (Figure 9.11).

This value shows the formation of all possible organic and inorganic deposits inside a heat exchanger tube: microorganisms, oil and other hydrocarbons, scale of carbonates, phosphates, silicates, corrosion products, silt, dirt, etc. More precisely, deposition monitoring measures fouling (accumulation of deposits) and is standardized [29, 30]. Thus, we do not know the exact contribution of microorganisms to *heat transfer resistance.* Even if we would be able to define only the biofouling formation, we would not be able to know which part and which kind of microorganisms is responsible for MIC. In spite of this substantial drawback, the *heat transfer resistance* method is used together with microbiological analysis and CM. Biofouling formation takes longer time, and we would not know what happened during this period. Electrochemical methods found application in the MIC online monitoring.

9.5.1.2 Electrochemical Methods

Electrochemical methods of MIC monitoring in real time are based on the principle that MIC is an electrochemical process and therefore influences such

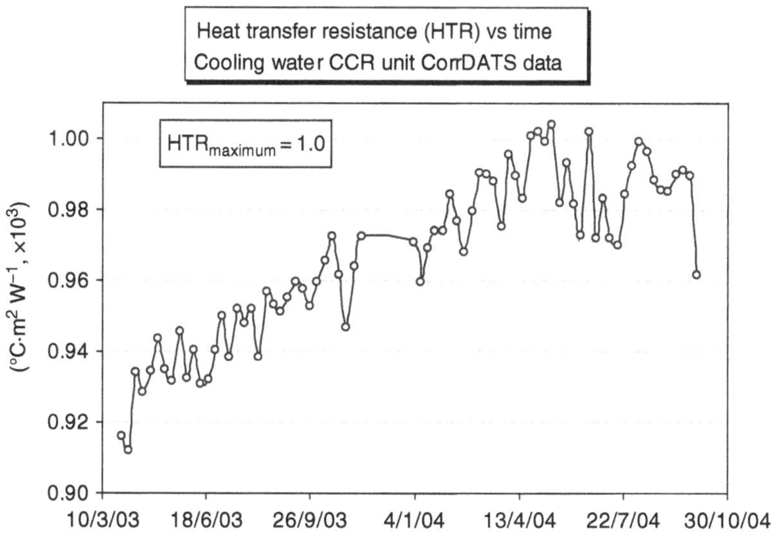

Figure 9.11 *Heat transfer resistance* (HTR) vs. time (cooling water system). *Source:* Groysman [1]. Reproduced with permission of Springer.

electrochemical parameters of corrosion such as electrical current, potential, and resistance of metals. Open circuit potential (OCP), LPR, EIS, and ENM are used for this purpose ([342], pp. 265–287; [380, 382, 383]) (see Section 9.6).

The *BIoGEORGE* electrochemical monitor was developed to provide online real-time indications of biofilm activity on metallic surfaces, namely, *sessile* microorganisms [350, 351, 382, 384–386]. This monitor is an electrochemical device that detects the activity of a biofilm by measuring the amount of electrical current necessary to offset the decrease in polarization resistance when a biofilm forms on a nonactive metal (e.g. titanium or stainless steel) and measures the current that flows between two nominally identical electrodes when no potential is applied [351]. In another configuration, one electrode (set of disks) is polarized relative to another for a short period of time each day [375, 376, 387–389]. During this polarization cycle, the environments at each disk surface are modified slightly so that biofilm formation on the electrodes is encouraged. For example, microbes that prefer electrochemical conditions associated with anodic sites will settle on the anode. Similarly, microbes that prefer cathodic conditions will settle on the cathode. Applied potential produces a current flow that is measured [375]. A *coupled multielectrode array sensor* was used for detection of MIC [390]. It is obvious that the *BIoGEORGE* probes work only in conductive media and cannot be used in crude oil, fuels, and other low conductive media.

A special device was developed for measuring *in situ* the pitting bacterial corrosion of carbon steel in oil and geothermal environments [391]. The principle is based on the generation of a pit on a circular carbon steel electrode by anodic polarization and measurement of the corrosion current between this electrode and the pipeline or the probe body used as a cathodic pole. The current flowing between the anode and the cathode is monitored and used to estimate if the pitting artificially created by an electric pulse may be maintained or not by the activity of microorganisms that may have developed on the surface of the electrode [391]. Electrochemical methods provide general information about all electrochemical corrosion processes occurring on the metal surface. In other words, the changes of the measured electrochemical parameters occur as a result of all corrosion processes, even which do not relate to MIC, and it is impossible to differentiate the only MIC from other electrochemical corrosion processes.

9.5.2 Scale Deposition Monitoring

It is important to monitor calcium carbonate (and other salts, such as calcium phosphates and sulfates, ferrous silicates, and others, not dissolved in water) crystallization from natural, industrial, and synthetic waters. Scale deposition leads to decrease of heat exchange in tubes and obstruction to flowing water in pipelines. Heat transfer resistance method described in Section 9.5.1.1 can be used for this purpose. Optical microscope was used for *in situ* scale deposition monitoring [392]. The device includes *QCM* and microscope for scale deposition monitoring [393]. The QSM allows to simultaneously measure the mass of scale deposit on transparent electrode.

BK Giulini developed a monitoring device called "Gilumat S3," which allows to measure scaling and fouling tendency separately [394]. This device is based on a dual channel measurement of relative differential heat transfer coefficients. At the start (clean surface), the heat transfer coefficient is set by definition to zero. One channel of the measuring device determines the scaling tendency and the other the fouling tendency. The channel for scaling measurement is permanently heated to a predefined temperature (default value is 60 °C). If a scale develops on the surface, the heat transfer coefficient changes and gives the relative scaling factor of the system. The channel for the fouling is not heated; therefore normally no scale can develop. A change in the heat transfer coefficient therefore indicates the presence of biofouling.

9.5.3 Hydrogen Monitoring

It is important to detect hydrogen absorption, hydrogen diffusion, and hydrogen damages at the early stage of use of metallic equipment and structures. Different analytical, mechanical, and metallurgical *monitoring methods* were developed for detection of possible *hydrogen damage* of metallic constructions and equipment. We can differentiate *hydrogen monitoring methods* into two groups: *analytical methods* (*detection of hydrogen gas*) and *physical methods* (*detection of physical discontinuities in metals* – fissures, cracks, blisters, and decarburization appeared because of hydrogen penetration into metals, resulting in changes of their mechanical properties). These methods may be destructive and NDT. Hydrogen gas can be detected either in intrusive or nonintrusive devices called *hydrogen probes* [395–400]. These probes are based on the following three principles:

1) Pressure increase with time in a controlled chamber, as hydrogen passes through the metallic thickness into the probe chamber (manometric or vacuum method) [401, 402]. Vacuum extraction at 400 °C [271] and hydrogen effusion [403, 404] are also based on this principle.
2) An electrochemical current resulting from the oxidation of hydrogen under an applied potential, namely, hydrogen ionization from H^0 atoms into H^+ ions (*electrolytical,* also named *electrochemical* method) [405–407]. Evaluation of hydrogen uptake, permeation, and transport in metals by an *electrochemical technique* is standardized [50, 408].
3) Current flow in an external circuit, based on a fuel cell principle, whereby hydrogen entering the miniature fuel cell causes the current flow.

There are two types of hydrogen pressure or vacuum probes: insertion and clamp-on (patch) [402, 405]. Intrusive probes (often called *finger* probes) are inserted into a vessel or pipe (pressure of 245 bar). They consist of a steel sensing element that has a hollow space inside connected to a pressure device that monitors the buildup (or vacuum loss) of hydrogen pressure. Intrusive corrosion probes are prone to provide false positive indications of corrosion when first installed, as they initially have clean surfaces [397]. Particularly in sour (H_2S) service, steel tends to be passivated by an iron sulfide film only after some period of exposure. Therefore, nonintrusive methods in such cases may be more effective than intrusive methods.

Vacuum hydrogen probes operate at pressures below atmospheric pressure. As hydrogen gas is collected at the outer surface of the vessel, a decrease in vacuum occurs within the probe. By collecting the hydrogen gas in a known volume and monitoring the rate of decrease in vacuum, a hydrogen flux or corrosion rate can be calculated [402]. Clamp-on-style hydrogen probes are nonintrusive and thus offer an advantage in that they can be installed without penetrating the pressure boundary. The main advantage of probes that operate under vacuum compared with those that operate above atmospheric pressure is sensitivity. Sensitivity is enhanced because smaller amounts of permeating hydrogen cause larger percentage of changes in the pressures of probes operating at lower absolute pressures. *Hydrogen effusion probes* are clamp-on pressure probes, can measure hydrogen flux and corrosion rates for the surfaces to 500 °C as the cup portion of the clamp is made of silver [402], and can operate under vacuum; their limitation is that the measured corrosion rates can only be considered semiquantitative and the accuracy is not clear, as the metal and fluid dictate what percentage of hydrogen goes through the wall and what percentage of hydrogen is dissolved into solution and remained in the metal [404].

Nonintrusive hydrogen probes utilize an externally applied cell, patch, or gauge to monitor the rate of hydrogen egress from the outer surface of a vessel or pipe wall [409]. Nonintrusive online *hydrogen flux monitoring* (*hydrogen collection method*) was developed [410–415] and used for assessing the possibility of hydrogen damage in crude oil processing equipment [416] such as naphthenic acid corrosion [417, 418] and corrosion rate measurements on HIC-susceptible and HIC-resistant carbon steels (ASTM A516) in sour environment [419]. The device consisting of a collector is attached to the outside of the pipe, vessel, or reactor. A low-voltage pneumatic pump draws airstream from the collector plate across the sensing face of the detector that generates a very small current. This current is amplified and is routed to a digital display and data storage unit. Hydrogen flux data is usually expressed in terms of $pL\,cm^{-2}\,s^{-1}$ [411–414] and then converted to hydrogen activity [414, 420]. This method allows to measure corrosion rates $<0.001\,mm\,yr^{-1}$ up to 500 °C with resolution time of 2 min [397, 421–423]. The dimension of a *hydrogen flux monitoring* probe is 15 cm [397]. *Hydrogen flux monitoring* at high temperatures has been carried out at different units since 2002 [414]. These devices are used for online real-time CM for industrial boiler tubes. Hydrogen partial pressure after permeation through a metal may be measured using a mass spectrometer (residual gas analyzer) [401]. A conventional *scanning Kelvin probe* (SKP), operating on a macroscopic scale, was used for determination of the egress of hydrogen at the surface of pure iron, because this egress produced a marked Volta potential depression (of several hundred millivolts) [424].

Electrochemical cell (*electrochemical permeation method*) is also used where hydrogen reacts electrochemically, producing a current signal [425–430]. This hydrogen diffusion measurement is based on the Devanathan–Stachurski method [431] and is standardized [50, 408]. Intrusive electrochemical probes are developed [50, 406]. Electrochemical sensor based on Nb/Nb_2O_5 system was developed for detection of hydrogen permeation [432].

Hydrogen permeation technique allows evaluation of the state of the metal degradation [433]. Correlation between the hydrogen uptake and the rates of corrosion or deposition in oil refinery units was determined [271, 419].

Finite element analysis is used to calculate the hydrogen concentration and hydrogen diffusion flux (hydrogen attack) in steels and metallic glasses [434].

Hydrogen content in metals can be determined by analytical methods: in hot glycerol bath, by the LECO hydrogen analyzer [435] or by H_2A 2002 Leybold–Heraeus hydrogen analyzer [436]. The LECO hydrogen analyzer RH600 is based on destructive method measuring hydrogen in ~1 g of metal employing the inert gas fusion principle. A pre-weighed sample, contained in a single-use graphite crucible, is fused under a flowing helium gas stream. Hydrogen present in the sample is released as molecular hydrogen into the flowing gas stream. This hydrogen (H_2) is passed over heated copper oxide, converting it to water (H_2O). Water is measured by an IR detector. This method allows to determine 0.1–2500 ppm of hydrogen with precision of 0.05 ppm [437].

The following physicochemical methods are used for hydrogen evolution analysis in amorphous and quasicrystalline alloys [438]: thermal desorption spectroscopy (TDS) [436], DSC, AES, elastic recoil detection analysis (ERDA), and transmission electron microscopy (TEM). SIMS is used in recognizing improper baking of Cd-electroplated steel that results in hydrogen embrittlement [439].

The benefits of *hydrogen monitoring* are as follows:

1) It provides information about potential hydrogen attack.
2) In some cases, the method provides online real-time monitoring of hydrogen permeation.

The limitations of *hydrogen monitoring* are as follows:

1) It provides *qualitative* information in most cases.
2) Hydrogen probe data usually do not correlate with other CM methods, for instance, coupon (WL) method.
3) The hydrogen probes can be subjected to error or delays in reading process transients, since many processes with hydrogen participation can be prone to rapid changes in hydrogen charging severity with plant operating conditions.

Hydrogen monitoring is applicable to the oil and gas, oil refining, and petrochemical industries with hydrocarbon process streams. *Hydrogen monitoring* is used in wet sour (H_2S) environments, control of HTHA, corrosion by hydrofluoric acid (HF), and high temperature corrosion by naphthenic acids [239]. Various inspection methods are used to monitor HIC, SOHIC, and blister damage growth. Common methods are *straight beam UT* for HIC and blisters and *angle beam UT* for HIC and SOHIC damage [440]. Various hydrogen probes (internal or external) are used to monitor hydrogen charge. The inspection monitoring interval can be adjusted on the basis of the measured hydrogen charging levels. Inspection and monitoring of HTHA is described in literature [441]. HTHA damage may occur in welds, weld HAZs, and base metals. It is difficult to detect the early stages of HTHA with fissures and small cracks. Special NDT techniques (two *ultrasonic testing*), such as *high frequency shear wave* and *angle beam spectrum analysis*, are used in the fissuring stage for welds. Conventional *shear wave UT* and *time-of-flight diffraction* (*TOFD*) techniques can be used to detect HTHA when there is significant cracking. The other UT, namely, *frequency-dependent backscatter method* in combination with the *velocity ratio and spectral analysis techniques*, is used for detecting HTHA damage in the fissuring stage of base metal. *Wet fluorescent magnetic particle testing* (*WFMT*) is used for HTHA detection in the form of surface-breaking cracks. Hot spot monitoring and sophisticated UT technique (velocity ratio, attenuation, and spectral analysis method) are also used for detecting HTHA [442]. Thus, most methods of detection and quantification of *hydrogen damage* are based on US: ultrasonic echo attenuation method, amplitude-based backscatter, velocity ratio, creeping waves/time-of-flight measurements, pitch-catch mode shear wave velocity, and advanced ultrasonic backscatter technique. The principle, equipment required, and advantages and limitations of these methods are described in literature [443].

In situ field *metallography* and *replication* are used for detecting the early stages of HTHA (decarburization and fissuring) at the surface of steel as well as for differentiating between HTHA and other forms of cracking (see Section 9.4.1.12). Conventional *radiography* is used for inspection and detection of cracks in welds and can detect HTHA only after cracks have formed.

9.6 Electrochemical Methods

Scientists study the world as it is;
Engineers create the world that has never been.

Theodore von Kármán (1881–1963), Hungarian-American engineer and physicist

Electrochemical CM methods are based on the established fact that corrosion of metals in solutions of electrolytes is an electrochemical process. This means that electrochemical methods are suitable only for corrosion of metals in media with high enough electrical conductivity. Thus, if corrosion reactions are of electrochemical nature, *electrical potential, current,* and *resistance* (as derivative of the two first parameters) formed on the metal surface are main characteristics of corrosion. If we measure these three parameters, we are able to characterize and monitor corrosion of metals in conductive media. Electrical potential of a metal is a thermodynamic characteristic of corrosion process, and one can use the formula for its definition: $\Delta G = -nFE$ (ΔG is the change of Gibbs energy of corrosion reaction, n is the number of moles of electrons involved in the metal dissolution, F is Faraday constant, and E is the difference of electromotive forces of anodic and cathodic reactions). We should be careful as usually we measure the *corrosion potential* (*irreversible potential*) and not a *reversible electrical potential* used in this formula. The difference between these potentials is explained in literature [1]. The *electrical current* shows the kinetic characteristic of corrosion and is connected with a physical loss (in gram) of a metal by Faraday's law:

$$I \cdot t = n \cdot F \cdot \frac{m}{M} \tag{9.3}$$

where I is the corrosion current, t is a period of current flow, n is the number of moles of electrons involved in the metal dissolution, F is the Faraday constant, m is the mass of corroded metal, and M is the molecular weight of a metal. Thus, corrosion rate of a metal is proportional to electrical or corrosion current. Based on three electrochemical parameters (*electrical potential, current,* and *resistance*), electrochemical methods may be qualitative or quantitative. Qualitative methods include the identification of corrosiveness of media (determination of *redox potential*) and determination of *corrosion (electrode) potential* of corroded metal surface. Quantitative methods include the identification of *corrosion current* on a metal surface. Any electrochemical method that gives rise to the calculation of corrosion *current* and as a result a *corrosion rate* can be related to a quantitative method.

The advantages of the *electrochemical methods* are as follows:

1) Electrochemical methods allow receiving the quickest dynamic information about corrosion rate in real time. Therefore, we can follow the fast changes in the corrosiveness of a system (a system includes a metal and an electrolytic environment).
2) From an engineering point of view, the electrochemical methods are convenient as they give electrical signal that allows to react quickly and store the data in electronic devices.

3) Some electrochemical methods use polarization of a metal in the environment. Such methods allow receiving useful information about active and passive regions at the metal surface, mechanism of corrosion and inhibition, and a corrosion rate (see Section 9.6.3) [1].

The limitations of the *electrochemical methods* are as follows:

1) Electrochemical methods are based on measurement of electric current flowing through the solution. Therefore, there are limitations for the use of electrochemical methods in solution with low electrical conductivity: in nonelectrolytes (organic solvents, fuels), in two-phase electrolyte–nonelectrolyte system (with large quantity of the latter), or in very pure water.
2) Corrosion rate measurements assume uniform corrosion.
3) If used in multiphase systems, nonconductive phases may quickly isolate the metal electrodes and invalidate the readings until the electrodes are cleaned again.
4) Noncorrosive electrochemical reactions occurring on metal surface result in errors. For example, hydrogen sulfide (H_2S), hydrogen peroxide (H_2O_2), or other species in solutions take part in the oxidation/reduction reactions on metal surface. Such noncorrosive electrochemical reactions give rise to additive electrical current that does not relate to corrosion current.
5) Not always all metallic surfaces corrode, but calculations relate to a current density ($A\,cm^{-2}$) and to a WL of all metallic surfaces ($g\,cm^{-2}$). Therefore, we do not receive a real corrosion rate.
6) So many factors influence the electrochemical processes on a metal surface, and they measure such small changes in the electrical current and potential that the electrochemical methods are not as reliable as WL method, and their repeatability is poor. But "trend" of the electrical parameters may provide useful information about corrosion behavior of metals.

The results of the electrochemical methods must be compared with the WL data. Electrochemical methods use analysis of the metal–electrolyte interface with direct current (DC), with alternating current (AC), and without electrical current. The measurements of oxidation/reduction (redox) potential and electrode (corrosion) potential relate to the technique without electrical current. DC technique uses potentiodynamic polarization (LPR and Tafel extrapolation) and ZRA. AC technique uses EIS, ENM, and HDA. Analysis of studies on LPR, EIS, and ENM before 2000 was conducted [21].

9.6.1 Measurement of Oxidation/Reduction (Redox) Potential

Atoms of different elements may be present in solutions in various oxidized states, for instance, Fe^{2+} and Fe^{3+}, Mn^{2+} and Mn^{4+}, Co^{2+} and Co^{3+}, Cu^{2+} and Cu^{+}, S^{2-} and SO_4^{2-}, O_2 and OH^-, Cl_2 and Cl^-, etc. If such ions and molecules are present in solution in significant quantities, and when electron exchange with the electrode is sufficiently fast, *oxidation/reduction* (or *redox*) equilibrium is established at some inert electrode (for example, platinum), giving it a well-defined potential, or reversible *oxidation/reduction potential* (ORP):

$$E = E^{o}_{Ox/Red} + \frac{RT}{nF} \cdot \ln\left(\frac{a_{Ox}}{a_{Red}}\right) \tag{9.4}$$

where $E^{o}_{Ox/Red}$ is standard electrode potential (at $a_{Ox} = a_{Red} = 1$); a_{Ox} and a_{Red} are activities of oxidized and reduced forms; R is 8.314 J·mol^{-1}·K^{-1}, gas constant; T is temperature in Kelvin; n is number of moles of electrons participating in redox reaction; and F is 96 500 C·mol^{-1} (Faraday constant).

The redox potential E is a voltage that is proportional to the ratio of concentrations of oxidized to reduced states (forms) of specific substances in liquid solutions and is measured on the platinum electrode in relation to some reference electrode (for instance, calomel). Atoms of different elements that are present in various oxidized states (oxidizers and reducers) in electrolytic media may take part in cathodic and anodic corrosive reactions and may influence microbiological activity. Therefore, ORP is a qualitative parameter for identification of corrosiveness of a medium. It is useful to monitor the "trend" of ORP in order to define the changes in the corrosiveness of an environment. The presence of O_2, Fe^{3+}, and other oxidizing agents in a solution results in a deficiency of electrons on a metal surface and will attempt to acquire electrons (oxidizing conditions). Likewise, the presence of H_2S, Fe^{2+}, Mn^{2+}, and other reducing agents in a solution results in the situation that electrons are available and will attempt to give up electrons (reducing conditions). The polarity and value of ORP depend on the concentration of oxidizing and reducing species in the environment. In an oxidizing environment, higher ORP values exist (usually positive, for instance, +30 mV), whereas lower ORP values exist in a more reducing environment (usually negative, for instance, −30 mV). This method was used for microbiological control (pump feeding of oxidizing biocides) [33], corrosivity (reducing or oxidizing environment) of water condensate after distillation of crude oil [315], corrosivity of gasoline cuts containing water and peroxides [444], and detection of corrosion in coated vessels and tubes (monitoring defects in coatings) [445]. The following are a few *benefits* of the ORP method:

1) It is very simple, is quick, and can be used to control the injection of oxidizing biocides or other oxidizing additives.
2) It is used for control of microbiological activity.
3) It can serve as a rough indication of the corrosiveness of the solution.

The main drawback of this method is that the measured ORP is referred to as the buck conditions in the solution and is different from the ORP formed at the surface of the equipment. Some ions that are present in various oxidized states do not take part in corrosive anodic and cathodic reactions, but influence the ORP. Deposits (for example, biofilm) may be formed on the platinum electrode and do not allow some aggressive species (e.g. oxygen molecules, chloride anions, etc.) to penetrate the electrode's surface. Such situations may give rise to errors in the determination of real ORP. This method is applicable for corrosion control in BFW and microbiological growth in cooling water. This is the only electrochemical method concerned with the determination of corrosive parameter in the environment. Other electrochemical methods deal with corrosive parameter of the metal–environment border. Now, we shall consider the electrochemical methods designated to determine corrosive properties of metal equipment in contact with solutions of electrolytes connected to the metal–environment border. These methods are the measurement of corrosion potential, potentiodynamic polarization curves (Tafel extrapolation and LPR), ENM, ZRA, EIS, and HDA.

9.6.2 Measurement of Corrosion Potential of Metallic Equipment

Corrosion potential is an electrical (electrode) potential of a metal surface formed in contact with an electrolyte solution when an electrical current does not flow through the metal. A voltmeter with high inner ER (above $10^6\,\Omega$) is used for this purpose. Corrosion potential measured at open electrical circuit is called an *open circuit potential* and is used as a *qualitative* characteristic of a corrosion process on a metal surface. The *corrosion potential* is measured in comparison with reference electrodes such as calomel, silver/silver chloride, or copper/copper sulfate electrodes. This method is similar to the redox potential measurement, but instead of inert platinum electrode, the metallic equipment or some metallic specimen of the same equipment is used. If corrosion potential of a metal increases with time (changes in a positive direction), passivation occurs, and corrosion slows down. However, pitting corrosion can occur. If corrosion potential of a metal decreases with time (changes in a negative direction), corrosion process intensifies. However, it is impossible to answer how quick corrosion can be decreased or intensified. This general rule is not always correct, as many factors can shift the potential toward more positive or negative values, but these shifts may not necessarily be related to the occurrence of corrosion of metals. For example, decrease of oxygen concentration on the metallic surface causes more negative corrosion potential values, but they may not necessarily be associated with a high probability of steel corrosion. The presence of cathodic (or mixed anodic and cathodic) inhibitors causes the corrosion potential to shift toward a more negative value with a corresponding reduction in the severity of the steel corrosion. It is used to assist prediction of corrosion behavior by comparison with polarization data obtained from laboratory or site polarization scan [51]. This method is used to control cathodic and anodic protection systems [61, 280] (Table 9.2), which hold the metal at

Table 9.2 Cathodic protection potentials (volts) depending on soil conditions.

Soil type	Soil conditions	Reference electrode	
		$Cu/CuSO_4$	Ag/AgCl/ seawater
Water and low resistivity soil (<100 Ωm)	Aerobic $T < 40\,°C$	−0.850	−0.800
	Aerobic $T > 60\,°C$	−0.950	−0.900
	Anaerobic	−0.950	−0.900
High resistivity aerated sandy soil regions	Resistivity: 100–1000 Ωm	−0.750	−0.700
	Resistivity: >1000 Ωm	−0.650	−0.600

Source: From EN 12954:2001 [61].

controlled immune or passive potential, respectively [15], as a criterion of cathodic protection of underground constructions [446] and reinforcement in concrete [447], standardized for measurement of the corrosion potential of rebars in concrete [448–450] (see Section 9.8), wireless potential mapping of reinforced concrete structures [451], measuring corrosion potential in concrete structures by means of the graphite pseudoreference electrode [452], and evaluating corrosion and inhibition under sludge deposits in crude oil pipelines [453].

The measurement of potential is used for identifying active and passive zones and for prediction of corrosion rates in reinforced concrete (including bridges) [454–457], for monitoring corrosivity of soil [281, 458], for water chemistry improvement in pressurized water reactor (PWR) (as electrode potential is very sensitive to changes of dissolved oxygen in water) [459], in nuclear power plants for controlling IGSCC of SS 304 [460], for determination of the presence of biofilm on stainless steels [380], for the determination of corrosivity of water condensate after distillation of crude oil [315], and for monitoring corrosion of steel sheet piles and their protection by paints [461]. The multi-probe CM system containing reference electrodes for measuring potentials of tank material was used in high level nuclear waste tanks for determination of tendency to SCC and pitting corrosion [462]. Continuous OCP monitoring was used for capturing transient pitting events [113, 463]. In this case, pitting corrosion was characterized by a rapid drop in corrosion potential followed by repassivation. Pit depths were calculated by converting potential transients into current transients. The resulting current transients were integrated to obtain the total charge transferred in the event, which was used to calculate pit depth [113].

The measurements of potentials of epoxy (or other paints) coated and galvanized rebars as well as structures under cathodic protection are not suitable [450].

The *advantages* of *corrosion potential* measurements are the same as that for the *redox* method: it is very simple, fast detection of active/passive transition and can be used for online monitoring. There are more *drawbacks*:

1) Changes in corrosion potential with time do not allow defining which process, cathodic or anodic, is responsible for increase or decrease of corrosion of metals.
2) It is impossible to identify the corrosion phenomenon type such as pitting corrosion, crevice corrosion, etc.
3) Contamination in the environment can influence results and stability of reference electrode.
4) The method is qualitative and does not provide information about corrosion kinetics.

It was developed an "electrochemical tongue" (sensor) based on the measurement of the electrochemical potentials of some metals (Fe, Cu, Pb, Ag, Pt) and oxides of iridium and titanium for the assessment of the environment corrosivity in electrolytes, which is defined by pH, conductivity, concentrations of dissolved oxygen, chlorides, and sulfates [464]. It was suggested to use this "electrochemical tongue" (qualitative characteristic) with measurement of LPR (quantitative characteristic) for assessment of corrosivity of electrolytes [464].

SKP measuring potential is used for estimation of protective properties of coatings on metals [92].

Other electrochemical methods based on polarization curves, Tafel extrapolation and LPR, are used (and the latter is widely used in industry) for quantitative determination of corrosion rate of metallic equipment in aqueous solutions of electrolytes.

9.6.3 Tafel Extrapolation

Tafel extrapolation (*Tafel polarization*) technique is based on *polarization curves* (also named *Evans diagram*) (Figure 9.12) ([1], pp. 305–306).

Two main parameters are defined from polarization curves: corrosion potential E_{corr} and corrosion current (i_{corr}) (corrosion rate). The latter is obtained from extrapolating the currents in the two Tafel regions (AB and CD; Figure 9.12) to the corrosion potential E_{corr}. Both Tafel extrapolation and LPR techniques are based upon application of specific potential followed by monitoring of the current. The main difference between these two methods is that the change in potential must be kept less than ±10 to 30 mV for the LPR technique (see Section 9.6.4), while the change of potential can go up to ±250 mV (and more) for the Tafel extrapolation technique. Another difference between Tafel extrapolation and LPR is in the interpretation of testing results for the calculation of corrosion rate. In Tafel extrapolation, corrosion rate is

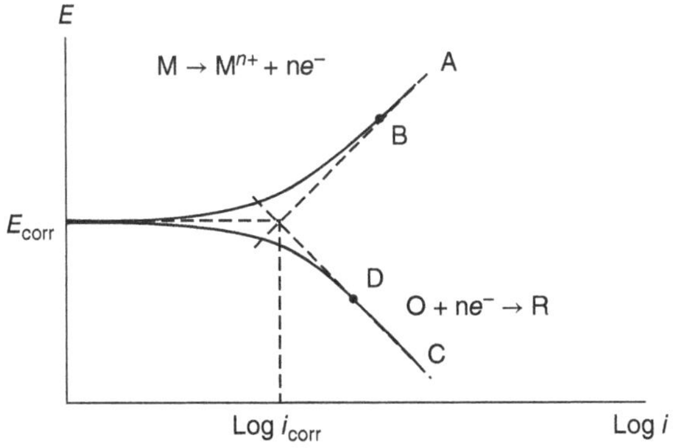

Figure 9.12 Polarization curves (Evans diagram) and Tafel extrapolation for corrosion process $M_{(s)} + O_{(aq)} \rightarrow M^{n+}_{(aq)} + R_{(aq)}$. M, metal; O, oxidizer (O_2, H_3O^+, Fe^{3+}); R, reduced species. *Source:* Groysman [1]. Reproduced with permission of Springer.

calculated using straightforward substitution of Tafel slope values (β_a and β_c) to obtain the corrosion current and is examined in Eq. (9.5):

$$i = i_{corr}\left\{\exp\left[a_1\left(E - E_{corr}\right)\right] - \exp\left[-a_2\left(E - E_{corr}\right)\right]\right\} \tag{9.5}$$

where a_1 is a slope of the anodic curve equaled to $2.303/\beta_a$, a_2 is a slope of the cathodic curve equaled to $2.303/\beta_c$, β_a is an anodic Tafel constant, β_c is a cathodic Tafel constant, and E and i are the potential and current density at any time. This method was used for evaluation of corrosion inhibition of brass in seawater by benzotriazole [465], for determination of corrosivity of soils [280], and for definition of mechanism of corrosion of metals in various media [315], including gasoline cuts containing water and peroxides [444].

The benefits of Tafel extrapolation method are as follows:

1) Tafel plots provide a direct measure of the corrosion current, which relates to corrosion rate.
2) Rapid determination of corrosion rates with Tafel plots can be advantageous for such studies as choice of inhibitors and alloys, as well as studying of influence of different factors on the efficiency of inhibitors, behavior of alloys, and corrosion mechanisms.

9.6.4 Linear Polarization Resistance Method

This method is based on the measurement of electrical current formed between two similar electrodes (Figure 9.13) when electrical potential scan between them is about ±10 to 30 mV. The propensity of the metal ions of an alloy to pass into solution, or corrode, is inferred by the relationship between a small change

Figure 9.13 Linear polarization resistance probe with replaceable electrodes.

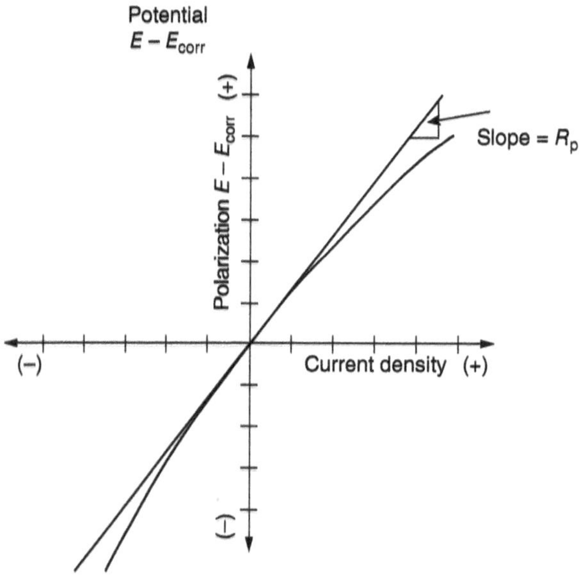

Figure 9.14 Electrical potential versus electrical current between two similar electrodes (linear polarization resistance method).

in electrical potential (typically 10–30 mV) across the corrosion interface and the increase in current density, which results from a consequent change in the flow of metal ions into solution.

If two metallic identical electrodes as in Figure 9.13, put into solution of electrolytes, begin to change electrical potential from −10 mV regarding the OCP ($E_{ocp} = E_{corr}$) of corroding electrodes with constant rate to +10 mV, we receive nearly straight line "electrical potential versus electrical current" (Figure 9.14) [49]. In other words, when the corroding electrode is polarized within a few millivolts around E_{corr}, the measured electrical current density change Δi is proportional to the corrosion current density i_{corr}. The ratio of potential change ($\Delta E = E - E_{corr}$) to current density change (Δi) at the point of zero current ($i_o = 0$) gives the *polarization resistance* R_p to corrosion reaction, that is, the slope of the linear curve corresponds to the *polarization resistance*:

$$R_p = \left(\frac{\Delta E}{\Delta i}\right)_{i_0} = 0 \tag{9.6}$$

Milton Stern, one of the inventors of the LPR method, noted [466] that this is not a "resistance" in the usual sense. Qualitatively, we may emphasize that the greater the polarization resistance, the smaller is the corrosion rate. For example, the efficiency of corrosion inhibitors in aqueous solutions of electrolytes may be evaluated according to R_p values.

The theory of this method was developed by M. Stern and A.L. Geary [466, 467] who showed in 1957 that the corrosion current density i_{corr} (mA cm^{-2}) can be calculated if we know the polarization resistance R_p:

$$i_{corr} = \frac{B}{R_p} \tag{9.7}$$

where B is the *Stern–Geary constant* (mV) depending on the type of metal and environment. The principle of this technique has a basis only in the evaluation of steady-state uniform corrosion. The B values must be calculated from other electrochemical experiments or determined empirically and are generally taken to be in the range of 26–30 mV for most metal–environment systems [468].

The use of *polarization resistance* for measuring corrosion rates has one particularly important advantage. The potential range investigated is close to the corrosion potential, and the applied currents are generally smaller than the corrosion current. Thus, the nature of the corroded surface does not change significantly, and the reactions that proceed during polarization are those that actually occur during the corrosion [466]. The LPR method has been reviewed [469–471] and standardized [24, 28, 47–49].

If the electrical conductivity of a solution is above 100 μS cm^{-1} ($R < 10^4$ Ohm·cm), two electrodes are used for LPR measurements. For a two-electrode measurement, the two electrodes are of the material being studied, and twice the value of R_p for a single electrode (plus the resistance of the solution between the electrodes) is measured [472, 473]. If the electrical conductivity of the solution is between 1 and 100 μS cm^{-1} ($10^4 < R < 10^6$ Ohm·cm), three electrodes are used. The third electrode is used for compensation of the IR drop and for the measurement of real electric potential of the electrode. The three electrodes are normally identical. LPR measurements can be made both manually and automatically to produce data from 5 min to 24 h [474].

LPR method is used to determine internal CM of subsea production flowlines [475], corrosion rate of reinforcement in concrete [12, 31, 476–479], assessing soil corrosiveness [32], corrosion in furnaces [480], CM of nuclear waste tanks [481], for evaluation of the efficacy of corrosion inhibitors in crude oil and gas pipelines [453, 482, 483], for measuring corrosion rates of outer surfaces of pipes at dig sites [484], online monitoring of the corrosive attack on superheater tubes in waste to energy and biomass incineration plants [485, 486], for determination of the initiation time and the mass loss due to crevice corrosion [487, 488], for evaluation of corrosion rate of copper-nickel alloys in seawater and marine sediments [489], for evaluation of corrosion inhibition of brass in seawater by benzotriazole [465], for optimization of corrosion inhibition in cooling water [490], for CM in wash water pump outlet in the overhead crude oil distillation column [275], in district heating systems [278, 491], the effect of oxygen on the inhibition of slightly sour corrosion [279], high salt concentration effects on CO_2 and H_2S corrosion [121], CM in cooling water systems [492] and in high pressure (up to 250 atm.) water lines [78]. If corrosion rate is low (for instance, in district heating pipelines), it is important to monitor changes in the corrosivity of the water than the corrosion rate itself [491]. Good correlation was shown between LPR and AE monitoring of uniform corrosion of mild steel in acid and neutral media [180]. Guidelines for online CM in plant equipment using electrochemical techniques are described in literature [47]. Special LPR sensors were developed for environments with high resistivity (hydrocarbon/water, atmosphere) [21], but they did not receive widespread attention. Localized CM technique based on combination of LPR and a calibrated white light interferometer (optical microscope) was developed for detecting pitting corrosion [493].

Using Faraday's law, one can convert the results of electrochemical measurements (i_{corr}, $\mu A\,cm^{-2}$) to rates of uniform corrosion (CR, $mm\,yr^{-1}$) [48]:

$$CR = K \cdot i_{corr} \cdot \rho^{-1} \cdot EW \tag{9.8}$$

where CR is corrosion rate, $mm\,yr^{-1}$; K is 3.27×10^{-3}, mm·g $(\mu A{\cdot}cm{\cdot}yr)^{-1}$; ρ is density of metal, $g\,cm^{-3}$; i_{corr} is corrosion current density, $\mu A\,cm^{-2}$; EW is equivalent weight of metal; and EW is W/n (W is the atomic weight of the metal and n the valence of the metal).

The *benefits* of the *LPR* method [1] are as follows:

1) It allows online real-time measurement of the rate of uniform corrosion and provides data within a few minutes.
2) LPR method is particularly well suited to the application of aqueous electrolyte solutions so that process upsets or other corrosion conditions can be detected very quickly to enable remedial action to be taken almost immediately. The method is used for the fast estimation of performance of corrosion inhibitors in a conductive environment.
3) Some systems combine this method with electrochemical current noise measurements (see Section 9.6.5) to determine the initiation of localized corrosion or as an indication of the stability of the inhibitor film on the metal surface.

The *limitations* of the *LPR* method [1] are as follows:

1) It is impossible to measure localized corrosion. We shall show later that two-electrode system is used for measuring pitting tendency by means of ENM (see Section 9.6.5).
2) Scale and other deposits formed on the electrode surface give rise to increase in the ER non-referring to the change of corrosion rate of the metal.
3) Redox reactions, different from corrosion reactions, contribute to the increase of corrosion current and can result in errors.
4) The constant *B* in formula (9.7) may vary with time, and this change may result in wrong results. The method requires prior knowledge of Tafel constants (see Section 9.6.3) for the calculation of corrosion rates.
5) It is very important to correctly mount the electrodes in a flow stream and to prevent "shading" of one electrode by another (Figure 9.15).
6) The main disadvantage is that the LPR method does not work in solutions of low conductivity, and it is impossible to monitor localized corrosion. ENM have been introduced in order to overcome the abovementioned drawbacks.

9.6.5 Electrochemical Noise Measurements

Analysis of corrosion failures shows that 70–90% of them result from localized corrosion. Data from the oil refining industry show that these values are between 40 and 65%. Ninety percent of these failures occur in 10% of the time that the equipment is in service [494]. ENM allow monitoring of localized

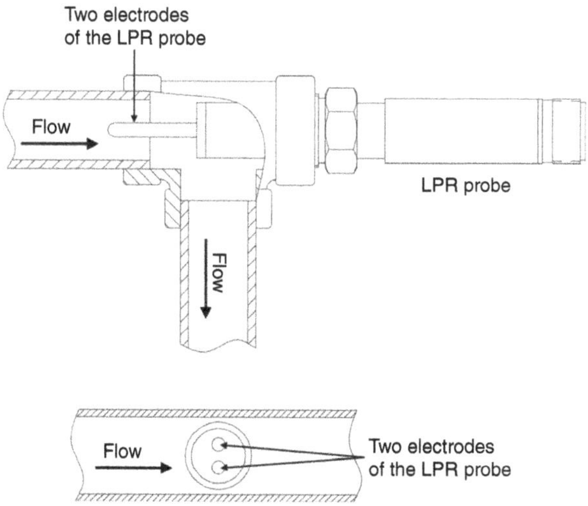

Figure 9.15 Correct installation of the LPR electrodes. *Source:* Groysman [1]. Reproduced with permission of Springer.

corrosion. First of all, we should explain what *electrochemical noise* means; it is a term coined in 1979. In an "ideal" world of corrosion, the electrode potential and current would be constant, metals would corrode uniformly, and all measurements would be easy. In the real world of corrosion, metals tend to corrode nonuniformly and corrosion potentials and currents vary with time. All this is connected with *electrochemical noise.* Everything in the universe is under *noise. Noise* is everywhere, from atom to planetary orbits. All particle trajectories exhibit chaotic, fractal, and turbulent behavior. *Noise* is opposite to *silence.* As it is impossible to reach the *absolute zero temperature,* it is impossible to reach the *absolute silence. Noise* in many cases is undesirable phenomenon. People learned to use *noise* in some cases. If a metal is "ill," a hole or crack is formed on the metal surface or inside, high energy goes outside (leaves) the active region, and we may "hear" the energy flow effects. The *AE* method is based on this principle (see Section 9.4.1.2). For example, tear a piece of a paper or break a tree. We will hear some *noise* – a crash, a crackle. This means that we hear how the bonds between atoms and molecules break. A similar situation occurs in a metallic corrosion. Metal ions leave their places in a crystal and cross into a solution in the anodic process. A metal corrodes, that is, some bonds are destroyed, and some bonds are formed. Some cations return to a crystal from a solution. The elementary charge transfer process (e.g. 2 for the oxidation of iron to ferrous ions or 4 for the reduction of a molecule of oxygen) is a random process with a binomial distribution known as "shot noise" [472]. Corrosion potential and current (important characteristics of corrosion) are not constant. Their fluctuations occur. *Electrochemical noise* in corrosion refers to the naturally occurring fluctuations in corrosion potential and corrosion current. Adsorption/desorption processes on a metal surface, pitting, and crack initiation may be the sources of *electrochemical noise.* If corrosion is an electrochemical phenomenon, it occurs according to a random or stochastic process. *Electrochemical noise* may be explained by the uncertainty principle of quantum mechanics. Some uncertainties or fluctuations exist in the positions and velocities of the particles (ions, electrons, and molecules) taking part in the electrochemical process. Up to the 1970s, scientists thought that the sources of the *electrochemical noise* were bias and error diminishing the reliability of electrochemical results. Therefore, they did not use the *electrochemical noise* information. Warren P. Iverson was the first researcher to use the *electrochemical potential noise* measurements for the analysis of metallic corrosion [495].

The electrochemical noise technique is defined as the analysis of the spontaneous fluctuations of the electrical current and potential of a corroding electrode. Such fluctuations are generated by spontaneous variations in corrosion kinetics and mechanism, commonly indicated as surface micro-events in the low frequency field and with amplitude from 10^{-6} to 0.1 V for potential and 10^{-11} to 10^{-3} A for current fluctuations [496]. In 1992, a method for the detection of localized corrosion using *current noise measurements* was patented [497].

Two techniques are used for measuring electrochemical potential and current noise. In the first technique, two "identical" electrodes are connected through a ZRA. Nominally "identical" means that the statistical properties of the two electrodes are the same [472]. There are a number of artifacts that can be introduced during the measurement process, including instrument noise, aliasing, and quantization noise [498]. Changes of electrical potential and current between two "identical" electrodes are measured in regard to the reference electrode with time. It is explicit that it is impossible to create two ideally "identical" electrodes. Therefore, another technique was developed with one working electrode, which is connected with a counter electrode (usually made of platinum) and a reference electrode of the potentiostat. This technique is called the *potentiostatic electrochemical noise technique* where both potential and current noises are measured at the same working electrode at an OCP. Current noise relates to current variations between working and counter electrodes, whereas potential noise refers to the variations in potential between a working electrode and a reference electrode. One characteristic of ENM is that corrosion that appears on the electrodes can be measured without generating any electrochemical disturbance to the electrodes. Various software programs were developed for mathematical treatment of measured standard deviation of the natural *potential (voltage) noise* (σ_E) and standard deviation of the natural *current noise* (σ_I) fluctuations generated in corrosion cells.

The ER noise (R_n) (or *electrochemical noise resistance*) can be calculated as a derivative of a potential and a current noise according to Ohm's law ($R_n = \sigma_E/\sigma_I$), and these values are in accord with the LPR (R_p) data [499–502]. Assuming that values σ_E and σ_I are based on the measured potential (i.e. σ_I has not been divided by the specimen area), R_n should be normalized through multiplying by the specimen area to give the normal polarization resistance units of Ohm·m^2 [472]. Sometimes, *spectral noise resistance* (R_{sn}) is calculated from Fourier transforms. The R_{sn} values are equivalent to R_n and R_p under a few circumstances [471, 502–504] and are very important for analysis of the efficiency of protective coatings on metals [500, 504, 505]. ENM is generally able to estimate the corrosion rate using R_n. Correlation between R_n and DC resistance measurements (R_{dc}) on coated metals [506] was seen.

Both electrochemical noise techniques with two "identical" and one electrode are used for monitoring both general and localized corrosion, especially pitting and crack initiation. The main benefit of this method is that it is used for online measurement of all types of localized corrosion in real time. The theory and practice of ENM is described in literature [472, 473, 498, 507, 508]. The pitting index (PI), or pitting factor (PF), is calculated according to literature [472, 473, 508, 509]:

$$\mathrm{PI} = \frac{\sigma_I}{I_{\mathrm{corr}}} \tag{9.9}$$

where σ_I is the current recorded from ECN (mA) and I_{corr} is the corrosion current recorded by LPR (mA) or harmonic analysis (see Section 9.6.9). Mixed quantitative–qualitative assessment is used:

1) PI < 0.01 (general corrosion).
2) 0.01 < PI < 0.1 (intermediate zone, but still predominantly general corrosion).
3) PI > 0.1 (localized corrosion).

Karel Hladky, one of the early workers on EN for corrosion studies, said, "the problem with noise is that it looks pretty much like noise"; in other words, it is difficult, if not impossible, to separate real EN from other sources of noise once a measurement has been made [472]. In spite of this, ENM are used in industry and laboratory investigations. The industrial devices are built on this principle for the *pitting tendency* CM, that is, transform raw electrochemical noise data into a process parameter that can be understood and used by operation personnel similar to a temperature or pressure reading (Figure 9.16).

ENM are unique among other electrochemical monitoring techniques in that they can be used to detect and distinguish between different localized corrosion phenomena such as pitting, crevice corrosion, and SCC [71, 472, 488, 496, 509–520]. Good correlation between ENM and AE measurements for monitoring SCC in high temperature (300 °C) and pressure (2.7 MPa) water [519] was exhibited. ENM were applied together with the reversed *direct current potential drop* (DCPD) technique for detection of SCC initiation in high purity water [71, 521]. ENM are used in combination with AE, EIS, and direct and alternate current potential drop methods (DCPD, ACPD) for CM in

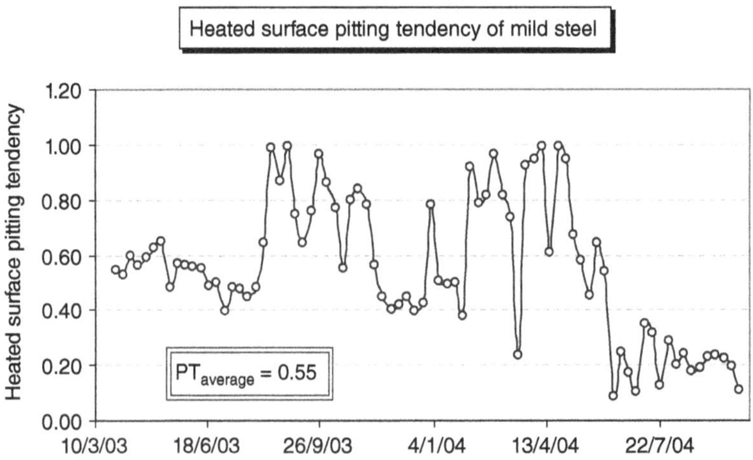

Figure 9.16 Pitting tendency of heated carbon steel surface vs time in cooling water. (Data were measured by means of CorrDATS system of Rohrback Cosasco; CorrDATS – Corrosion and Deposit Monitoring System.)

typical light water reactor service and nuclear waste disposal environments [71]. EN probes are used for CM in tanks containing high level radioactive wastes [522, 523] and in sour oil processing facilities [524], determination of pitting tendency in cooling water systems [525], CM in BFW, and monitoring of quality of BFW and optimizing of inhibition in BFW [526]. ENM are used for evaluation of general corrosion [21], for CM in reinforced concrete structures [12], for monitoring of exfoliation corrosion [447, 527, 528] and intergranular corrosion of aluminum alloys [529], for the estimation of corrosion mechanisms [530], for monitoring of localized corrosion and corrosion inhibitor performance [295] and protective properties of organic coatings [21, 500, 504, 506, 531], for assessing MIC [355, 496, 532, 533], for the estimation of protective properties of the Al–Zn galvannealed coatings in automotive industry [534], for continuous monitoring of inhibitor film performance and persistency [535], for screening (ranking) of corrosion inhibitors for aluminum alloys [536], for continuous Kraft digester [537], for CM in wet gas environment [538], for oxidizing hydrochloric acid solutions at high temperature [539], for CM in steam turbines [540], for CM in furnaces [541], and for CM in fuel combustion plants [503, 505, 542] and industrial boilers (both in furnaces and superheaters) [137, 543]. The measuring systems require three electrodes: two working and a reference electrode. In aqueous solutions, a standard reference electrode is used; however, at high temperatures, a pseudoreference electrode of the same material as the working electrode is employed. The conductive aqueous medium is replaced in the high temperature corrosion cell by molten salts formed in the surface deposit [503] or conductive fly ash at ~500 °C [137]. Noise data are obtained by measuring, at one second intervals, the instantaneous current between two working electrodes. The resolution is $\sim 10^{-9}$ A for current noise and $\sim 10^{-6}$ V for voltage noise [503]. Modified ENM technique was developed for monitoring uniform and localized corrosion [544]. Short-time monitoring using ENM with "no contact to substrate" setup is being developed [545]. EN device is used to measure corrosion, to analyze the causes of corrosion, to establish countermeasures, and to verify the effectiveness of corrosion control at chemical plants [546]. The actual limitations of ENM lie in the quantification of corrosion rate, and it is not simple to identify the type of corrosion phenomenon [547]. The theory and application of ENM is constantly developing [71, 548].

9.6.6 Galvanostatic Pulse Technique

Galvanostatic pulse technique (named also *galvanostatic pulse transient method*) is an NDT method that determines corrosion potential, corrosion current, and ER, mainly in reinforced concrete structures [12, 549]. An anodic current pulse is applied in a short time (~10 s) to reinforcement galvanostatically from a counter electrode placed on the concrete surface together with a reference electrode.

The applied current is normally in the range of 5–400 μA and the typical pulse duration is up to 10 s. The small anodic current results in a change of reinforcement potential, which is recorded as a function of polarization time. When the constant current is applied, the polarized potential of reinforcement at a given time can be calculated as a function of R_p, ohmic resistance, and double layer capacitance. After the R_p value is determined by this analysis, the corrosion current can be calculated from the formula (9.7). The authors [549] reported that this method allowed CM in marine environment. The limitation of *galvanostatic pulse technique* is difficulty in assessment of corroded area of steel rebars embedded in concrete.

9.6.7 Zero Resistance Ammetry

Zero resistance ammetry (ZRA) is a current-to-voltage converter that provides a voltage that is proportional to the current flowing between the input terminals while imposing a "zero" voltage drop to the external circuit. In *ZRA technique*, a macro cell current is measured between two different or the same metals (sensor) in an electrolytic environment. A ZRA measures the short-circuit current between the electrodes by an ammeter with zero resistance in order to avoid reducing the short-circuit current [24]. *ZRA technique* provides online real-time measurements. ZRA is used for measuring the galvanic corrosion current between two dissimilar or the same electrodes. The galvanic current is a corrosion rate value that can be obtained through Faraday's law (9.8). Thin film microsensors based on *ZRA technique* were developed for evaluation of corrosivity of environment (especially atmospheric corrosion), coating performance, properties of sealant and adhesives, hidden corrosion detection, and composite degradation in aircraft systems [21]. This method is standardized [28, 550] and is used also for measuring the coupling current between two "identical" electrodes to monitor electrochemical current noise [507, 508]. In reality, these electrodes will be slightly different, and a small coupling current will exist. Such measurements are relevant to detecting the breakdown of passivity and the early stages of corrosion. The restriction is, if extensive corrosion is occurring on both electrodes, that the macro cell current measured will not accurately reflect the severity of attack. *ZRA technique* is used in galvanic sensors developed for aircraft application for CM of the environment and better prediction of the structural health in wide inaccessible areas [551].

The LOCORR-CELL based on measurements of the galvanic current flow between the steel anode covered by a porous glass frit and the surrounding steel casing was developed for CM in district heating systems [491, 552]. The electrical current is measured by a ZRA. Galvanic measurements are used simultaneously with ER and optical monitoring of corrosion in district heating systems [72].

The *differential flow cell method* uses *ZRA technique* for online real-time prediction of localized corrosion (pitting, crevice, and underdeposit corrosion) in cooling water systems [553–558]. In this method, two or three identical electrodes are used for measuring the electrical current through a ZRA. Usually, two small identical anodes are placed in the slow flow conditions, and a large cathode is placed in the fast flow conditions. Three electrodes are connected electrically through a ZRA. The monitor obtains the localized corrosion rates on the two anodes [556, 557].

9.6.8 Electrochemical Impedance Spectroscopy

The concept of *electrical resistance* – the ability of a circuit element to resist the flow of electric current, is known as Ohm's law:

$$R = \frac{E}{I} \tag{9.10}$$

This relationship is restricted by an ideal resistor that has the following properties: it follows Ohm's law at all current and voltage values; its resistance value is independent of frequency; an AC and voltage signals through a resistor are in phase with each other. The real world contains circuit elements that exhibit much more complex behavior. Instead of *resistance* (R) we use *impedance* (Z), which is a more general electric circuit parameter. Similar to resistance, *impedance* is a measure of the ability of a circuit to resist the flow of electric current. Unlike resistance, *impedance* is not limited by the simplifying properties listed above.

Electrochemical impedance (Z) is the frequency-dependent, complex-valued proportionality factor, $\Delta E/\Delta I$, between the applied alternating potential (or current) and the response current (or potential) in an electrochemical cell [559]. In other words, an alternating voltage (usually 10–20 mV) is applied to a metal, and the resultant current and phase angle are measured for various frequencies. This factor becomes the *impedance* when the perturbation and response are related linearly and the response is caused only by the perturbation. The value may be related to the corrosion rate when the measurement is made at the corrosion potential.

Electrochemical impedance spectroscopy (also known as *AC impedance spectroscopy*, or the *electrochemical impedance technique*, or *impedance spectroscopy*) is an electrochemical technique in which a low amplitude alternating potential (or current) wave is imposed on top of a potential (usually the corrosion potential and zero imposed current) [560]. In other words, EIS monitors the electric response of the metal–environment border to the applied AC signal over a frequency spectrum. EIS can be represented as an extension of the LPR method [23, 472]. Instead of DC signals, AC signals are applied. As the different corrosion steps (diffusion, adsorption, surface reaction) have various time constants, these steps can be measured separately by varying the frequency of the AC signal.

Electrochemical impedance is normally measured using a small excitation signal. The EIS method applies a series of potential perturbations in the form of sinusoidal waves with a peak-to-peak potential of ±5 to 15 mV about corrosion potential and measures the resulting current [21]. The relationship between the voltage and current is used to make a judgment about the corrosion rate and other corrosion characteristics. The theory of EIS is described and analyzed in literature [468, 507, 561–565]. The goal of the EIS technique is to measure the *impedance* as a function of frequency by proper procedures and then analyze the resulting spectrum to estimate the corrosion rates and the mechanism that might give rise to the spectra. The *impedance* is affected by the interaction between frequency and all of the physical and chemical processes that respond to that frequency change within the electrochemical cell and across the corroding interface. In the limit of zero frequency, the impedance approaches the DC resistance of the corroding system. EIS can determine the polarization resistance (R_p), which is related to corrosion rate [21]. Thus, the low frequency limit of EIS is the same as the LPR method. The difference is that only a few LPR instruments correct for the solution resistance (R_s) and EIS provides R_p and R_s as well as mechanistic information and has the ability of extrapolating to steady-state conditions, which in some cases provide better corrosion rate estimation compared with LPR results. However, if CM is to be continuous, there is little advantage of EIS over LPR [21]. EIS was developed for lifetime establishment of external pipe coatings that require long lifetime (50 years) and standardized for determination of protective properties of coatings [566, 567]. EIS is a method used for online monitoring of uniform and localized corrosion [71, 568], studying uniform and pitting corrosion, corrosion of metals covered with a thin layer of electrolyte, protective properties of organic coatings [93, 500, 504, 569, 570] and degradation of composite, detection of SCC [571], inhibitor performance [21, 572], determination of corrosivity of soils [280], quantifying corrosion of steel rebars embedded in concrete [12], measuring of the exfoliation corrosion of aluminum alloys [573–575], online monitoring of furnace-wall corrosion in power generation boiler [576], evaluation of corrosion rate of copper nickel alloys in seawater and marine sediments [489], CM in acid gas pipelines [577], in district heating systems [278], and CM of carbon steel in pasty clayey mixture at 50–150 °C [125]. Local EIS was used for the determination of pitting corrosion and coating defects [578]. It was shown that EIS measurements compare well with the R_n values obtained from analysis of the ENM [500, 579]. A novel *dynamic EIS* (which is used under nonstationary conditions) was suggested for evaluation of corrosion inhibition of brass by benzotriazole in seawater [465]. The EIS, ENM, and OCP were used for assessing protective properties of metallic and organic coatings in cyclic salt spray tests [580].

The *benefits* of the *EIS* technique are as follows:

1) Very small excitation amplitudes (usually 5–15 mV) minimally disturb the metal, attached corrosion products, and other adsorbed species.
2) Rapid estimation of general corrosion rates (0.5–24 h after immersion).
3) It can provide more information than LPR method, but it can be more time consuming than LPR method.
4) It can provide detailed information about the mechanism and kinetics of electrochemical reactions.
5) Estimation of corrosion rates in slowly changing systems and low corrosion rates (<0.01 mpy).
6) Corrosion rate estimation in low conductivity media (organic liquids, soil, and concrete) for evaluation of cathodic protection efficiency. In these environments, the solution resistance can be separated from the actual polarization resistance.
7) Rapid assessment of efficiency of corrosion inhibitors. The EIS technique can be more effective than the LPR method.
8) Evaluation of corrosion under coatings and quality assurance of coatings (pore resistance and film capacitance).

The *shortcomings* of the *EIS* technique are as follows:

1) Measuring of an EIS spectrum takes time (usually many hours).
2) The system (metal–electrolyte) should be at a steady state. Otherwise, inaccurate results will be received. In practice, steady state can be difficult to achieve as the system can change through adsorption of solution impurities, growth of an oxide layer, buildup of reaction products in solution, temperature changes, coating degradation, etc.
3) Similar to the LPR method, the assumption of uniform corrosion has to be made in the calculation of corrosion rates.
4) Analysis of the EIS data is relatively complex and difficult for interpretation compared with the commonly used ER or LPR techniques. The EIS method requires benchmarking with other CM techniques.

9.6.9 Harmonic Distortion Analysis or Harmonic Analysis

The LPR technique assumes that the response of the electrochemical system under investigation is linear (in essence, this means that the amplitude of the response is directly proportional to the amplitude of the perturbation) [472]. This is rarely true for electrochemical systems. The basis of the HDA is an extension of the Stern–Geary analysis and impedance method. This method is based on measuring the fundamental, second, and third harmonic currents from the current response at corrosion potential by perturbing a corroding system with a non-distorted sinusoidal signal of low amplitude [581, 582]. The HDA is a low frequency impedance measurement and, similar to the LPR and EIS, relies on a steady-state approximation, but requires more mathematical treatment than the LPR and EIS techniques. A simple amplitude sinusoidal voltage of 10 mV is applied to perturbate the working electrode in the same manner as the EIS method. However, only a single frequency is employed, and the current response is measured in terms of the fundamental, first, and second harmonics. The current response to a low frequency voltage sine wave is distorted due to the nonlinearities of the charge transfer process. This distortion is analyzed in terms of the higher harmonics to provide values for corrosion current, the Tafel constants, and hence the Stern–Geary constant B [583]. This technique relates to the EIS in that an alternating potential perturbation is applied to one sensor in a three-element probe, with a resultant current response. The theory of EIS has been developed in literature [564, 581, 582, 584–586], whereby all kinetic parameters (including the Tafel constants) can be calculated explicitly. No other technique offers this facility. The HDA is based on evolving and improving the performance of the LPR technique. By applying a low frequency sine wave to the measurement current, the resistance of the corrosive solution can be calculated through a harmonic analysis of the resulting signals. With both polarization resistance and solution resistance, a more accurate general corrosion rate can be determined. Both the EIS and HDA are used for measuring corrosion rates of metals in acids [587], neutral solutions of electrolytes [588], rebars in concrete [12, 589], and MIC [533] and for evaluation of corrosion inhibition of brass in seawater by benzotriazole [465].

The *advantages* of *HDA* are as follows:

1) This technique is quick, and mathematical data treatment facilitates direct computation of the Tafel constants, Stern–Geary constant B, and the corrosion rate from the system under test. All these data are stored in the transmitter [590].

The *restrictions* of *HDA* are as follows:

1) As in the LPR and EIS techniques, the assumption of uniform corrosion has to be made.
2) If localized corrosion is taking place, the data are of a qualitative nature, indicating the breakdown of passivity and the possibility of localized attack.

9.6.10 Electrochemical Frequency Modulation

The *electrochemical frequency modulation* (EFM), or *intermodulation distortion* (IMD), is an extension of the HDA that uses a pair of frequencies and monitors the harmonics and the intermodulation distortion [591–594]. The theory was developed [591, 595] and apparatus was patented [596]. In the EFM method, two sinusoidal signals having different frequencies are applied to a portion of the system to distort a response current in the system. Analysis of this current generated in response to the applied perturbation signal provides the corrosion rate and Tafel constants [596]. In this method, a single-electrode probe is used for online CM [594]. This method was used for measuring corrosion rate of copper-nickel alloys in seawater and marine sediments [489], in boiler tubes [594], and in constructions under cathodic protection [597].

9.6.11 Other Electrochemical Methods

The *direct potentiostatic coulometry* (DPC) is a direct coulometry at controlled potential on glassy carbon electrode that allows studying corrosion kinetics of metals [598]. This method was used for studying corrosion kinetics of welds and protective properties of polymer coatings and inhibitors [598].

9.7 CM in the Atmosphere

All that exists in the universe is developed, and flows one into another.
Paulo Coelho (born 1947), a Brazilian lyricist and novelist

In many cases, we must know the corrosiveness of the atmosphere at the plant, in marine atmosphere, in storage areas, or in rooms with electronic and electric equipment. Many aggressive variables, such as sulfur oxides, nitrogen oxides, chlorine, ammonia, hydrogen sulfide, ozone, water vapors, salts, dust, and microorganisms contained in air, influence corrosion of metals in the atmosphere. Climatic factors such as temperature and its changes, relative humidity, sun irradiation, rainfall, and time of wetness are vital factors of corrosiveness of the atmosphere. Failure may occur because of corrosion of electric contacts made of gold, silver, copper, zinc, or their alloys. The WL and ER methods together with control of physicochemical variables and climatic factors are used for defining the corrosiveness of the atmosphere [599]. The WL method allows the classification of the corrosive level of atmosphere (Table 9.3).

The ER method allows online real-time monitoring and to learn about corrosiveness of the atmosphere and is realized in the sensor permitting to measure precisely very low corrosion rates of 0.0025 mm yr^{-1} [601]. CORRLOG ER corrosion sensors were developed for real-time monitoring of atmospheric corrosion of cultural heritage, namely, corrosion rate of steel, zinc, copper, bronze, brass, tin, lead, silver, and nickel in air [284, 602, 603]. These ER sensors can be used also for pollution monitoring, monitoring of corrosivity during transportation and storage of vehicles and goods, better management of corrosion protection of ships and vehicles, and monitoring of atmospheric corrosion

Table 9.3 Corrosion rates of carbon steel and zinc (μm yr^{-1}) in atmospheric environments [600].

	Carbon steel		Zinc		
Corrosivity	**First year**	**Steady state**	**First year**	**Steady state**	**Typical environment**
Very low	<1.3	<0.1	<0.1	<0.05	Dry indoors
Low	1.3–25	0.1–1.5	0.1–0.7	0.05–0.5	Desert or nonpolluted urban
Medium	25–50	1.5–6	0.7–2.1	0.5–2	Mild marine or mild industrial
High	50–80	6–20	2.1–4.2	2–4	Marine (calm sea)
Very high	80–200	20–90	4.2–8.4	4–10	Marine (surf beach)

Table 9.4 Sensors of atmospheric corrosion, their use, and thickness of the metallic track.

	Indoor		Outdoor	
Metal/alloy	**High sensitivity**	**Long lifetime**	**High sensitivity**	**Long lifetime**
Copper	50 nm	500 nm	5 μm	
Bronze	200 nm		5 μm	
Brass	—	10 μm		
Silver	50 nm	500 nm		—
Iron/steel	200 nm		25 μm	250 μm
Zinc	—	25 μm		50 μm
Tin	—	5 μm		
Lead	200 nm		5 μm	

Source: From Kouřil et al. [603].

of electronic equipment and structures in civil engineering. The sensitivity of ER measurements varies from 1 to 10 nm, and the lifetime is ~2 years in medium corrosive environments (Table 9.4) [602, 603].

The *QCM* is used for online real-time CM of the environment for electronic equipment [22, 604, 605], but it is affected by humidity and particles or films deposited on the surface [603].

The *CLSM* is used for determination of initial pitting process of magnesium alloys under atmospheric conditions [606].

Atmospheric sensors for defining corrosion rates of zinc, nickel, and galvanized steel were developed [607]. Electrochemical sensors for high-resolution *in situ* atmospheric corrosivity monitoring were used in field environments [608].

Galvanic sensors use condensed moisture and pollutants as an electrolyte, generating electric current that relates to the corrosivity of the condensed film [551]. These sensors comprise a series of interdigitated strips of two alternating different materials (gold and cadmium), deposited on a nonconducting polymer with a little gap that serves as an insulator between the two metallic strips. At relative humidity values >50%, the condensed moisture and pollutants create a thin liquid film on the strips, which serves as an electrolyte. The magnitude of this galvanic current provides a measure of the corrosivity of the condensed film. These *galvanic sensors* are used in aircraft application for CM of the environment and better prediction of the structural health in wide inaccessible areas [551].

Multielectrode array sensor (see Section 9.9) and *coating degradation sensor* systems measuring the corrosion behavior of carbon steel, impedance, relative humidity, and temperature allow CM in dehumidified storage areas [609].

Corrosion coulometer was developed for measuring general corrosion of bridges in atmosphere and evaluating corrosive atmosphere at chemical plants and aeronautic launch facilities having corrosive exhaust emissions [610–612].

A *metal piece test kit* was developed [613], in which several metal strips are placed in the atmosphere for a month. Corrosion products are defined by XRF, and corrosiveness of the atmosphere is determined according to the presence of sulfur, chloride, and oxygen in corrosion products.

AE sensor, magnetic flux leakage, and magnetostrictive sensor technologies are used for CM of bridge cables in the atmosphere [614]. For CM of vehicles in the road environments, corrosion panels of different metals (carbon steel, zinc, and samples with crevices) together with measurements of relative humidity, temperature (air and metal surface, dew point), and collectors for road mud (for measurements pH, anions, and cations) are used [615].

The following methods are used for evaluating efficiency of vapor corrosion inhibitors (VCI): electrochemical, ER, QCM, radiochemical, mass spectroscopy, and IR spectroscopy [616–619].

9.8 NDT and CM of Reinforced Concrete Structures

If somebody understands nothing except chemistry,
then this person barely understands chemistry (paraphrase).

Georg Christoph Lichtenberg (1742–1799), a German scientist and satiric writer

The NDT and electrochemical techniques of CM of reinforced concrete structures have been reviewed [12]. Among these methods are NDT (*visual inspection, IR thermography, X-ray/gamma radiography measurements,*

US pulse velocity technique, cover thickness measurements) and electrochemical methods (*OCP, surface potential, concrete resistivity, LPR, Tafel extrapolation, galvanostatic pulse transient method, ENM, EIS, and harmonic analysis*). In addition to these methods, the use of *embedded CM sensor* (monitoring OCP, LPR, resistivity, chloride ion concentration, and temperature), vibrating wire and electrical strain gauges, and optical fiber sensors is described [12]. Electrochemical methods (OCP, LPR, and EIS) together with the measuring of physicochemical parameters (porosity, pore size distribution, and alkalinity of the cementitious materials) were used for the monitoring of reinforced corrosion [620]. *Visual inspection* covers data about cracks, rust stains, spalled concrete parts, and quality of concrete. *IR thermography* provides information about chloride content on the surface of concrete. The *radiographic technique* is used for determining internal cracks, voids, variations in density of concrete, and whether the prestressing strands are snapped or intact. *US pulse velocity technique* involves the measuring of the speed of sound through concrete in order to detect internal flaws (cracks and voids) and to determine the strength of concrete. *Cover thickness measurements* (a cover meter or profometer) is used for measuring concrete cover, namely, for the detection of rebar size and its position. Cover thickness is an important parameter that preserves the electrochemical stability of steel in chloride-contaminated concrete. *Optical fiber sensors* are used for the CM, strain, displacement, and detection of cracks in concrete. The *vibrating wire strain gauge* is used to detect the formation of internal cracks and delamination in concrete. Electrochemical methods, except *surface potential* and *concrete resistivity*, are described in Section 9.6.

The *surface potential measurement* is used for identifying anodic and cathodic regions in concrete structures and indirectly detecting the probability of rebar corrosion in concrete. During corrosion process, an electric current flow between the cathodic and anodic sites through the concrete and this flow is detected by measurement of potential drop in the concrete. Two reference electrodes are used for surface potential measurements [12]. No electrical connection to the rebar is necessary in this technique. One electrode is kept fixed on the structure at a symmetrical point. The other electrode called moving electrode is moved along the structure on the nodal points of the grid. The potential of movable electrode, when placed at nodal points, is measured against the fixed electrode. A more positive potential reading represents anodic area where corrosion is possible. The greater the potential difference between anodic and cathodic areas, the greater is the probability of corrosion.

Electrical resistivity measurements of concrete are used for determining the risk of reinforcing steel corrosion and evaluating the effectiveness and durability of cathodic protection of concrete structures. Two different techniques, AC and DC measurements, are used for determination of electrical resistivity. In these measurements, both surface and embedded probes are applied. It was suggested to assess possible corrosion (corrosiveness) in concrete structures according to electrical resistivity (Table 9.5). Such qualifiable differentiation is similar to defining of corrosiveness of soils according to their resistivity [1].

The *OCP* technique is based on the measurements of corrosion potential of rebar with respect to a standard reference electrode. The tendency of any metal to react with an environment is indicated by the potential it develops in contact with the environment (see Section 9.6.2). In reinforced concrete structures, concrete acts as an electrolyte, and the reinforcement will develop a potential depending on the concrete environment, which may vary from place to place [12]. Potential mapping is widely used for detection of corrosion in reinforced concrete structures [621]. OCP measurements determine the probability of reinforcement corrosion (Table 9.6). Similar to *electrical resistivity of concrete*, OCP is also a qualifiable parameter.

The *anodic polarization voltage gradient* (*APVG*) method was suggested for monitoring the corrosion state of the steel core in reinforced concrete pipes [622]. This method is based on the *direct current voltage gradient* (*DCVG*) method consisting of an anodic polarization of the steel core generating current leaks through the isolation defects of the passive layer and potential gradients in the soil. These gradients are detected and measured by moving of the two $Cu/CuSO_4$ reference electrodes on the ground along the pipe. Thus, dimensions of defects are located and evaluated from the ground, without

Table 9.5 Electrical resistivity of concrete and possible corrosion.

Electrical resistivity (Ω·cm)	Corrosion risk
Greater than 20 000	Negligible
10 000–20 000	Low
5 000– 10 000	High
Less than 5 000	Very high

Source: From Song and Saraswathy [12].

Table 9.6 OCP (mV) of concrete structures determining corrosion conditions.

OCP vs. SCE	OCP vs. CSE	Corrosion condition
<–426	<–350	Severe corrosion
<–276	<–350	High (90% risk of corrosion)
–126 to – 275	–350 to –200	Intermediate corrosion risk
>–125	>–200	Low (10% risk of corrosion)

Source: From ASTM C876-09 [450].
Note: SCE, saturated calomel electrode; CSE, copper/copper sulfate electrode.

excavation and service disruption (without emptying the pipe) [622]. Benefits and limitations of all these methods are described [12].

9.9 Online Real-time CM

Time is a space for development of abilities.
Karl Marx (1818–1873), a German philosopher

The ER and LPR methods are standardized and used for online CM of metallic equipment in plants [47]. The ER technique was introduced in the early 1950s. The LPR technique entered into use for CM in industry in the 1960s. The problem is that the constant *B* in Eq. (9.7) are generally taken to be in the range of 26–30 mV for most metal–environment systems and are regarded by most manufacturers of LPR instrumentation to be a constant that is configured into an instrument at the factory [583]. The constant *B*, however, is not constant. Change from 1 to more than 100 mV for various systems can vary from time to time and even within the same system [623, 624]. Therefore, for accurate corrosion rate measurements, variable constants *B* should be used. The HDA technique overcomes this limitation and allows the determination of the Stern–Geary constant *B*. Usually an average constant *B* is analyzed with time, and then the LPR corrosion rate is post-corrected. Good correlation was shown between the constant *B* corrected LPR corrosion rate, the WL method, and ultrasonic thickness measurements [625].

Online real-time CM must be integrated with the process control system, i.e. we should monitor technological parameters together with corrosion parameters [285, 306, 626]. CM allows rapid determination of changes in water or other process stream quality, physicochemical changes (chemical content of phases), environmental parameters (temperature, pressure, flow rate, and chemical feed rate), and inhibitor performance. The pertinent operating and analytical data are entered into the risk matrix analysis where the data are used to develop the relative risk and consequence of the corrosion [627]. CM technology can determine uniform and localized (pitting, crack, etc.) corrosion, even when the uniform corrosion rate is low. The CorrDATS system manufactured by Rohrback Cosasco works well for monitoring uniform corrosion, pitting tendency, and heat transfer resistance (criterion of biofouling and other deposits) in water cooling systems (see Section 9.5.1.1) [525].

Three electrochemical methods (LPR, HDA, and ENM) described in literature [300, 306, 499, 509, 533, 628–633] are realized by *SmartCET technology* (Honeywell International, Inc.) and are used for online real-time CM of general and localized corrosion in the oil refining, chemical, oil, and gas industries [494, 623, 634], influence microbiological activity [623, 635–637], in the trickle-bed bioreactors [638], in district heating systems [289], in water injection systems and suitable optimization of corrosion inhibitors selection [583] and for determining inhibitor optimal dosage requirements [623, 639], in hot organic streams with small concentrations (1–2%) of water [499, 623, 637], in multiphase oil–water systems with corrosive gases in liquid and vapor condensing phases [640], in inhibited multiphase and dehydrated transport lines [509], in natural gas transmission pipelines [640], to monitor fireside corrosion in a waste to energy plant [637, 641], and for evaluation of fireside corrosion of coal-fired boiler materials [633]. Measurements using the three electrochemical techniques (LPR, HDA, and ENM) only require a few minutes [623]. This multi-technique electrochemical monitoring allows following the general corrosion rate, pitting tendency, constants B, and corrosion mechanism. The plant DCS provides the plant personnel access to time-trended general and pitting corrosion and direct correlation with plant process information [300].

The *resistance corrosion monitoring* (RCM) device was developed for continuous monitoring of pipe wall thickness in high temperature areas where corrosion is suspected [642–645]. The RCM operates on the same underlying principle as traditional ER probes except it utilizes the pipe wall as the active sensor element and it provides much more accurate and precise data. The RCM is an array of pins welded directly onto the pipe, covering ~1 m^2 area of pipe to be monitored.

The *field signature method* (FSM) developed by CorrOcean Company (Norway) is used in subsea pipelines, nuclear power plants, refineries, and storage tanks [239, 646, 647]. The FSM method is based on feeding an electric DC through the selected sections of the structure to be monitored and sensing the pattern of the electrical field by measuring small potential differences set up on the surface of the monitored object [648]. Sensing pins are used to measure the voltage response to an induced current (Figure 9.17). This type of measurement between two pins is closely related to an ER measurement. Proper interpretation of these potential differences (voltage distribution) can lead to conclusion about wall thickness reduction. The FSM provides online information on wall thickness loss, erosion, cracking, or pitting and is used up to 450 °C [649].

CEION (Cormon, UK) and *Microcor* (RCS, USA), comprising ENM and ER probes, are used in subsea pipelines and onshore sour service flowlines for monitoring localized corrosion, inhibitor effectiveness, and its optimization [295, 297]. The data are transferred via wireless systems to central locations. CEION is a ratiometric nanometer metal loss and electrochemical technology with resolution of 100 times better than existing ER-based devices. CEION yields real-time performance in the sub-nanometer ($<10^{-9}$ m) domain with the resolutions equivalent to the atomic radii of iron (125 picometer): corrosivity, erosivity, deposition, films, and solution conductivity.

Microcor/Corrater online systems intelligent interface unit (RCS) provides instantaneous corrosion rate and pitting tendency. RCS developed

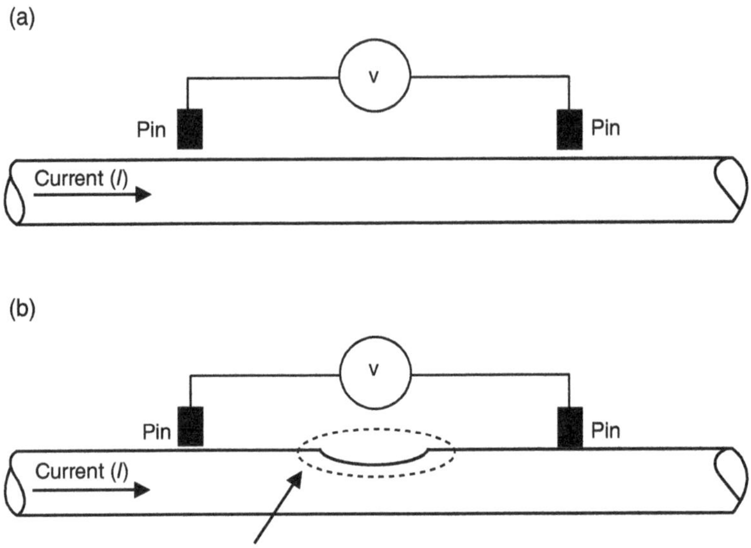

Figure 9.17 Scheme of the FSM technique. (a) No corrosion. (b) Corrosion occurs.

the *Corrater* LPR soil corrosion rate monitor for detecting soil corrosivity rate around pipelines.

CorrTran MV (Pepperl+Fuchs, Inc.) evaluates uniform and localized (pitting) corrosion.

Coupled multielectrode array sensor (CMAS) consists of an array of metal electrodes, all electrically insulated from each other, but connected through a network of resistors and arranged so that a small area of each electrode contacts a corrosive environment [650, 651]. The anodic current from each electrode is the signal that indicates localized corrosion. In a localized corrosion environment, anodic current flows into the more corroding electrodes, and cathodic current flows out of the less corroding or noncorroding electrodes. Such currents allow estimating the rate of localized corrosion using Faraday's law [650]. *CMAS* were used for the measurement of uniform and localized corrosion [66, 67, 651–656], MIC monitoring [390], evaluation of corrosion inhibitors in cooling water [657, 658], coating integrity, CM of pitting and crevice corrosion, and correlation with process parameters (conductivity, pH, temperature, dissolved oxygen, and corrosion potential) in water cooling system in order to understand the causes of corrosion and determine preventive anticorrosion measures [655], monitoring corrosion rates of Alloy 22 and titanium in aqueous solutions to 150 °C and pressure of 51–82 atm [650, 659], and monitoring corrosion rates of carbon steels, stainless steels, Ni–Cr, Ni–Cr–Mo, and Cu–Ni alloys in both liquid and vapor phases up to 300 °C and 140 atm [660, 661]. This sensor allows measuring the corrosion rate from

10^{-5} mm yr^{-1} (stainless steels in passive conditions) till localized penetration rates for these alloys up to 10 mm yr^{-1}. A *node electrode* system containing *CMAS, LPR,* and *ZRA* was developed for real-time CM of rebars in concrete and buried pipes in soil [662]. Brilliant review on *CMAS* was performed in 2006 with summary of use for different applications [651].

The *Concerto* system developed by CAPSIS (UK) is used for monitoring uniform and localized corrosion in nuclear waste tanks, sour gas pipelines, downholes, and concrete.

The *computerized CM system based on micropotentiostat/ZRA combination* was introduced [663]. Electrochemical corrosion probes are used in the chemical process industry [664].

KEMA manufactures the *KEMCOP* (*KEMa COrrosion Probe*) corrosion and fouling monitoring system for measuring corrosion rates and composition of corrosion products as a result of the firing conditions in power and waste incineration plants [137, 543, 665].

The *structural health monitoring* (SHM) was introduced in military aircraft for monitoring corrosion damage at "hot spots" in airframes [382]. The SHM system includes a combination of ER, galvanic, and microclimatic sensors, amounting under the lavatories and galley areas, doors, and its surroundings, etc. Microclimatic sensors measure the local conditions at the surface: temperature, humidity, surface wetness, pH, and conductivity of surface moisture. Galvanic corrosion sensor measures the corrosion-induced electric current between mated strips of copper and the aluminum alloy from which the monitored structure is fabricated. ER sensor is based on the aluminum alloy strip element measuring its corrosion in atmosphere. These multi-sensors measure real corrosion of "hot spots" in airframes and local microclimates, deduce corrosion activity, and are compared with actual corrosion (WL coupons and maximum pit depth) [111].

Advanced ER online corrosion probes with fiber-optic communication links are used for measuring the effectiveness of the chemical treatment program in gas, oil, and waterflooding plants [285]. This system is connected to 10 standalone servers linked to more than 300 transmitters/probes. The distance between the furthest monitoring site and the server is ~150 km. Remote seats are also provided with the software to allow users to access the corrosion server remotely. This system is ~256 times more sensitive to corrosion rate changes and less sensitive to temperature fluctuations than traditional ER systems [458]. The LPR and modified electrochemical potential measurements are used for CM of boilers [666].

During the construction of the Copenhagen Metro from 1996 until 2007, *corrosion monitoring system* (CMS) including 170 sets of probes was installed and embedded in the concrete structures [667]. This system includes the half-cell potential measurements, ZRA, impedance, LPR measurements with IR drop compensation, and temperature and allows monitoring uniform corrosion and stray current corrosion. The structures built for the Copenhagen Metro are designed for a 100-year service life, and existing CMS is used to update the probability-based service life assessment.

A *reinforced concrete corrosivity monitor* (*RCCM*) is an embeddable CM instrument that is capable of measuring the LPR, OCP, resistivity, temperature, and a potential related to chloride ion concentration [668]. The *RCCM* has many applications in the construction and maintenance of various structures such as high rise buildings, parking garages, bridges, dams, spillways, flood control channels, piers, pylons, and erosion control structures. During construction, engineers, builders, and supervisors can monitor parameters such as chloride concentration, resistivity, and temperature. These parameters can identify errors at an early stage of construction. One error that may be detectable is the use of seawater or contaminated water during mixing of the concrete ($[Cl^-]$). The moisture content and temperature of the structure can be monitored during the curing process to ensure maximum strength of the concrete. Once construction is complete, the instrument can be used to conduct long-term monitoring of corrosion conditions over time [668]. The RCCM is used in the Seohae Grand Bridge in South Korea [668]. The wireless CM system including ER probes, impedance coating-based degradation, and time-of-wetness sensors was developed for monitoring the US Marine Corps vehicles [94]. The OCP and LPR methods have been used over 20 years for CM of the Vevera Viaduct in Italy [669]. The level of sophistication of CM systems has improved significantly over the last 5 years due to the increased use of microprocessor-based instrumentation [285, 625, 639, 670–672]. We should take into consideration that CM is a very dynamic field in which new companies and devices appear every year.

9.10 NDT and CM of Pipelines

Gentlemen want to show their education,
therefore speak about incomprehensible things.
Anton Pavlovich Chekhov (1860–1904), a Russian writer

Both external and internal surveys are used for evaluating the condition of pipelines. The *ILI* tools are a common method to evaluate pipelines and pinpoint damage. The ILI surveys can provide information on wall thickness changes caused by corrosion, welds, branch connections, valves, and other imperfections in the pipe. The internal corrosion in pipelines is assessed by following NDT [2, 27, 41, 78, 246, 673–681]:

1) The use of "smart pigs" and verification by excavation and detailed examination at selected sites. "Pigs" (scrapers) are devices that are inserted into the pipeline and perform dedicated functions as they travel through the line. They can detect corrosion, cracks, and leaks in pipelines [2, 27, 682–686].

2) The use of *magnetic and electromagnetic techniques, flux leakage, ultrasonic testing, eddy currents, and EMAT tools* [27, 143, 680, 687, 688]. *Magnetic flux leakage* is based on induction of a magnetic flux density into the pipe wall. Pipe wall imperfections or metal loss in a magnetized section of pipe causes some of the magnetism to "leak" outside the pipe wall in the vicinity of defects. These flux leakage fields are detected by sensors on the instrumented tool as it moves through the pipeline [27, 689]. The "smart pigs" can use *magnetic flux leakage* tools [683]. Systems use either permanent magnets or electromagnets. The *EMAT* consists of a coil in a magnetic field at the internal surface of the pipe wall. The AC placed through the coil induces a current in the pipe wall Lorentz forces, which in turn generates ultrasound. In the *eddy current* method, a coil generating an alternating magnetic field is used to create the induced flux. The *eddy current* technology was combined with *magnetic flux leakage* method for monitoring inner corrosion of pipelines [688]. The *guided wave ultrasonic testing* (also known as the *long-range UT, GWUT,* or *GUL*) over the last decade has provided the industry with a new technique for ascertaining the integrity of pipelines [690, 691].
3) ISI of unpiggable lines based on the fluid flow modeling and determination of the critical pipeline inclination angles that may be likely sites for water accumulation [692, 693] and by means of high-resolution defectoscopes [694, 695].
4) Use of electrochemical probes based on the LPR, EIS, ENM, HDA, and the *multielectrode array sensor method* [78, 623, 631, 639, 640, 675, 696].
5) The *FSM* (see Section 9.9) [27, 646, 648]. It uses an induction coil that feeds an electrical current into the pipe at one location on the exterior near the test area and out at another location [697–699]. The current feed points are located 1–3 m apart. For a metal with an even wall thickness, without flaws or defects, a uniform electrical field is set up, reflecting the wall thickness. The uniform corrosion, pits, and cracks distort this electrical field, reflecting the reduction in the pipe wall thickness [27].
6) Use of *carbon–silver galvanic couple thin film sensors* for the indication of water accumulation possessing a corrosion risk in pipelines [673].
7) *Fiber-optic sensors* are used as a noninvasive tool for monitoring uniform and pitting corrosion in pipelines in real time [700, 701].
8) Use of galvanically *coupled multielectrode array sensor* for CM inside of high pressure (up to 10.34 MPa) natural gas pipelines [675, 676].
9) For the detection of *erosion* inside of pipelines, three physical methods are used: the finger probes with pressure gauges [702], the ER probes [287, 702, 703], and acoustic devices [703, 704].
10) Corrosion coupons, ER probes, and US thickness measurements are used for CM of internal surfaces of gas pipelines [681].

Most of these techniques are used for the monitoring of internal and external corrosion of pipelines. The main limitation of most of these techniques is the need to have a prior knowledge of the optimum location for sensors. Integrity management includes all data that supports the estimation of corrosion situation in the pipelines, namely, direct data (corrosion rate) and indirect data (chemical composition of gases and liquids, presence of microorganisms, operating conditions, flow rate, flow regime, etc.) [705]. Current technologies and software models of CM in dry gas and multiphase pipelines have been intensively developed during the last decade [706–709]. These software models can predict corrosion rate for any point along a pipeline.

Contactless immersion ultrasound modules (UT method) based on the clamp concept are used for real-time corrosion and/or erosion monitoring of pipelines [143]. If one distributes a number of transducers around the pipeline, it will be possible to detect localized corrosion [687]. The ER probes are used for CM of external corrosion of underground pipelines [294].

The systems of the Cormon company for subsea monitoring cover both intrusive probe applications for pressure, temperature, erosion, and corrosion and in-line spool-based flow assurance tools based on metal loss measurement [710].

The corrosion sensor utilizing the *resistive strain gauge* or an *optic fiber* instrument for measuring *diaphragm deflection* as an indication of sustained corrosion damage inside pipelines was developed [711, 712]. The sensing element (test diaphragm) is constructed of a material similar to the monitored pipe wall. As the test diaphragm thins due to corrosion, it becomes more sensitive to pipeline pressure and deflects accordingly. Minute changes in the diaphragm thickness are detected by monitoring the diaphragm's deflection under the internal pipeline pressure. One variation of CM system employs *resistive strain gauges* mounted directly on the diaphragm to observe diaphragm deflection. Another variation uses a *fiber-optic* measuring technique known as *extrinsic Fabry–Perot interferometry* (EFPI). In this variant, an EFPI gap is established between the end of an optic fiber and the reflective inner surface of the diaphragm. The instrument measures this gap distance to determine *diaphragm deflection* with sub-nanometer resolution. The sensor diaphragms are designed to have a corrosion span of 250 μm and a pressure rating of 689 kPa for use in potable water distribution pipelines. Thickness resolution 0.01 μm allows to measure corrosion rates of ~0.1 mm yr^{-1} with response time of 1 h [711, 712].

The *system including fiber-optic sensors* was developed to monitor internal corrosion, cracking and crack propagation, temperature and pressure changes, pH, and dissolved CO_2 concentration, as well to detect leaks, buckling, bending, and soil movement (environmental condition changes of a pipe) [674, 713, 714]. Thus, combined physical, chemical, and electrochemical methods are used for CM of pipelines.

The *probes* are used to detect erosion in flowlines caused by abrasive particles such as sand, corrosion products, scale, solids introduced into the well such as drilling mud and frac sand, and any other solids. Three physical methods are used for detection of *erosion: finger probes with pressure gauges* [702], the *ER probes* [287, 702, 703], and *acoustic* devices [703, 704, 715]. The *finger probes* are designed with a hollow cylindrical probe (sensing element) that is inserted into the flow. As a particulate impinges on the surface of the sacrificial sensing element, a hole is eventually eroded through the element, and the pressure from the system enters the inside of the probe itself. The pressure gauge detects that the element has been breached. In many cases, finger probes failed with no measurable erosion on any other piece of production equipment [702]. The *ER probes* measure the resistance of the sensing element exposed to the flow and corrosion, and usually it is impossible to distinguish between erosion and corrosion or erosion–corrosion. The *acoustic* nonintrusive (passive ultrasonic measurement) monitors detect high frequency noise from outside of a pipe wall. These monitors detect the noise that is generated by solid particle impact as well as noise that is generated by turbulence and bubble and droplet impact on the pipe wall [704]. The acoustic monitors require calibration.

Corrosion sensing system operating on a diaphragm-based principle, where the diaphragm is a surrogate element that corrodes similarly to the pipeline, was developed [716]. Small changes in diaphragm thickness can be detected by monitoring the diaphragm response to the applied pressure intrinsic to the pipeline process. The primary sensing element is mechanical in nature, not relying on process conductivity, and is suited to a wide range of processes such as potable water, oil and gas, chemical, and electric power applications [716].

A new device based on *eddy current array* for measuring, documenting, and assessing of external corrosion on steel pipelines was described [107]. Software produces a color contour map of the corrosion, which identifies the depth and location of the deepest pit and the axial length of the corrosion and utilizes the industry's widely accepted critical engineering assessment methods to calculate a safe operating pressure.

9.11 NDT and CM Under Insulation

Ultrasonic guided waves, pulsed eddy current, radiography, and *IR thermography* methods are used in detecting corrosion under insulation (CUI) [235, 252, 717–719]. A mathematical model was developed for prediction of the number of CUI-susceptible locations that needs to be refurbished [720].

9.12 Monitoring of Cathodic Protection

Whatever begins also ends.
Seneca (54 BC–39 AD), a Roman rhetorician and writer

The effectiveness of CP, especially in the presence of stray currents, is difficult to measure. Monitoring the efficiency of CP of underground and submerged metallic constructions and structures has been described [19, 42, 721–725], as well as automated remote monitoring system to control the stray current drainage [726] and CP [727], wireless remote monitoring [728], real-time monitoring of the IR-free polarization (E_{off}) potentials of multiple coupon test station buried in close proximity to the monitored feeder [729], and new devices with modified permanent Cu/$CuSO_4$ reference electrode coupled with corrosion coupons and an electronic alternating switch device [730]. Monitoring of pipes within cased crossings [37, 39, 731, 732] and the use of coupons for the monitoring of CP of prestressed concrete cylinder pipe [733] have been described. Visual inspection inside shipboard tanks demands drainage, opening, venting, and certifying as gas-free by a chemist. Monitoring systems inside shipboard tanks including sensors for the measuring corrosion potentials of metallic tank surface and electrical currents between anodes and tank surface were developed [734]. Evaluation of underground pipeline coating condition is described in standard [103, 735] and brilliant reviews [736].

9.13 Selecting a CM Method

The less people know,
they think that the more their knowledge.
Jean-Jacques Rousseau (1712–1778), a Genevan philosopher

Pros and cons of commonly available CM techniques have been analyzed [5, 6, 300]. Combined electrochemical techniques are used widely. The ENM are generally able to estimate the corrosion rate using R_n. However, the resulting measurement is inherently noisy, and other electrochemical methods (LPR, EIS, HDA, and EFM) provide better results [472]. Electrochemical techniques (OPC, LPR, ENM, and EIS) are used for CM of casing steel during underground storage of carbon dioxide [737] and detection of corrosion of post-tensioned strands in grouted assemblies [299, 738]. Corrosion rate sensors based on LPR and galvanic current measurements are embedded in concrete for monitoring corrosion in reinforced concrete structures [739]. Combined methods consisting of UT, ENM (for detection SCC), ER, and LPR probes are used for CM in high level waste storage tanks [740].

Many CM techniques are available, but relatively few are in widespread use. Comparison of properties and possibilities of some *CM* methods appropriate for the environmental conditions and for identification of particular corrosion phenomena are shown in Tables 9.7 and 9.8. The choice of CM method depends on the industry, type of equipment and structure, conditions, and objectives. Therefore, there is no one absolute recommendation for all industries. You should try to use as more methods as you can.

Table 9.7 Comparison of possibilities of CM methods used in the industry.

	CM method					
Corrosion phenomenon	**Ultrasonic**	**Coupons (WL method)**	**ER probes**	**LPR**	**Chemical analytical methods**	**ENM + HDA**
General (uniform) corrosion	Excellent	Excellent	Excellent	Good	Good	Good
Pitting corrosion	Fair	Excellent	NA	NA	NA	Excellent
Galvanic corrosion	NA	Excellent	NA	NA	NA	Good
MIC	Fair	Good	NA	NA	NA	Good
Erosion–corrosion	Good	Excellent	Excellent	NA	NA	Excellent
SCC	Fair	Good	NA	NA	NA	Excellent
Intergranular corrosion	NA	Good	NA	NA	NA	Excellent
Hydrogen-induced corrosion	Fair	Fair	NA	NA	NA (good for H_2 determination)	Excellent
Dealloying	Fair	Excellent	NA	NA	Good	Good
Crevice corrosion and underdeposit corrosion	NA	Good	NA	NA	NA	Excellent

Source: From Groysman [6].
Note: ER, electrical resistance; LPR, linear polarization resistance; ENM, electrochemical noise measurements; HDA, harmonic distortion analysis; MIC, microbiologically induced corrosion; SCC, stress corrosion cracking; NA, not applicable.

Table 9.8 Comparison of characters of some CM methods used in the industry.

	Corrosion control method					
Character	**Ultrasound**	**Coupons (WL method)**	**ER probes**	**LPR**	**Chemical analytical methods**	**ENM + HDA**
Period of measurements needed	6–24 months	1–6 months	One to several days	Several minutes	Minutes to hours	Several minutes
Periodical inspection, online or offline measurements	Inspection	Offline measurements	Online measurements	Online measurements	Online or offline measurements	Online measurements
Real-time information ability	No	No	Yes	Yes	Yes	Yes

Source: From Groysman [6].

Note:

[1] *Online* measurements are continuous control of metal loss, or corrosion rate, or other parameters in a system. Data are obtained without disrupting the monitoring process.

[2] *Offline* measurements are periodical control of corrosion rate, or other parameters, in a system with disruption of monitoring process.

[3] *Real-time* information is corrosion rate data that can be received instantaneously or with more frequency than the changes in the parameter being investigated. These data are usually received during several minutes.

9.13.1 Conclusion

I believe that getting to know more and more,
a man acquires infinite power, and if not happiness,
at least clearness (lucidity) of spirit....

Émile François Zola (1840–1902), a French writer

1) The NDT technology is constantly being improved and new methods are being developed. Advances in the use of lasers and imaging technology (including fiber-optic technique, video, holography, and thermography) have made noncontact NDT more viable in many situations. Most NDT techniques are used for detection of corrosion damages.
2) The NDT and CM are used to pursue, detect, predict, and prevent corrosion failures. The NDT and CM are an important part of anticorrosion management in any plant, including anticorrosion control program. The NDT and CM are a multidisciplinary issue, covering a wide range of measurement techniques, sensors, devices, instrumentation, data analysis, and standards. The general philosophy of NDT and CM is that multiple techniques are used to both complement and check each other.
3) We should differentiate *CM* from *inspection* and *survey*. Even though the latter is planned and organized periodically, CM is conducted online; and we can receive information about corrosion situation in real time.
4) Use of the NDT and CM can effectively reduce the failure risk to nearly zero. The NDT and CM are widely used for assessment of corrosion damage in materials. In some cases, visual inspection will not observe anything until a significant damage has developed, which causes a high cost of repair in the short term.

   There is no single technology that can detect and monitor all types of corrosion damage. Why are there so many methods of *NDT* and *CM*? To follow up corrosion is more complicated than to monitor such processing parameters as temperature, pressure, flow rate of streams, concentrations of reagents, etc., because of the following:
   a) There are many corrosion phenomena. One can see from Table 9.7 that not every *CM* method is suitable for monitoring all corrosion phenomena.
   b) Corrosion may be uniform or localized (pitting or crack). Specific methods are suitable for monitoring localized corrosion.

5) The *NDT* and *CM* must be planned and carried out at the stage of design and project. It is very important to correctly select *NDT* and *CM* type and inspection points. For NDT, ultrasonic thickness measurements must be taken in the vicinity of the CM points. Corrosion rate even for uniform corrosion may be significantly changed at different distances. The more places we monitor, the more complete and the more real corrosion situation may be determined.

6) The most sophisticated CM techniques are electrochemical measurements, and they are used today not only for corrosion trending measurements but also for online real-time corrosion information. Thus, ENM have proven to be most successful for CM in wide conditions (from ambient ~20 °C to high temperatures ~500 °C), and corrosion rates are collected for a few minutes.
7) NDT (US technique) allows direct *in situ* measurements of accurate corrosion rates of steel structures and used for determining their lifetime.
8) Probes or their sensing elements should be replaced when they reach 90–95% of their nominal life. They should be changed immediately if localized corrosion has occurred. In some cases, the sensing elements need to be cleaned if the data received appears questionable.
9) Use of only one method of *CM* provides some definite but not full information. The more methods of CM are used, the more we cover corrosion situations and as a result increased production, injection of chemicals, and technological parameters. It is important to use automated multi-technique monitoring capabilities. CM should be integrated with other plant programs designed to optimize the process conditions, injection of chemicals, and inspection. Only then we can successfully manage plant operation. CM is used with assessing the operational conditions, and thus the causes of corrosion can be identified.
10) Commercially online real-time monitoring systems based on electrochemical and physicochemical techniques are developed and introduced into practice. Some engineers have found these devices to be useful, while others remain skeptical about their benefits. Therefore, different NDT and CM integrated systems that combine more than one NDT and CM technique (in order to increase reliability and confidence in the data) must be used often and compared in various industries.

9.13.2 Instead the End

How pleasant to know,
That you learned something new!

Jean-Baptiste Poquelin Moliere (1622–1673), a French playwright and actor

Corrosion monitoring is practiced in many ways and occupies only ~0.05% of all corrosion cost in some industries. Therefore, it is out of place to talk about the *high* cost of CM techniques.

When corrosion occurs, cost is incurred. There is no doubt that we should monitor corrosion. However, we must be careful to use CM, as R.A. Cottis points out: "CM does not, of itself, reduce costs, increase reliability or even give greater assurance of safe operation; it just provides an indication of what the corrosion rate may be" [741]. Any statement (and this too) can be refuted as there are many examples from various industries (including personal experience of author of this chapter on oil refining industry) how NDT and CM helped to know about the corrosion situation, to stop using equipment with high corrosion rate, to know the reason for sudden appearance of corrosion, to optimize chemical injection or change to better chemicals, etc. Universal NDT and CM systems (device) for all corrosion phenomena and cases may not be created. However, we have witnessed that NDT and CM parameters (thicknesses and corrosion rates) are entered as many technological parameters and appear on monitors in many industries (chemical, oil refining, petrochemical, oil and gas) as other technological parameters: temperature, pressure, flow rate, concentration, etc. Therefore, we will observe in the nearest future that different NDT and CM methods will be increasingly incorporated in general systems with other tools measuring both technological and corrosive (corrosion) parameters.

The same NDT and CM techniques can be used in different industries. Some industries have special corrosion problems under specific conditions, for instance, industrial boilers or airplanes. Therefore, it will be important to write a chapter or review in the nearest future about NDT and CM in specific industries or how to monitor particular corrosion phenomena.

Corrosion thermodynamics is understood well, but corrosion kinetics (similar to chemical kinetics) is in the stage of alchemy in the Middle Ages and is based on experimental data. The NDT and CM, based on kinetics, should use the achievements of general kinetic theories for their own development. Only in few cases *corrosion rate* and *remaining lifetime* can be predicted on basis of the NDT and CM data.

The key factor of successful NDT and CM is the determination of the "areas of interest" (dangerous places or points), and this demands close cooperation from plant inspection departments, corrosion engineers, plant operators, and processing engineers. We should remember that any corrosion sensor measures corrosion only in a local place, and we do not know exactly what happens in the vicinity. This defines some vagueness in general philosophy of NDT and CM. Some corrosion phenomena (intergranular attack and dezincification) cannot be monitored in real time; only coupons can be used. Information about NDT and CM increases exponentially. Therefore, do not be surprised that this chapter does not encompass all literature concerning NDT and CM. As Kozma Prutkov (the fictional author invented by Aleksey Konstantinovich Tolstoy and three Zhemchuzhnikov brothers who lived during Russia's authoritarian reign by Nicholas I) said, "One cannot embrace unembraceable."

There is resemblance between corrosion rates and the results of stock exchanges; *results from the past are not guarantee for the future* [137]. Any changes in processing (temperature, installation additional pumps, changes of diameter of pipelines, fuel or crude oil type, inhibitors, neutralizers, and other chemicals) can result in sudden increase of corrosion. Therefore, we should analyze any changes and how they can influence the corrosion situation. In this light, NDT and CM play important roles in keeping metallic constructions in good conditions and preservation of people and environment from catastrophes.

References

1 Groysman, A. (2010). *Corrosion for Everybody*, 368. Dordrecht: Springer.
2 Raj, B., Jayakumar, T., and Sharma, G.K. (2008). NDE techniques for assessment of corrosion damage in materials and components. In: *Corrosion Science and Technology* (ed. U. Kamachi Mudali and B. Raj), 416–467. New Delhi, India: Narosa Publishing House.
3 API 570 (2003). *Piping Inspection Code: Inspection, Repair, Alteration, and Rerating of In-service Piping Systems*, 54. Washington, DC: API.
4 Hellier, C. (2001). *Handbook of Nondestructive Evaluation*, 603. New York: McGraw-Hill Professional.
5 Groysman, A. (2009). Corrosion monitoring in the oil refining industry: review. In: *Proceedings of the International Conference on Corrosion CORCON2009*, Mumbai, India, (28 September–1 October 2009), 85. European Federation of Corrosion.
6 Groysman, A. (2009). Corrosion monitoring. *Corrosion Reviews* 27 (3–4): 205–343.
7 Groysman, A. (2009). Corrosion monitoring in the oil refining industry. In: *Proceedings EUROCORR 2009, the European Corrosion Congress*, Nice, France (6–10 September 2009), 75. European Federation of Corrosion.
8 Roberge, P.R. (2000). *Handbook of Corrosion Engineering*, 371–483. New York: McGraw-Hill.
9 Davis, J.R. (ed.) (2000). *Corrosion: Understanding the Basics*, 467–474. Materials Park, OH: ASM International.
10 Winston Revie, R. (ed.) (2006). *Uhlig's Corrosion Handbook*, 2e, 1179–1238. Hoboken, NJ: Wiley-Interscience.
11 Canale, G., De Marco, M., Pinca, S., and Liratzis, T. (2010). The role of welded joint vulnerability and various damage mechanisms active in process & petrochemical plants reliability analysis with Risk Based Inspection (RBI) approach. In: *Proceedings EUROCORR 2010, the European Corrosion Congress*, Moscow, Russia (10–14 September 2010), Paper No. 9442, 24. European Federation of Corrosion.
12 Song, H.-W. and Saraswathy, V. (2007). Corrosion monitoring of reinforced concrete structures – a review. *International Journal of Electrochemical Science* (2): 1–28.

13 Groysman, A. (2008). Corrosion monitoring in industry. In: *Corrosion Science and Technology* (ed. U. Kamachi Mudali and B. Raj), 500–550. New Delhi, India: Narosa Publishing House.

14 NDT Encyclopedia. https://www.ndt.net/ndtaz/ndtaz.php (accessed 13 November 2018).

15 Rothwell, N. and Tullmin, M. (2000). *The Corrosion Monitoring Handbook*, 1e, vol. 180. Southampton, UK: Coxmoor Publishing Company.

16 Corrosion Testing and Monitoring CICS (2001). *Laboratory Techniques for In-situ Corrosion Monitoring*. Houston, TX: NACE International.

17 Marcus, P. and Mansfield, F. (eds.) (2005). *Analytical Techniques in Corrosion Science and Engineering*, 776. Baton Roca, FL: CRC Press, Taylor & Francis Group.

18 Yang, L. (ed.) (2008). *Techniques for Corrosion Monitoring*, 712. Cambridge, UK: Woodhead Publishing Ltd.

19 Roberge, P.R. (2007). *Corrosion Inspection and Monitoring*, vol. 383. Hoboken, NJ: Wiley-Interscience.

20 Tan, Y. et al. (2010). Review of critical issues in carbon dioxide corrosion testing and monitoring techniques. In: *NACE International Conference, CORROSION 2010*, Paper No. 10155, 15. Houston, TX: NACE International.

21 Agarwala, V.S. and Ahmad, S. (2000). Corrosion and monitoring – a review. In: *NACE International Conference, CORROSION 2000*, Paper No. 00271, 19. Houston, TX: NACE International.

22 Sridhar, N., Dunn, D.S., Brossia, C.S., and Moghissi, O. (2000). Corrosion monitoring techniques for thermally driven wet and dry conditions. In: *NACE International Conference, CORROSION 2000*, Paper No. 00283, 22. Houston, TX: NACE International.

23 Notten, G. (2008). Corrosion examination, inspection and monitoring, quality control of materials of construction. *Stainless Steel World* (October): 65–70.

24 NACE International Publication 3T 199 (1999). *Techniques for Monitoring Corrosion and Related Parameters in Field Applications*, Item No. 24203, 41. Houston, TX: NACE International.

25 NACE Standard RP0775-2005 (2005). *Preparation, Installation, Analysis, and Interpretation of Corrosion Coupons in Oilfield Operations*, 10. Houston, TX: NACE International.

26 NACE Standard RP0192-98 (1998). *Monitoring Corrosion in Oil and Gas Production with Iron Counts*, Item No. 21053, 13. Houston, TX: NACE International.

27 NACE International Publication 1D199 (1999). *Internal Corrosion Monitoring of Subsea Production and Injection Systems*, Item No. 24202, 13. Houston, TX: NACE International.

28 NACE Publication 1C187 (2005). *Use of Galvanic Probe Corrosion Monitors in Oil and Gas Drilling and Production Operations*, Item 24003, 13. Houston, TX: NACE International.
29 NACE Standard RP0189-2002 (2002). *On-Line Monitoring of Cooling Waters*, Item No. 21041, 19. Houston, TX: NACE International.
30 NACE Standard TM0286-2001 (2001). *Cooling Water Test Unit Incorporating Heat Transfer Surfaces*, Item No. 21219, 9. Houston, TX: NACE International.
31 NACE International Publication 05107 (2007). *Report on Corrosion Probes in Soil or Concrete*, Item No. 24234, 21. Houston, TX: NACE International.
32 NACE Standard RP0497-2004 (2004). *Field Corrosion Evaluation Using Metallic Test Specimens*, Item No. 21083, 26. Houston, TX: NACE International.
33 NACE International Publication 46107 (2007). *Control of Corrosion, Deposition, and Microbiological Growth in Recirculating Water Systems in Buildings*, Item No. 24233, 43. Houston, TX: NACE International.
34 NACE Standard TMP0497-2002 (2002). *Measurement Techniques Related to Criteria for Cathodic Protection on Underground or Submerged Metallic Piping Systems*, Item No. 21231, 31. Houston, TX: NACE International.
35 NACE SP0169-2007 (2007). *Control of External Corrosion on Underground or Submerged Metallic Piping Systems*, 35. Houston, TX: NACE International.
36 NACE Standard TM0194-2004 (2004). *Field Monitoring of Bacterial Growth in Oil and Gas Systems*, 20. Houston, TX: NACE International.
37 NACE SP0502-2008 (2008). *Standard Practice – Pipeline External Corrosion Direct Assessment Methodology*, 58. Houston, TX: NACE International.
38 NACE TM0101-2001 (2001). *Measurement Techniques Related to Criteria for Cathodic Protection on Underground or Submerged Metallic Tank Systems*, 27. Houston, TX: NACE International.
39 NACE SP0200-2008 (formerly RP0200-2000) (2008). *Standard Practice Steel-cased Pipeline Practices*, Item No. 21091, 23. Houston, TX: NACE International.
40 NACE SP0286-2007 (formerly RP0286) (2007). *Standard Practice Electrical Isolation of Cathodically Protected Pipelines*, Item No. 21032, 20. Houston, TX: NACE International.
41 NACE SP0206-2006 (2006). *Standard Practice Internal Corrosion Direct Assessment Methodology for Pipelines Carrying Normally Dry Natural Gas (DG-ICDA)*, Item No. 21112, 24. Houston, TX: NACE International.
42 NACE SP0207-2007 (2007). *Standard Practice Performing Close-interval Potential Surveys and DC Surface Potential Gradient Surveys on Buried or Submerged Metallic Pipelines*, Item No. 21121, 51. Houston, TX: NACE International.

43 NACE SP0274-2004 (2004). *Standard Recommended Practice High-voltage Electrical Inspection of Pipeline Coatings*, Item No. 21010, 6. Houston, TX: NACE International.

44 ASTM G4-01 (2008). *Standard Guide for Conducting Corrosion Tests in Field Applications, Annual Book of ASTM Standards*, vol. 03.02, 9. West Conshohocken, PA: ASTM International.

45 ASTM NACE/ASTM G31-12A (2012). *Standard Guide for Laboratory Immersion Corrosion Testing of Metals, Annual Book of ASTM Standards*, vol. 03.02, 8. West Conshohocken, PA: ASTM International.

46 ASTM G1-03 (2008). *Standard Practice for Preparing, Cleaning and Evaluating Corrosion Test Specimens, Annual Book of ASTM Standards*, vol. 03.02, 9. West Conshohocken, PA: ASTM International.

47 ASTM G96-90(2008) (2008). *Standard Guide for Online Monitoring of Corrosion in Plant Equipment (Electrical and Electrochemical Methods), Annual Book of ASTM Standards*, vol. 03.02, 10. West Conshohocken, PA: ASTM International.

48 ASTM G102-89(2015) (2015). *Standard Practice for Calculation of Corrosion Rates and Related Information from Electrochemical Measurements, Annual Book of ASTM Standards*, vol. 03.02, 7. West Conshohocken, PA: ASTM International.

49 ASTM G59-97(2014) (2015). *Standard Test Method for Conducting Potentiodynamic Polarization Resistance Measurements, Annual Book of ASTM Standards*, vol. 03.02, 4. West Conshohocken, PA: ASTM International.

50 ASTM G148-97(2003) (2003). *Standard Practice for Evaluation of Hydrogen Uptake, Permeation, and Transport in Metals by an Electrochemical Technique*, 10. West Conshohocken, PA: ASTM International.

51 ASTM G5-94(2004) (2006). *Standard Reference Test Method for Making Potentiostatic and Potentiodynamic Anodic Polarization Measurements, Annual Book of ASTM Standards*, vol. 03.02, 12. West Conshohocken, PA: ASTM International.

52 ASTM G51-95 (2005) (2006). *Standard Test Method for Measuring pH of Soil for Use in Corrosion Testing, Annual Book of ASTM Standards*, vol. 03.02, 3. West Conshohocken, PA: ASTM International.

53 ASTM G57-06 (2006). *Standard Test Method for Field Measurement of Soil Resistivity Using the Wenner Four-Electrode Method, Annual Book of ASTM Standards*, vol. 03.02, 5. West Conshohocken, PA: ASTM International.

54 ASTM G187-05 (2006). *Test Method for Measurement of Soil Resistivity Using the Two-Electrode Soil Box Method, Annual Book of ASTM Standards*, vol. 03.02, 4. West Conshohocken, PA: ASTM International.

55 API Standard 1163 (2005). *In-line Inspection Systems Qualification Standard*, 1e, 37. Washington, DC: American Petroleum Institute.

56 API Standard 653 (2003). *Tank Inspection, Repair, Alteration, and Reconstruction*, 116. Washington, DC: American Petroleum Institute.
57 API RP 45 (1998). *Recommended Practice for Analysis of Oilfield Waters*, 3e, 60. Washington, DC: American Petroleum Institute.
58 ASME (2010). *Boiler and Pressure Vessel Code, Section V. Nondestructive Examination*, 700. New York: ASME.
59 ASNT Recommended Practice No. SNT-TC-1A (2006). *Personnel Qualification and Certification in Nondestructive Testing.* Columbus, OH: ASNT.
60 ANSI/NACE Standard RP0104-2004 (2004). *The Use of Coupons for Cathodic Protection Monitoring Applications*, Item No. 21105, 29. Houston, TX: NACE International.
61 BS EN 12954:2001 (2001). *Cathodic Protection of Buried or Immersed Metallic Structures. General Principles and Application for Pipelines*, 34. London: BSI.
62 Brossia, C.S. and Dunn, D.S. (2005). Apparatus and method for detecting the degradation of a coating using embedded sensors. US Patent 6,911,828.
63 Pope, D.H., Lin, Y.P.J., St. Martin, E.J., and Frank, J.R. (1999). In-situ process for the monitoring of localized pitting corrosion. US Patent 5,888,374.
64 Martinchek, G.A. and Yaffe, M.R. (2000). Detection of pitting corrosion. US Patent 6,015,484.
65 Lin, Y.P.J., St. Martin, E.J., Frank, J.R., and Pope, D.H. (2001). Electrode design for corrosion monitoring using electrochemical noise measurements. US Patent 6,294,074.
66 Yang, L. and Sridhar, N. (2004). Sensor array for electrochemical corrosion monitoring. US Patent 6,683,463.
67 Yang, L. and Sridhar, N. (2006). Sensor array for electrochemical corrosion monitoring. US Patent 6,987,396.
68 Schweitzer, P.A. (2006). *Corrosion of Linings & Coatings: Cathodic and Inhibitor Protection and Corrosion Monitoring*, 568. Boca Raton, FL: CRC Press, Taylor & Francis Group.
69 Dean, S.W. (1986). Overview of corrosion monitoring in modern industrial plants. In: *Corrosion Monitoring in Industrial Plants Using Nondestructive Testing and Electrochemical Methods*, STP 908 (ed. G.C. Morgan and P. Labine), 197–220. West Conshohocken, PA: ASTM International.
70 Dean, S.W. (1987). In-service monitoring. In: *NACE International Conference, CORROSION*, Metals Handbook, 9e, vol. 13, 197–203. Materials Park, OH: ASM International.
71 Ritter, S. and Molander, A. (eds.) (2010). *Corrosion Monitoring in Nuclear Systems: Research and Applications*, European Federation of Corrosion Series, vol. 56, 280. London, UK: Maney Publishing.
72 Hemmingsen, T., Hilbert, L., and Nielsen, L.V. (2003). Assessment of sulphur and H_2S corrosion by use of simultaneous ER, galvanic and optical

measurements. In: *Proceedings EUROCORR 2003, the European Corrosion Congress*, Budapest, Hungary (28 September–2 October 2003), 10. European Federation of Corrosion.

73 NDT.net. *Proceedings of the 10th Asia-Pacific Conference on Non-Destructive Testing*, Brisbane, Australia (17–21 September 2001). http://www.ndt.net/apcndt2001/index.htm (accessed 14 November 2018).

74 ASNT. www.asnt.org. (accessed 14 November 2018).

75 Muravin, G.B. (2000). *Inspection, Diagnostics and Monitoring of Construction Materials and Structures by the Acoustic Emission Method*, 480. London: Minerva Press.

76 Thompson, D.O. and Chimenti, D.E. (eds.) (2006). *Review of Progress in Quantitative Nondestructive Evaluation*, vol. 26. New York: Springer.

77 Groysman, A. and Groysman, O. (2010). Interdisciplinary relationships in corrosion teaching and education at school, college and university. In: *Proceedings EUROCORR 2010, the European Corrosion Congress*, Moscow, Russia (10–14 September 2010), Paper No. 9148, 23. European Federation of Corrosion.

78 Faritov, A.T., Gumerov, A.G., and Hudyakova, L.P. Corrosion monitoring system and oil field pipelines database software. In: *Proceedings EUROCORR 2010, the European Corrosion Congress*, Moscow, Russia (13–17 September 2010), Paper No. 9663, 12. European Federation of Corrosion.

79 Bauge, A., Wold, K., and Petrova, J. (2010). Benchmarking of the modern methods of internal corrosion monitoring of the oil and gas product facilities. In: *Proceedings EUROCORR 2010, the European Corrosion Congress*, Moscow, Russia (13–17 September 2010), Book of Abstracts, 683. European Federation of Corrosion.

80 Munger, C.G. and Vincent, L.D. (1999). *Corrosion Protection by Protective Coatings*, 2e, 510. Houston, TX: NACE International.

81 Keane, J.D. (ed.) (1989). *Good Painting Practice*, 2e, vol. 1, 580. Pittsburgh, PA: Steel Structures Painting Council.

82 Peabody, A.W. (2001). *Control of Pipeline Corrosion*, 2e (ed. R.L. Bianchetti), 347. Houston, TX: NACE International.

83 Parker, M.E. and Peattie, E.G. (1999). *Pipe Line Corrosion and Cathodic Protection*, 3e, 166. Houston, TX: Gulf Publishing Company.

84 Morgan, J. (1987). *Cathodic Protection*, 2e, vol. 519. Houston, TX: NACE.

85 Martinez, S., Valek, L., and Kapor, F. (2008). Assessment of CP effectiveness on PE tape coated buried pipeline by external corrosion direct examination. In: *Proceedings EUROCORR 2008, the European Corrosion Congress*, Edinburgh, Scotland (7–11 September 2008), 8. European Federation of Corrosion.

86 Makama, Z. et al. (2008). Surface preparation and adhesion in the prevention of cathodic delamination. In: *Proceedings EUROCORR 2008, the European*

Corrosion Congress, Edinburgh, Scotland (7–11 September 2008). European Federation of Corrosion.

87 Wan, H., Song, D., Li, X. et al. (2017). Failure mechanisms of the coating/metal interface in waterborne coatings: the effect of bonding. *Materials* 10: 397.

88 Nielsen, L.V. and Petersen, M.B. (2008). Consequences of AC/DC interference on pipeline cathodic protection criteria. In: *Proceedings EUROCORR 2008, the European Corrosion Congress*, Edinburgh, Scotland (7–11 September 2008). European Federation of Corrosion.

89 NACE International Publication 10A392 (2006). *Effectiveness of Cathodic Protection on Thermally Insulated Underground Metallic Structures*, Item No. 24156, 8. Houston, TX: NACE International.

90 Brazy, J.-L., Roche, M., and Tobiassen, P. (2004). Feed-back from long-term experience with offshore cathodic protection. In: *Proceedings EUROCORR 2004, the European Corrosion Congress*, Nice, France (12–16 September 2004), 13. European Federation of Corrosion.

91 Festy, D. (2004). Monitoring of cathodic protection for design and surveillance. In: *Proceedings EUROCORR 2004, the European Corrosion Congress*, Nice, France (12–16 September 2004), 8. European Federation of Corrosion.

92 Feser, R., Schnier, K., and Haarmann, S. (2010). New interpretation of corrosion behaviour of coated metals by combination of impedance and scanning Kelvin probe measurements. In: *NACE International Conference, CORROSION 2010*, Paper No. 10158, 12. Houston, TX: NACE International.

93 Tanabe, H. (2010). Evaluation of protective performance of organic coatings using current interrupter technique. In: *NACE International Conference, CORROSION 2010*, Paper No. 10163, 8. Houston, TX: NACE International.

94 Caseres, L. and Jakab, M.A. (2010). Field deployment of wireless sensors to monitor coating degradation and environmental corrosivity. In: *NACE International Conference, CORROSION 2010*, Paper No. 10175, 15. Houston, TX: NACE International.

95 Weldon, D.G. (2009). *Failure Analysis of Paints and Coatings*, 362. Chichester, UK: Wiley.

96 de Wacht, M. (2010). Field survey results of a recently developed cathodic protection survey methods to improve the ECDA process. In: *NACE International Conference, CORROSION 2010*, Paper No. 10057, 11. Houston, TX: NACE International.

97 Picas, J. and Shoaf, W. (2010). Weather – outside force threat "lightning" to pipelines and coatings. In: *NACE International Conference, CORROSION 2010*, Paper No. 10058, 14. Houston, TX: NACE International.

98 Biagiotti, S.F., Biles, S.R., and Chaney, C.E. (2010). Using soil analysis and corrosion rate modeling to support ECDA and integrity management of pipelines and buried plant piping. In: *NACE International Conference, CORROSION 2010*, Paper No. 10059, 12. Houston, TX: NACE International.

99 Norsworthy, R. (2010). Finding and determining the cause of external corrosion on cathodically protected pipelines. In: *NACE International Conference, CORROSION 2010*, Paper No. 10069, 10. Houston, TX: NACE International.

100 Nielsen, L.V. and Baumgarten, B. (2010). Monitoring of average coating resistance with off-potential mode T/R's – operator experiences. In: *NACE International Conference, CORROSION 2010*, Paper No. 10065, 14. Houston, TX: NACE International.

101 Souto, R.M., González-García, Y., and González, S. (2009). Characterization of coating systems by scanning electrochemical microscopy: surface topology, blistering and debonding. In: *Proceedings EUROCORR 2009, the European Corrosion Congress*, Nice, France (6–10 September 2009), Paper No. 8278, 1. European Federation of Corrosion.

102 Fumei, O., Di Biase, L., Cigna, R. et al. (2010). Cathodic protection of LPG mounded vessels – specific solutions for particular problems. In: *Proceedings EUROCORR 2010, the European Corrosion Congress*, Moscow, Russia (13–17 September 2010), Paper No. 920, 11. European Federation of Corrosion.

103 NACE TM0109-2009 (2009). *Aboveground survey techniques for the evaluation of underground pipeline coating condition*, 35. Houston, TX: NACE International.

104 Nicholson, J.P. (2010). Combined close interval potential surveys and direct current voltage surveys for increased pipeline integrity. In: *Proceedings EUROCORR 2010, the European Corrosion Congress*, Moscow, Russia (13–17 September 2010), Paper No. 9145, 9. European Federation of Corrosion.

105 Augustyński, Ł. and Królikowska, A. (2010). Appraised dry film thickness – new concept for the interpretation of DFT data collected from coated structures. In: *Proceedings EUROCORR 2010, the European Corrosion Congress*, Moscow, Russia (13–17 September 2010), Paper No. 9639, 1. European Federation of Corrosion.

106 Al-Meshari, A., Abdelgalil, A., and Al-Enazi, S. (2010). Investigation of cracking in the shell of a reactor of MTBE plant Catofin unit. In: *Proceedings EUROCORR 2010, the European Corrosion Congress*, Moscow, Russia (13–17 September 2010), Paper No. 9135, 9. European Federation of Corrosion.

107 Laughlin, S. (2010). Conformable array corrosion mapping tool. In: *NACE International Conference, CORROSION 2010*, Paper No. 10171, 11. Houston, TX: NACE International.

108 Nelson, B.N., Lemieux, E.J., Slebodnick, P., and Groeninger, W. (2006). Tank coatings assessments using the Insertible Stalk Inspection System and Corrosion Detection Algorithm. In: *NACE International Conference, CORROSION 2006*, Paper No. 06323, 17. Houston, TX: NACE International.

109 Paul N. Gardner Company, Inc.. *Video Borescope*. https://www.gardco.com/pages/optical/ma/video_bore.cfm (accessed 27 October 2018).

110 GE Inspection Technologies. *10017 Corrosion-Erosion Brochure*, 28. https://www.offshore-technology.com/contractors/instrumentation/ge-inspection (accessed 27 October 2018).

111 Trego, A., Price, D., Hedley, M. et al. (2007). Development of a system for corrosion diagnostics and prognostics. *Corrosion Reviews* 25 (1–2): 161–177.

112 van Teylingen, C. (2003). Measuring and classification of corrosion using VIEEW (digital image analysis system). In: *Proceedings EUROCORR 2003, the European Corrosion Congress*, Budapest, Hungary (28 September–2 October 2003), 8. European Federation of Corrosion.

113 Stewart, S., Jovancicevic, V., Menendez, C. et al. (2010). Evaluation a new sour gas corrosion inhibitor for field applications via localized corrosion monitoring techniques. In: *NACE International Conference, CORROSION 2010*, Paper No. 10274, 17. Houston, TX: NACE International.

114 Papavinasam, S., Doiron, A., Li, J. et al. (2010). Sour and sweet corrosion of carbon steel: general or pitting or localized or all of the above? In: *NACE International Conference, CORROSION 2010*, Paper No. 10332, 23. Houston, TX: NACE International.

115 Polder, R. and Peelen, W. (2006). Risk based management of concrete structures using advanced corrosion monitoring. In: *Proceedings EUROCORR 2006, the European Corrosion Congress*, Maastricht, The Netherlands (25–28 September 2006), 11. European Federation of Corrosion.

116 Michvocik, M. and Zima, G. (2010). Effectiveness of NDT methods for corrosion/erosion monitoring and diagnostics, Appendix 15. In: *Minutes of EFC WP15 Corrosion in the Refinery Industry* Budapest, Hungary (22 June 2010), Slides Nos. 123–134. European Federation of Corrosion.

117 Mathiesen, L., Klingaard, S., Hilbert, L.R., and Andersen, A. (2010). Assessing the residual lifetime of district heating system, casestory from municipality of Kalundborg. In: *Proceedings EUROCORR 2010, the European Corrosion Congress*, Moscow, Russia (13–17 September 2010), Paper No. 9226, 5. European Federation of Corrosion.

118 Penkala, J.E., Fichter, J., and Ramachandran, S. (2010). Protection against microbiologically influenced corrosion by effective treatment and

monitoring during hydrotest shut-in. In: *NACE International Conference, CORROSION 2010*, Paper No. 10404, 17. Houston, TX: NACE International.
119 Rajasekar, A., Maruthamuthu, S., and Ting, Y.-P. (2008). Electrochemical behavior of *Serratia marcescens* ACE2 on carbon steel API 5L=X60 in organic/aqueous phase. *Industrial & Engineering Chemistry Research* 47: 6925–6932.
120 Strehblow, H.-H. and Foelske, A. (2003). The investigation of passive layers on Co with XPS and STM. In: *Proceedings EUROCORR 2003, the European Corrosion Congress*, Budapest, Hungary (28 September–2 October 2003), 1. European Federation of Corrosion.
121 Fang, H., Brown, B., and Nešić, S. (2010). High salt concentration effects on CO_2 corrosion and H_2S corrosion. In: *NACE International Conference, CORROSION 2010*, Paper No. 10276, 29. Houston, TX: NACE International.
122 Neufeld, A.K. and Cole, I.S. (1997). Using Fourier transform infrared analysis to detect corrosion products on the surface of metals exposed to atmospheric conditions. *Corrosion* 53 (10): 788–799.
123 Abraham, G.J., Kain, V., and Dey, G.K. (2009). MIC failure of type 316L seawater pipeline. *Materials Performance* 48 (1): 64–69.
124 Letelier, M.V., Lagos, G.E., and Reyes, A. (2007). Blue stains in domestic fixtures exposed to drinking water: chemical characterization. In: *Proceedings EUROCORR 2007, the European Corrosion Congress*, Freiburg, Germany (9–13 September 2007), 6. European Federation of Corrosion.
125 Bataillon, C., Martin, F.A., and Roy, M. (2007). Corrosion monitoring of carbon steel in pasty clayey mixture as function of temperature. In: *Proceedings EUROCORR 2007, the European Corrosion Congress*, Freiburg, Germany (9–13 September 2007), 16. European Federation of Corrosion.
126 Lyon, S.B., Bingham, R., and Mills, D.J. (2017). Advances in corrosion protection by organic coatings: what we know and what we would like to know. *Progress in Organic Coatings* 102: 2–7.
127 Videla, H.A. and Herrera, L.K. (2010). The use of synchrotron radiation in the characterization of corrosion products and materials: potential in MIC and biodeterioration. In: *NACE International Conference, CORROSION 2010*, Paper No. 10220, 11. Houston, TX: NACE International.
128 Xie, J., Simanzhenkov, V., Santos, B. et al. (2010). Corrosion of UNS S30400 stainless steel trays in an amine unit. In: *NACE International Conference, CORROSION 2010*, Paper No. 10187, 13. Houston, TX: NACE International.
129 Verona, C. and Higginson, R. (2009). Stress corrosion cracking of CrNi martensitic stainless steel for steam turbine blades. In: *Proceedings EUROCORR 2009, the European Corrosion Congress*, Nice, France (6–10 September 2009), Paper No. 7752, 7. European Federation of Corrosion.

130 El Hajj, H., Abdelouas, A., and Grambow, B. (2010). Biocorrosion of P235GH steel in column experiments simulating geological conditions of high-level radioactive waste disposal. In: *Proceedings EUROCORR 2010, the European Corrosion Congress*, Moscow, Russia (13–17 September 2010), Paper No. 4770, 1. European Federation of Corrosion.
131 El Kamel, M., Galtayries, A., Foulonneau, G. et al. High temperature corrosion of sulfided steels in naphthenic acid environments. In: *Proceedings EUROCORR 2010, the European Corrosion Congress*, Moscow, Russia (13–17 September 2010), Paper No. 9377, 22. European Federation of Corrosion.
132 Prasad, R. (2000). Assessment and control of MIC in the oil industry in the 20th century. In: *NACE International Conference, CORROSION 2000*, Paper No. 00405, 11. Houston, TX: NACE International.
133 Maxwell, S., Devine, C., Rooney, F., and Spark, I. (2004). Monitoring and control of bacterial biofilms in oilfield water handling systems. In: *NACE International Conference, CORROSION 2004*, Paper No. 04752, 16. Houston, TX: NACE International.
134 Little, B.J., Ray, R.I., and Wagner, P.A. (1999). Marine bacteria and localized corrosion on polymer coated steel: cause and effect. In: *NACE International Conference, CORROSION99*, Paper No. 183, 8. Houston, TX: NACE International.
135 Videla, H.A. and Herera, L.K. (2005). Microbiologically influenced corrosion: looking to the future. *International Microbiology* 8: 169–180.
136 Natarajan, K.A. (2008). Microbially-influenced corrosion. In: *Corrosion Science and Technology* (ed. U. Kamachi Mudali and B. Raj), 95–121. New Delhi, India: Narosa Publishing House.
137 de Jong, M.P. and Leferink, R.G.I. (2006). Corrosion monitoring on demand. In: *Proceedings EUROCORR 2006, the European Corrosion Congress*, Maastricht, The Netherlands (25–28 September 2006), 18. European Federation of Corrosion.
138 Buryak, A.K., Serdyuk, T.'y.M., and Uleanov, A.V. (2010). Mass-spectrometry for corrosion processes and corrosion inhibitors investigation. In: *Proceedings EUROCORR 2010, the European Corrosion Congress*, Moscow, Russia (13–17 September 2010), Paper No. 9238, 1. European Federation of Corrosion.
139 Groysman, A. and Groysman, O. (2002). Corrosion and philosophy. In: *Proceedings 15th International Corrosion Congress on Frontiers in Corrosion Science and Technology*, Granada, Spain (22–27 September 2002), Paper No. 482. Madrid: National Centre for Metallurgical Research.
140 Lavarde, C. (2010). Rightrax corrosion monitoring, Appendix 13. In: *Minutes of EFC WP15 Corrosion in the Refinery Industry*, Budapest, Hungary (22 June 2010), Slides Nos. 85–108. European Federation of Corrosion.
141 Rannou, G., Thierry, D., and LeBozec, N. (2010). Ultrasonic monitoring of steel corrosion during accelerated corrosion testing and outdoor field

exposures. In: *NACE International Conference, CORROSION 2010*, Paper No. 10170, 14. Houston, TX: NACE International.

142 Al-Barout, M., Al-Abdulhadi, A.A., Al-Mutairi, S. et al. (2010). Non-intrusive online multi-sensors for pipeline corrosion monitoring "field testing". In: *NACE International Conference, CORROSION 2010*, Paper No. 10174, 12. Houston, TX: NACE International.

143 Rommetveit, T., Johnsen, R., and Baltzersen, Ø. (2008). Using ultrasound measurements for real-time process control of pipelines and process equipment subjected to corrosion and/or erosion. In: *NACE International Conference, CORROSION 2008*, Paper No. 08285, 13. Houston, TX: NACE International.

144 Ho, K.S., Billson, D.R., and Hutchins, D.A. (2007). Ultrasonic Lamb wave tomography using scanned EMATs and wavelet processing. *Nondestructive Testing and Evaluation* 22 (1): 19–34.

145 Volker, A. and Mast, A. (2008). Permanent corrosion monitoring using guided waves. *The 17th World Conference on Nondestructive Testing*, Shanghai, China (25–28 October 2008), p. 8.

146 Pei, J., Yousuf, M.I., Degertekin, F.L. et al. (1995). Lamb wave tomography and its application in pipe erosion/corrosion monitoring. *Proceedings IEEE Ultrasonics Symposium* 755–758.

147 Mažeika, L., Kažys, R., Raišutis, R., and Šliteris, R. (2007). Ultrasonic guided wave tomography for the inspection of the fuel tanks floor. *4th International Conference on NDT*, Chania, Crete, Greece (11–14 October 2007), p. 8.

148 Leonard, K.R., Malyarenko, E.V., and Hinders, M.K. (2002). Ultrasonic Lamb wave tomography. *Inverse Problems* 18 (6): 1795.

149 Rivera, H., Carlos, C., Canto, J. et al. (2010). Guided waves methodology application in the analysis of pipeline integrity of docks along Pacific Coast and Mexico Gulf. In: *NACE International Conference, CORROSION 2010*, Paper No. 10067, 11. Houston, TX: NACE International.

150 Muravin, G., Muravin, B., and Lezvinsky, L. (2008). Acoustic emission non-destructive inspection method for cranes. In: *Proceedings of the International Conference Safety and Reliability of Welded Components in Energy and Processing Industry*, Graz, Austria (10–11 July 2008), Vol. 52, Special Issue – 2008, pp. 211–215.

151 BS EN 16018 (2011). *Non-destructive Testing. Terminology. Terms Used in Ultrasonic Testing with Phased Arrays*, 44. London: BSI.

152 NDT (2007). *Advances in Phased Array Ultrasonic Technology Applications*, 491. Olympus NDT.

153 GE Inspection Technologies. *Corrosion & Erosion*. https://www.gemeasurement.com/sites/gemc.dev/files/corrosion_erosion_inspection_brochure.pdf (accessed 27 October 2018).

154 Drinkwater, B.W. and Wilcox, P.D. (2006). Ultrasonic arrays for non-destructive evaluation: a review. *NDT&E International* 39 (7): 525–541.

155 Nondestructive Testing Handbooks. https://www.asnt.org/MajorSiteSections/Publications/NDT_Handbooks.aspx (accessed 27 October 2018).

156 Ciorau, P., Chartier, C., and Mair, K.(2008). A contribution of phased array ultrasonic technology (PAUT) to detection and sizing of stress corrosion cracks. *NDT.net – The e-Journal of Nondestructive Testing* 13(10): 9. http://www.ndt.net/article/v13n10/ciorau2.pdf (accessed 27 October 2018).

157 Ciorau, P., Pullia, L., Hazelton, T., and Daks, W. (2008). Phased array ultrasonic technology (PAUT) contribution to detection and sizing of microbially influenced corrosion (MIC) of service water systems, and shut down coolers heat exchangers in OPG CANDU stations. *NDT.net – The e-Journal of Nondestructive Testing* 13(10): 8. http://www.ndt.net/article/v13n10/ciorau1.pdf (accessed 27 October 2018).

158 Muravin, B. (2009). Acoustic emission science and technology (in Hebrew). *Journal of Building and Infrastructure Engineering of the Israeli Association of Engineers and Architects*, 10. http://www.docstoc.com/docs/10239348/Acoustic-Emission-Science-and-Technology (in English) (accessed 27 October 2018).

159 ASTM E1316-10a (2010). *Standard Terminology for Nondestructive Examinations*, 33. West Conshohocken, PA: ASTM International.

160 Huang, M., Jiang, L., Liaw, P.K. et al. (1998). Using acoustic emission in fatigue and fracture materials research. *JOM (Journal of Engineered Materials)* 50(11). https://www.tms.org/pubs/journals/jom/9811/huang/huang-9811.html (accessed 27 October 2018).

161 Muravin, G.B., Lezvinskaya, L.M., Makarova, N.O., and Pavlovskaya, G.S. (1990). *Problems in the Acoustic-Emission Diagnosis of the Corrosion Process (Review)*, 100–109. Plenum Publishing Corporation Translation from the journal *Defektoskopiya* (in Russian), 1990, No. 2, pp. 18–28.

162 Kim, Y.P., Fregonese, M., Mazille, H. et al. (2006). Study of oxygen reduction on stainless steel surfaces and its contribution to acoustic emission recorded during corrosion processes. *Corrosion Science* 48 (12): 3945–3959.

163 ASTM E1419-00 (2000). *Standard Test Method for Examination of Seamless, Gas-Filled, Pressure Vessels Using Acoustic Emission*, 6. West Conshohocken, PA: ASTM International.

164 ASNT (2005). *Acoustic Emission Testing, Nondestructive Testing Handbook*, 3e, vol. 6. Columbus, OH: American Society for Nondestructive Testing.

165 ANSI/ASNT CP-105-2006 (2006). *ASNT Standard Topical Outlines, for Qualification of Nondestructive Testing Personnel*. Columbus, OH: American Society for Nondestructive Testing.

166 ASNT (2001). *Recommended Practice No. SNT-TC-1A – Non-Destructuve Testing*, 165. Columbus, OH: American Society for Nondestructive Testing.

167 EN 1330-9:2009 (2009). *Non-destructive Testing. Terminology. Terms Used in Acoustic Emission Testing*, 30. London: BSI.

168 EN 13477-1:2001 (2001). *Non-destructive Testing. Acoustic Emission. Equipment Characterization. Equipment Description*, 12. London: BSI.
169 EN 13477-2:2001 (2001). *Non-destructive Testing. Acoustic Emission. Equipment Characterization. Verification of Operating Characteristic*, 18. London: BSI.
170 EN 14584:2005 (2005). *Non-destructive Testing – Acoustic Emission – Examination of Metallic Pressure Equipment During Proof Testing – Planar Location of AE Sources*, 15. London: BSI.
171 ASTM E 569-07 *Standard Practice for Acoustic Emission Monitoring of Structures during Controlled Stimulation*. West Conshohocken, PA: ASTM International.
172 ASTM E 1211-07 *Practice for Leak Detection and Location Using Surface-Mounted Acoustic Emission Sensors*. West Conshohocken, PA: ASTM International.
173 ASTM E 1932-07 *Guide for Acoustic Emission Examination of Small Parts*. West Conshohocken, PA: ASTM International.
174 Moran, G.C. and Labine, P. (eds.) (1986). *Corrosion Monitoring in Industrial Plants Using Nondestructive Testing and Electrochemical Methods*, ASTM STP 908, 515. West Conshohocken, PA: ASTM.
175 Rettig, T.W. and Felsen, M.J. (1976). Acoustic emission method for monitoring corrosion reactions. *Corrosion* 32 (4): 121–126.
176 Mansfield, F. and Stocker, P.J. (1979). Acoustic emission from corroded electrodes. *Corrosion* 35 (12): 541–544.
177 Rettig, T.W. and Felsen, M.J. (1976). Acoustic emission method for detection and monitoring of corrosion. US Patent 3,946,600.
178 Fenn, R. and Condello, C. (1986). Monitoring corrosion by acoustic emission. *Nondestructive Testing and Evaluation* 2 (5 & 6): 169–176.
179 Jaubert, L., Fregonese, M., Caron, D. et al. (2005). On the opportunity to use non intrusive acoustic emission recordings for monitoring uniform corrosion of carbon steel and austenitic stainless steel in acid and neutral media Insight – non-destructive testing and condition monitoring. *The Journal of the British Institute of Non-Destructive Testing* 47 (8): 465–472.
180 Jaubert, L., Fregonese, M., Caron, D. et al. (2003). Correlation between electrochemical determination of corrosion rate and non intrusive acoustic emission recordings for monitoring uniform corrosion of mild steel in acid and neutral media. In: *Proceedings EUROCORR 2003, the European Corrosion Congress*, Budapest, Hungary (28 September–2 October 2003), Paper No. 57. European Federation of Corrosion.
181 Lewis, M. (2002). Acoustic monitoring of a prestressed concrete tank. *Materials Performance* 41 (7): 56–59.
182 Young-Pil, K., Fregonese, M., Mazille, H. et al. (2003). Sensitivity of acoustic emission for detecting crevice corrosion on AISI 304L austenitic stainless steel. In: *Proceedings EUROCORR 2003, the European Corrosion Congress*,

Budapest, Hungary (28 September–2 October 2003), 9. European Federation of Corrosion.

183 Wevers, M., Van Dijck, G., Desadeleer, W. et al. (2004). Acoustic emission for on-line monitoring of damage in various application fields. In: DGZfP-Proceedings BB 90-CD, EAGAE, Lecture 59, 587–595.

184 Kharebov, V. (2010). Automated systems of complex corrosive monitoring and prospects of application of acoustic emission method in their structure. In: *Proceedings EUROCORR 2010, the European Corrosion Congress*, Moscow, Russia (13–17 September 2010), Paper No. 9241, 8. European Federation of Corrosion.

185 Lee, C.K., Scholey, J.J., Worthington, S.E. et al. (2008). Acoustic emission from pitting corrosion in stressed stainless steel plate. *Corrosion Engineering, Science and Technology* 43 (1): 54–63.

186 Fregonese, M., Idrissi, H., Mazille, H. et al. (2001). Initiation and propagation steps in pitting corrosion of austenitic stainless steels: monitoring by acoustic emission. *Corrosion Science* 43: 627–641.

187 Fregonese, M., Idrissi, H., Mazille, H. et al. (2001). Monitoring pitting corrosion of AISI 316L austenitic stainless steel by acoustic emission technique: choice of representative acoustic parameters. *Journal of Materials Science* 36: 557–563.

188 Fregonese, M., Idrissi, H., Mazille, H. et al. (2001). Monitoring pitting corrosion of 304L and 316L austenitic stainless steels by acoustic emission: influence of the pit morphology. In: *Marine Corrosion of Stainless Steels* (ed. D. Feron) EFC Publications No. 33, 292–298. London: IOM Communications.

189 Krakowiak, S. and Darowicki, K. (2009). Electrochemical and acoustic emission studies of aluminum pitting corrosion. *Journal of Solid State Electrochemistry* 13 (11): 1653–1657.

190 Kim, Y.P., Fregonese, M., Mazille, H. et al. (2005). Study of the respective roles of anodic and cathodic reactions in the acoustic emission recorded during crevice corrosion of 304L austenitic stainless steel. *Corrosion Engineering, Science and Technology* 40 (4): 301–312.

191 Kim, Y.P., Fregonese, M., Mazille, H. et al. (2003). Ability of acoustic emission for detection and monitoring of crevice corrosion of 304L austenitic stainless steel. *NDT&E International* 36 (8): 553–562.

192 Yonezu, A., Cho, H., Ogawa, T., and Takemoto, M. (2006). Simultaneous monitoring of acoustic emission and corrosion potential fluctuation for mechanistic study of chloride stress corrosion cracking. *Key Engineering Materials* 321–323), *Advanced Nondestructive Evaluation I*: 254–259.

193 Janík, P. and Sajdl, P. (2009). Acoustic emission during hydrogen charging and desorption process. In: *Proceedings EUROCORR 2009, the European Corrosion Congress*, Nice, France (6–10 September 2009), Paper No. 8148, 10. European Federation of Corrosion.

194 Carboni, C., Peyre, P., Fregonese, M. et al. (2001). Influence of laser-peening and shot-peening on the pitting corrosion behavior of 316L steel: acoustic emission monitoring of the pit propagation. In: *Proceedings of SMT14* (ed. T.S. Sudarshan and M. Jeandin), 381–388. Materials Park, OH: ASM International.
195 Smanio, V., Cassagne, T., Ropital, F. et al. (2008). Evaluation of hydrogen embrittlement induced damages in steels using acoustic emission. In: *NACE International Conference, CORROSION 2008*, Paper No. 08111, 14. Houston, TX: NACE International.
196 Smanio-Renaud, V., Cassagne, T., Ropital, F. et al. (2007). Evaluation of hydrogen induced cracking damages in steels using acoustic emission. In: *Proceedings EUROCORR 2007, the European Corrosion Congress*, Freiburg, Germany (9–13 September 2007). European Federation of Corrosion.
197 Smanio, V., Kittel, J., Fregonese, M. et al. (2008). Evaluation of hydrogen embrittlement induced damages in steels using acoustic emission. In: *NACE International Conference*, Paper No. 08111. Houston, TX: NACE International.
198 Amami, S., Marchand, P., Duval, S. et al. (2003). Detection by acoustic emission technique of the corrosion stress cracking within steel structures in H_2S containing mediums. In: *Proceedings EUROCORR 2003, the European Corrosion Congress*, Budapest, Hungary (28 September–2 October 2003), Paper No. 156. European Federation of Corrosion.
199 Smanio, V., Fregonese, M., Kittel, J. et al. (2010). Wet hydrogen sulfide cracking of steel monitoring by acoustic emission: discrimination of AE sources. *Journal of Materials Science* 45 (20): 5534–5542.
200 Smanio, V., Fregonese, M., Kittel, J. et al. (2008). Wet hydrogen sulfide cracking monitoring by acoustic emission. In: *Proceedings EUROCORR 2008, The European Corrosion Congress*, Edinburgh, Scotland (7–11 September 2008), 13. European Federation of Corrosion.
201 Yuyama, S. and Nishida, T. (2003). Acoustic emission evaluation of corrosion damages in buried pipes of refinery. *Journal of Acoustic Emission* 21: 187–196.
202 Miller, R.K. and McIntire, P. (eds.) (1987). *Nondestructive Testing Handbook*, vol. 5, 58–61. Columbus, OH: American Society for Nondestructive Testing, Inc., Acoustic Emission Testing.
203 Voronin, V.P., Shul'te, A.Y., Sukhinin, Y.V. et al. (1990). Acoustic emission inspection of stress corrosion of structural steels. *Materials Science* 26 (2): 230–233. (Translated from Fiziko-Khimicheskaya Mekhanika Materialov, March–April 1990, No. 2, pp. 115–117, in Russian).
204 Clough, R.B., Chang, J.C., and Travis, J.P. (1981). Acoustic emission signatures and source microstructure using indentation fatigue and stress corrosion cracking in aluminum alloys. *Scripta Metallurgica* 15: 417–422.

205 Munch, E., Duisabeau, L., Fregonese, M. et al. (2004). Acoustic emission detection of environmentally assisted cracking in Zircaloy-4 alloy. In: *Proceedings EUROCORR 2004, the European Corrosion Congress*, Nice, France (12–16 September 2004). European Federation of Corrosion.

206 Moorthy, V., Jayakumar, T., and Raj, B. (1996). Influence of microstructure on acoustic emission behavior during stage 2 fatigue crack growth in solution annealed, thermally aged and weld specimens of AISI type 316 stainless steel. *Materials Science and Engineering* A212: 273–280.

207 Almeida, A. and Hill, E.v.K. (1995). Neural network detection of fatigue crack growth in riveted joints using acoustic emission. *Materials Evaluation* 53 (1): 76–82.

208 Scala, M. and Cousland, S.M. (1983). Acoustic emission during fatigue crack propagation in the aluminium alloys 2024 and 2124. *Materials Science and Engineering* 61: 211–218.

209 Wang, Z.F., Zhu, Z., and Ke, W. (1991). Behavior of acoustic emission for low-strength structural steel during fatigue and corrosion fatigue. *Metallurgical and Materials Transactions A* 22A: 2677–2680.

210 Yuyama, S., Kishi, T., and Hisamatsu, Y. (1984). Fundamental aspects of AE monitoring on corrosion fatigue processes in austenitic stainless steel. *Journal of Materials for Energy Systems, American Society for Metals* 5 (4): 212–221.

211 Yuyama, S., Kishi, T., and Hisamatsu, Y. (1983). AE analysis during corrosion, stress corrosion cracking and corrosion fatigue processes. *Journal of Acoustic Emission* 2 (1/2): 71–93.

212 Bellenger, F., Mazille, H., and Idrissi, H. (2002). Use of acoustic emission technique for the early detection of aluminum alloys exfoliation corrosion. *NDT&E International* 35: 385–392.

213 Yuyama, S., Li, Z.W., Yoshizawa, M. et al. (2001). Evaluation of fatigue damage in reinforced concrete slab by acoustic emission. *NDT & E International* 34: 381–387.

214 Cole, P. and Watson, J. (2005). Acoustic emission for corrosion detection. In: *Proceedings of the 3rd Middle East Nondestructive Testing Conference & Exhibition*, Bahrain, Manama (27–30 November 2005), 7. NDT.net.

215 Proverbio, E. (2009). Evaluation of deterioration in reinforced concrete structures by AE technique. In: *Proceedings EUROCORR 2009, the European Corrosion Congress*, Nice, France (6–10 September 2009), Paper No. 8398, 14. European Federation of Corrosion.

216 Coulter, J.E., Robertson, M.O., and Stevens, D.M. (1996). Acoustic emission for detection of corrosion under insulation. US Patent 5,526,689.

217 Bakirov, M.B., Povarov, V.P., Gromov, A.F., and Levchuk, V.I. (2015). Development of a technology for continuous acoustic emission monitoring of in-service damageability of metal in safety-related NPP equipment. *Nuclear Energy and Technology* 1 (1): 32–36.

218 Muravin, G., Adams, C.W., Muravin, B., and Lezvinsky, L. (2000). Revealing, identifying, and evaluating failures in high pressure reactors by acoustic emission image recognition method. In: Progress in Acoustic Emission X, The Japanese Society for NDI, 243–248.

219 Muravin, G., Muravin, B., Kralj, S. et al. (2007). Investigation of weld joints of LPG pressure vessel made from A516 Grade 70 steel. In: *Proceedings of the IIW (International Institute of welding) International Conference Welding & Materials Technical, Economic and Ecological Aspects*, Dubrovnik&Cavtat, Croatia (1–8 July 2007), 465–474.

220 Muravin, G., Muravin, B., Mizrahi, I., and Lezvinsky, L. (2008). The possibility of revealing and assessment age related degradation and industrial flaws in high chromium steels piping and equipment by quantitative acoustic emission non-destructive method. *Scientific Israel – Technological Advantages* 10 (2): 102–118.

221 Muravin, G., Muravin, B., and Lezvinsky, L. Revealing, identifying, assessing and monitoring flaws in water cooling system of operating bwr reactors by quantitative acoustic emission NDI method. In: *Proceeding of the International Conference & Exhibition on Pressure Vessels and Piping, "OPE 2006 – CHENNAI"*, Chennai, India (7–9 February 2006), 13.

222 Muravin, G., Muravin, B., and Lezvinsky, L. (2009). Application of quantitative acoustic emission non-destructive inspection technology for testing welded structures and components. In: *Proceedings of the 7th European Congress on Joining Technology EUROJOIN 7*, Venezia Lido (21–22 May 2009), 1–9.

223 Muravin, G., Muravin, B., and Lezvinsky, L. (2008). Acoustic emission inspection of operating steam piping for detection and assessment of caustic corrosion cracking. In: Progress in Acoustic Emission XIV, The Japanese Society for NDI, 477–486.

224 Muravin, G., Adams, C.W., Muravin, B., and Lezvinsky, L. (2000). Acoustic emission revealing and danger level evaluation of stress corrosion cracking in stainless steel pipes. In: Progress in Acoustic Emission X, The Japanese Society for NDI, 135–140.

225 Muravin, G., Muravin, B., and Lezvinsky, L. (2006). Evaluating the remaining lifetime of tubes undergone stress corrosion cracking by quantitative acoustic emission NDI and numerical method. In: *Proceedings of the Fourth International Conference on NDE in Relation to Structural Integrity for Nuclear and Pressurized Components* (6–8 December 2004), 715–723. London, UK: Institute for Energy.

226 Murakami, Y. and Yuyama, S. (1996). Acoustic emission evaluation of structural integrity in reinforced concrete beams deteriorated due to corrosion of reinforcement. *Progress in AE VIII (JSNDI), Proceedings of 13th International AE Symposium*, Nara, Japan (27–30 November 1996), pp. 217–224.

227 Muravin, G., Muravin, B., Lezvinsky, L. et al. (2001). Acoustic emission monitoring of water-cooling system of BWR reactor. In: *Proceedings 3rd International Conference on NDE in Relation to Structural Integrity for Nuclear and Pressurised Components*, Seville, Spain (14–16 November 2001), 9.

228 Muravin, G., Adams, C.W., Muravin, B., and Lezvinsky, L. (2004). Revealing systems of flaws in high pressure reactors operating under dynamic pulse loading. In: *Proceedings of SPIE, Noise and Information in Nanoelectronics, Sensors, and Standards II*, Maspalomas, Gran Canaria, Spain (26–28 May 2004), vol. 5472, 409–419.

229 Seah, K.H.W., Lim, K.B., Chew, C.H., and Teoh, S.H. (1993). The correlation of acoustic emission with rate of corrosion. *Corrosion Science* 34 (10): 1707–1713.

230 Yanting, X., Ding, S., Liu, F. et al. (2008). Acoustic emission in-service detection of cryogenic storage tank floors. In: *17th World Conference on Nondestructive Testing*, Shanghai, China (25–28 October 2008), 4.

231 Baeté, C., Straetmans, S., Buelens, C., and Surgeon, M. (2004). Non intrusive inspection of aboveground storage tanks. In: *Proceedings EUROCORR 2004, the European Corrosion Congress*, Nice, France (12–16 September 2004), 1. European Federation of Corrosion.

232 Ropital, F., Kittel, J., and Dascotte, P. (2006). Method combining acoustic emission and thermogravimetry for an improved evaluation of the high temperature corrosion. In: *Proceedings EUROCORR 2006, the European Corrosion Congress*, Maastricht, The Netherlands (25–28 September 2006), 9. European Federation of Corrosion.

233 Daniel Efird, K. (1986). Testing coatings using acoustic emission technology. In: *Corrosion Monitoring in Industrial Plants Using Nondestructive Testing and Electrochemical Methods* (ed. G.C. Moran, P. Labine and ASTM STP 908), 515. West Conshohocken, PA: ASTM.

234 Fregonese, M., Jaubert, L., and Cètre, Y. (2007). Contribution of Acoustic Emission technique for monitoring damage of rubber coating on metallic surfaces. Comparison with electrochemical measurements. *Progress in Organic Coatings* 59 (3): 239–243.

235 EFC (European Federation of Corrosion) (2004). *Working Party 15 Meeting: Corrosion Under Insulation, Guideline* (15 September 2004).

236 Shaikh, H., Sivaibharasi, N., Sasi, B. et al. (2006). Use of eddy current testing method in detection and evaluation of sensitization and intergranular corrosion in austenitic stainless steels. *Corrosion Science* 48: 1462–1482.

237 Almadani, F., Zaid, M., Gaydecki, P., and Miller, G. (2010). Signal profiles and images of corroded steel bars in concrete generated by inductive scanning technique using heterodyning sensor. In: *NACE International Conference, CORROSION 2010*, Paper No. 10167, 11. Houston, TX: NACE International.

238 Crouzen, P.C.N., Verstijnen, W., Munns, I.J., and Hulsey, R.C. (2006). Application of pulsed eddy current corrosion monitoring in refineries and oil production facilities. In: *NACE International Conference, CORROSION 2006*, Paper No. 06312, 9. Houston, TX: NACE International.

239 Mathers, R. (2005). Treatment of high acid crudes and the methods used in refineries to mitigate naphthenic acid corrosion. In: *Proceedings EUROCORR 2005, the European Corrosion Congress*, Lisbon, Portugal (4–8 September 2005), 11. European Federation of Corrosion.

240 van Andel, P.W. (2001). Eddy current inspection technique. US Patent 6,291,992.

241 Crouzen, P.C.N. (2003). Method for inspecting an object of electrically conducting material. US Patent 6,570,379.

242 Wikswo, J.P. (1997). Squid magnetometers for studying corrosion and corrosion protection in aircraft aluminum. In: *NACE International Conference, CORROSION/97*, Paper No. 97293, 17. Houston, TX: NACE International.

243 Goldfine, N. and Greig, N.A. (1994). Using electromagnetic sensors (magnetometers and dielectrometers) to detect corrosion beneath and moisture within paint coatings on aircraft. In: *NACE International Conference, CORROSION/94*, Paper No. 94353. Houston, TX: NACE International.

244 Atherton, L., Laursen, P., and Siebert, M.A. (1993). Small – diameter MFL detector overcoming technical hurdles. *Pipe Line Industry* 76: 69–73.

245 Csizinszky, L. (2003). New NDT technologies in pipeline and tank inspection. In: *Proceedings EUROCORR 2003, the European Corrosion Congress*, Budapest, Hungary (28 September–2 October 2003), vol. 5. European Federation of Corrosion.

246 Borjailah, W., Al-Oadah, A., Beuker, T. et al. (2007). In-line inspection with high resolution electro-magnetic acoustic transducer (EMAT) technology crack detection and coating disbondment. In: *NACE International Conference, CORROSION 2007*, Paper No. 07131, 7. Houston, TX: NACE International.

247 Amir, N., Barzelay, O., Yefet, A., and Pechter, T. (2008). Condenser tube examination using acoustic pulse reflectometry. In: *Proceedings of POWER2008, ASME Power 2008*, Orlando, FL (22–24 July 2008), 5.

248 Sheppard, K., Zagrai, A., and Donskoy, D. (2007). A non-linear acoustic, vibro-modulation, technique for the detection and monitoring of contact-type defects, including those associated with corrosion. *Corrosion Reviews* 25 (1–2): 81–96.

249 Meo, M., Zumpano, G., and Polimeno, U. (2007). Corrosion identification on an aluminium plate-like structure by monitoring the wave propagation phenomena. *Corrosion Reviews* 25 (1–2): 213–232.

250 Poonguzhali, A., Shaikh, H., Dayal, R.K., and Khatak, H.S. (2008). A review on degradation mechanism and life estimation of civil structures. *Corrosion Reviews* 26 (4): 215–294.

251 Niccolls, E.H., Gallon, A.E., and Yamamoto, K. (2008). Systematic integration of advanced NDE and corrosion monitoring for improved refinery reliability. In: *NACE International Conference, CORROSION 2008*, Paper No. 08280, 11. Houston, TX: NACE International.

252 Twomey, M. (1997). Inspection techniques for detecting corrosion under insulation. *Materials Evaluation* 55: 129–132.

253 Yoshikawa, H. and Matsui, T. (2008). A sampling method and data evaluation of archaeological samples to support long-term corrosion prediction. In: *NACE International Conference, CORROSION 2008*, Paper No. 08RTS07, 10. Houston, TX: NACE International.

254 McConnell, M.D. (2010). Application of thermography in diagnosing corrosion and material issues in today's refinery. In: *NACE International Conference, CORROSION 2010*, Paper No. 10362, 14. Houston, TX: NACE International.

255 Han, J.-S. and Park, J.-H. (2004). Detection of corrosion steel under organic coating by infrared photography. *Corrosion Science* 46: 787–793.

256 Jönsson, M., Bo, R., and Annergren, I. (2009). The use of infrared thermography in the corrosion science area. In: *Proceedings EUROCORR 2009, the European Corrosion Congress*, Nice, France (6–10 September 2009), Paper No. 7983, 13. European Federation of Corrosion.

257 ASM Handbook (2018). *Nondestructive Evaluation of Materials*, vol. 17. USA: ASM International.

258 Vagn Hansen, J. (1998). Non-destructive metallography used on-site, possibilities and experiences. *NDT* 3(10). http://www.ndt.net/article/ecndt98/chemical/107/107.htm#0 (accessed 30 October 2018).

259 Zuljan, D. and Grum, J. (2005). Non-destructive metallographic analysis of surface and microstructures by means of replicas. In: *The 8th International Conference of the Slovenian Society for Non-destructive Testing "Application of Contemporary Non-Destructive Testing in Engineering"*, Portorož, Slovenia (1–3 September 2005), 359–368.

260 Artamonov, V.V. and Artamonov, V.P. (2007). Aluminum replicas for optical metallography. *Journal of Materials Engineering and Performance* 16 (1): 52–54.

261 Groysman, A., Feldman, B., Kaufman, A., and Balali, R. (2011). Hydrogen damage and prevention in oil refinery. In: *NACE International Conference, CORROSION 2011*, Paper No. 11295, 15. Houston, TX: NACE International.

262 Helgesen, T., Tjernæs, A., Heiberg, G., and Heier, E. (2005). Failure investigation and condition assessment using field metallography. *Engineering Failure Analysis* 12 (6): 974–985.

263 ASTM E 1351-01 (2006) (2006). *Standard Practice for Production and Evaluation of Field Metallographic Replicas*, Book of Standards, vol. 03.01, 6. West Conshohocken, PA: ASTM.

264 ASTM A956-06 (2006). *Standard Test Method for Leeb Hardness Testing of Steel Products*, Book of Standards, vol. 01.05, 12. West Conshohocken, PA: ASTM.

265 Kuzbozhev, A.S., Aleksandrov, Y.V., and Solovej, V.O. (2010). System of monitoring of gas pipelines on stress corrosion cracking. In: *Proceedings EUROCORR 2010, the European Corrosion Congress*, Moscow, Russia (13–17 September 2010), Paper No. 4700, 7. European Federation of Corrosion.

266 Mears, D. (2009). Handheld x-ray technology identifies alloy composition of critical metal piping in petrochemical plants. *Materials Performance* 48 (12): 18–20.

267 Oxford Instruments. https://www.oxinst.com/products (accessed 30 October 2018).

268 Zhou, Z., Zhou, K., Hou, X., and Luo, H. (2005). Arc/spark optical emission spectrometry: principles, instrumentation, and recent applications. *Applied Spectroscopy Reviews* 40 (2): 165–185.

269 Rigual, Y., Hernández, S., and Biomorgi, J. (2004). New device for corrosion monitoring and flow effect evaluation in oil field. In: *Proceedings EUROCORR 2004, the European Corrosion Congress*, Nice, France (12–16 September 2004), 9. European Federation of Corrosion.

270 Groysman, A. (2004). Corrosion problems and their solutions in the oil refining industry. In: *Proceedings EUROCORR 2004, the European Corrosion Congress*, Nice, France (12–16 September 2004), 9. European Federation of Corrosion.

271 Lunarska, E. and Nikiforov, K. (2008). Hydrogen degradation of refinery and electric power installations. *Corrosion Reviews* 26 (2–3): 173–213.

272 NACE International Publication 35201 (2001). *Technical Report on the Application and Interpretation of Data from External Coupons Used in the Evaluation of Cathodically Protected Metallic Structures*, Item No. 24213, 20. Houston, TX: NACE International.

273 Groysman, A. (1994). Corrosion monitoring at the oil refinery. In: *Second NACE Asian Corrosion Conference*, Paper No. 1005. Singapore: NACE.

274 Groysman, A. and Hiram, A. (1997). Corrosion monitoring and control in refinery process units. In: *NACE International Conference, CORROSION/97*, Paper No. 512, 25. Houston, TX: NACE.

275 Eaton, P. (2006). Effect of non-neutralizer corrosion inhibitors and iron sulfide on corrosion/erosion in refinery overheads. In: *Proceedings EUROCORR 2006, the European Corrosion Congress*, Maastricht, The Netherlands (25–28 September 2006), 12. European Federation of Corrosion.

276 Groysman, A. (1996). Corrosion monitoring in the oil refining industry. In: *Proceedings of the 13th International Corrosion Congress*, Melbourne, Australia (25–29 November 1996). Australasian Corrosion Association, Inc.
277 Groysman, A. (1995). Corrosion monitoring in the oil refinery. *International Conference on Corrosion in Natural and Industrial Environments: Problems and Solutions*, Italy (23–25 May 1995), Paper No. 7.
278 Nørskov, L., Richter, S., and Thorarinsdottir, R.I. (2003). Comparative studies of electrochemical methods in sulfide containing water distribution systems. In: *Proceedings EUROCORR 2003, the European Corrosion Congress*, Budapest, Hungary (28 September–2 October 2003), 8. European Federation of Corrosion.
279 Wang, H. and Wylde, J. (2010). Corrosion inhibitor development for slight sour environment with oxygen intrusion issue. In: *NACE International Conference, CORROSION 2010*, Paper No. 10275, 21. Houston, TX: NACE International.
280 Wang, S., Du, C., Li, X. et al. (2015). Field corrosion characterization of soil corrosion of X70 pipeline steel in a red clay soil. *Progress in Natural Science: Materials International* 25 (3): 242–250.
281 Freitas, D.S. et al. (2010). Methodology for evaluation of soil corrosivity in buried pipelines for different regions of Rio de Janeiro. In: *Proceedings EUROCORR 2010, the European Corrosion Congress*, Moscow, Russia (13–17 September 2010), Paper No. 9328, 13. European Federation of Corrosion.
282 Nielsen, L.V. and Neilsen, K.V. (2003). Differential ER-technology for measuring degree of accumulated corrosion as well as instant corrosion rate. In: *NACE International Conference, CORROSION 2003*, Paper No. 03443, 13. Houston, TX: NACE International.
283 CORMON (2004). *Development in Real-time Corrosion Monitoring – Recent Advances in Techniques and Management, Display and Correlation of Data*. Report by CORMON, Business Briefing: Exploration & Production: The Oil & Gas Review, pp. 1–4.
284 Prosek, T., Kouril, M., Dubus, M. et al. (2013). Real-time monitoring of indoor air corrosivity in cultural heritage institutions with metallic electrical resistance sensors. *Studies in Conservation* 58 (2): 117–128.
285 Abdulhadi, A., Al-Subaie, M.F., Al-Zahrani, A.M., and Al-Qarni, M.M. (2007). Field and laboratory experiences with advanced electrical resistance online corrosion monitoring system. In: *NACE International Conference, CORROSION 2007*, Paper No. 07265, 10. Houston, TX: NACE International.
286 Carl Bovankovich, J. (1994). On-line corrosion monitoring for process plant control. *Materials Performance* 33 (11): 57–60.
287 Braaten, N.A. and Johnsen, R. (2000). Six years experience with an intrusive erosion monitoring system used subsea and topside. In: *NACE International*

Conference, CORROSION 2000, Paper No. 00097, 10. Houston, TX: NACE International.
288 Yang, L., Sun, X., and Steward, F. (1999). An on-line electrical resistance corrosion monitor for studying flow assisted corrosion of carbon steel under high-temperature and high-pressure conditions. In: *NACE International Conference, CORROSION99*, Paper No. 459, 16. Houston, TX: NACE International.
289 Thorarinsdottir, R.I. and Hilbert, L.R. (2004). Online real-time, corrosion monitoring in district heating systems. News from DBDH (February), 3.
290 Groysman, A. (2005). Anti-corrosion management and environment at the oil refining industry. In: *Proceedings of the International Conference on Corrosion CORCON2005*, Chennai, India (28–30 November 2005), 18. NACE International.
291 Brown, G.K., Davies, J.R., and Hemblade, B.J. (2000). Real time metal loss internal monitoring. In: *NACE International Conference, CORROSION2000*, Paper No. 278, 16. Houston, TX: NACE International.
292 Paiva, G.J.M. and Cobucci, J.C.S. (2007). Monitoring naphthenic acid corrosion on-line. In: *NACE International Conference, CORROSION 2007*, Paper No. 07269, 18. Houston, TX: NACE International.
293 Welsh, R.A. and Benfield, J. (2006). Environmental protection through automated remote monitoring of fuel storage tank bottoms using electrical resistance probes. *Materials Performance* 45 (3): 38–40.
294 Marshakov, A., Petrunin, M., and Ignatenko, V. (2010). Monitoring of external corrosion of underground pipelines. In: *Proceedings EUROCORR 2010, the European Corrosion Congress*, Moscow, Russia (13–17 September 2010), Paper No. 9673, 19. European Federation of Corrosion.
295 Esaklul, K.A. (2007). Perspective on the needs of online, real-time monitoring to meet future asset integrity management requirements. In: *NACE International Conference, CORROSION 2007*, Paper No. 07264, 8. Houston, TX: NACE International.
296 Ridd, B., Blakset, T.J., and Queen, D. (1998). Field trails for corrosion inhibitors selection and optimization, using a new generation of electrical resistance probes. In: *NACE International Conference, CORROSION/98*, Paper No. 78, 6. Houston, TX: NACE International.
297 Esaklul, K. and Ballard, A. (2007). Challenges in the design of corrosion and erosion monitoring for a deepwater subsea equipment – stretching the limits of technology. In: *NACE International Conference, CORROSION 2007*, Paper No. 07338, 8. Houston, TX: NACE International.
298 Farrell, D.M. (1993). On-line monitoring of fireside corrosion in power plant. In: *Proceedings 12th International Corrosion Congress*, vol. 12, 4131–4140. Houston, TX: NACE International.
299 Taveira, L.V., Sagüés, A.A., Lopez-Sabando, J., and Joseph, B. (2008). Detection of corrosion of post tensioned strands in grouted assemblies.

In: *NACE International Conference, CORROSION 2008*, Paper No. 08398, 19. Houston, TX: NACE International.

300 Hernandez, J., Kane, R., Briegel, K., and Clark, R. (2008). Real-time, online corrosion monitoring & process optimization through the chemical plant control system. In: *NACE International Conference, CORROSION 2008*, Paper No. 08281, 14. Houston, TX: NACE International.

301 Farrell, D.M. and Robbins, B.J. (2003). On-line corrosion and thermal mapping of industrial plant by non-intrusive means. In: *Proceedings EUROCORR 2003, the European Corrosion Congress*, Budapest, Hungary (28 September–2 October 2003), 1. European Federation of Corrosion.

302 Niblock, T., Ooi, T.K., and Marotta, S.A. (2007). Optimizing sensors, systems and performance. *Corrosion Reviews* 25 (1–2): 195–212.

303 Niblock, T., Surangalikar, H.S., Morse, J. et al. (2004). AMSN (arrayed multiple sensor networks) for material and structural prognostics. *Proceedings of the Materials Science & Technology*, New Orleans, LA (September 2004).

304 Niblock, T., Surangalikar, H.S., Morse, J. et al. (2002). Development of a commercial micro corrosion monitoring system. *Proceedings from SPIE Smart Materials 2002, Melbourne, Australia* (September 2002).

305 Richez, M. (2010). Crack on the effluent nozzle of a HDS reactor, Appendix 7. In: *Minutes of EFC WP15 Corrosion in the Refinery Industry*, Budapest, Hungary (22 June 2010), Slides Nos. 47–52. European Federation of Corrosion.

306 Kane, R., Eden, D., Amidi, S., and Delve, D. (2008). Implementation of real-time corrosion monitoring with industrial process control and automation. In: *NACE International Conference, CORROSION 2007*, Paper No. 07268, 16. Houston, TX: NACE International.

307 Johnson, D.A., Townsend, G.S., and Abraham, T.K. (2006). Causes, consequences and remedies of operational variability in cooling tower operation and treatment. In: *Proceedings EUROCORR 2006, the European Corrosion Congress*, Maastricht, The Netherlands (25–28 September 2006), 20. European Federation of Corrosion.

308 ASTM Standards (2008). *Water and Environmental Technology*, vol. 11.01, 1082. West Conshohocken, PA: ASTM.

309 Eaton, A.D., Clesceri, L.S., Rice, E.W., and Greenberg, A.E. (eds.) (2005). *Standard Methods for the Examination of Water and Wastewater*, 21e, 1368. Washington, DC: American Water Work Association.

310 BETZ (1980). *Handbook of Industrial Water Conditioning*, 8e, 349–420. Trevose, PA: BETZ Laboratories, Inc.

311 NACE Standard TM0173-2005 (2005). *Methods for Determining Quality of Subsurface Injection Water Using Membranes Filters*, Item No. 21205, 19. Houston, TX: NACE International.

312 NACE Standard MR 0175/ISO 15156-1 (2001). *Petroleum and Natural Gas Industries – Materials for Use in H_2S-Containing Environments in Oil and Gas Production – Part 1: General Principles for Selection of Cracking-resistant Materials*, Item No. 21306, 16. Houston, TX: NACE International.

313 Panova, A.A., Pantano, P., and Walt, D.R. (1997). In-situ fluorescence imaging of localized corrosion with a pH-imaging sensitive fiber. In: *NACE International Conference, CORROSION/97*, Paper No. 314, 4. Houston, TX: NACE International.

314 Panova, A.A., Pantano, P., and Walt, D.R. (1997). In-situ fluorescence imaging of localized corrosion with a pH-sensitive imaging fiber. *Analytical Chemistry* 69: 1635–1641.

315 Tomin, V.P., Silinskaya, Y.N., and Korchevin, N.A. (2010). Peculiarities of the corrosion in cat cracker and delayed coker gasoline distillates. In: *Proceedings EUROCORR 2010, the European Corrosion Congress*, Moscow, Russia (13–17 September 2010), Paper No. 9242, 11. European Federation of Corrosion.

316 Latyshev, A.A. and Suranov, G.I. (2010). The analysis of oils, fuel and operating fluids spectrometric methods. In: *Proceedings EUROCORR 2010, the European Corrosion Congress*, Moscow, Russia (13–17 September 2010), Paper No. 4673, 12. European Federation of Corrosion.

317 Varga, Á., Nagy, L., Izquierdo, J. et al. (2011). Development of solid contact micropipette Zn-ion selective electrode for corrosion studies. *Analytical Letters* 44 (18): 2876–2886.

318 Nie, M., Wharton, J.A., Cranny, A.W. et al. (2008). Crevice corrosion solution analysis by capillary electrophoresis. In: *Proceedings EUROCORR 2008, the European Corrosion Congress*, Edinburgh, Scotland (7–11 September 2008). European Federation of Corrosion.

319 Pucić, I., Jakšić, M., Madžar, T., and Rogan, I. (2008). PIXE spectroscopy for determination of volatile corrosion inhibitor concentration in anticorrosion polymer films. In: *NACE International Conference, CORROSION 2008*, Paper No. 08364, 12. Houston, TX: NACE International.

320 Estevao, L.R.M. and Nascimento, R.S.V. (2001). Modifications in the volatilization rate of volatile corrosion inhibitors by means of host–guest systems. *Corrosion Science* 43 (6): 1133–1153.

321 Mišković, L., Madzar, T., Vignetti, A.M. et al. (2005). Preservation of high-performance military diesel engines using volatile corrosion inhibitors. *Materials Performance* 44 (1): 44–51.

322 Godfrey, M.R. and Chen, T.Y. (1995). Monitoring corrosion in boiler systems with colorimetric tests for ferrous and total iron. *Corrosion* 51 (10): 797–804.

323 Chen, T.Y., Godfrey, M.R., and Avallone, S.C. (1994). Corrosion and corrosion product transport monitoring in boiler condensate systems.

In: *NACE International Conference, CORROSION/94*, Paper No. 336. Houston, TX: NACE International.
324 Johnson, R.E., Agarwala, V.S., Rajan, K.S. et al. (1999). Fluorescent materials utilized as early warning sensors for corrosion of aluminum alloys. In: *NACE International Conference, CORROSION/99*, Paper No. 505, 10. Houston, TX: NACE International.
325 Johnson, R.E. and Agarwala, V.S. (1994). The use of fluorescent compounds and complexes of metals as early warning detectors for corrosion. In: *NACE International Conference, CORROSION/94*, Paper No. 332. Houston, TX: NACE International.
326 Agarwala, V.S. (1993). Sensors for corrosion detection and monitoring in hidden areas. In: *Proceedings of Symposium on Chemical Sensors II*, vol. 93–97 (ed. M. Butler, A. Ricco and N. Yamazoe), 698–708. Pennington, NJ: The Electrochemical Society.
327 Georgieva, I. and Petkova, N. (2004). Application of inhibiting protection at the crude oil atmospheric distillation units. In: *Proceedings EUROCORR 2004, the European Corrosion Congress*, Nice, France (12–16 September 2004), 9. European Federation of Corrosion.
328 Chambers, B.D., Srinivasan, S., Yap, K.-m., and Kane, R.D. (2009). Corrosion in crude distillation unit overhead operations: a comprehensive review. In: *Proceedings EUROCORR 2009, the European Corrosion Congress*, Nice, France (6–10 September 2009), Paper No. 8232, 11. European Federation of Corrosion.
329 Claesen, C., Thornthwaite, P., and Lordo, S. (2009). Changing crude oil quality and refinery corrosion inhibition. In: *Proceedings EUROCORR 2009, the European Corrosion Congress*, Nice, France (6–10 September 2009), Paper No. 7847, 22. European Federation of Corrosion.
330 Jackson, T., Craig Winslow, M., and Wilson, M. (2005). Prolonged experience processing high acid crude cross oil & refining company. In: *Proceedings EUROCORR 2005, the European Corrosion Congress*, Lisbon, Portugal (4–8 September 2005), 13. European Federation of Corrosion.
331 Cosentino, P., Grossman, B., and Shieh, C. (1995). Fiber-optic chloride sensor development. *Journal of Geotechnical Engineering* 121: 610–617.
332 Payer, J.H., Fink, K., and Song, I. (1997). Metal–metal oxide pH sensor: the effect of anions and aeration on pH measurements. In: *NACE International Conference, CORROSION/97*, Paper No. 308, 14. Houston, TX: NACE International.
333 Fink, K. and Payer, J.H. (1994). Corrosion sensors for condensate corrosivity in heat exchangers. In: *NACE International Conference, CORROSION/94*, Paper No. 330. Houston, TX: NACE International.
334 Sridhar, N. and Dunn, D.S. (1994). Effect of applied potential on changes in solution chemistry inside crevices on type 304L stainless steel and alloy 825.

In: *NACE International Conference, CORROSION/94*, Paper No. 347. Houston, TX: NACE International.

335 Ferreira, C.A.M. and Ponciano, J.A.d.C. (2006). Determination of the soil corrosivity of samples from southeastern Brazilian region. In: *Proceedings EUROCORR 2006, the European Corrosion Congress*, Maastricht, The Netherlands (25–28 September 2006), 9. European Federation of Corrosion.

336 Kichenko, S., Petrunin, M., and Marshakov, A. (2010). New methods of ensuring integrity of TNK-BP field pipelines. In: *Proceedings EUROCORR 2010, the European Corrosion Congress*, Moscow, Russia (13–17 September 2010), Paper No. 4698, 1. European Federation of Corrosion.

337 Correia, M.J., Pereira, E.V., Salta, M.M., and Fonseca, I.T.E. (2005). Sensor for oxygen evaluation in concrete. In: *Proceedings EUROCORR 2005, the European Corrosion Congress*, Lisbon, Portugal (4–8 September 2005), 9. European Federation of Corrosion.

338 NACE Standard TM0399-2005 (2005). *Standard Test Method for Phosphonate in Brine*, Item No. 21238, 11. Houston, TX: NACE International.

339 Port, R.D. and Herro, H.M. (1991). *The Nalco Guide to Boiler Failure Analysis*, 293. New York: McGraw-Hill, Inc.

340 ASTM Standards (2008). *Pesticides; Environmental Assessment; Hazardous Substances and Oil Spill Responses*, vol. 11.05, 1326. West Conshohocken, PA: ASTM.

341 Flemming, H.C. and Geesey, G.G. (eds.) (1991). *Biofouling and Biocorrosion in Industrial Water Systems*, 220. Berlin: Springer-Verlag.

342 Geesey, G.G., Lewandowski, Z., and Flemming, H.C. (eds.) (1994). *Biofouling and Biocorrosion in Industrial Water Systems*, 297. Boca Raton, FL: Lewis Publishers, CRC Press Inc.

343 Videla, H.A. (1996). *Manual of Biocorrosion*, 273. Boca Raton, FL: Lewis Publishers, CRC Press Inc.

344 Zhu, X.Y., Modi, H., Ayala, A., and Kilbane, J.J. II (2006). Rapid detection and quantification of microbes related to microbiologically influenced corrosion using quantitative polymerase chain reaction. *Corrosion* 62 (11): 950–955.

345 Flynn, D.J. (2009). *The Nalco Water Handbook*, 1280. New York: The McGraw-Hill Companies, Inc.

346 Roy Cullimore, D. (2010). *Practical Atlas for Bacterial Identification*, 2e, 327. Boca Raton, FL: Lewis Publishers, CRC Press.

347 ASTM D6469-08e1 (2008). *Standard Guide for Microbial Contamination in Fuels and Fuel Systems*, ASTM Book of Standards, vol. 05.03, 11. West Conshohocken, PA: ASTM.

348 Borenstein, S.W. and Licina, G.J. (1994). An overview of monitoring techniques for the study of microbiologically influenced corrosion. In: *NACE International Conference, CORROSION94*, Paper No. 611. Houston, TX: NACE International.

349 Hall-Stoodley, L., Rayner, J.C., Stoodley, P., and Lappin-Scott, H.M. (1999). Establishment of experimental biofilms using the modified robins device and flow cells. *Environmental Monitoring of Bacteria, Methods in Biotechnology* 12: 307–318.

350 Kesner, D. (2010). Microbiologically influenced corrosion: association with biofilm, monitoring and removal, Appendix 14. In: *Minutes of EFC WP15 Corrosion in the Refinery Industry*, Budapest, Hungary (22 June 2010), Slide Nos. 109–122. European Federation of Corrosion.

351 Jenneman, G.E., Webb, R., Holle, E. et al. (2004). Evaluation of an on-line biofilm detector and bio-traps for monitoring MIC in produced oilfield brine. In: *NACE International Conference, CORROSION2004*, Paper No. 04758, 27. Houston, TX: NACE International.

352 Nalepa, C.J., Ceri, H., and Stremick, C.A. (2000). A novel technique for evaluating the activity of biocides against biofilm bacteria. In: *NACE International Conference, CORROSION 2000*, Paper No. 00347, 19. Houston, TX: NACE International.

353 Nivens, D.E., Palmer, R.J., and White, D.C. (1995). Continous nondestructive monitoring of microbial biofilms: a review of analytical techniques. *Journal of Industrial Microbiology* 15: 263–376.

354 Krooneman, J., Appeldoorn, P., and Trompert, R. (2006). Detection, prevention and control of microbial corrosion. In: *Proceedings EUROCORR 2006, the European Corrosion Congress*, Maastricht, The Netherlands (25–28 September 2006), 9. European Federation of Corrosion.

355 Heyer, A., D'Suoza, F., Ferrari, G. et al. (2009). Combining microscopy and electrochemical studies on microbial influenced corrosion (MIC) in seawater environments. In: *Proceedings EUROCORR 2009, the European Corrosion Congress*, Nice, France (6–10 September 2009), Paper No. 7851, vol. 6. European Federation of Corrosion.

356 Flemming, H.C. (2002). Biofouling in water systems – cases, causes, countermeasures. *Applied Microbiology and Biotechnology* 59: 629–640.

357 Schmid, T., Panne, U., Haisch, C. et al. (2002). A photoacoustic technique for depth-resolved in situ monitoring of biofilms. *Environmental Science & Technology* 36 (19): 4135–4141.

358 Bischoff, P., Krolla-Sidenstein, P., Gerdes, A., and Obst, U. (2006). Characterization of microbial biocoenosis of concrete coated surfaces of drinking water reservoirs by PCR-DGGE. In: *Proceedings EUROCORR 2006, the European Corrosion Congress*, Maastricht, The Netherlands (25–28 September 2006), 2. European Federation of Corrosion.

359 Tanji, Y., Handa, T., and Miyanaga, K. (2010). Microbiologically influenced corrosion of carbon steel in aqueous natural gas plant. In: *NACE International Conference, CORROSION 2010*, Paper No. 10219, 11. Houston, TX: NACE International.

360 Price, A., Álvarez, L.A., Whitby, C., and Larsen, J. (2010). Detection of SRP activity by quantification of mRNA for the dissimilatory (bi) sulfite reductase gene (dsrA) by reverse transcription quantitative PCR. In: *NACE International Conference, CORROSION 2010*, Paper No. 10253, 15. Houston, TX: NACE International.

361 Larsen, J., Rasmussen, K., and Sørensen, K. (2010). Consortia of MIC bacteria and archaea causing pitting corrosion in top side oil production facilities. In: *NACE International Conference, CORROSION 2010*, Paper No. 10252, 12. Houston, TX: NACE International.

362 Nichols, P., Henson, J.M., Guckert, J.B. et al. (1985). Fourier transform-infrared spectroscopic methods for microbial ecology: analysis of bacteria, bacteria-polymer mixtures and biofilms. *Journal of Microbial Methods* 4: 79–94.

363 Schmitt, J. and Flemming, H.-C. (1998). FTIR-spectroscopy in microbial and material analysis. *International Biodeterioration and Biodegradation* 41: 1–11.

364 Quintelas, C., Ferreira, E.C., Lopes, J.A., and Sousa, C. (2018). An overview of the evolution of infrared spectroscopy applied to bacterial typing. *Biotechnology Journal* 13: 10.

365 Razban, B., Chen, W., Wang, D. et al. (2010). Preliminary assessment on bacterial deterioration of asbestos reinforced concrete pipes for water distribution. In: *NACE International Conference, CORROSION 2010*, Paper No. 10212, 13. Houston, TX: NACE International.

366 Port, R.D. and Herro, H.M. (1993). *The NALCO Guide to Cooling Water System. Failure Analysis*, 127–129, 157. New York: McGraw-Hill, Inc.

367 Setareh, M. and Javaherdashti, R. (2003). Precision comparison of some SRB detection methods in industrial systems. *Materials Performance* 42 (5): 60–63.

368 Philip Yu, F., Dillon, J.J., and Henry, T. (2010). Identification and control of microbiologically influenced corrosion in a power plant. In: *NACE International Conference, CORROSION 2010*, Paper No. 10211, 20. Houston, TX: NACE International.

369 Little, B.J. and Lee, J.S. (2007). *Microbiologically Influenced Corrosion*. Hoboken, NJ: Wiley.

370 Haile, T., Sooknah, R., Papavinasam, S. et al. (2010). Simultaneous online monitoring of SRB activity and corrosion rate. In: *NACE International Conference, CORROSION 2010*, Paper No. 10255, 14. Houston, TX: NACE International.

371 Horn, J., Carrillo, C., and Dias, V. (2003). Comparison of the microbial community composition at Yucca Mountain and laboratory test nuclear repository environments. In: *NACE International Conference, CORROSION 2003*, Paper No. 03556, 12. Houston, TX: NACE International.

372 Chattoraj, M., Fehr, M.J., Hatch, S.R., and Allain, E.J. (2002). Online measurement and control of microbiological activity in industrial water systems. *Materials Performance* 41 (4): 40–45.

373 Scott, P.J.B. (2000). Microbiologically influenced corrosion monitoring: real world failures and how to avoid them. *Materials Performance* 39 (1): 54–63.

374 Kharchenko, U., Beleneva, I., and Karpov, V. (2010). Estimation of the seawater biocorrosiveness. In: *Proceedings EUROCORR 2010, the European Corrosion Congress*, Moscow, Russia (13–17 September 2010), Paper No. 4715, 1. European Federation of Corrosion.

375 Licina, G.J. (2010). Biocide optimization: performance improvements and reduced chemical usage. In: *NACE International Conference, CORROSION 2010*, Paper No. 10052, 15. Houston, TX: NACE International.

376 Little, B.J., Lee, J.S., and Ray, R.I. (2006). Diagnosing microbiologically induced corrosion: a state-of-art review. *Corrosion* 62 (11): 1006–1017.

377 Little, B.J., Pope, R.K., Daulton, T.L., and Ray, R.I. (2001). Application of environmental cell transmission electron microscopy to microbiologically influenced corrosion. In: *NACE International Conference, CORROSION 2001*, Paper No. 01266, 18. Houston, TX: NACE International.

378 Lewandowski, Z. (2000). MIC and biofilm heterogeneity. In: *NACE International Conference, CORROSION/2000*, Paper No. 00400, 7. Houston, TX: NACE International.

379 A. Heyer (2013). Microbiologically influenced corrosion in ship ballast tanks. Master of Science, The Netherlands, p. 194.

380 Marconnet, C., Dagbert, C., Roy, M., and Féron, D. (2006). Microbially influenced corrosion of stainless steels in the Seine River. In: *Proceedings EUROCORR 2006, the European Corrosion Congress*, Maastricht, The Netherlands (25–28 September 2006), 17. European Federation of Corrosion.

381 Boffardi, B.P. (1995). Water treatments. In: *Corrosion Tests and Standards: Application and Interpretation*, ASTM Manual Series: MNL, vol. 20 (ed. R. Baboian), 697–704. West Conshohocken, PA: ASTM.

382 Dorsey, M.H., Licina, G.L., Saldanha, B.J., and Ebersole, R.C. (2002). Monitoring for corrosion and microbiological activity in a cooling water system. In: *NACE International Conference, CORROSION/2002*, Paper No. 02009, 28. Houston, TX: NACE International.

383 Veazey, M.V. (2003). A unique approach to fighting microbiologically influenced corrosion. *Materials Performance* 42 (12): 16–18.

384 Licina, G.J., Nekoksa, G., and Howard, R.L. (1994). An electrochemical method for on-line monitoring of biofilm activity in cooling water using the biogeorge probe. In: *Microbially Influenced Corrosion Testing*, ASTM STP 1232 (ed. J.R. Kearns and B.J. Little), 118–127. Philadelphia, PA: ASTM International.

385 Licina, G.J. and Carney, C.S. (1999). Monitoring biofilm formation and incipient MIC in real time. In: *NACE International Conference, CORROSION99*, Paper No. 175, 14. Houston, TX: NACE International.
386 Licina, G.J. (2001). Monitoring biofilms on metallic surfaces in real time. In: *NACE International Conference, CORROSION 2001*, Paper No. 01442, 23. Houston, TX: NACE International.
387 Licina, G.J. (2004). Optimizing biocide additions via real time monitoring of biofilms. In: *NACE International Conference, CORROSION 2004*, Paper No. 04582, 15. Houston, TX: NACE International.
388 Zintel, T.P. and Kostuck, D.A. (2003). Evaluation of chemical treatments in natural gas systems vs. MIC and other forms of internal corrosion using carbon steel coupons. In: *NACE International Conference, CORROSION 2003*, Paper No. 03574, 13. Houston, TX: NACE International.
389 Eckert, R.B., Aldrich, H.C., Edwards, C.A., and Cookingham, B.A. (2003). Microscopic differentiation of internal corrosion initiation mechanisms in a natural gas system. In: *NACE International Conference, CORROSION 2003*, Paper No. 03544, 13. Houston, TX: NACE International.
390 Sean Brossia, C. and Yang, L. (2003). Studies of microbiologically influenced corrosion using a coupled multielectrode array sensor. In: *NACE International Conference, CORROSION 2003*, Paper No. 03575, 8. Houston, TX: NACE International.
391 Cotiche, C. and Daumas, S. (2004). Development of an equipment for the detection and measurement of localized corrosion of carbon steel induced by sulfidogenic bacteria. In: *Proceedings EUROCORR 2004, the European Corrosion Congress*, Nice, France (12–16 September 2004), 5. European Federation of Corrosion.
392 Euvrard, M., Filiatre, C., and Crausaz, E. (2000). A cell to study in situ crystallization of calcium carbonate. *Journal of Crystal Growth* 125: 1461.
393 Devos, O., Gabrielli, C., and Tribollet, B. (2003). Scale deposition analysis by electrochemical and optical techniques at a potential corresponding to cathodic protection. In: *Proceedings EUROCORR 2003, the European Corrosion Congress*, Budapest, Hungary (28 September–2 October 2003), 8. European Federation of Corrosion.
394 Haug, R., de Wrangel, A., Weindel, E., and Paping, L. (2008). Reuse of water and still maintaining reliability by avoiding scale, biological slime and corrosion effects even more effectively. In: *Proceedings EUROCORR 2008, the European Corrosion Congress*, Edinburgh, Scotland (7–11 September 2008), 21. European Federation of Corrosion.
395 http://www.alspi.com/hy4000.htm; https://www.cosasco.com/corrosion-hydrogen-induced-cracking.html; http://www.caproco.com/catalog/pdf/Probes-Instruments/Hydrogen/Hydrogen-Patch-Saddle-Probes.pdf (accessed 3 November 2018).

396 Dean, F. (2004). Continuous active corrosion and HIC risk measurement. *EUROCORR 2004, Working Party 15 (Corrosion in the Refining Industry)*, Nice, France (15 September 2004).
397 Dean, F. (2008). The utility of hydrogen flux measurement in refineries. In: *Proceedings EUROCORR 2008, the European Corrosion Congress*, Edinburgh, Scotland (7–11 September 2008), 9. European Federation of Corrosion.
398 Shreir, L.L., Jarman, R.A., and Burstein, G.T. (eds.) (1994). *Corrosion*, 3e, vol. 2, 19:154–19:178. London, UK: Butterworth Heinemann.
399 Messer, B., Oprea, V., Beaulieu, K., and Wright, A. (2008). Role of nascent hydrogen in refinery corrosion. In: *NACE International Conference, CORROSION 2008*, Paper No. 08549, 21. Houston, TX: NACE International.
400 Groysman, A., Kaufman, A., Feldman, B., and Balali, R. (2010). Hydrogen damage, monitoring, and prevention in the oil refinery. In: *Proceedings EUROCORR 2010, the European Corrosion Congress*, Moscow, Russia (13–17 September 2010), Paper No. 9167, 30. European Federation of Corrosion.
401 Barakat, N. and Reuben, R.L. (2008). Hydrogen permeation through low alloy steels. In: *Proceedings EUROCORR 2008, the European Corrosion Congress*, Edinburgh, Scotland (7–11 September 2008), 1. The European Corrosion Congress.
402 McKeen, K., Justason, A., and Lalonde, M. (2010). Sealing of non-intrusive vacuum hydrogen probe at temperatures up to 500°C. In: *NACE International Conference, CORROSION 2010*, Paper No. 10177, 13. Houston, TX: NACE International.
403 Bloom, M. and Krulfeld, M. (1957). A hydrogen effusion method for the determination of corrosion rates in aqueous systems at elevated temperature and pressure. *Journal of the Electrochemical Society* 104 (5): 264.
404 McKeen, K., Lalonde, M., Scott, A., and Leblanc, K. (2008). Corrosion monitoring at the Coleson Cove generating station. In: *NACE International Conference, CORROSION 2008*, Paper No. 08294, 11. Houston, TX: NACE International.
405 NACE International Publication 1C184 (2008). *Monitoring Internal Corrosion in Oil and Gas Production Operations with Hydrogen Probes*, Item No. 24002, vol. 11. Houston, TX: NACE International.
406 Lewis, A.L. and Boah, J.K. (2003). Hydrogen permeation probe. US Patent 6,554,981.
407 Park, G.T., Kim, S.J., and Kim, K.Y. (2010). Quantitative measurement of hydrogen diffusion and trapping parameters for linepipe steel under constant load in sour environment. In: *NACE International Conference, CORROSION 2010*, Paper No. 10180, 10. Houston, TX: NACE International.

408 ISO 17081:2004 (2004). *Method of Measurement of Hydrogen Permeation and Determination of Hydrogen Uptake and Transport in Metals by an Electrochemical Technique*, 16. Geneva, Switzerland: ISO.

409 Aikens, J. (2009). Device and system for corrosion detection. US Patent 7,552,643.

410 Dean, F.W.H. and Ling, A.E. (2003). Hydrogen collection and detection. US Patent 6,637,253.

411 Tems, R.D. and Dean, F.W.H. (2000). Field application of new, portable, non-intrusive hydrogen monitor for sour service. In: *NACE International Conference, CORROSION/2000*, Paper No. 00471, 16. Houston, TX: NACE International.

412 Tems, R.D., Lewis, A.L., and Abdulhadi, A.I. (2002). Field and laboratory measurements with hydrogen permeation measurement devices. In: *NACE International Conference, CORROSION/2002*, Paper No. 02345, 20. Houston, TX: NACE International.

413 Dean, F.W.H. and Powell, S.W. (2004). A laboratory study of types of corrosion inducing hydrogen permeation through steel. In: *NACE International Conference, CORROSION/2004*, Paper No. 04472, 14. Houston, TX: NACE International.

414 Dean, F.W.H. (2002). Corrosion trends as discerned by the hydrogen collection method. In: *NACE International Conference, CORROSION/2002*, Paper No. 02344, 15. Houston, TX: NACE International.

415 Dean, F.W.H. (2005). Measurement of hydrogen permeation through steel sections of varying thickness at 19°C. *Journal of Materials Science and Technology* 21 (3): 347–351.

416 Al-Sulaiman, S.A., Al-Mithin, A.W., Al-Shamari, A.R. et al. (2010). Assessing of possibility of hydrogen damage in crude oil processing equipment. In: *NACE International Conference, CORROSION 2010*, Paper No. 10176, 14. Houston, TX: NACE International.

417 O'Kane, J.M., Rudd, T.F., Harrison, J.H. et al. (2010). Correlation of hydrogen flux and corrosion rate measurements carried out during a severe episode of corrosion-erosion attributable to naphthenic acid. In: *NACE International Conference, CORROSION 2010*, Paper No. 10178, 9. Houston, TX: NACE International.

418 O'Kane, J.M., Rudd, T.F., Cooke, D. et al. (2010). Detection and monitoring of naphthenic acid corrosion in a visbreaker unit using hydrogen flux measurements. In: *NACE International Conference, CORROSION 2010*, Paper No. 10351, 15. Houston, TX: NACE International.

419 Dean, F.W.H., Fowler, C.M., Farnell, R.T., and Mishael, S.J. (2010). Hydrogen flux and corrosion rate measurements on hydrogen induced cracking susceptible and resistant A516 steels in various sour environments. In: *NACE International Conference, CORROSION 2010*, Paper No. 10179, 19. Houston, TX: NACE International.

420 Mishael, S.J., Dean, F.W.H., and Fowler, C. (2004). Practical applications of hydrogen permeation monitoring. In: *NACE International Conference, CORROSION/2004*, Paper No. 04476, 12. Houston, TX: NACE International.

421 Dean, F.W.H. and Fray, D.J. (2000). An ultrasensitive hydrogen detection method. *Journal of Materials Science and Technology* 16: 41–49.

422 Dean, F.W.H. and Powell, S.W. (2006). Hydrogen flux and high temperature acid corrosion. In: *NACE International Conference, CORROSION 2006*, Paper No. 06436, 9. Houston, TX: NACE International.

423 Al-Aqeer, F. and Addington, F. (2008). Hydrogen permeation application to crude unit overhead corrosion monitoring. In: *NACE International Conference, CORROSION 2008*, Paper No. 08548, 9. Houston, TX: NACE International.

424 Williams, G., Neil McMurray, H., and Newman, R.C. (2010). Detection of hydrogen permeation through iron foil using a scanning kelvin probe. In: *Proceedings EUROCORR 2010, the European Corrosion Congress*, Moscow, Russia (13–17 September 2010), Paper No. 9322, 1. European Federation of Corrosion.

425 Cheng, Y.F. and Du, Y.L. (1993). Design of an electrochemical probe for monitoring susceptibility of steel in pickling to hydrogen-induced cracking. *Corrosion* 49 (9): 776–781.

426 Serna, S., Campillo, B., Ramirez, M. et al. (2008). Hydrogen permeability behaviour of a high strength microalloyed steel developed for sour service. In: *NACE International Conference, CORROSION 2008*, Paper No. 08405, 12. Houston, TX: NACE International.

427 Taylor, M.L. and McMillion, L.G. (2008). Hydrogen permeation of alloy 22. In: *NACE International Conference, CORROSION 2008*, Paper No. 08407, 15. Houston, TX: NACE International.

428 Smirnova, A., Johnsen, R., and Nisancioglu, K. (2010). Effect of tensile stress on hydrogen permeation in 13% Cr super martensitic stainless steel. In: *Proceedings EUROCORR 2010, the European Corrosion Congress*, Moscow, Russia (13–17 September 2010), Paper No. 9474, 12. European Federation of Corrosion.

429 Frappart, S., Aulnoye-Aymeries, Feaugas, X. et al. Study of the hydrogen diffusion and trapping mechanisms into quenched and tempered HSLA steels using electrochemical permeation technique. In: *Proceedings EUROCORR 2010, the European Corrosion Congress*, 12. European Federation of Corrosion, Moscow, Russia (13–17 September 2010), Paper No. 9580.

430 Zhou, X. and Bo, H. (2010). Methods of detecting hydrogen stress corrosion cracking and its protection in tribochemical environments. In: *Proceedings EUROCORR 2010, the European Corrosion Congress*, Moscow, Russia

(13–17 September 2010), Paper No. 9156, 1. European Federation of Corrosion.

431 Devanathan, M.A.V. and Stachurski, Z. (1962). The adsorption and diffusion of electrolytic hydrogen in palladium. *Proceedings of Royal Society of London A* 270 (1340): 90–102.

432 da Silva, A.G.S.G., Ponte, H.A., Pashchuk, A., and Marino, C.E.B. (2007). A study of the electrochemical behavior of niobium in relation to hydrogen charging. In: *Proceedings EUROCORR 2007, the European Corrosion Congress*, Freiburg, Germany (9–13 September 2007), Paper No. 1399. European Federation of Corrosion.

433 Charca, S.M., Uwakweh, O.N.C., and Agarwala, V.A. (2007). Comparative study of hydrogen transport behaviors in armco-Fe and AF1410 steels. *Corrosion Reviews* 25 (3–4): 377–400.

434 Eliaz, N., Banks-Sills, L., Ashkenazi, D., and Eliazi, R. (2004). Modeling failure of metallic glasses due to hydrogen embrittlement in the absence of external loads. *Acta Materialia* 52: 93–105.

435 Bolzoni, F., Fassina, P., Fumagalli, G., and Lazzari, L. (2010). Hydrogen charging of carbon and low alloy steel by electrochemical methods. In: *Proceedings EUROCORR 2010, the European Corrosion Congress*, Moscow, Russia (13–17 September 2010), Paper No. 9503, 16. European Federation of Corrosion.

436 Todoshchenko, O., Yagodzinskyy, Y., and Hänninen, H. (2010). Hydrogen diffusion and trapping in austenitic stainless steels studied with thermal desorption spectroscopy. In: *Proceedings EUROCORR 2010, the European Corrosion Congress*, Moscow, Russia (13–17 September 2010), Paper No. 4742, 9. European Federation of Corrosion.

437 RH600 Hydrogen (2010). *Specification Sheet of LECO Corporation*, 2. St. Joseph, MI: LECO Corporation.

438 Eliaz, N., Eliezer, D., Abramov, E. et al. (2000). Hydrogen evolution from Zr-based amorphous and quasicrystalline alloys. *Journal of Alloys and Compounds* 305: 272–281.

439 Kossoy, E., Khoptiar, Y., Cytermann, C. et al. (2008). The use of SIMS in quality control and failure analysis of electrodeposited items inspected for hydrogen effects. *Corrosion Science* 50 (5): 1481–1491.

440 API 579-1/ASME FFS-1 2007 (2007). *Fitness-for-Service*, 7–15. Washington, DC: API.

441 API Recommended Practice 941 (2008). *Steels for Hydrogen Service at Elevated Temperatures and Pressures in Petroleum Refineries and Petrochemical Plants*, 7e, 56. Washington, DC: American Petroleum Institute.

442 Harjac, S., Keen, D., and Macraild, A. (2008). Hot spot related high temperature hydrogen attack of secondary reformer vessel in ammonia gas plant. In: *Proceedings EUROCORR 2008, the European Corrosion Congress*,

Edinburgh, Scotland (7–11 September 2008), Paper No. 1281. European Federation of Corrosion.

443 Kot, R. (2001). Hydrogen attack, detection, assessment and evaluation. In: *Proceedings 10th Asia-Pacific Conference on Non-destructive Testing*, Brisbane, Australia (17–21 September 2001), 17. NDT.

444 Tomin, V.P., Silinskaya, Y.N., and Belyaeva, E.V. (2010). Impact of the primary oxidation products of gasoline cuts and oxygen-containing compounds on corrosion behavior of carbon steel. In: *Proceedings EUROCORR 2010, the European Corrosion Congress*, Moscow, Russia (13–17 September 2010), Paper No. 9243, 8. European Federation of Corrosion.

445 Schäfer, G. (2007). Fast, easy and reliable corrosion detection with DC monitoring techniques. In: *Proceedings EUROCORR 2007, the European Corrosion Congress*, Freiburg, Germany (9–13 September 2007), 6. European Federation of Corrosion.

446 Lindemuth, D., Carroll, J., Kroon, D. et al. (2010). PHMSA – sponsored research: improvements to ECDA process potential measurements in paved areas. In: *NACE International Conference, CORROSION 2010*, Paper No. 10055, 16. Houston, TX: NACE International.

447 Mork, J.H., Mayer, S., and Rosenbom, K. (2007). Cathodic protection of reinforcement in concrete. In: *Proceedings EUROCORR 2007, the European Corrosion Congress*, Freiburg, Germany (9–13 September 2007), 6. European Federation of Corrosion.

448 Elsener, B. and Bohni, H. (1990). Potential mapping and corrosion of steel in concrete. In: *Corrosion Rates of Steel in Concrete*, ASTM STP 1065, 200. West Conshohocken, PA: American Society for Testing and Materials.

449 Broomfield, J.P., Langford, P.E., and Ewins, A.J. (1990). The use of a potential wheel to survey reinforced concrete structures. In: *Corrosion Rates of Steel in Concrete*, ASTM STP 1065, 200. West Conshohocken, PA: American Society for Testing and Materials.

450 ASTM C876-09 (2009). *Standard Test Method for Half-cell Potentials of Uncoated Reinforcing Steel in Concrete*, Annual Book of ASTM Standards, vol. 03.02, 7. West Conshohocken, PA: ASTM.

451 Reihling, K. and Raupah, M. (2010). Wireless potential mapping of reinforced concrete structures. In: *NACE International Conference, CORROSION 2010*, Paper No. 10153, 8. Houston, TX: NACE International.

452 Cunha, M.P., Brito, P.S.D., and Rodrigues, L.F. (2009). A promising potential pseudo-reference electrode for concrete corrosion monitoring. In: *Proceedings EUROCORR 2009, the European Corrosion Congress*, Nice, France (6–10 September 2009), Paper No. 8381, 1. European Federation of Corrosion.

453 Been, J., Place, T.D., and Holm, M. (2010). Evaluating corrosion and inhibition under sludge deposits in large diameter crude oil pipelines.

In: *NACE International Conference, CORROSION 2010*, Paper No. 10143, 11. Houston, TX: NACE International.

454 Gonzales, J.A., Benito, M., and Feliu, S. (1995). Suitability of assessment methods for identifying active and passive zones in reinforced concrete. *Corrosion* 51 (2): 145–152.

455 Glass, G., Hassanein, A.M., and Buenfeld, N.R. (1997). Monitoring the passivation of steel in concrete induced by cathodic protection. *Corrosion Science* 39: 1451–1458.

456 Duffó, G.S., Farina, S.B., and Giordano, C.M. (2008). Characterization of embeddable reference electrodes for corrosion monitoring in reinforced concrete structures. In: *Proceedings EUROCORR 2008, the European Corrosion Congress*, Edinburgh, Scotland (7–11 September 2008), 18. European Federation of Corrosion.

457 Elsener, B. (2003). The importance of monitoring in corrosion control of highway bridges. In: *Proceedings EUROCORR 2003, the European Corrosion Congress*, Budapest, Hungary (28 September–2 October 2003), 1. European Federation of Corrosion.

458 Miyata, Y. and Asakura, S. (1997). Corrosion monitoring of metals in soils by electrochemical and related methods: part I – monitoring of actual field buried metal structures and electrochemical simulation with corrosion probes and pilot pieces. *Japan Society for Corrossion Engineering* 46 (9): 451–455.

459 Molander, A. and Ullberg, M. (2007). A new technique for in-plant high mass transfer rate ECP monitoring. In: *Proceedings EUROCORR 2007, the European Corrosion Congress*, Freiburg, Germany (9–13 September 2007), 9. European Federation of Corrosion.

460 Molander, A. (2004). Corrosion potential monitoring in nuclear power environments. In: *Proceedings EUROCORR 2004, the European Corrosion Congress*, Nice, France (12–16 September 2004). European Federation of Corrosion.

461 Bethencourt, M., Sánchez-Amaya, J.M., González-Rovira, L., and Botana, F.J. (2007). Monitoring the corrosion of steel sheet piles by means of ultrasound and corrosion potential measurements. In: *Proceedings EUROCORR 2007, the European Corrosion Congress*, Freiburg, Germany (9–13 September 2007), 8. European Federation of Corrosion.

462 Anda, V., Edgemon, G., Dahl, M. et al. (2010). *Performance of Multi-Probe Corrosion Monitoring Systems at the Hanford Site*, Paper No. 10243, 20. Houston, TX: NACE International.

463 Menendez, C.M., Jovancicevic, V., and Mok, W.Y. (2007). Quantitative analysis of localized corrosion in the presence of corrosion inhibitors. In: *NACE International Conference, CORROSION 2007*, Paper No. 07619, 15. Houston, TX: NACE International.

464 Ferreira Gorjao, P., Bogaerts, W., Hubrecht, J. et al. (2003). An artificial intelligence sensor to assess environment corrosivity: the sensing and pattern recognition system. In: *Proceedings EUROCORR 2003, the European Corrosion Congress*, Budapest, Hungary (28 September–2 October 2003), 10. European Federation of Corrosion.

465 Darowicki, K., Gerengi, H., Bereket, G., and Slepski, P. (2009). Evaluation of corrosion inhibition of brass-118 alloy in artificial seawater by benzotriazole using dynamic EIS. In: *Proceedings EUROCORR 2009, the European Corrosion Congress*, Nice, France (6–10 September 2009), Paper No. 7908, 18. European Federation of Corrosion.

466 Stern, M. (1958). A method for determining corrosion rates from linear polarization data. *Corrosion* 14 (9): 60–64.

467 Stern, M. and Geary, A.L. (1957). Electrochemical polarization: I. A theoretical analysis of the shape of polarization curves. *Journal of the Electrochemical Society* 104 (1): 56–63.

468 Scully, J.R., Silverman, D.C., and Kendig, M.W. (eds.) (1993). *Electrochemical Impedance: Analysis and Interpretation*. West Conshohocken, PA: ASTM.

469 Mansfeld, F. (1976). The polarization resistance techniques for measuring corrosion currents. In: *Advances in Corrosion Science and Technology*, vol. 6 (ed. M.G. Fontana and R.W. Staehle), 163–262. New York: Plenum Press.

470 Scully, J.R. (1998). The polarization resistance method for determination of instantaneous corrosion rates: a review. In: *NACE International Conference, CORROSION/98*, Paper No. 98304, 18. Houston, TX: NACE International.

471 Scully, J.R. (2000). Polarization resistance method for determination of instantaneous corrosion rates. *Corrosion* 56 (2): 199–218.

472 Cottis, R.A. (2006). An evaluation of electrochemical noise for the estimation of corrosion rate and type. In: *NACE International Conference, CORROSION 2006*, Paper No. 06432, 11. Houston, TX: NACE International.

473 Cottis, R.A. (2005). Electrochemical noise as a technique for corrosion monitoring. In: *Proceedings International Conference on Corrosion, CORCON2005*, Chennai, India (28–30 November 2005), 12.

474 Kane, R.D. (1996). *Selection of Corrosion Monitoring Methods for Refinery Application, Corrosion in the Oil Refining Industry* (26–27 September 1996), 14/1–14/17. Houston, TX: NACE International.

475 Joosten, M.W., Kolts, J., Humble, P.G. et al. (1998). Internal corrosion monitoring of subsea production flowlines – probe design and testing. In: *NACE International Conference, CORROSION/98*, Paper No. 77, 24. Houston, TX: NACE International.

476 Gowers, K.R., Millard, S.G., Gill, J.S., and Gill, R.P. (1994). Programmable linear polarization meter for determination of corrosion rate of reinforcement in concrete structures. *British Corrosion Journal* 29 (1): 25–32.

477 Flis, J., Pickering, H.W., and Osseo-Asare, K. (1995). Assessment of data from three electrochemical instruments for evaluation of reinforcement corrosion rates in concrete bridge components. *Corrosion* 51 (8): 602–609.

478 Kelly, R.G., Yuan, J., Jones, S.H. et al. Embeddable microinstruments for corrosion monitoring. In: *NACE International Conference, CORROSION97*, Paper No. 97294, 12. Houston, TX: NACE.

479 Andrade, C. and Martinez, I. (2010). Non contacting corrosion method applied to the steel corrosion rate determination in different solution and reinforced concrete specimens. In: *NACE International Conference, CORROSION 2010*, Paper No. 10125, 13. Houston, TX: NACE International.

480 Linjewile, T.M., Valentine, J., Davis, K.A. et al. (2003). Prediction and real-time monitoring techniques for corrosion characterization in furnaces. *Materials at High Temperature* 20 (2): 175–183.

481 Mickalonis, J.I. (1997). Coupon corrosion monitoring of a high level waste tank. In: *NACE International conference, CORROSION/97*, Paper No. 97129. Houston, TX: NACE International.

482 Papavinasam, S., Revie, R.W., Attard, M. et al. (2003). Comparison of laboratory methodologies to evaluate corrosion inhibitors for oil and gas pipelines. *Corrosion* 59 (10): 897–912.

483 Perdomo, J.J., Ramirez, M., and Viloria, A. (2000). Test spot inconsistencies. *Oil & Gas Journal* 98 (1): 46–48.

484 Yunovich, M., Tossey, B.M., and Mendez, C. (2006). Measuring corrosion growth rates for determining integrity verification reassessment intervals by in-situ LPR technique. In: *NACE International Conference, CORROSION 2006*, Paper No. 06311, 14. Houston, TX: NACE International.

485 Waldmann, B., Haider, F., Horn, S., and Warnecke, R. (2008). Corrosion monitoring in waste-to energy plants. In: *Proceedings EUROCORR 2008, the European Corrosion Congress*, Edinburgh, Scotland (7–11 September 2008), 13. European Federation of Corrosion.

486 Waldmann, B., Schrupp-Heidelberger, D., Stöcker, B. et al. (2009). Online corrosion monitoring in waste to energy (WTE) and biomass incineration plants – application of a corrosion probe allows quantifying the gases' corrosive effects. In: *Proceedings EUROCORR 2009, the European Corrosion Congress*, Nice, France (6–10 September 2009), Paper No. 8287, 1. European Federation of Corrosion.

487 Inman, M., Jennings Taylor, E., Rawat, A. et al. (1997). Detection of corrosion in natural seawater using polarization resistance measurements. In: *NACE International Conference, CORROSION/97*, Paper No. 97296, 12. Houston, TX: NACE International.

488 Inman, M., Rawat, A.K., Jennings Taylor, E., and Moran, P.J. (1997). Detection of crevice corrosion under an O-ring by polarization resistance measurements using electrodes embedded in the O-ring. In: *NACE International Conference, CORROSION/97*, Paper No. 312, 5. Houston, TX: NACE International.

489 Lenard, D.R. (2007). Electrochemical frequency modulation measurements on the corrosion of copper nickel alloys in clean seawater and marine sediments. In: *NACE International Conference, CORROSION 2007*, Paper No. 07246, 19. Houston, TX: NACE International.

490 Eden, D.C., Eden, D.A., Delve, D., and Mehdawe, A. (2007). *Performance Improvement for Cooling Water Systems*, 15. Houston, TX: Honeywell Process Solutions. https://www.automation.com/pdf_articles/CoolingWaterWhitePaper.pdf (accessed 5 November 2018).

491 Pehkonen, A., Forsén, O., Kärkkäinen, K., and Leino, J. (2003). Development of electrochemical measurement system for corrosion monitoring in district heating systems. In: *Proceedings EUROCORR 2003, the European Corrosion Congress*, Budapest, Hungary (28 September–2 October 2003), 9. European Federation of Corrosion.

492 Fofano, S., Chaves Neto, A., and Ponte, H.A. (2003). Determining corrosion rates by electrochemical techniques in an industrial water cooling system. *Materials Performance* 42 (2): 62–65.

493 Molony, J.J., Mok, W.Y., and Menendez, C.M. (2010). The in-situ assessment of pitting corrosion and its inhibition using LCM. In: *NACE International Conference, CORROSION 2010*, Paper No. 10156, 35. Houston, TX: NACE International.

494 Eden, D.C. and Kane, R.D. (2004). Achieving improved process control and real-time corrosion monitoring. In: *EUROCORR 2004, Working Party 15 (Corrosion in the Refining Industry)*, Nice, France (15 September 2004). European Federation of Corrosion.

495 Iverson, W.P. (1968). Transient voltage changes produced in corroding metals and alloys. *Journal of the Electrochemical Society* 115 (6): 617–618.

496 Cafissi, A., Invernizzi, A.J., and Trasatti, S.P. (2007). Electrochemical noise as a powerful tool for studying corrosion processes. *Corrosion Reviews* 25 (1–2): 233–246.

497 Eden, D.A., John, D.G., and Dawson, J.L. (1992). Corrosion monitoring. US Patent 5,139,627.

498 Cottis, R.A. (2001). Interpretation of electrochemical noise data. *Corrosion* 57 (3): 265–285.

499 Eden, D.C. and Kintz, J.D. (2004). Real-time corrosion monitoring for improved process control: a real and timely alternative to upgrading of materials of construction. In: *NACE International Conference, CORROSION/2004*, Paper No. 04238, 14. Houston, TX: NACE International.

500 Mills, D.J. and Woodcock, C.P. (2008). Use of electrochemical impedance spectroscopy for investigation of a set of organic coatings on steel. In: *Proceedings EUROCORR 2008, the European Corrosion Congress*, Edinburgh, Scotland (7–11 September 2008), 10. European Federation of Corrosion.

501 Mabbutt, S.J. and Mills, D.J. (1998). Technical note: novel configurations for electrochemical noise measurments. *British Corrosion Journal* 33 (2): 158–160.

502 Torres-Islas, A., Uruchurtu, J., Serna, S., and Gonzálea-Rodríguez, J.G. (2007). Correlation analysis between electrochemical noise resistance in time and frequency domain for mild steel and stainless steel in solutions within all pH scale. In: *Proceedings EUROCORR 2007, the European Corrosion Congress*, Freiburg, Germany (9–13 September 2007), 10. European Federation of Corrosion.

503 Mabbutt, S., Simms, N., and Oakey, J. (2008). High temperature corrosion monitoring by electrochemical noise techniques. In: *Proceedings EUROCORR 2008, the European Corrosion Congress*, Edinburgh, Scotland (7–11 September 2008), 19. European Federation of Corrosion.

504 Bos, W.M., Homborg, A., Ferrari, G.M., and de Wit, J.H.W. (2006). Complementary use of EIS and ENM for qualification of organic coatings. In: *Proceedings EUROCORR 2006, the European Corrosion Congress*, Maastricht, The Netherlands (25–28 September 2006), 8. European Federation of Corrosion.

505 Greisiger, H. and Schauer, T. (2001). On the interpretation of the electrochemical noise data for coatings. *Progress in Organic Coating* 39: 31–36.

506 Mills, D.J. and Broster, M. (2006). Continuing work to enable electrochemical methods to be used to monitor the performance of organic coatings in the field. In: *Proceedings EUROCORR 2006, the European Corrosion Congress*, Maastricht, The Netherlands (25–28 September 2006), 7. European Federation of Corrosion.

507 Cottis, R. and Turgoose, S. (1999). *Electrochemical Impedance and Noise*, 149. Houston, TX: NACE International.

508 Kearns, J.R., Scully, J.R., Roberge, P.R. et al. (eds.) (1996). *Electrochemical Noise Measurement for Corrosion Applications*, 476. West Conshohocken, PA: ASTM.STP 1277

509 Kane, R.D., Cayard, M.S., and Maldonado, J.G. (2003). Technology for improving pipeline integrity: online, real-time corrosion monitoring. In: *Proceedings EUROCORR 2003, the European Corrosion Congress*, Budapest, Hungary (28 September–2 October 2003), 10. European Federation of Corrosion.

510 Báez, V.B., Vera, J.R., and Villalba, R. (2001). Electrochemical noise for evaluating pitting resistance of CRA materials under simulated well conditions. In: *NACE International Conference, CORROSION/2001*, Paper No. 01103, 15. Houston, TX: NACE International.

511 González-Rodriguez, J.G., Salinas-Bravo, V.M., García-Ochoa, E., and Diaz-Sánchez, A. (1997). Use of electrochemical potential noise to detect initiation and propagation of stress corrosion cracks in a 17-4 PH steel. *Corrosion* 53 (9): 693–699.

512 Edgemon, G.L., Danielson, M.J., and Bell, G.E.C. (1997). Detection of stress corrosion cracking and general corrosion of mild steel in simulated defense nuclear waste solutions using electrochemical noise analysis. *Journal of Nuclear Materials* 245 (2–3): 201–209.

513 Watanabe, Y., Shoji, T., and Kondo, T. (1998). Electrochemical noise characteristics of IGSCC in stainless steels in pressurized high-temperature water. In: *NACE International Conference, CORROSION/98*, Paper No. 98129, 7. Houston, TX: NACE International.

514 Edgemon, G.L., Ohl, P.C., Bell, G.E.C., and Wilson, D.F. (1996). Detection of general and localized corrosion of mild steel in simulated defense nuclear waste solutions using electrochemical noise analysis. In: *NACE International Conference, CORROSION/96*, Paper No. 96094, 29. Houston, TX: NACE International.

515 Aballe, A., Newman, R.C., and Cottis, R. (2003). Electrochemical noise study of stress corrosion cracking of sensitized 304H in thiosulphate. In: *NACE International Conference, CORROSION/2003*, Paper No. 03403, 13. Houston, TX: NACE International.

516 Hickling, J., Taylor, D.F., and Andersen, P.L. (1998). Use of electrochemical noise to detect stress corrosion crack initiation in simulated BWR environments. *Materials and Corrosion* 49 (9): 651–658.

517 Stewart, J., Wells, D.B., Scott, P.M., and Williams, D.E. (1992). Electrochemical noise measurements of stress corrosion cracking of sensitized austenitic stainless steel in high-purity oxygenated water at 288°C. *Corrosion Science* 33 (1): 73–88.

518 Klapper, H., Heyn, A., and Goellner, J. (2008). Influence of the cathodic process on the electrochemical noise from pitting of stainless steel. In: *NACE International Conference, CORROSION 2008*, Paper No. 08401, 9. Houston, TX: NACE International.

519 Sajdl, P., Kucera, P., Novotny, R. et al. (2003). Electrochemical noise measurement of stress corrosion cracking. In: *Proceedings EUROCORR 2003, the European Corrosion Congress*, Budapest, Hungary (28 September–2 October 2003), 11. European Federation of Corrosion.

520 Breimesser, M., Ritter, S., Seifert, H.P. et al. (2010). Electrochemical noise of SCC in austenitic stainless steels: a combined macro- and micro-electrochemical approach. In: *Proceedings EUROCORR 2010, the European Corrosion Congress*, Moscow, Russia (13–17 September 2010), Paper No. 9534, 14. European Federation of Corrosion.

521 Ritter, S. and Seifert, H.-P. (2008). Detection of stress corrosion cracking in simulated BWR environment by combining the electrochemical noise and DCPD technique. In: *Proceedings EUROCORR 2008, the European Corrosion Congress*, Edinburgh, Scotland (7–11 September 2008), 11. European Federation of Corrosion.

522 Mickalonis, J.I., Tshishiku, E.M., and Edgemon, G.L. (2006). Development of a movable electrochemical noise corrosion probes for nuclear waste tanks, p. 14. Report WSRC-MS-2001-00734. U.S. Department of Energy.

523 Edgemon, G.L. (2001). Electrochemical noise based corrosion monitoring at the Hanford site: third generation system. In: *NACE International*

Conference, CORROSION/2001, Paper No. 01140, 18. Houston, TX: NACE International.

524 Barr, E.M., Goodfellow, R.B., and Rosenthal, L.M. (2000). Noise monitoring in Canada's Simonette sour oil processing facility. In: *NACE International Conference, CORROSION/2000*, Paper No. 00414, 15. Houston, TX: NACE International.

525 Groysman, A. and Schwarz, I. (2006). Study of efficiency of industrial corrosion inhibitors for cooling water systems at oil refining industry. In: *NACE International Conference, CORROSION 2006* (12–16 March 2006), Paper 06097, 10. Houston, TX: NACE International.

526 Ohtsu, T. and Miyazawa, M. (2010). Study of the boiler water quality in chemical plants using the electrochemical method. In: *NACE International Conference, CORROSION 2010*, Paper No. 10154, 17. Houston, TX: NACE International.

527 Sánchez-Amaya, J.M., Bethencourt, M., González-Rovira, L., and Botana, F.J. (2007). Monitoring exfoliation corrosion of AA2024 by means of electrochemical noise. In: *Proceedings EUROCORR 2007, the European Corrosion Congress*, Freiburg, Germany (9–13 September 2007), 10. European Federation of Corrosion.

528 Jamali, S.S. and Mills, D.J. (2016). A critical review of electrochemical noise measurement as a tool for evaluation of organic coatings. *Progress in Organic Coatings* 95: 25–37.

529 Sánchez-Amaya, J.M., Gonzalez-Rovira, L., El Amrani, K. et al. (2010). Electrochemical noise of AA2024 in solutions promoting IGC. In: *Proceedings EUROCORR 2010, the European Corrosion Congress*, Moscow, Russia (13–17 September 2010), Paper No. 9488, 12. European Federation of Corrosion.

530 Cao, F.H., Zhang, Z., Su, J.X. et al. (2006). Electrochemical noise analysis of LY12-T3 in EXCO solution by discrete wavelet transform technique. *Electrochimica Acta* 51 (7): 1359–1364.

531 Mansfeld, F., Han, L.T., and Lee, C.C. (1997). Analysis of electrochemical impedance and noise data for polymer coated metals. *Corrosion Science* 39: 255–279.

532 Amaya, M., Romero, J.M., Garcia, E., and Martinez, L. (2001). Electrochemical noise in assessing MIC in offshore injection systems. In: *NACE International Conference, CORROSION/2001*, Paper No. 01269, 9. Houston, TX: NACE International.

533 Eden, D.A. (2002). Control of microbiologically induced corrosion in sea water injection systems using electrochemical noise technology. In: *NACE International Conference, CORROSION 2002*, Paper No. 02338, 15. Houston, TX: NACE International.

534 Halama, M., Jerolitsch, D., Linhardt, P., and Fafilek, G. (2009). Active corrosion management in automotive industry: hyphenation of electrochemical noise analysis with artificial neural networks – feasibility study. Part I. In: *Proceedings EUROCORR 2009, the European Corrosion Congress*, Nice, France (6–10 September 2009), Paper No. 7854, 8. European Federation of Corrosion.

535 Tan, Y.J., Bailey, S., and Kinsella, B. (1996). The monitoring of the formation and destruction of corrosion inhibitor films using electrochemical noise analysis (ENA). *Corrosion Science* 38: 1681–1695.

536 Plagemann, P., Kaune, M., Schneider, M., and Yezerska, O. (2006). Electrochemical current noise measurements for corrosion screening purposes. In: *Proceedings EUROCORR 2006, the European Corrosion Congress*, Maastricht, The Netherlands (25–28 September 2006), 13. European Federation of Corrosion.

537 Townley, D.W., Pawel, S.J., Wilson, D.F., and Roy, D.E. (2003). Investigation of the use of electrochemical noise to monitor corrosion in kraft pulp digesters. In: *NACE International Conference, CORROSION/2003*, Paper No. 03411, 17. Houston, TX: NACE International.

538 Wang, H.B., George, K., Hong, T. et al. (2000). Corrosion monitoring of top flow in wet gas environment using electrochemical noise measurement (ECN). In: *NACE International Conference, CORROSION/2000*, Paper No. 00045, 16. Houston, TX: NACE International.

539 Inoue, H., Mori, S., and Maeda, Y. (2003). Electrochemical noise measurement on UNS 10276 in oxidizing hydrochloric acid solutions at high temperature. In: *NACE International Conference, CORROSION/2003*, Paper No. 03417, 9. Houston, TX: NACE International.

540 Bignold, G.J. and Quirk, G.P. (2002). Electrochemical noise measurements in a 500 MW steam turbine to maximize lifetime under changing operational demands. In: *NACE International Conference, CORROSION 2002*, Paper No. 02333, 20. Houston, TX: NACE International.

541 Linjeville, T.M., Valentine, J., Davis, K.A. et al. (2003). Prediction and real-time monitoring techniques for corrosion characterization in furnaces. *Materials at High Temperatures* 20 (2): 175–184.

542 Covino, B.S., Bullard, S.J., Cramer, S.D. et al. (2004). High temperature electrochemical noise corrosion sensors for fossil fuel combustion. In: *NACE International Conference, CORROSION 2004*, Paper No. 04528, 9. Houston, TX: NACE International.

543 de Jong, M.P. and Leferink, R.G.I. (2003). Probes to monitor in-plant material degradation. *Materials at High Temperatures* 20 (2): 169–173.

544 Chen, J.F. and Bogaerts, W.F. (1996). Electrochemical emission spectroscopy for monitoring uniform and localized corrosion. *Corrosion* 52 (10): 753–759.

545 Halama, M., Jerolitsch, D., Linhardt, P. et al. (2010). Improvement of ENA – NOCS technique using Artificial Neural Networks approach for the detection of corrosion. In: *Proceedings EUROCORR 2010, the European*

Corrosion Congress, 8. European Federation of Corrosion, Moscow, Russia (13–17 September 2010), Paper No. 9504.

546 Ohtsu, T. and Miyazawa, M. (2007). Application of corrosion analysis and corrosion control by using corrosion monitoring at chemical plants. In: *NACE International Conference, CORROSION 2007*, Paper No. 07266, 15. Houston, TX: NACE International.

547 Cafissi, A. and Trasatti, S. (2005). *Minutes of EFC WP 15 Corrosion in the Refinery Industry*, Lisbon, Portugal (6 September 2005), 7 p.

548 Ritter, S., Huet, F., and Cottis, R.A. (2009). Reliability of electrochemical noise measurements – part 2: first ECG-COMON guideline for an assessment of electrochemical noise measurement devices. In: *Proceedings EUROCORR 2009, the European Corrosion Congress*, Nice, France (6–10 September 2009), Paper No. 8046, 12. European Federation of Corrosion.

549 Bjegovic, D., Stipanovic, I., Skazlic, M. et al. (2003). Case study – corrosion monitoring in marine environment in croatia. In: *Proceedings EUROCORR 2003, the European Corrosion Congress*, Budapest, Hungary (28 September–2 October 2003), 11. European Federation of Corrosion.

550 ASTM G71-81 (2003) (2006). *Standard Guide for Conducting and Evaluating Galvanic Corrosion Tests in Electrolytes*, Annual Book of ASTM Standards, vol. 03.02, 5. West Conshohocken, PA: ASTM International.

551 Aiello, L. and Agarwala, V. (2007). Galvanic sensor for monitoring structural damage. *Corrosion Reviews* 25 (1–2): 39–50.

552 Andersen, A., Hilbert, L., Jansen, P., and Thorarinsdottir, R. (2003). A localized corrosion cell for industrial applications. In: *Proceedings EUROCORR 2003, the European Corrosion Congress*, Budapest, Hungary (28 September–2 October 2003), 9. European Federation of Corrosion.

553 Yang, B. (1994). Method and apparatus for measuring underdeposit localized corrosion rate or metal rate under tubercles corrosion in cooling water systems. US Patent 5,275,704.

554 Bo, Y. (1994). A novel method for on-line determination of underdeposit corrosion rate in cooling water systems. In: *NACE International Conference, CORROSION/94*, Paper No. 335. Houston, TX: NACE International.

555 Bo, Y. (1995). Localized corrosion monitoring in cooling water systems. In: *NACE International Conference, CORROSION/95*, Paper No. 541. Houston, TX: NACE International.

556 Yang, B. (1999). Minimizing localized corrosion via new chemical treatments and performance based treatment optimization and control. In: *NACE International Conference, CORROSION99*, Paper No. 307, 27. Houston, TX: NACE International.

557 Yang, B. (1995). Method for on-line determination of underdeposit rates in cooling water systems. *Corrosion* 51 (2): 153.

558 Fu, S.-L., Griffin, A.M., Garcia, J.G., and Yang, B. (1996). A new localized corrosion monitoring technique for the evaluation of oilfield inhibitors. In: *NACE International Conference, CORROSION96*, Paper No. 346, 21. Houston, TX: NACE International.

559 ASTM G15-08 (2008). *Standard Terminology Relating to Corrosion and Corrosion Testing*, Annual Book of ASTM Standards, vol. 03.02, 5. West Conshohocken, PA: ASTM International.

560 Silverman, D.C. (1987). Primer on the AC impedance technique. In: *Electrochemical Techniques for Corrosion Engineering* (ed. R. Baboian), 73. Houston, TX: NACE.

561 Macdonald, J.R. (1987). *Impedance Spectroscopy*. New York: Wiley.

562 Silverman, D.C. and Carrico, J.E. (1988). Electrochemical impedance technique – a practical tool for corrosion prediction. *Corrosion* 44 (5): 280.

563 Barsoukov, E. and Macdonald, J.R. (eds.) (2005). *Impedance Spectroscopy: Theory, Experiment, and Applications*, 2e, vol. 616. New York: Wiley Interscience Publications.

564 Bard, A.J. and Faulkner, L.R. (2000). *Electrochemical Methods, Fundamentals and Applications*. New York: Wiley.

565 Orazem, M.E. (2005). Interpretation of impedance spectroscopy. In: *Proceedings EUROCORR 2005, the European Corrosion Congress*, Lisbon, Portugal (4–8 September 2005), 9. European Federation of Corrosion.

566 ISO 16773-1:2016 (2016). *Electrochemical Impedance Spectroscopy (EIS) on Coated and Uncoated Metallic Specimens – Part 1: Terms and Definitions*, 7. Geneva, Switzerland: ISO.

567 ISO 16773-2:2016 (2016). *Electrochemical Impedance Spectroscopy (EIS) on Coated and Uncoated Specimens – Part 2: Collection of Data*, 24. Geneva, Switzerland: ISO.

568 Roberge, P.R. and Sastri, V.S. (1994). On-line corrosion monitoring with electrochemical impedance spectroscopy. *Corrosion* 50 (10): 744–754.

569 Sauvant-Moynot, V., Kittel, J., and Lefébvre, X. (2006). Use of EIS for the monitoring and modelling of multi-layer anticorrosion coatings. In: *Proceedings EUROCORR 2006, the European Corrosion Congress*, Maastricht, The Netherlands (25–28 September 2006), 15. European Federation of Corrosion.

570 Davis, G.D., Dunn, R.C., and Ross, R.A. (2010). Wireless, battery-powered coating-health monitor (CHM). In: *NACE International Conference, CORROSION 2010*, Paper No. 10091, 9. Houston, TX: NACE International.

571 Adib, B., Neshati, J., and Sardashti, A. (2010). Stress corrosion cracking detection of sensitized stainless steel 304 in chloride media by using electrochemical impedance spectroscopy (EIS). In: *Proceedings EUROCORR 2010, the European Corrosion Congress*, 8. European Federation of Corrosion, Moscow, Russia (13–17 September 2010), Paper No. 9162.

572 Hamduri, H. and Nasrazadani, S. (2008). Evaluation of DBU as corrosion inhibitor for secondary cycle pressurized water reactor. In: *NACE International Conference, CORROSION 2008*, Paper No. 08363, 7. Houston, TX: NACE International.

573 Conde, A. and De Damborenea, J.J. (2000). Evaluation of exfoliation susceptibility by means of the electrochemical impedance spectroscopy. *Corrosion Science* 42: 1363–1377.

574 Conde, A. and De Damborenea, J.J. (1998). Electrochemical modeling of exfoliation corrosion behaviour of 8090 alloy. *Electrochimica Acta* 43 (8): 849–860.

575 Keddam, M., Kuntz, C., Takenouti, H. et al. (1997). Exfoliation corrosion of aluminum alloys examined by electrode impedance. *Electrochimica Acta* 42 (1): 87–97.

576 Farrell, D.M., Mok, W.Y., and Pinder, L.W. (1989). On-line monitoring of furnace-wall corrosion in a 125 MW power generation boiler. *Materials Science and Engineering* A121: 651–659.

577 Neshati, J., Fardi, M.R., and Ghassem, H. (2004). Studying of acid-gas pipelines corrosion with impedance spectroscopy. In: *Proceedings EUROCORR 2004, the European Corrosion Congress*, Nice, France (12–16 September 2004), 8. European Federation of Corrosion.

578 Shukla, P.K. and Dasgupta, B. (2010). Modeling of local electrochemical impedance spectroscopy using boundary element method. In: *NACE International Conference, CORROSION 2010*, Paper No. 10152, 13. Houston, TX: NACE International.

579 Nagiub, A.M. (2003). Evaluation of localized corrosion for Al-2024 with electrochemical impedance spectroscopy (EIS) and electrochemical noise analysis (ENA). In: *Proceedings EUROCORR 2003, the European Corrosion Congress*, Budapest, Hungary (28 September–2 October 2003), 1. European Federation of Corrosion.

580 Jerolitsch, D., Krenn, K., Schnattinger, H. et al. (2009). Monitoring of electrochemical parameters during cyclic salt spray test. In: *Proceedings EUROCORR 2009, the European Corrosion Congress*, Nice, France (6–10 September 2009), Paper No. 8386, 8. European Federation of Corrosion.

581 Rangarajan, S.K. (1975). Random process in electrochemistry: Part 1. Noise as an input signal. *Journal of Electroanalytical Chemistry* 62: 62.

582 Devay, J. and Meszaros, L. (1979). Study of the rate of corrosion of metals by a faradaic distortion method. *Acta Chimica Academiae Hungaricae* 100: 183–202.

583 Eden, D.C., Eden, D.A., Winning, I.G., and Fell, D. (2006). On-line, real-time optimization of corrosion inhibitors in the field. In: *NACE International Conference, CORROSION/2006*, Paper No. 06321, 14. Houston, TX: NACE International.

584 Gamry Instruments. *Basics of Electrochemical Impedance Spectroscopy*. https://www.gamry.com/application-notes/EIS/basics-of-electrochemical-impedance-spectroscopy (accessed 10 November 2018).
585 Meszaros, L., Meszaros, G., and Lengyel, B. (1994). Application of harmonic analysis in the measuring technique of corrosion. *Journal of the Electrochemical Society* 141 (8): 2068–2071.
586 Darowicki, K. and Krakowiak, A. (2003). The application of Gabor transformation in the harmonic analysis of corrosion processes. *Anti-Corrosion Methods and Materials* 50 (3): 193–200.
587 Sathiyanarayanan, S. and Balakrishnan, K. (1994). Critique of harmonis analysis for corrosion rate measurements. *British Corrosion Journal* 29 (2): 152–155.
588 Gill, J.S., Callow, L.M., and Scantlebury, J.D. (1983). Corrosion measurements derived from small perturbation non-linearity: part 1 – harmonic analysis. *Corrosion* 39 (2): 61–65.
589 Vavpetič, P. (2008). Seminar corrosion in concrete steel (April 2008), p. 16. University of Ljubljana, Kamnik. http://mafija.fmf.uni-lj.si/seminar/files/2007_2008/Corrosion_Seminar.pdf (accessed 10 November 2018).
590 Hohenstein, D. (2005). Corrosion as a real-time process variable. *Control Engineering* (1 March): 4. https://www.controleng.com/single-article/corrosion-as-a-real-time-process-variable.html (accessed 10 November 2018).
591 Bosch, R.W., Hubrecht, J., Bogaerts, W.F., and Syrett, B.C. (2001). Electrochemical frequency modulation: a new electrochemical technique for online corrosion monitoring. *Corrosion* 57 (1): 60–70.
592 Bosch, R.W., Hubrecht, J., Bogaerts, W.F., and Syrett, B.C. (2001). Electrochemical frequency modulation: a new electrochemical technique for online corrosion monitoring. In: *NACE International Conference, CORROSION 2001*, Paper No. 01010060. Houston, TX: NACE International.
593 Eden, D.A. (2005). Practical measurements using non-linear analysis techniques – harmonic distortion and intermodulation distortion. In: *NACE International Conference, CORROSION 2005*, Paper No. 05335, 9. Houston, TX: NACE International.
594 Eden, D.A. and Kane, R.D. (2005). Single electrode probes for on-line electrochemical monitoring. In: *NACE International Conference, CORROSION 2005*, Paper No. 05373, 11. Houston, TX: NACE International.
595 Obot, I.B. and Onyeachu, I.B. (2018). Electrochemical frequency modulation (EFM) technique: theory and recent practical applications in corrosion research. *Journal of Molecular Liquids* 249: 83–96.
596 Bosch, R.W. and Bogaerts, W.F. (2001). Apparatus and method for electrochemical corrosion monitoring. US Patent 6,320,395.
597 Bosch, R.W., Bogaerts, W.F., and Syrett, B. (2004). Corrosion rate measurement under cathodic protection by the electrochemical frequency modulation (EFM) technique. *Electrochimica Acta* 49: 3122–3133.

598 Kuzmak, A.E. and Kozheurov, A.V. (2010). Coulometry at controlled potential for evaluation of corrosion kinetics. In: *Proceedings EUROCORR 2010, the European Corrosion Congress*, Moscow, Russia (13–17 September 2010), Paper No. 9679, 14. European Federation of Corrosion.
599 Lebozec, N., Pautasso, J.-P., Le Calve, P., and Thierry, D. (2009). Assessment of the aggressiveness and environmental conditions on ships. In: *Proceedings EUROCORR 2009, the European Corrosion Congress*, Nice, France (6–10 September 2009), Paper No. 7745, 10. European Federation of Corrosion.
600 a ISO 9223 (2012). *Corrosion of Metals and Alloys – Classification of Corrosivity of Atmospheres*, 15. Geneva, Switzerland: ISO.b ISO 9224 (2012). *Corrosion of Metals and Alloys – Corrosivity of Atmospheres – Guiding Values for the Corrosivity Categories*, 13. Geneva, Switzerland: ISO.c ISO 12944-2 (2017). *Paints and Varnishes – Corrosion Protection of Steel Structures by Protective Paint Systems, Part 2: Classification of Environments*, 11. Geneva, Switzerland: ISO.
601 Kempe, P., Denzine, A., and Tiefnig, E. (2000). An advanced atmospheric corrosion measurement system. In: *NACE International Conference, CORROSION 2000*, Paper No. 00405, 8. Houston, TX: NACE International.
602 Prosek, T., Thierry, D., Kouril, M., and Degres, Y. (2008). Automated corrosion loggers for corrosion monitoring in the atmosphere. In: *NACE International Conference, CORROSION 2008*, Paper No. 08296, 6. Houston, TX: NACE International.
603 Kouřil, M., Prosek, T., Scheffel, B. et al. (2010). Real time corrosion monitoring system for protection of cultural heritage. In: *Proceedings EUROCORR 2010, the European Corrosion Congress*, Moscow, Russia (13–17 September 2010), Paper No. 9676, 11. European Federation of Corrosion.
604 Sjögren, L. (2002). *On-line Corrosion Monitoring of Environments for Electronic Equipment*, Report 2002:2. Stockholm, Sweden: Swedish Corrosion Institute.
605 Sjögren, L. and Le Bozec, N. (2004). On-line corrosion monitoring of indoor atmospheres. In: *Proceedings EUROCORR 2004, the European Corrosion Congress*, Nice, France (10–14 September 2004), Paper No. 224. European Federation of Corrosion.
606 Jönsson, M., Persson, D., and Thierry, D. (2006). The initial steps of atmospheric corrosion on magnesium alloy AZ91D studied by confocal microscopy. In: *Proceedings EUROCORR 2006, the European Corrosion Congress*, Maastricht, The Netherlands (25–28 September 2006), 8. European Federation of Corrosion.
607 Mansfeld, F. (1994). Determination of time-of-wetness and corrosion rates with atmospheric corrosion rate monitors. In: *NACE International Conference, CORROSION/94*, Paper No. 333. Houston, TX: NACE International.

608 Forslund, M., Majoros, J., and Leygraf, C. (1997). A sensor system for high resolution in-situ atmospheric corrosivity monitoring in field environments. *Journal of the Electrochemical Society* 144: 2637–2642.

609 Trillo, E. and Caseres, L. (2010). Corrosion monitoring in dehumidified storage. In: *NACE International Conference, CORROSION 2010*, Paper No. 10165, 15. Houston, TX: NACE International.

610 White, M.L. and Leidheiser Jr., H. (1991). Environmental corrosion monitor system incorporating deposited contaminants and corrosion products. US Patent 5,045,775.

611 Granata, R.D., Fisher, J.W., and Wilson, J.C. (1997). Update on applications of corrosion coulometers. In: *NACE International Conference, CORROSION/97*, Paper No. 307, 11. Houston, TX: NACE International.

612 White, M.L. and Leidheiser, H. Jr. (1990). The corrosion coulometer – a new corrosion monitor for steel structures. *Corrosion* 46 (8): 653–661.

613 Iikawa, T., Udo, Y., and Nakajima, E. (1993). Assessment of specific environments by metal piece test kits. *Corrosion Science* 35 (1–4): 735–742.

614 (2009). Corrosion simulation tests bridge cable sensors. *Materials Performance* 48 (3): 18–21.

615 Jönsson, M., Bo, R., and Annergren, I. (2010). Monitoring the road environment for on-vehicle testing. In: *Proceedings EUROCORR 2010, the European Corrosion Congress*, Moscow, Russia (13–17 September 2010), Paper No. 9502, 1. European Federation of Corrosion.

616 Vasanth, K.L. (2008). Volatile corrosion inhibitors. In: *Corrosion Science and Technology* (ed. U. Kamachi Mudali and B. Raj), 284–300. New Delhi, India: Narosa Publishing House.

617 Jaeger, P. (1997). Characterization of volatile corrosion inhibitors using the quartz crystal microbalance and supporting techniques. In: *NACE International Conference, CORROSION/97*, Paper No. 180, 6. Houston, TX: NACE International.

618 Vasanth, K.L. and Dacres, C.M. (1997). Vapor phase corrosion inhibitors for navy applications. In: *NACE International Conference, CORROSION/97*, Paper No. 179, 12. Houston, TX: NACE International.

619 Jaeger, P., Garfias-Mesias, L.F., and Büchler, M. (1999). A novel technique to studying the effect of a VCI. In: *NACE International Conference, CORROSION/99*, Paper No. 132, 9. Houston, TX: NACE International.

620 Shi, J., Sun, W., Chen, H., and Zhang, Q. (2009). Influence of porosity of the corrosion of reinforcement in cementitious materials. In: *Proceedings EUROCORR 2009, the European Corrosion Congress*, Nice, France (6–10 September 2009), Paper No. 7758, 12. European Federation of Corrosion.

621 Kessler, S., Gehlen, C., and Menzel, K. (2010). Detectability of active corroding areas in reinforced concrete structures with potential mapping. In: *Proceedings EUROCORR 2010, the European Corrosion Congress*,

Moscow, Russia (13–17 September 2010), Paper No. 9492, 1. European Federation of Corrosion.

622 Marconnet, C., Djafer, M., and Buttoudin, S. (2010). Investigation of an electrochemical non destructive technique to control the degradation of reinforced concrete pipes in drinking water distribution systems. In: *Proceedings EUROCORR 2010, the European Corrosion Congress*, Moscow, Russia (13–17 September 2010), Paper No. 9560, 14. European Federation of Corrosion.

623 Kane, R.D., Eden, D.C., and Eden, D.A. (2005). Real-time solutions integrate corrosion monitoring with process control. *Materials Performance* 44 (2): 36–41.

624 Kane, R.D. and Trillo, E. (2004). Evaluation of multiphase environments for general and localized corrosion. In: *NACE International Conference, CORROSION/2004*, Paper No. 04656, 20. Houston, TX: NACE International.

625 Veazey, M.V. (2004). A paradigm shift for process control? *Materials Performance* 43 (12): 16–19.

626 Eden, D.A. and Srinivasan, S. (2004). Real-time, on-line and on-board: the use of computers, enabling corrosion monitoring to optimize process control. In: *NACE International Conference, CORROSION/2004*, Paper No. 04059, 16. Houston, TX: NACE International.

627 Morgan, N., Winslow, M.C., and Howard, C. (2005). *Development and Implementation Strategies for Safe & Profitable Opportunity Crude Processing*, Technical Paper GE Betz TP1005EN 0503, 9 p.

628 Eden, D.C., Cayard, M.S., Kintz, J.D. et al. (2003). Making credible corrosion measurements – real corrosion, real time. In: *NACE International Conference, CORROSION/2003*, Paper No. 03376, 22. Houston, TX: NACE International.

629 Teevens, P. (2000). Defeating internal corrosion of petroleum pipelines in remote locations: winning strategies integrating real-time electrochemical noise corrosion surveillance with recent advancements in power and communications technologies. *Corrosion and Prevention, 2000, Publication No. 064, Australasian Corrosion Association.*

630 Eden, D.A. (1998). Electrochemical noise – the first two octaves. In: *NACE International Conference, CORROSION/98*, Paper No. 98386, 31. Houston, TX: NACE International.

631 Teevens, P. (1998). Electrochemical noise – a potent weapon in the battle against sour gas plant corrosion: over three years operating and turnaround inspection experiences in two Canadian plants. *Corrosion and Prevention, 1998, Publication No. 038, Australasian Corrosion Association.*

632 Eden, D.A. and Leiper, G.A. (2001). On-line electrochemical monitoring using distributed hardware and data analysis envelopes. In: *NACE International Conference, CORROSION 2001*, Paper No. 01284, 11. Houston, TX: NACE International.

633 Eden, D.A. and Breene, B. (2003). On-line electrochemical corrosion monitoring in fireside applications. In: *NACE International Conference, CORROSION 2003*, Paper No. 03361, 10. Houston, TX: NACE International.

634 Eden, D.C., Eden, D.A., and Kane, R. (2005). Take corrosion monitoring on-line. *Chemical Processing* 68 (8): 37–41.

635 Kane, R.D. and Campbell, S. (2004). Real-time monitoring of steel influenced by microbial activity (SRB) under controlled seawater injection conditions. In: *Proceedings EUROCORR 2004, the European Corrosion Congress*, Nice, France (12–16 September 2004), 13. European Federation of Corrosion.

636 Kane, R.D. and Campbell, S. (2004). Real-time corrosion monitoring of steel influenced by microbial activity (SRB) in simulated seawater injection environments. In: *NACE International Conference, CORROSION/2004*, Paper No. 04579, 15. Houston, TX: NACE International.

637 Kane, R.D. (2004). On-line, real-time monitoring for petrochemical and pipeline process control applications. In: *Proceedings EUROCORR 2004, the European Corrosion Congress*, Nice, France (12–16 September 2004), 12. European Federation of Corrosion.

638 Kus, S., Bartelmus, G., Kane, R.D. et al. (2009). Corrosion in continuous biodegradation of volatile organic compounds. In: *Proceedings EUROCORR 2009, the European Corrosion Congress*, Nice, France (6–10 September 2009), Paper No. 8088, 13. European Federation of Corrosion.

639 Kane, R.D., Eden, D.A., and Eden, D.C. (2003). Online, real-time corrosion monitoring for improving pipeline integrity – technology and experience. In: *NACE International Conference, CORROSION/2003*, Paper No. 03175, 18. Houston, TX: NACE International.

640 Guo, X., Dong, Z., Gu, T. et al. (2012). Research on the intelligent safety monitoring system of pipeline corrosion in acidic oil and gas fields – II. *Procedia Engineering* 27: 1664–1670.

641 Matthes, S.A., Covino, B.S., Bullard, S.J., and Williamson, K.M. (2008). Field test of high temperature corrosion sensors in a waste to energy plant. In: *NACE International Conference, CORROSION 2008*, Paper No. 08294, 20. Houston, TX: NACE International.

642 Lasiuk, B., Wilson, M., and Winslow, C. (2005). *Advances in Optimizing Refinery Profitability*, Technical Report GE Betz AM-05-13 (January 2005), p. 12.

643 Winslow, M.C., Wilson, M., Lasiuk, B. et al. (2005). Solutions for processing opportunity crudes. *ERTC (European Refining Technology Conference) 10th Annual Meeting*, Vienna, Austria (November 2005), p. 14.

644 Jackson, T., Winslow, M.C., and Wilson, M. (2004). Prolonged experience processing high acid crude – cross oil & refining company. *ERTC 9th Annual Meeting*, Prague, Czech Republic (November 2004).

645 Vanhove, A. (2006). Advances in corrosion monitoring while processing high acidic opportunity crudes. *The 7th Israel Conference on Corrosion and Electrochemistry*, Bar-Ilan University, Israel (10 May–11 2006).

646 Horn, H., Sivertsen, S.T., and Pedersen, A.E. (2003). Evaluation of the accuracy of a subsea system for internal corrosion monitoring based on a retrieved pipe. In: *NACE International Conference, CORROSION/2003*, Paper No. 03425, 16. Houston, TX: NACE International.

647 Claesen, C. and Kulic, M. (2010). Progress in petroleum refinery high temperature sulfidic corrosion inhibition. In: *Proceedings EUROCORR 2010, the European Corrosion Congress*, Moscow, Russia (10–14 September 2010), Paper No. 9595, 10. European Federation of Corrosion.

648 Strommen, R.D., Horn, H., and Wold, K.R. (1993). New technique monitors pipeline corrosion cracking. *Oil & Gas Journal* (27 December): 88–92. https://www.ogj.com/articles/print/volume-91/issue-52/in-this-issue/pipeline/new-technique-monitors-pipeline-corrosion-cracking.html (accessed 14 November 2018).

649 Morton, K. (2010). Reliable solutions for corrosion monitoring in high temperature plant. In: *Proceedings EUROCORR 2010, the European Corrosion Congress*, Moscow, Russia (10–14 September 2010), Paper No. 9542, 1. European Federation of Corrosion.

650 Chiang, K.-T.K. and Yang, L. (2010). High-temperature electrochemical sensor for online corrosion monitoring. In: *NACE International Conference, CORROSION 2010*, Paper No. 10168, 21. Houston, TX: NACE International.

651 Yang, L. and Sridhar, N. (2006). Coupled multielectrode array systems and sensors for real-time corrosion monitoring – a review. In: *NACE International Conference, CORROSION 2006*, Paper No. 06681, 46. Houston, TX: NACE International.

652 Budiansky, N.D., Hudson, J.L., and Scully, J.R. (2004). Origins of persistent interaction among localized corrosion sites on stainless steel. *Journal of the Electrochemical Society* 151 (4): B233–B243.

653 Dorsey, M.H., Demarco, D.R., Fisher, G.A. et al. (2005). Laboratory evaluation of a multi-array sensor for detection of underdeposit corrosion and/or microbiologically influenced corrosion. In: *NACE International Conference, CORROSION 2005*, Paper No. 05371, 17. Houston, TX: NACE International.

654 Cong, H., Budiansky, N.D., Scully, J.R., and Michels, H.T. (2007). Use of coupled electrode arrays to elucidate copper pitting as a function of potable water chemistry. In: *NACE International Conference, CORROSION 2007*, Paper No. 07392, 24. Houston, TX: NACE International.

655 Colbert, R. and Reich, R. (2008). Corrosion monitoring of a water based rolling facility with coupled multielectrode array sensors and the correlations with other process variables: conductivity, pH, temperature, dissolved oxygen

and corrosion potential. In: *NACE International Conference, CORROSION 2008*, Paper No. 08295, 15. Houston, TX: NACE International.
656 Anderko, A., Sridhar, N., Brossia, C.S. et al. (2003). An electrochemical approach to predicting and monitoring localized corrosion in chemical process streams. In: *NACE International Conference, CORROSION 2003*, Paper No. 03375, 24. Houston, TX: NACE International.
657 Duke, D. and Yang, L. (2008). Electrochemical studies with coupled multielectrode array sensors on corrosion inhibition of copper and affected metals in the presence of ammonia and other corrosive ions in zero liquid discharge cooling water. In: *NACE International Conference, CORROSION 2008*, Paper No. 08372, 17. Houston, TX: NACE International.
658 Yang, L. and Dunn, D.S. (2002). Evaluation of corrosion inhibitors in cooling water systems using a coupled multielectrode array sensor. In: *NACE International Conference, CORROSION 2002*, Paper No. 02004, 17. Houston, TX: NACE International.
659 Chiang, K.T. and Yang, L. (2008). Development of crevice-free electrodes for multielectrode array sensor for applications at high temperatures. *Corrosion* 64 (10): 805–812.
660 Yang, L., Sridhar, N., and Cragnolino, G. (2002). Comparison of localized corrosion of Fe-Ni-Cr-Mo alloys in concentrated brine solutions using a coupled multielectrode array sensor. In: *NACE International Conference, CORROSION 2002*, Paper No. 02545, 14. Houston, TX: NACE International.
661 Yang, L. and Sridhar, N. (2003). Coupled multielectrode online corrosion sensor. *Materials Performance* 42 (9): 48–52.
662 Yang, L., Chiang, K.T., and Shukla, P. (2010). Distributed node electrodes for corrosion monitoring of concrete rebars and buried pipes. In: *NACE International Conference, CORROSION 2010*, Paper No. 10166, 17. Houston, TX: NACE International.
663 Kelly, R.G., Yuan, J., Jones, S.H. et al. (1997). Embeddable microinstruments for corrosion monitoring. In: *NACE International Conference, CORROSION/97*, Paper No. 97294, 12. Houston, TX: NACE International.
664 Van Orden Chidester, A. (1994). Monitor on-line corrosion electrochemically. *Chemical Engineering Progress* 90: 67–73.
665 Zhan, H., de Jong, M., Goole, S., and Versluis, P. (2009). Practical approach on optimizing the co-firing conditions at the coal fired power plant of EPZ. In: *Proceedings EUROCORR 2009, the European Corrosion Congress*, Nice, France (6–10 September 2009), Paper No. 8183, 23. European Federation of Corrosion.
666 Laurila, T. and Pulliainen, M. (1999). Computerized on-line corrosion monitoring & corrosion control systems in process industry. In: *NACE International Conference, CORROSION99*, Paper No. 244, 10. Houston, TX: NACE International.
667 Jensen, B.B., Schultz, K., and Grønvold, F. (2010). Management system for corrosion monitoring in the Copenhagen Metro. In: *NACE International*

Conference, CORROSION 2010, Paper No. 10160, 12. Houston, TX: NACE International.
668 Dunn, R.C., Davis, G.D., and Ross, R.A. (2010). Corrosion monitoring of steel reinforced concrete structures using embedded instrumentation. In: *NACE International Conference, CORROSION 2010*, Paper No. 10173, 10. Houston, TX: NACE International.
669 Mercalli, A., Cigna, R., and Viaduct, V. (2009). Twenty years of corrosion monitoring. In: *Proceedings EUROCORR 2009, the European Corrosion Congress*, Nice, France (6–10 September 2009), Paper No. 8425, 8. European Federation of Corrosion.
670 Quirk, G.P., Eden, D.A., Eden, D.C., and Hobbs, G. (2000). Experience of using multiple-technique corrosion monitoring in real time. *Proceedings of the 40th Annual Conference of the Australasian Corrosion Association*, Australia (19–22 November 2000), Paper No. 65.
671 Simon Thomas, M.J.J., Pots, B.F.M., and Hendriksen, E.L.J.A. (2001). Field corrosivity measurements – an essential component of the corrosion control process. In: *NACE International Conference, CORROSION 2001*, Paper No. 01038, 16. Houston, TX: NACE International.
672 Johnson, B. V. (2004). Advanced electrical resistance on-line corrosion monitoring system at a gas gathering facility. *10th Middle East Corrosion Conference*, Bahrain (7–10 March 2004).
673 Sridhar, N., Tormoen, G., Sean Brossia, C., and Sabata, A. (2006). Development and application of mobile sensor network to monitor corrosion in pipelines. In: *NACE International Conference, CORROSION2006*, Paper No. 06322, 17. Houston, TX: NACE International.
674 Gu, P., Zheng, W., and Revie, R.W. (2007). Intelligent pipeline monitoring system. In: *NACE International Conference, CORROSION 2007*, Paper No. 07267, 11. Houston, TX: NACE International.
675 Sridhar, N., Yang, L., and Song, F. (2006). Application of multielectrode array sensor to study dewpoint corrosion in high pressure natural gas pipeline environments. In: *NACE International Conference, CORROSION 2006*, Paper No. 06673, 13. Houston, TX: NACE International.
676 Yang, L., Sridhar, N., Pensado, O., and Dunn, D.S. (2002). An in-situ galvanically coupled multielectrode array sensor for localized corrosion. *Corrosion* 58 (12): 1004–1014.
677 ANSI/ASME B31G-1991 (R2004) (2004). *Manual for Determining the Remaining Strength of Corroded Pipelines, a Supplement to ASME B31 Code for Pressure Piping*, 55. New York: ASME.
678 ASTM E317-16 (2016). *Standard Practice for Evaluating Performance Characteristics of Ultrasonic Pulse-Echo Testing Instruments and Systems Without the Use of Electronic Measurement Instruments*, ASTM Book of Standards, vol. 03.03, 13. West Conshohocken, PA: ASTM International.

679 NACE SP0102-2010 (2010). *Standard Practice In-Line Inspection of Pipelines*, 42. Houston TX: NACE International.

680 NACE International Publication 35100 (2000). *In-Line Nondestructive Inspection of Pipelines*, Item No. 24211, 35. Houston, TX: NACE International.

681 Bushkovskiy, A.L., Thuy, L.C., Ivanov, A.N., and Galljamov, T.N. (2010). Internal corrosion of pipelines and the gas preparation equipment of JV Vietsovpetro. In: *Proceedings EUROCORR 2010, the European Corrosion Congress*, Moscow, Russia (13–17 September 2010), Paper No. 9340, 1. European Federation of Corrosion.

682 Greenwood, C. (2006). Pigging the diesel pipeline between Hawaii's red hill facility and pearl harbor. *Materials Performance* 45 (3): 16–19.

683 Al-Maslamani, M.J., Chaudhuri, P.K., and Queenan, E.D. (2003). Validation of intelligent PIG inspection data for management of pipeline corrosion. In: *Proceedings EUROCORR 2003, the European Corrosion Congress*, Budapest, Hungary (28 September–2 October 2003), 10. European Federation of Corrosion.

684 Gunaltun, Y., Punprak, S., Thammachart, M., and Tanaprasertsong, P. (2010). Worst case top of the line corrosion: cold spot corrosion. In: *NACE International Conference, CORROSION 2010*, Paper No. 10097, 9. Houston, TX: NACE International.

685 Gunaltun, Y. and Payne, L. (2003). A new technique for the control of top of the line corrosion: TLCC-PIG. In: *NACE International Conference, CORROSION 2003*, Paper No. 03344, 9. Houston, TX: NACE International.

686 Wikipedia. *Pigging*. https://en.wikipedia.org/wiki/Pigging (accessed 10 November 2018).

687 Baltzersen, Ø., Waag, T., Johnsen, R. et al. (2007). Wall thickness monitoring of new and existing subsea pipelines using ultrasound. In: *NACE International Conference, CORROSION 2007*, Paper No. 07333, 9. Houston, TX: NACE International.

688 Stawicki, O., Beuker, T., Ahlbrink, R., and Brown, B. (2010). Monitoring of top of line corrosion with eddy current technology combined with magnetic flux leakage method. In: *NACE International Conference, CORROSION 2010*, Paper No. 10094, 7. Houston, TX: NACE International.

689 Madi, M.S. (2003). Detection of internal & external corrosion using guided wave UT and MFL inspection techniques in wafra field. In: *Proceedings EUROCORR 2003, the European Corrosion Congress*, Budapest, Hungary (28 September–2 October 2003), 10. European Federation of Corrosion.

690 Van Deventer, L. (2009). Advancements in guided wave UT equipment regarding sensitivity. *Materials Performance* 48 (2): 56–58.

691 Van Deventer, L. (2009). Advancements in GUL G-3 guided wave UT equipment regarding sensitivity. *Materials Performance* 48 (3): 58–61.

692 Burwell, D., Sridhar, N., Moghissi, O.C., and Perry, L. (2004). Internal corrosion direct assessment of dry gas transmission pipelines – validation. In: *NACE International Conference, CORROSION/2004*, Paper No. 04195, 17. Houston, TX: NACE International.

693 Moghissi, O.C., Perry, L., Cookingham, B., and Sridhar, N. (2004). Internal corrosion direct assessment of dry gas transmission pipelines – application. In: *NACE International Conference, CORROSION/2003*, Paper No. 03204, 18. Houston, TX: NACE International.

694 Kanaykin, V.A. and Steklov, O.I. (2010). Modern trunk pipeline in-line diagnostics technologies to detect corrosion-related defects. In: *Proceedings EUROCORR 2010, the European Corrosion Congress*, Moscow, Russia (3–17 September 2010), Paper No. 4684, 1. European Federation of Corrosion.

695 Kharinovsky, V.V. (2010). Forecasting of the lifecycle of gas pipeline section with the corrosion defects. In: *Proceedings EUROCORR 2010, the European Corrosion Congress*, Moscow, Russia (13–17 September 2010), Paper No. 9655, 1. European Federation of Corrosion.

696 Covino, B.S., Cramer, S.D., Bullard, S.J. et al. (2005). Electrochemical corrosion rate sensors for detecting internal corrosion of natural gas transmission pipelines. In: *NACE International Conference, CORROSION 2005*, Paper No. 05133, 14. Houston, TX: NACE International.

697 Duesso, F., Horn, H., and Wold, K.R. (1995). Qualification of the FSM technology for corrosion monitoring of a subsea pipeline at the Froy field. In: *NACE International Conference, CORROSION/95*, Paper No. 27. Houston, TX: NACE.

698 Clarida, D.R., Scanlan, R.J., and Boothman, R.M. Corrosion monitoring experience in the refining industry using the FSM technique. In: *NACE International Conference, CORROSION 2003*, Paper No. 03655, vol. 2003. Houston, TX: NACE International.

699 Strommen, R., Horn, H., and Wold, K.R. (1992). FSM – a unique method for monitoring corrosion pitting, erosion and cracking. In: *NACE International Conference, CORROSION/92*, Paper No. 7. Houston, TX: NACE.

700 Tennyson, R.C., Morison, D., and Miesner, T. (2006). Fiber optic monitoring focuses on bending, corrosion. *Oil & Gas Journal* (20 February): 55–60. https://www.ogj.com/articles/print/volume-104/issue-7/transportation/fiber-optic-monitoring-focuses-on-bending-corrosion.html (accessed 3 November 2018).

701 Morison, D. (2008). Remote monitoring of pipeline corrosion using fiber optic sensors. In: *NACE International Conference, CORROSION 2008*, Paper No. 08290, 9. Houston, TX: NACE International.

702 Brown, G.K., Davies, J.R., and Hemblade, B.J. (2000). Solids and sand monitoring – an overview. In: *NACE International Conference, CORROSION 2000*, Paper No. 00091, 28. Houston, TX: NACE International.

703 Salama, M.M. (2000). Performance of sand monitors. In: *NACE International Conference, CORROSION 2000*, Paper No. 00085, 18. Houston, TX: NACE International.

704 Shirazi, S.A., McLaury, B.S., and Ali, M.M. (2000). Sand monitor evaluation in multiphase flow. In: *NACE International Conference, CORROSION 2000*, Paper No. 00084, 19. Houston, TX: NACE International.

705 Eckert, R.B., Cookingham, B., and Bensman, L. (2006). Optimizing internal corrosion monitoring and response through integration of direct and indirect data. In: *NACE International Conference, CORROSION 2006*, Paper No. 06307, 13. Houston, TX: NACE International.

706 Srinivisan, S., Lagad, V., and Kane, R.D. (2007). Internal corrosion assessment for dry gas and multiphase pipelines using corrosion prediction models. In: *EUROCORR 2007*, Freiburg, Germany (9–13 September 2007), 21. European Federation of Corrosion.

707 Jangama, V. and Srinivisan, S. (1997). A computer model for prediction of corrosion of carbon steels. In: *NACE International Conference, CORROSION/97*, Paper No. 97318, 16. Houston, TX: NACE International.

708 Srinivisan, S. and Kane, R.D. (2003). Critical issues in the application and evaluation of a corrosion prediction model for oil and gas systems. In: *NACE International Conference, CORROSION 2003*, Paper No. 03640, 18. Houston, TX: NACE International.

709 Lagad, V., Srinivisan, S., and Kane, R.D. (2004). Software system for automating internal corrosion direct assessment of pipelines. In: *NACE International Conference, CORROSION 2004*, Paper No. 04197, 16. Houston, TX: NACE International.

710 Teledyne Marine. http://www.teledynemarine.com/news/mariner-september (accessed 4 November 2018).

711 Larsen, K.R. (2009). Pipeline sensors measure corrosion rate and cumulative corrosion damage. *Materials Performance* 48 (3): 24–25.

712 Brown, N.K. and Friedersdorf, F.J. (2008). Corrosivity monitoring system for pipelines. In: *NACE International Conference, CORROSION 2008*, Paper No. 08203, 17. Houston, TX: NACE International.

713 Tennyson, R.C., Morison, W.D., Manuelpillai, G., and Revie, W. (2004). Application of fiber optic sensors to monitor pipeline corrosion. In: *NACE International Conference, CORROSION 2004*, Paper No. 04739, 18. Houston, TX: NACE International.

714 Tennyson, R.C., Morison, W.D., and Manuelpillai, G. (2005). Monitoring pipeline integrity using fiber optic sensors. In: *NACE International Conference, CORROSION 2005*, Paper No. 05134, 8. Houston, TX: NACE International.

715 Wold, K., Jakobsen, T., Erik, S. et al. (2010). New sand management system integrating erosion based and acoustic sensors in new generation

management software. In: *NACE International Conference, CORROSION 2010*, Paper No. 10142, 14. Houston, TX: NACE International.
716 Brown, N.K. and Friedersdorf, F.J. (2010). Pipelines process corrosivity measurement using mechanical diaphragm sensors. In: *NACE International Conference, CORROSION 2010*, Paper No. 10172, 12. Houston, TX: NACE International.
717 Scanlan, R.J., Valbuena, R.R., Harrison, I.M., and Rengifo, R. (2008). A refinery approach to address corrosion under insulation. In: *NACE International Conference, CORROSION 2008*, Paper No. 08558, 35. Houston, TX: NACE International.
718 McKinney, K.E., Busch, F.J.M., Blaauw, A., and Etheridge, A. (2010). Development of risk assessment and inspection strategies for external corrosion management. In: *NACE International Conference, CORROSION 2005*, Paper No. 05557, 13. Houston, TX: NACE International.
719 Pechacek, R.W. (2003). Advanced NDE methods of inspecting insulated vessels and piping for ID corrosion and corrosion under insulation (CUI). In: *NACE International Conference, CORROSION 2003*, Paper No. 03031, 9. Houston, TX: NACE International.
720 Erickson, T.H., Christopher Dash, L., Murali, J.J., and Ayers, C.R. (2010). Predicting the progression of wetness and corrosion under insulation damage in aboveground pipelines. In: *NACE International Conference, CORROSION 2010*, Paper No. 10373, 10. Houston, TX: NACE International.
721 Mierau, T. (2006). Remote monitoring & control systems as a solution for corrosion monitoring at airports with elevated security standards. In: *NACE International Conference, CORROSION 2006*, Paper No. 06309, 12. Houston, TX: NACE International.
722 Lukacs, Z., Gabor, L., and Fodor, G. (2003). Equipment and computerized evaluation technology for close interval potential survey (CIPS) on cathodic protection of pipelines. In: *Proceedings EUROCORR 2003, the European Corrosion Congress*, Budapest, Hungary (28 September–2 October 2003), 8. European Federation of Corrosion.
723 Segall, S.M., Reid, R.G., and Gummow, R.A. (2006). Use of an integrated CIPS/DCVG survey in the ECDA process. In: *NACE International Conference, CORROSION 2006*, Paper No. 06193, 12. Houston, TX: NACE International.
724 Segall, S.M., Gummow, R.A., Shore, J., and Reid, R.G. (2010). Ensuring the accuracy of indirect inspections data in the ECDA process. In: *NACE International Conference, CORROSION 2010*, Paper No. 10061, 13. Houston, TX: NACE International.
725 López, C., Knight, C., Godoy, A. et al. (2010). Numerical modeling of cathodic protection system looking for present condition evaluation and improvement of pipeline network at Manzanillo, Mexico. In: *NACE*

International Conference, CORROSION 2010, Paper No. 10068, 8. Houston, TX: NACE International.
726 Rothman, P.S., Nikolakakos, S., and Szeliga, M.J. (2006). Innovative approach to stray current monitoring at the site of the World Trade Center in New York City. In: *NACE International Conference, CORROSION 2006*, Paper No. 06310, 14. Houston, TX: NACE International.
727 Song, H.S., Kim, Y.G., and Park, K.W. (2008). Experiences of remote monitoring and controlling systems in CP. In: *NACE International Conference, CORROSION 2008*, Paper No. 08292, 7. Houston, TX: NACE International.
728 Blankenstein, L. (2003). Wireless remote monitoring for oil and gas corrosion protection assets. *Materials Performance* 42 (8): 26–29.
729 Yunovich, M. and Thompson, N.G. (2002). Coupon monitoring for cathodic protection optimization: real-time automated remote control of dynamic stray current on pipe-type cables. In: *NACE International Conference, CORROSION 2002*, Paper No. 06310, 18. Houston, TX: NACE International.
730 Panossian, Z., Filho, S.A., Almeida, N. et al. (2010). A new thermodynamic criterion and a new field methodology to verify the probability of AC corrosion in buried pipelines. In: *NACE International Conference, CORROSION 2010*, Paper No. 10115, 15. Houston, TX: NACE International.
731 Pikas, J. (2009). Direct assessment monitoring of pipes within cased crossings – part1. *Materials Performance* 48 (2): 30–32.
732 Pikas, J. (2009). Direct assessment monitoring of pipes within cased crossings – part 2. *Materials Performance* 48 (3): 40–44.
733 Du, Y., Zhang, L., Xu, L., and Lu, M. (2010). Cathodic protection monitoring of prestressed concrete cylinder pipe. In: *NACE International Conference, CORROSION 2010*, Paper No. 10121, 14. Houston, TX: NACE International.
734 Larsen, K.R. (2009). Monitoring system measures in situ corrosion condition of shipboard tanks. *Materials Performance* 48 (3): 22–23.
735 NACE Standard Recommended Practice SP0502-2010 (formerly RP0502) (2010). *Pipeline External Corrosion Direct Assessment Methodology*, Item No. 21097, 57. Houston, TX: NACE International.
736 Papavinasam, S. and Doiron, A. (2010). A 5-M approach to control external pipeline corrosion. In: *NACE International Conference, CORROSION 2010*, Paper No. 10062, 28. Houston, TX: NACE International.
737 Xu, L., Xu, J., Xu, M.-B. et al. (2018). Corrosion behavior of 3% Cr casing steel in CO_2-containing environment: a case study. *The Open Petroleum Engineering Journal* 18: 13. https://benthamopen.com/FULLTEXT/TOPEJ-11-1 (accessed 10 November 2018).
738 Taveira, L.V., Joseph, B., and Sagüés, A.A. Electrochemical noise corrosion monitoring of post tensioned tendons, 210th ECS Meeting, 29 October–3 November 2006, Cancun, Mexico. *ECS Transactions* 13 (13): 129–138.

739 Pereira, E.V., Figueira, R.B., Salta, M.M., and Fonseca, I.T.E. (2008). Non destructive measurement of reinforcement corrosion rate in existing RC structures. In: *Proceedings EUROCORR 2008, the European Corrosion Congress*, Edinburgh, Scotland (7–11 September 2008). European Federation of Corrosion.

740 Edgemon, G.L., Anda, V.S., Berman, H.S. et al. (2008). History and operation of the Hanford high-level waste storage tanks. In: *NACE International Conference, CORROSION 2008*, Paper No. 08RTS01, 71. Houston, TX: NACE International.

741 Cottis, R.A. (2007). Corrosion monitoring – what's the point? In: *Proceedings EUROCORR 2007, the European Corrosion Congress*, Freiburg, Germany (9–13 September 2007). European Federation of Corrosion.

Index

Non-Destructive Evaluation of Corrosion and Corrosion-assisted Cracking, First Edition.
Edited by Raman Singh, Baldev Raj, U. Kamachi Mudali and Prabhakar Singh.

e

f